Fundamentals of Physics II

THE OPEN YALE COURSES SERIES is designed to bring the depth and breadth of a Yale education to a wide variety of readers. Based on Yale's Open Yale Courses program (http://oyc.yale.edu), these books bring outstanding lectures by Yale faculty to the curious reader, whether student or adult. Covering a wide variety of topics across disciplines in the social sciences, physical sciences, and humanities, Open Yale Courses books offer accessible introductions at affordable prices.

The production of Open Yale Courses for the Internet was made possible by a grant from the William and Flora Hewlett Foundation.

TITLES IN THIS SERIES

Paul H. Fry, *Theory of Literature*
Roberto González Echevarría, *Cervantes' "Don Quixote"*
Christine Hayes, *Introduction to the Bible*
Shelly Kagan, *Death*
Dale B. Martin, *New Testament History and Literature*
Giuseppe Mazzotta, *Reading Dante*
R. Shankar, *Fundamentals of Physics: Mechanics, Relativity, and Thermodynamics*
R. Shankar, *Fundamentals of Physics II: Electromagnetism, Optics, and Quantum Mechanics*
Ian Shapiro, *The Moral Foundations of Politics*
Steven B. Smith, *Political Philosophy*

Fundamentals of Physics II

Electromagnetism, Optics, and Quantum Mechanics

R. SHANKAR

Yale

UNIVERSITY PRESS

New Haven and London

Yale University Press books may be purchased in quantity for educational,
business, or promotional use. For information, please e-mail sales.press@
yale.edu (U.S. office) or sales@yaleup.co.uk (U.K. office).

Set in Minion type by Newgen North America.
Printed in the United States of America.

ISBN: 978-0-300-21236-5 (pbk. : alk. paper)
Library of Congress Control Number: 2015956862

A catalogue record for this book is available from the British Library.

This paper meets the requirements of ANSI/NISO Z39.48-1992
(Permanence of Paper).

10 9 8 7 6 5 4 3 2 1

*To Stella, Vesper, Matteo,
and the rest of G3.*

JOSIAH WILLARD GIBBS LL.D.
PROFESSOR OF MATHEMATICAL PHYSICS
IN YALE COLLEGE MDCCCLXXI TO
MCMIII   DISCOVERER AND
INTERPRETER OF THE LAWS
OF CHEMICAL EQUILIBRIUM

Deep and original, but also humble and generous, the physicist
Josiah Willard Gibbs spent much of his life at Yale University. His father was
a professor of sacred languages at Yale, and Gibbs received his bachelor's
and doctorate degrees from the university before teaching there until his death in
1903. The sculptor Lee Lawrie created the memorial bronze tablet pictured
above, which was installed in Yale's Sloane Physics Laboratory in 1912. It now
resides in the entrance to the J. W. Gibbs Laboratories, Yale University.
His life and work continue to inspire the author as they do everyone who is
familiar with them.

Contents

Preface

This is the companion volume to *Fundamentals of Physics: Mechanics, Relativity, and Thermodynamics*. It is the second half of an introductory course taught at Yale and covers electromagnetism, optics, and quantum mechanics. Like Volume I, it is based on the lectures given at Yale to a diverse class. The two volumes could be used for a year-long course in introductory physics that covers all the major topics. It may also be used for self-study. Some instructors may prescribe it as a supplement to another text.

The chapters in the book more or less follow the Yale lectures with a few minor modifications. The style preserves the classroom atmosphere. Often I introduce the questions asked by the students or the answers they give when I believe they will be of value to the reader. The problem sets and exams, without which one cannot learn or be sure one has learned the physics, may be found along with their solutions at the Yale website, http://oyc.yale.edu/physics, free and open to all. The lectures may also be found at venues YouTube, iTunes (https://itunes.apple.com/us/itunes-u/physics-video/id341651848?mt=10), and Academic Earth, to name a few.

In the lectures I sometimes refer to my *Basic Training in Mathematics*, published by Springer and intended for anyone who wants to master the undergraduate mathematics needed for the physical sciences.

This book, like its predecessor, owes its existence to many people. Peter Salovey, now president, then dean of Yale College, persuaded me to be part of the first batch of Open Yale Courses, funded by the Hewlett Foundation. Diana E. E. Kleiner, Dunham Professor, History of Art and Classics, encouraged and guided me in many ways. She was also the one who persuaded me to write both these books. At Yale University Press, Joe Calamia has been an invaluable guide, making countless suggestions to improve the book's contents. He has also lent his name to many subatomic particles that appear in this book. Once again Ann-Marie Imbornoni skillfully shepherded the book through various stages of production. I am delighted that Liz Casey was once again able to apply her editorial magic to the manuscript, greatly improving not only the punctuation, syntax, and grammar but also the clarity. She made sure my intended sense was captured by the words used.

I thank Professor Ganpathy Murthy (University of Kentucky) and Branislav Djordjevic (George Mason University) for thoughtful comments. My very special thanks go to Phil Nelson of the University of Pennsylvania for his detailed and insightful comments on many parts of the book.

The writing of this book started a year ago and ended August 2015 at the Aspen Center for Physics (ACP). I am most grateful for the climate provided by the ACP where both the scientist and author in me found intellectual nourishment. The ACP is supported by the National Science Foundation (NSF) Grant number 1066293.

A large portion of the book was written at the Kavli Institute for Theoretical Physics (KITP) in Santa Barbara, where I was fortunate to receive a Simons Distinguished Visiting Scholar award for Fall 2014. The KITP is supported in part by the National Science Foundation under Grant number NSF PHY11-25915. I am especially grateful to Professor Lars Bildsten for making this possible.

The day I find I cannot write books at either of these marvelous places, I will switch to another line of work.

Barry Bradlyn and Alexey Shkarin were two exceptional graduate students who proofread the book, caught bugs, and suggested stylistic changes.

My family, all three generations of it, was very supportive as always. The final check was provided by Stella, who left many unsolicited notes in the margins and inside using her crayons. She is responsible for all remaining errors.

CHAPTER 1

Electrostatics I

We begin the second half of the course with an introduction to a new force: electromagnetism. Then we will study optics. We will conclude with a study of quantum mechanics. Now, quantum mechanics is not like a new force. It's a whole different ball game. It's not about what forces are acting on this or that object that determine its trajectory. The question there is: should we be even thinking about particles going on *any* trajectory? The answer will be negative. You will find out that most of the cherished ideas from Newtonian mechanics get overthrown. But the good news is that you need quantum mechanics only to study very small things like atoms or molecules. Of course, the big question is, where do you draw the line? How small is small? Some people even ask me, "Do you need quantum mechanics to describe the human brain?" And the answer is, "Yes, if it is small enough." I've gone to parties where after a few minutes of talking to a person I'm thinking, "Okay, this person's brain needs a full-fledged quantum mechanical treatment." But most of the time everything is macroscopic, and you can describe it with Newtonian mechanics and classical electrodynamics.

1.1 Review of $F = ma$

Before we start with electromagnetism, let us recall the interplay between the ideas of force, mass, acceleration, and $F = ma$ discussed at length in the prequel to this book, referred to as Volume I. The only thing everyone knows from the nursery is that *a* stands for acceleration, and we all know how to measure it. You find the position now and the position slightly later, take the difference, divide by the time, and get the velocity. Even though velocity requires two successive position measurements, we talk of velocity "right now," because you can make those two successive measurements arbitrarily close to each other, and in the limit in which the time difference between them goes to zero, you can talk about the velocity right now. If the speedometer in your car points to 60 miles per hour, that's your velocity right now. Likewise, find the velocity now, find the velocity a little later, divide the difference by time, and you get the acceleration. It's also an instantaneous quantity. If you step on the gas and feel the seat pushing you, that reflects your acceleration right now.

Given that we know how to measure acceleration, how should we determine the mass of anything? First of all you need an arbitrarily chosen standard of mass. The Bureau of Standards has a block of some material that *defines* a kilogram. Using that, can you find the value of another mass? Surely you know that using a weighing scale is not the correct answer because that measures the weight of the object due to earth's gravity, while the mass of an object is defined anywhere, even far from the earth. Now you might say, "Well, take a known force and divide by the acceleration it produces," but we haven't talked about how to measure the force either. All you have is this equation $F = ma$.

One correct option is to use $F = ma$ itself, as follows. Take a spring, attach one end to a wall, hook the known 1 *kg* mass to the other end, pull it by some amount, release it, and measure a_1, the acceleration. Now pick any object whose mass you want, say an elephant. You detach that 1 *kg* mass, attach the elephant, pull the spring by the *same amount*, and measure a_E, the acceleration of the elephant. Since you pulled the spring by the same amount, the force is the same in both cases. *You don't know and don't have to know what it is, just that it is the same.* Therefore we know

$$1 \cdot a_1 = m_E \cdot a_E, \tag{1.1}$$

which determines m_E, the mass of the elephant.

So imagine that the masses of all objects can be determined by this process. Can we now use $F = ma$ to find the trajectory of bodies? No, we still need to know what forces will be acting on a body in any given situation. We need to know the F in $F = ma$ in the given context. Newton does not tell you that in general. For example, for the spring, *you* have to determine what force it exerts when it's pulled by various amounts. To this end, you pull it by some amount x, attach it to a known mass, find the acceleration, and then the product ma gives the force as a function of x, namely Hooke's law $F = -kx$. So this is an example of your finding out the left-hand side of Newton's law by measuring the acceleration of known masses. Newton *did*, however, give the left-hand side in one famous case, the law of universal gravity between masses M and m in terms of their separation r and the gravitational constant G:

$$F = \frac{GMm}{r^2}. \tag{1.2}$$

Using this law we have been able to do some very impressive celestial mechanics, right up to the present.

Unlike the spring force, *there is no real contact between the earth and the object that it is pulling*, whether it be the apple or the moon. This is an example of *action at a distance*. It was a great abstraction to believe that things can reach out and pull (or push) other things without touching them. Gravity was the first formally described force of this kind.

Remember the distinction between $F = ma$ and $F = -kx$. The first is always true and relates the force on a body to the acceleration it produces, but does not tell us what force F will act in any given situation. It is our job to find out every time what forces might be acting on a body. If it's connected to a spring, we have to study the spring experimentally to find out that $F = -kx$.

So $F = ma$ is good for three things: to define mass, to determine the forces acting on bodies of known mass by seeing how they accelerate, and to find the acceleration of bodies given the forces.

Every time a body accelerates, we must be able to relate its acceleration to the sum of all the forces acting on it. But now and then we will not be able to do this. We can either abandon $F = ma$, or, putting our faith in the correctness of $F = ma$, *provided all forces are included*, we can go on to discover and characterize the new force behind the discrepancy.

1.2 Enter electricity

Now I will describe an experiment that reveals a new force. I take a comb and vigorously brush my hair and and then touch the comb to a small piece of paper. I find it sticks to the comb and I can lift it. But when I shake the comb vigorously, the paper falls down. What can we learn from this?

Clearly, the force between the comb and paper is not the force of gravity, because gravity doesn't care if you comb your hair or not. We may concede that there is a new force, but we may conclude it is feeble compared to gravity because it eventually yields to gravity when we shake the comb. It would be a mistake to think so. In fact, this new force is roughly 10^{40} times stronger than the gravitational force, as determined by a criterion that I will explain shortly. But first, let us grasp this fact intuitively.

Look at Figure 1.1. You see me holding the comb, which is holding up the piece of paper. What is trying to pull it down? The entire planet! The Himalayas are pulling it down, the Pacific Ocean is pulling it down, even the Loch Ness Monster is pulling it down. *Everything* is pulling it down. I am one of these people generally convinced the world is acting against me, but this time I'm right. Everything *is* against me and my comb, and yet we are able to triumph against all of that. And that is how you compare the electric force with the gravitational force. It takes the entire planet to compensate whatever force I created between the comb and the piece of paper. Later we will see how the number 10^{40} quantifies this fact.

Something has happened to the comb when I rubbed it against my hair, something that allowed it to attract the paper. We describe that

Figure 1.1   The comb is pulling the paper electrostatically, and all of the world is pulling the other way gravitationally.

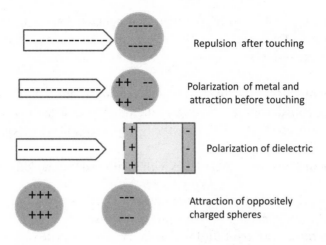

Figure 1.2   Top: A negatively charged rod imparts some negative charge to the sphere upon contact and they repel each other. Second: The negatively charged rod attracts a neutral sphere by polarizing it without touching it. Third: A charged rod polarizes a dielectric. The light region in the middle is the overlap of rectangles with positive (dotted boundary) and negative charges (solid boundary). The light region is neutral and the edges carry the uncanceled charges. Bottom: The charged spheres attract because they have been charged oppositely.

condition of the comb by saying "The comb is charged." If the comb is briefly dipped in water and removed, we find it no longer attracts the piece of paper. We say the comb is now *discharged*.

I am going to describe in detail the microscopic theory that can explain this experiment and many more, qualitatively and quantitatively. But, first, let us consider, at a qualitative level, a few more such experiments (depicted in Figure 1.2) and their explanations in terms of these ideas.

Experiment 1: Wearing silk gloves, take an aluminum rod and rub it against a passing furry animal, say a Yeti. Briefly touch an uncharged metallic sphere, isolated from everything else. The rod and sphere will repel.

Experiment 2: Move the charged rod near an isolated and uncharged sphere. Before they touch, they will attract each other. The same happens when the metallic sphere is replaced by a piece of paper.

Experiment 3: Take two more uncharged spheres. Repeat the previous two experiments after rubbing the rod on a piece of polyester and you will find the same results.

Experiment 4: Take two uncharged spheres; touch one with the rod that has been rubbed against Yeti fur and the other with a rod that has been rubbed against polyester. This time the two spheres will attract each other.

Experiment 5: Connect the spheres with a wire and they no longer attract.

Now we turn to the underlying theory, which is a result of centuries of investigation. We first consider some qualitative facts.

The most important idea is that everything is made of atoms. The atom has a nucleus consisting of protons and neutrons. The nucleus is surrounded by some very light particles called electrons. Normally the number of electrons and protons in an atom is equal. Two protons will repel each other, as will two electrons, but the proton and electron will attract each other. A neutron will not interact with, i.e., attract or repel, another neutron, proton, or electron. (Here I refer to electrical interactions, not nuclear interactions. These are much stronger, but significant only at very short distances [$\simeq 10^{-15}$ m]. The neutron fully participates in nuclear interactions and the electron does not.) Objects like electrons and protons that take part in electrical interactions are said to be charged or to carry a charge, while neutrons are said to be (electrically) neutral.

Just as mass is the reason particles experience the force of gravity, charge is the reason they interact electrically. But there are differences. There are no gravitationally neutral particles—everything has a positive mass. Second, gravity is always attractive but the electric forces can go either way. This is described by saying there are two kinds of charge, positive and negative, which can cancel each other out just like positive and negative numbers can. By convention the proton has positive charge and the electron has negative charge. Like charges repel and unlike charges attract. A system made of an equal amount of positive and negative charges will appear neutral, at least from a distance, when the internal structure is irrelevant. There is no such way to neutralize gravity.

An atom with equal number of protons and electrons is neutral. This is due to a remarkable fact that the *electron and proton have exactly equal and opposite charges*. This equality is quite a mystery since the two particles are otherwise very dissimilar: the proton is about 1836 times as heavy and experiences forces that the electron does not.

Here is how we understand Experiments 1 through 5 in terms of the preceding facts.

Experiment 1: Upon rubbing, electrons flow from the Yeti to the rod, leaving the rod negative and Yeti positive. The protons stay where they are. The silk gloves keep the electrons in the rod from jumping on to your body and then to the ground: silk is an *insulator*. When the rod touches the sphere, some electrons migrate to the sphere in order to get away from each other. The sphere and the rod are both negatively charged and repel each other.

Experiment 2: When the negatively charged rod goes near the neutral metallic sphere, the electrons in the sphere are repelled by the extra electrons in the rod, and they preferentially occupy the far side, leaving a positive region near the rod. The positive region is attracted to the rod and the negative one repelled by it, but the attraction wins since the positive part is closer. *Such free motion of electrons can take place in a conductor.* If we replace the metallic sphere by a piece of paper, it too gets attracted, but by a more complicated mechanism. The paper is a *dielectric.* The electrons in it are not free to run off to one end, because paper is also an insulator, but they can move a little from their orbits centered on the nuclei if coaxed. Think of a rectangular piece of paper as made of two superposed layers, one positive (bounded by the dotted line in the figure) and made of the nuclei and one negative, made of the electrons (bounded by the solid line). Initially the two layers overlap completely and neutralize each other everywhere. When the negative rod comes near one edge, the electronic layer is displaced by a tiny amount (of atomic dimensions) away from the rod, while the positive nuclei stay put. The bulk of the paper (solid region of overlap in the middle) is still neutral, but the edge near the rod has a strip of unbalanced protons and the edge far from it has a strip of unbalanced electrons. This process, in which the positive and negative charges are displaced relative to each other by a small amount, is called *polarization.* Again, the attraction of the nearby positive strip beats the repulsion of the distant negative one.

Experiment 3: When the rod is rubbed against polyester, the electrons flow the other way: from the rod to the polyester, leaving the rod positively charged. We can repeat the arguments from Experiments 1 and 2, simply reversing the sign of all charges.

Experiment 4: Now one sphere is positively charged (by polyester) and one is negatively charged (thanks to Yeti) and they attract.

Experiment 5: The wire, a conductor, allows electrons to flow from the negative to the positive sphere till both become neutral (assuming they had equal and opposite charges).

Observe that in all cases, *it is only the electrons that do the moving.* Consider in particular Experiment 3, when a positively charged rod touches the neutral sphere, and both end up positive. The protons do not flow from the rod to the sphere. Instead, the rod starts out with *a deficit of electrons* it lost to the polyester. It is hungry for electrons, some of which it takes from the sphere when it touches it. The sphere then becomes positive and the rod slightly less positive. It is *as if* positive charge had migrated from the rod to the sphere. Likewise, the electric current in a wire is assigned a direction conventionally associated with the flow of positive charge, while in reality it is the electrons that are moving in the opposite direction. We will run into one exception: within a cell or battery, current is carried by positive and negative ions (non-neutral atoms with an excess or deficit of electrons).

1.3 Coulomb's law

We now progress from a qualitative description of charges to a quantitative one. How do we measure or quantify q, the charge? What precisely is the force between two *static* charges q_1 and q_2 as a function of their positions? All the answers are contained in one formula called *Coulomb's law*, after Charles-Augustine de Coulomb (1736–1806). Even though only Coulomb's name is on it, his work was the culmination of many previous efforts. He did, however, give the law its final and direct verification, which is why the unit of charge, denoted by C, is called a coulomb.

Coulomb's law says that the force between two charges q_1 and q_2, located at points $\mathbf{r}_1$ and $\mathbf{r}_2$ (as shown in Figure 1.3), is

$$\mathbf{F}_{21} = -\mathbf{F}_{12} = \frac{q_1 q_2}{4\pi \varepsilon_0 r_{12}^2} \mathbf{e}_{12} \quad \text{where} \tag{1.3}$$

$$r_{12} = |\mathbf{r}_2 - \mathbf{r}_1| \quad \text{and} \tag{1.4}$$

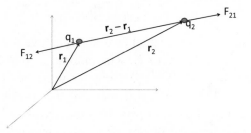

Figure 1.3   The forces between two charges q_1 and q_2 located at $\mathbf{r}_1$ and $\mathbf{r}_2$. The force $\mathbf{F}_{12}$ acts on q_1 due to q_2 and is equal and opposite to $\mathbf{F}_{21}$, similarly defined.

$$\mathbf{e}_{12} = \frac{\mathbf{r}_2 - \mathbf{r}_1}{|\mathbf{r}_2 - \mathbf{r}_1|} \quad \text{is a unit vector from } \mathbf{r}_1 \text{ to } \mathbf{r}_2 \text{ and} \qquad (1.5)$$

$$\frac{1}{4\pi\,\varepsilon_0} = 9 \cdot 10^9 \frac{N \cdot m^2}{C^2}. \qquad (1.6)$$

In the formula $\mathbf{F}_{21}$ $(= -\mathbf{F}_{12})$ is the force on charge 2, due to charge 1. The figure corresponds to the case when the charges are of the same sign and hence repel. If they are of opposite sign, the forces will be reversed and describe mutual attraction.

We will spend considerable time unearthing the numerous implications of this formula.

First, notice that regardless of how q is measured, the formula shows the charges pushing each other away if they have the same sign and attracting if they have opposite signs.

Next, the formula defines a charge of one coulomb: if two charges, 1 coulomb each, are separated by 1 meter, the repulsive force between them will be $9 \cdot 10^9$ N.

That's an enormous force (the weight of about 10,000,000 adults), and normally you don't run into 1 coulomb of unneutralized charge. A coulomb arises more commonly when we consider currents: an ampere (denoted by A) is the flow of one coulomb per second and that is not unusual. (Remember the wire is still neutral: the flowing electrons are neutralized by a static nuclear background.)

In these units the charge of the proton, denoted by e, is $1.6 \cdot 10^{-19}$ C and that of the electron is $-e$.

1.4   Properties of charge

Now we consider two fundamental facts about charge that are not part of Coulomb's law: *it is conserved and it is quantized.*

As you know, "conserved" is a physics term for saying "does not change with time." Electrical charge may migrate from body to body or place to place, but the total charge is conserved, provided you keep track of the signs. In a chemical reaction or in particle accelerators where all kinds of new particles are produced in a collision, the total charge of the final products always equals the total charge of the incoming products.

Charge is not merely conserved: it is conserved *locally.* I will illustrate what I mean by considering a conservation law that is not local. Suppose I say the number of students in the class is conserved. That means that if you count them at any time, you will get the same number. But suppose Joe suddenly disappears from the back of the room and instantaneously reappears at the front. The number of Joes is conserved. This is, however, not local conservation because Joe disappears in one part of the world and appears in another, without following an interpolating trajectory. Such non-local conservation laws do not seem to exist and do not interest us, since they cannot survive relativity: the disappearance and reappearance of Joe, simultaneous in one frame, need not be simultaneous in another frame. There we could have a period with no Joe anywhere or two Joes. If you want conservation laws that hold in all frames, they have to be local.

The conservation of electrical charge is local. So charge doesn't just disappear at one place and reappear somewhere else; it just moves around. As it moves we can follow this motion continuously. We can employ this notion to restate the local conservation of charge as follows. Suppose you mark off a closed region of space and (i) count all the charge inside and (ii) keep track of all charge entering or leaving the region via the boundary. You will find that the increase (decrease) of the enclosed charge is precisely accounted for by the charge flowing in (out) across the boundary. This would not have been the case with Joe: if you had counted the number of Joes inside a region in the back of the class and another region in front of the class, both numbers would have jumped abruptly with no accompanying flow of Joe at either boundary.

The conservation of charge had been assumed from Coulomb's time and played a big part in the explanation of the electrostatic experiments described earlier.

The second feature of charge is that it is *quantized*. That means the electrical charge does not take a continuum of possible values, unlike, say, the x-coordinate of any object, which can be any number you like. All the charges we have ever seen are integral (positive or negative) multiples of a basic unit of charge, $e = 1.6 \cdot 10^{-19}$ C. (Quarks are an exception, but they are always trapped inside particles like protons and neutrons. Their charge is also quantized but as a fraction of e. For example, the proton is made up of 2 quarks of charge $\frac{2}{3}e$ and one of charge $-\frac{1}{3}e$ and a cloud of quark-antiquark pairs of net charge 0.)

Paul Dirac (1902–1984) has provided a possible explanation of charge quantization using two ideas I have not discussed yet: magnetic monopoles and quantum mechanics. I nonetheless digress here to describe Dirac's work because by the time I cover these two topics, you may have forgotten the question we are discussing. Briefly, a magnetic monopole, if it existed, would possess an attribute called magnetic charge that comes in two signs, just like electric charge. Monopoles of like charge would repel and monopoles of unlike charge would attract with an inverse square law. All the magnetic phenomena we see, like with bar magnets, are associated with magnetic dipoles, which have net zero magnetic charge and are actually produced by electric currents. A magnetic monopole will be like a bar magnet with just the north pole, something we have not seen yet. So far we have not had direct and reproducible evidence of even a single monopole, let alone a macroscopic manifestation in the form of a magnet with just one pole. Some grand unified theories, however, predict magnetic monopoles. They are expected to be fairly heavy and to interact more strongly than electric charges. Dirac showed that if quantum theory is to consistently describe the interaction of electric charges with monopoles, *all electric charges have to be multiples of some basic unit, inversely proportional to the monopole's magnetic charge.* Thus even a single monopole, anywhere in the universe, guarantees electric charge quantization. If you believe that anything that *can* exist *will* exist, you can hope that one day these monopoles will be seen.

Let us briefly consider a few facts you may have known but not wondered about at any length.

Every electron, anywhere in the universe, is identical to every other one: it has exactly the same charge and exactly the same mass. Now, you might say, "Look, that's a tautology, because if it hadn't the same charge and the same mass, you would simply call it something else." But what makes my sentence non-empty is that there *are* many, many, many electrons that *are* absolutely identical. This never happens macroscopically. Even identical twins are not identical, and cars that are supposed to be identical are not. But at the microscopic level, elementary particles like electrons are identical to other electrons anywhere in the universe, even if they were produced in collisions in different parts of the universe. That is a mystery, at least in classical mechanics, though relativistic quantum field theory gives an explanation. (Relativistic quantum field theory is a description of fields, like the electromagnetic field, satisfying the laws of relativity and quantum mechanics. It forms the basis of all modern particle theory.) The fact that they are absolutely identical makes our life easy, because if every particle were different from every other particle, we could not make many useful predictions. For example, assuming that the hydrogen atom on a receding galaxy is identical to the hydrogen atom on the earth and observing that the light from it has a shifted frequency, we deduce the galaxy's velocity from the Doppler shift, instead of simply saying the "hydrogen" in the other galaxy is a different atom. This identity of atoms and molecules is also why structures like DNA are stable and reproducible.

Why is the charge of the electron exactly equal and opposite to the charge of the proton, given that they have very different masses and non-electric interactions? The standard model of strong, weak, and electromagnetic interaction can explain this based on a consistency condition called "anomaly cancellation." This equality of charge is the key to the neutrality of atoms and the reason behind our existence. It is also why we can detect gravity despite its relative weakness, a point we will explore in greater detail shortly.

1.4.1 Superposition principle

We now pass to an application of Coulomb's law when there are three charges q_1, q_2, and q_3 at $\mathbf{r}_1, \mathbf{r}_2$, and $\mathbf{r}_3$ as shown in Figure 1.4. What will be the force on q_3 due to the other two? Most students answer right away that it is the vector sum of $\mathbf{F}_{31}$ and $\mathbf{F}_{32}$ in our notation, i.e., the sum of the force q_1 by itself would exert on q_3 and what q_2 would exert by itself. While this is indeed correct, it is not simply a consequence of Coulomb's law. The law only says what happens when we have just one pair of charges, while the students' answer assumes that $\mathbf{F}_{31}$, the force on q_3 due to q_1, is unaffected by the presence of q_2. This is not a logical necessity or a consequence of Coulomb's law, and it is not even true

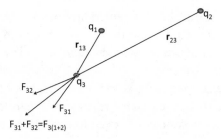

Figure 1.4 The force on q_3 due to charges q_1 and q_2 is the sum of the forces each would have exerted on q_3 in the absence of the other. This is the superposition principle.

if effects of relativistic quantum mechanics are included. We then find that when there are three charges present, certain new forces appear that cannot be described in terms of the pairwise "two-body" interactions. In other words, studying pairs of particles in isolation will not tell us everything we need to know when more than one pair is present. *However, in classical electrodynamics, which we focus on here, we may add the force q_1 would have exerted on q_3 in the absence of q_2 to the force q_2 would have exerted on q_3 in the absence of q_1 to find the force on q_3 when all three are present.* This is called the *superposition principle.* I repeat: this is not a logical consequence of Coulomb's law, but an empirically established feature of classical electrodynamics that simplifies our life enormously.

1.5   Verifying Coulomb's law

Suppose I give you Coulomb's law and ask you to verify it. How will you confirm the dependence of the force on q_1, q_2, and $r = |\mathbf{r}_1 - \mathbf{r}_2|$? Think about it, before reading the answers given by my class. An idea my class generated was that we keep q_1 and q_2 fixed and vary r, and measure the mutual force as a function of r. Here are two ideas my students came up with for measuring the force. One was to connect the charges to the two ends of a spring and watch how much it expanded (or contracted) to balance the electrical force. The other was to tie one of the charges down, let the other accelerate, and use $F = ma$ to find the force (as a function of the starting separation r).

I should point out that I accepted any procedure that was right in principle and did not require that they corresponded to what experimentalists, who are devilishly clever, would employ in practice.

Notice that to confirm the $1/r^2$ dependence, we don't have to know what q_1 and q_2 are, as long as we keep them fixed. Double (or triple) the r and see if the force (measured as described above) falls to one fourth (one ninth) of the initial. Of course, you need to consider a lot of values of r to truly nail down the r dependence.

Next, you want to verify that the force goes as the first power of q_1 and the first power of q_2. Consider the following suggestion: "Take two metal spheres, put a fixed charge on one (this will be the fixed q_1) and vary the charge on the other (q_2), and track the force. For example, if you halve q_2, the force should drop to half the old value." To halve the charge on sphere 2, you cannot simply say "Halve the number of electrons dumped on it," because the existence of electrons was unknown at Coulomb's time, and you have to play by the rules of that pre–atomic theory period. After some discussion the following acceptable strategy was generated.

Take two charged spheres numbered 1 and 2, and find the force between them. Do not touch number 1. Take sphere number 2 and bring it in contact with an identical uncharged sphere and separate them. *By symmetry*, they should each end up with $\frac{1}{2}q_2$. Even though we did not know what q_2 was, we know we have halved it in the process. Put 2 at the old location and see if the force has halved.

We physicists love these symmetry arguments, which transcend physics and border on philosophy: When two identical spheres are made to share some charge, there is no reason why nature would not give each exactly half the total.

Returning to the spheres, by another splitting, you can get a sphere with charge $\frac{1}{4}q_2$. By making a sphere with $\frac{1}{4}q_2$ share its charge with an identical one carrying $\frac{1}{2}q_2$, you can get one with $\frac{3}{8}q_2$ and so on.

That's how we can verify that the force depends linearly on q_2. Of course, it must then also depend linearly on q_1, because it's up to us to decide which one we want to call q_2.

Here is another challenge. I give you a charged sphere and I want you to find how much charge it carries, in coulombs. What will you do? When a student said: "Put it in the vicinity of a reference charge and then measure the acceleration," I asked her how to get a known reference charge. Her answer (correct, but by no means unique) was as follows. Take these two identical spheres, each with the same unknown charge q

(say by making them share $2q$ equally), place them at a known separation, say a meter, and measure the force needed to keep them where they are. Use Coulomb's law to extract q^2.

If you constantly think about how you would measure anything you work with, you'll understand physics more deeply and also find solving problems a lot easier. If instead you are busy pushing symbols around and chasing factors of 2π, you will eventually be lost.

1.6   The ratio of gravitational to electric forces

Recall the claim that F_g/F_e, the ratio of gravitational to electric forces, is of the order 10^{-40}. We have to specify how we got this number. Our task is not like selling toothpaste where one can glibly say it makes teeth 3.14 times whiter: that is a different game, not subject to any rules.

We *do* have to explain how we come up with 10^{-40}. It turns out the answer does depend slightly on what comparison method we choose. There will be some variations, but they will be tiny compared to the enormous ratio, i.e., the number of zeros may range from 37 to 43 depending on the comparison method.

Consider two particles of mass m_1 and m_2 and charges q_1 and q_2, a distance r apart. We find

$$\frac{F_g}{F_e} = \left(\frac{Gm_1 m_2}{r^2} \right) \cdot \left(\frac{q_1 q_2}{4\pi \varepsilon_0 r^2} \right)^{-1} = \frac{Gm_1 m_2 \cdot 4\pi \varepsilon_0}{q_1 q_2}. \qquad (1.7)$$

Fortunately the ratio does not depend on the separation r we choose for comparison since both forces fall as $1/r^2$. It does, however, depend on the charges and masses. If there were only one kind of particle (and its antiparticle) in the universe, we could plug in its mass and charge. But there are of course many. However, we can focus on the two key players out of which everything we see is made, the proton and electron.

If we take two electrons we get

$$\frac{F_g}{F_e} = \frac{6.7 \cdot 10^{-11} \cdot (9 \cdot 10^{-31})^2}{(1.6 \cdot 10^{-19})^2 \cdot 9 \cdot 10^9} \simeq 2.3 \cdot 10^{-43}. \qquad (1.8)$$

If we take a proton and an electron, the ratio will be of order 10^{-40}, and if we take two protons it will be of the order 10^{-36}. Gravity is incredibly weaker than electricity, no matter how you slice it.

If gravity is so weak, how did anyone discover it? Suppose we knew only about electricity and didn't know about gravitation. One way to find out that there is an extra force is to measure the force between two particles to a fantastic accuracy and find some discrepancy in the 40th decimal place. But that's not how it was done, of course. Everyone seems to know the reason: the electric force, even though it's very strong, comes with opposite charges. Consider the planet Earth. It has lots and lots of atoms and lots of charges in each atom, but every atom is neutral. The moon too has lots and lots of atoms, but they're also neutral. So all the powerful electric forces amount to nothing, due to internal cancellations. But the *mass of the electron does not cancel the mass of the proton in determining the mass of the atoms.* So mass can never be hidden, whereas charge can be hidden. That's the reason why, in spite of the incredible amount of electrical forces they're potentially capable of exerting, the earth and moon see each other as neutral entities. In most cosmological calculations you can forget the electric force. The remaining (gravitational) force plays a dramatic role in the structure of the universe.

It is this feature of gravity, that mass cannot be hidden, that allowed us to infer existence of *dark matter*. Let us recall how we know of its existence in our own galaxy. If a star is orbiting the center of our galaxy, just by using Newtonian gravity, by knowing the velocity of the object as it goes around, you can calculate how much mass is enclosed by the orbit. In case you forgot, for a circular orbit, the velocity at radius r is constrained by

$$v^2 r = GM \tag{1.9}$$

where M is the enclosed mass. If you take orbits of bigger and bigger radius, you will find more and more enclosed mass, until you reach orbits as big as the visible galaxy. So far so good. But you find that as you consider bigger orbits, you still keep picking up more mass, out to some great distance. That is the dark matter halo of our galaxy. Dark matter is made of hard-to-detect particles, but its gravitational effects cannot be hidden. It occurs everywhere, even in galaxy clusters. Physicists around the world, including here at Yale, are trying to find dark matter. The problem is, we don't know exactly what particles dark matter is made of. They are not any of the usual suspects, which would have interacted with other particles and been detected already. You have to build detectors that will detect that unknown species. And you're hoping that one of these dark matter particles will collide with the stuff in your detector and trigger a reaction. Of course, there will be lots of reactions due to other particles. That's called *background*. You've got to throw the background events out and hope that whatever is left over is due to dark matter. One diagnostic is that while normal particles will typically collide multiple times in the detector, we weep for joy if the dark matter particles collide even once. The particles that form dark matter are very interesting to astrophysicists and particle physicists, and there are many candidates.

1.7 Coulomb's law for continuous charge density

We conclude with one final variant of Coulomb's law. We have seen how to use the superposition principle to add up the pairwise forces on any one charge due to many others. But often we consider problems where the charges are continuous. (In real life everything is discrete, made of protons and electrons, but at some macroscopic scale, it will look like charge is continuous, just as water, which is made of molecules, appears to be a continuous fluid.) We tackle this variation just as we did the problem of gravity due to a continuous mass distribution: we replace the sums by integrals.

As an example, consider a circular wire of radius R with λ coulombs per meter, lying in the xy-plane with its center at the origin, as shown in Figure 1.5. I want to find the electric force it exerts on a point charge q

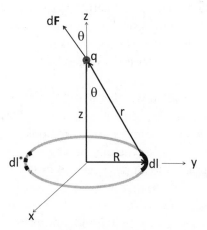

Figure 1.5   The electric force due to a loop in the xy-plane, on a charge located on the z-axis. The highlighted segment of length dl has charge λdl and exerts a force $d\mathbf{F}$. We keep only the vertical part along the z-axis since the diametrically opposite segment dl^* (shown by a dotted curve) will cancel the horizontal part.

located at a height z on the z-axis. I divide the loop into tiny segments of length dl. Consider the tiny highlighted segment of length dl perpendicular to the yz-plane, as shown in Figure 1.5. It can be treated as a point charge λdl. It exerts a force $d\mathbf{F}$ of magnitude

$$|d\mathbf{F}| = \frac{q \cdot \lambda dl}{4\pi \varepsilon_0 (R^2 + z^2)}. \tag{1.10}$$

The force vector lies in the yz-plane. We need keep only the vertical part, pointing up the z-axis, since the horizontal part will be canceled by the segment dl^* at the diametrically opposite point, shown as a dotted curve. The total vertical force has a magnitude given by integration:

$$F_z = \int dF_z = \int \frac{q \cdot \lambda dl}{4\pi \varepsilon_0 (R^2 + z^2)} \times \frac{z}{\sqrt{R^2 + z^2}} \tag{1.11}$$

$$= \frac{q \cdot \lambda \cdot 2\pi Rz}{4\pi \varepsilon_0 (R^2 + z^2)^{3/2}}, \tag{1.12}$$

where the factor

$$\frac{z}{\sqrt{R^2 + z^2}} = \cos\theta \tag{1.13}$$

projects out the vertical part of dF and $2\pi R$ is the integral over dl.

Once you've done such a calculation you must think of ways to test the result. Here are two good tests. First, if you set $z = 0$, i.e., find the force at the center of the loop, you should get zero since every segment that exerts a force toward the center is countered by the diametrically opposite one. This is true of our answer.

The second test is to go very far away from the loop, when it should look like a point charge $\lambda \cdot 2\pi R$. How far is far? Any one length, say the diameter of my head, can be made to look impressively large or depressingly small by choosing the unit of length to be a micron or a light year. Only ratios of lengths can be described as large or small, and to be useful, the ratios should be relative to some intrinsic length in the problem. For my head to appear point-like I should be seen from a distance many times the size of my head. For the loop to appear point-like, it should be viewed from a distance $z \gg R$. You may verify that in this limit the formula indeed reduces to the force between q and a point charge $2\pi R\lambda$, separated by a distance z.

CHAPTER 2

The Electric Field

I begin with a review of a subset of ideas from the last chapter that you will need going forward.

2.1 Review of key ideas

Several species of particles, such as protons and electrons, have an attribute called electric charge or simply charge. Others like the neutron do not. Objects with charge exert forces on other objects with charge. The coulomb is the unit for measuring charge. It is denoted by C and is defined by Coulomb's law, which I repeat for convenience:

$$\mathbf{F}_{21} = -\mathbf{F}_{12} = \frac{q_1 q_2}{4\pi\,\varepsilon_0 r_{12}^2}\mathbf{e}_{12} \quad \text{where} \tag{2.1}$$

$$r_{12} = |\mathbf{r}_2 - \mathbf{r}_1| \quad \text{and} \tag{2.2}$$

$$\mathbf{e}_{12} = \frac{\mathbf{r}_2 - \mathbf{r}_1}{|\mathbf{r}_2 - \mathbf{r}_1|} \quad \text{and} \tag{2.3}$$

$$\frac{1}{4\pi\,\varepsilon_0} = 9 \cdot 10^9 \frac{N \cdot m^2}{C^2}. \tag{2.4}$$

In the formula, q_1 and q_2 are charges of the particles located at $\mathbf{r}_1$ and $\mathbf{r}_2$, and $\mathbf{F}_{12}\,(=-\mathbf{F}_{21})$ is the force on charge 1, due to charge 2.

If two charges, each equal to 1 C, are placed one meter apart, they will experience a force equal to $9 \cdot 10^9$ N. Once such a reference charge (or a known fraction of it) is given, any other charge may be measured using Coulomb's law. (Here is a way to create a reference charge. We take two identical uncharged spheres, charge one by an unknown amount q and let it share its charge with the other. Each then has $q/2$ and the force between them at a known separation then determines $q^2/4$.)

Charge can be positive or negative. From Coulomb's law we see that like charges repel (i.e., $\mathbf{F}_{12}$ and $\mathbf{F}_{21}$ point away from each other) while opposite charges attract. The proton and electron have charges $e = 1.6 \cdot 10^{-19}$ C and $-e$ respectively. The neutron has no charge. Finally, we need the superposition principle to go beyond a pair of charges. This principle allows us to compute the force on any one charge due to many others by adding their individual contributions. The force between a pair of charges is indifferent to the presence of other charges.

The total charge of a collection of charges is the algebraic sum of the charges of the constituents. As a result, an atom with an equal number of electrons and protons is electrically neutral. This is the reason we can detect the gravitation force between the earth and the moon: given their electrical neutrality, only gravity remains and is detectable despite being 10^{40} times weaker.

2.2   Digression on nuclear forces

Now for a brief digression. We can understand the atom as resulting from the attraction between the protons in the nucleus and the electrons. But what are the protons doing, so close to each other inside a nucleus of size 10^{-15} m? Why doesn't the Coulomb repulsion make the nucleus explode? The answer, which you might already know, is that protons experience another force, the *nuclear or strong force*, which is attractive and much stronger than their Coulomb repulsion. If that is so, how did we manage to detect the relatively tiny electrical force hiding underneath this nuclear force? The answer has to do with the fact that the nuclear force F_n *has a very different distance dependence compared to the electrical force F_e*. It varies with distance roughly as

$$F_n = A\frac{e^{-r/r_0}}{r^2} \tag{2.5}$$

where $r_0 \simeq 10^{-15}$ m is called the *range of the nuclear force*. The electric force of course behaves as

$$F_e = \frac{k}{r^2} \qquad (2.6)$$

where k includes q, ε_0, etc. As a result of the different r-dependences, *the ratio F_n/F_e, unlike F_g/F_e, is distance dependent*:

$$\frac{F_n}{F_e} = \frac{Ae^{-r/r_0}}{k}. \qquad (2.7)$$

Deep inside the nucleus, i.e., $r \ll r_0$, $e^{-r/r_0} \simeq 1$ and

$$\frac{F_n}{F_e} = \frac{A}{k} \qquad (2.8)$$

and the nuclear force dominates because $A \gg k$. As we go to distances $r \gg r_0$, the exponential e^{-r/r_0} completely suppresses the factor $\frac{A}{k}$ and the Coulomb repulsion wins. Of course, the crossover between the two forces is not abrupt, but occurs over the rough dimension of the nucleus.

Now for the role of the neutron in the nucleus. What are the neutrons doing here? Whereas they are nobodies with respect to Coulomb interactions, the attractive nuclear force between two neutrons or between a neutron and a proton is as strong as the nuclear force between two protons. (This is one reason protons and neutrons are collectively called *nucleons*.) As the nuclei get bigger, the exponential suppression of the attractive nuclear force really kicks in, while the Coulomb repulsion between protons lives on. So additional protons eventually cause instability, while neutrons contribute to the stability: they bring in nuclear attraction without the Coulomb repulsion that necessarily accompanies protons and tries to blow up the nucleus. There are far more neutrons than protons as the nuclei get heavier. (For example, $^{235}_{92}$ U has 92 protons and 143 neutrons.) But neutrons can only do so much: the laws of quantum mechanics force the added neutrons to have more and more kinetic energy, and beyond some size nuclei are unstable and decay into stable nuclei, say by emitting α particles, which are He nuclei, made of two protons and two neutrons.

This ends the brief digression on the complicated subject of nuclear physics.

2.3   The electric field E

Now for the main business of this chapter: the seminal notion of the electric field.

Let us rewrite the force on q_2 due to q_1 as follows:

$$\mathbf{F}_{21} = \frac{q_1 q_2}{4\pi\,\varepsilon_0 r_{12}^2}\mathbf{e}_{12} \tag{2.9}$$

$$= \frac{q_1}{4\pi\,\varepsilon_0 r_{12}^2}\mathbf{e}_{12} \cdot q_2 \tag{2.10}$$

$$\equiv \mathbf{E}(\mathbf{r}_2)q_2. \tag{2.11}$$

What we have done is to write the force on q_2 due to q_1 as a product of q_2 and $\mathbf{E}(\mathbf{r}_2)$, which is called *the electric field at the location of q_2*.

Where does this cosmetic factorization of F_{21} into $\mathbf{E}(\mathbf{r}_2)$ and q_2 lead us?

First we will say that the interaction between q_1 and q_2 is a two-step process:

Step 1. The charge q_1 produces a field $\mathbf{E}(\mathbf{r}_2)$ at the location of q_2 given by

$$\mathbf{E}(\mathbf{r}_2) = \frac{q_1}{4\pi\,\varepsilon_0 r_{12}^2}\mathbf{e}_{12}. \tag{2.12}$$

Step 2. The charge q_2 responds to the field by feeling a force $\mathbf{F}_{21} = q_2\mathbf{E}(\mathbf{r}_2)$.

Thus we have split the simple Coulomb interaction into two parts: the creation of the field by one charge and the response to that field of the other. Of course, we could just as well factorize $\mathbf{F}_{12}$ as $\mathbf{E}(\mathbf{r}_1)$, the field produced by q_2 at the location of q_1, times q_1.

It will be a while before you can appreciate the cleverness behind this factorization. For now, just understand the terminology and the procedure.

Notice two things.

Thing 1: While it takes two charges to feel a force, it takes only one charge to produce a field. A charge q at the origin produces the following field at point **r**:

$$E(\mathbf{r}) = \frac{q}{4\pi\,\varepsilon_0 r^2}\mathbf{e}_r \tag{2.13}$$

where $\mathbf{e}_r = \mathbf{r}/r$ is a unit vector in the radial direction, from the origin (where q is) to **r** where **E** is being computed.

Thing 2: The field due to q is non-zero everywhere, not just where there is another charge to feel the field.

We think of $\mathbf{E}(\mathbf{r})$ as a condition in space, produced by the presence of q. Something is different at **r** when q is around, compared to when it isn't: with q present, any charge placed at **r** will feel a force, while without it, it will just sit there.

A field is a force waiting to happen: just put a test charge there and you will see it in action. The field of a charge is felt only by other charges.

If there are many charges, we invoke the superposition principle: the field at some **r** due to many charges will be the (vector) sum of the fields due to each one. You have to perform this possibly very complicated vector sum to *calculate* the field there. To *measure* it is easier: put a known test charge q at **r**, equate the force it experiences to $q\mathbf{E}$. If $q = 1$ C, the force and **E** are numerically equal *but dimensionally different*. This is why one says the *field is the force on a unit charge*.

Let us get some practice by computing **E** at the corner (a, a) of a square with charges q at the other three corners $(0,0), (a,0)$, and $(0,a)$ as shown in Figure 2.1. Once you get this, you can add twists: make the square

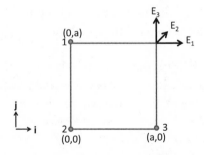

Figure 2.1   The electric fields $\mathbf{E}_1$, $\mathbf{E}_2$, and $\mathbf{E}_3$ at (a, a) in terms of the unit vectors **i** and **j**, due to three equal charges q located at $(0,a), (0,0)$, and $(a,0)$. The total field at (a, a) is the vector sum of the three pieces.

into a rectangle, make the charges unequal and of different signs, and so forth.

The figure shows separately the three contributions $\mathbf{E}_1$, $\mathbf{E}_2$, and $\mathbf{E}_3$ at (a, a) due to three equal charges q located at $(0, a), (0, 0)$, and $(a, 0)$:

$$\mathbf{E}_1 = \frac{q}{4\pi\,\varepsilon_0 a^2}\mathbf{i} \tag{2.14}$$

$$\mathbf{E}_2 = \frac{q}{4\pi\,\varepsilon_0 \cdot 2a^2}\frac{\mathbf{i}+\mathbf{j}}{\sqrt{2}} \tag{2.15}$$

$$\mathbf{E}_3 = \frac{q}{4\pi\,\varepsilon_0 a^2}\mathbf{j}. \tag{2.16}$$

I hope you can see why the field $\mathbf{E}_2$ is half as big as the other two, and points in the 45° direction, i.e., along $\mathbf{i}+\mathbf{j}$. The corresponding *unit* vector is obtained by dividing $\mathbf{i}+\mathbf{j}$ by its length, $\sqrt{2}$. It is easy to add the three pieces to get the total field at (a, a):

$$\mathbf{E}(a, a) = \frac{q}{4\pi\,\varepsilon_0 a^2}\left(1 + \frac{1}{2\sqrt{2}}\right)(\mathbf{i}+\mathbf{j}). \tag{2.17}$$

We will soon be doing numerous versions of this problem, computing the field due to various charge distributions, discrete and continuous. But at the outset I must warn you that Coulomb's law, as stated, violates relativity. Suppose you and I hold two positive charges q_1 and q_2, and I am one light-year away from you. You hold your charge in place by pushing against the repulsive force mine exerts. Now I suddenly move mine away from you by a bit. You will feel the reduced repulsion right away. I have managed to send you a signal instantaneously and this faster-than-light signaling is disallowed.

Does electrodynamics then violate relativity? No, we will see it is remarkably compatible with it. What happens in the *complete* theory is that if I wiggle my charge, the signal will reach you a year later, traveling at the speed of light. Until such time, the field at your location due to my charge will remain unaltered.

Of the two parts of the story, the computation of $\mathbf{E}$ in terms of the charges and the response of a test charge q to the field, only the former gets modified in the complete theory. The field at some space-time point, say $(\mathbf{r} = \mathbf{0}, t = 0)$, will receive contributions from all other charges based not on what they are doing now, but what they were doing at an earlier time. The amount by which we have to go back in time is just the time light would take to go from the source of the field to $(\mathbf{r} = \mathbf{0}, t = 0)$. A charge that was a light-year away a year ago will be contributing to the field at $(\mathbf{0}, 0)$. This is called the *retarded interaction*. We will discuss this briefly toward the end of chapter 15.

What is the role of Coulomb's law then? *In principle* it is to be used when none of the charges is moving. In this case, the delay does not matter: since every charge knows where every other charge is, all the signals have arrived and are unchanging. In *practice* we also use Coulomb's law provided the charges in question are near each other and moving at $\frac{v}{c} \ll 1$, and retardation effects are negligible, as in most electrical circuits.

Remarkably, the second part of the story, the equation giving the response to the field, $\mathbf{F} = q\mathbf{E}(\mathbf{r})$ remains unaltered in the final theory of electrodynamics. It is a *local* relation between the field at a space-time point and the charge at that point. The field at any point could be a very complicated function of every charge in the history of the universe, but the response (of test charge q) to it depends only on its current value at the location of the charge q. It does not care what went into producing $\mathbf{E}$. So the field concept is essential to making electromagnetic theory compatible with relativity.

2.4   Visualizing the field

Let us go back now to the simplest problem in the world: the electric field due to one charge. The formula is very simple. Let's put that charge q at the origin. The electric field is

$$\mathbf{E}(\mathbf{r}) = \frac{q}{4\pi\,\varepsilon_0 r^2}\mathbf{e}_r \tag{2.18}$$

where $\mathbf{e}_r$ is the radial unit vector, $\mathbf{r}/r$. Sometimes you see Eqn. 2.18 rewritten as

$$\mathbf{E}(\mathbf{r}) = \frac{q}{4\pi\,\varepsilon_0 r^3}\mathbf{r}. \tag{2.19}$$

If you encounter this version, do not get fooled into thinking the field is falling as r^{-3}. It's still r^{-2} because there's an extra $\mathbf{r}$ at the top.

So here you have this formula. If you're a person who likes to work with formulas this is all you need. You manipulate the stuff on paper, and you add different fields. But people like to visualize this. How do we visualize this? That's the real question. Suppose someone asks you, what's the height above sea level of a certain part of the United States? You've got some mountains. You've got some valleys. Somebody can give you a function that gives you the height at any point in the United States, but it's more revealing for most of us to have some kind of a contour map. Each contour is a different height. If you go hiking, you want this map, not the corresponding function. Similarly, you want a pictorial representation of this electric field. Unlike the height function, which is a scalar, i.e., just a number at each point, the electric field is a vector $\mathbf{E}(\mathbf{r})$ at each point $\mathbf{r}$.

Suppose I want to communicate to you pictorially the information contained in the function in Eqn. 2.18. I begin with the modest goal of describing $\mathbf{E}$ at just one point, 1, in Figure 2.2.

Like many of the figures I will show you, it is a two-dimensional cross section of a three-dimensional configuration.

I take that point and draw an arrow there to represent $\mathbf{E}(1)$. The length of the arrow gives the size of the field in some scale, so many centimeters of length for each *newton/coulomb* of field. That is the electric field at that point 1. Then I pick a few points, say eight in all, at the same radius. (The points lie on a circle in the plane of the paper, while real charges live in three dimensions. You should think of this as a cross section of what happens on a sphere of the same radius.) *The points are*

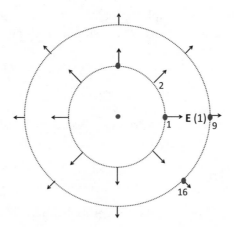

Figure 2.2   The electric fields $\mathbf{E}$, at a few representative points on circles of two different radii around a charge at the origin.

also uniformly distributed to reflect the isotropy of the electric field. The figure is already telling you something: the field is radially outward and same in magnitude at points with the same r. Be very careful about what it is *not* telling you. An arrow is not telling you what is happening throughout the length of the arrow. It's telling you what's happening at the starting point, the tail. You understand the arrow is in your mind. It's not really sticking out in space. It's a property or a condition at that starting point, but we've got to draw it somehow, so we draw it that way. (If **E** were the velocity of a fluid in a river, the arrow starting at some point **r** *would be the velocity at* **r** *only*, even if the arrow is a foot long and passes over regions where the actual water velocity is totally different in magnitude and direction.)

What happens when we go further out in r? If I put a test charge further away, it is still going to be repelled radially, but less. So I draw a few arrows at representative points 9–16 and make them shorter, to reflect the $1/r^2$ nature. I can draw a few more arrows and hope you get the picture from the few discrete sampling points.

Then someone had this clever idea: join all these arrows as in Figure 2.3. These are called *field lines.*

The actual charge and lines should be drawn in three dimensions but Figure 2.3 shows what happens in a representative plane, which I assume for illustrative purposes has 8 lines. What have I gained and what have I lost? Previously I knew the field direction only at the chosen

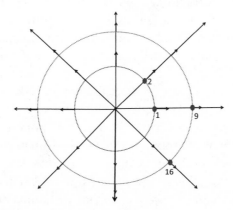

Figure 2.3   The electric field lines due to a charge at the origin. The actual charge and lines live in three dimensions, and the figure shows what happens in a representative plane, which I assume has 8 lines.

points at some radii, but now I know it throughout each line. On the other hand, I have lost information on the magnitude of the field: the arrows, whose lengths encoded $|\mathbf{E}|$ are gone and replaced by the field lines that go on forever. They merely tell me in which direction $\mathbf{E}$ points, but not how big it is. They just tell me that the charge is pushing every (positive) test charge out radially and that the force is isotropic.

But, thanks to a miraculous property of the Coulomb force, namely that it falls like $1/r^2$, there is information even on the strength of the electric field. That information is contained in the *density of electric field lines*. By density of lines, I mean the number of lines crossing a surface perpendicular to the lines, divided by the area of that surface.

To grasp this, let us pick some convention, that for every coulomb of charge, we will draw a certain number of lines emanating from it, say 64. If we draw a sphere of some radius surrounding the coulomb, 64 lines will cross that sphere, everywhere perpendicular to the surface *and of uniform density*, reflecting the isotropy of the electric field of a point charge. If I draw a bigger sphere, the same 64 lines will cross that sphere also, *but they will be less dense*, with fewer lines per unit area. Since the area of the sphere grows as r^2, the density of lines will fall as $1/r^2$. *This is exactly how the field strength* $|\mathbf{E}| \equiv E$ *falls with distance*.

This wonderful ability of field lines to encode the magnitude and direction of the field exists only because we are living in three dimensions (where the sphere surrounding the charge has an area that grows as r^2) and dealing with a field that falls as $1/r^2$. For example, if a radial field that falls as $1/r^3$ is represented by such lines, their direction will faithfully represent the direction of the field, but their density will not represent the field strength E.

The lines help you visualize the field strength. Wherever the lines are dense, the field is strong. Wherever the lines are spread apart, the field is weak. It is a very precise statement. The only thing not precise is how many lines you want to draw per coulomb. That is really up to you, but you must be consistent. Once you choose 64 lines per coulomb, and you are dealing with a charge of two coulombs, you should draw 128 lines coming out of it uniformly spread out. As long as you do that, the number of lines crossing per unit area will be proportional to the field.

Now, no matter how many lines you pick per coulomb, there will be spaces between the lines. That does not mean the field is zero between the lines. The field is continuous in space and not concentrated literally on these lines. You must read between the lines. For example, at a point

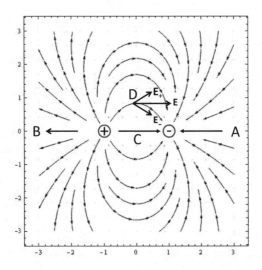

Figure 2.4   The electric field due to a dipole. The two vectors shown at point D are the contributions $\mathbf{E}_+$ and $\mathbf{E}_-$ to $\mathbf{E}$ from the two charges. Their vector sum will be horizontal.

midway between two adjacent lines, the field is pointing midway as well, with an intensity given by the density of the lines at that radius.

Clearly, if we consider the field of a negative charge, the lines will point inward, reflecting the attraction felt by the test charge.

The notion of field lines extends beyond the field due to just one charge. Figure 2.4 shows the field due to a pair of charges $\pm q$, called a *dipole*. The first thing I want you to notice is that very close to any one charge the lines point uniformly and radially out or in depending on its sign, no matter how many other charges there are. This is because as we approach any charge, the field it produces diverges as $1/r^2$ and swamps the finite contributions from the others.

Next consider the field lines labeled A, B, C and D. Look at line A. It is clear that a (positive) test charge placed anywhere on A (which goes from the minus charge all the way to infinity) would be attracted to the minus end of the dipole, which attracts it more than the plus end repels it. The reverse argument explains the outward pointing line B, on which repulsion wins. On the line C, pointing from the plus to the minus, both charges apply a force to the right. Finally, look at the line labeled D. It contains the point D, which lies on the perpendicular bisector of the line joining the two charges. Notice the field line at D is horizontal. This directionality follows from a symmetry argument. The minus charge attracts the test charge on a line from D to itself; the plus charge repels it along a line joining it to D. Both forces have the same magnitude (since D is equidistant from them), canceling vertical components, and additive horizontal components.

The figure also makes it clear that *any* closed surface enclosing only the plus (minus) charge will intercept 10 lines going outward (inward). If we draw any surface enclosing both charges, the net flow in or out will be zero. This is your qualitative introduction to *Gauss's law*, which relates the net (outgoing minus incoming) number of lines leaving a closed surface to the net enclosed charge. (If the surface is convoluted a field line may exit, reenter, and exit again for example. This will count as a *net* exit of one line.)

Figure 2.5 shows two identical positive charges. Far from both, it will look like the field of a point charge of double the strength. The number of lines crossing a closed surface enclosing both charges is the sum of the lines emanating from each. The closed surface I have shown is a nice ellipse, but the lines crossing it will not change if I distort it in any way that does not exclude either charge (once again an example of Gauss's law).

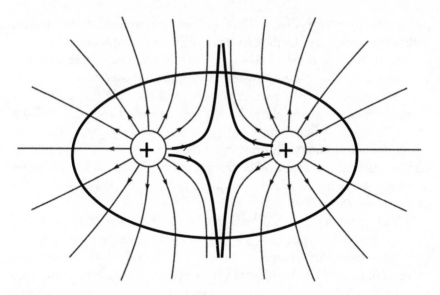

Figure 2.5   The electric field due to two positive charges. Far from both, it looks like the field of a point charge of double the strength. The number of lines crossing a closed surface enclosing both charges is the sum of the lines emanating from both.

What if we had two opposite but unequal charges, say 10 C and $-5C$, and associate 10 and 5 lines with each? You can draw the sketch yourself with the following features: near each charge you can forget the other, 5 lines will flow from the 10 C to the -5 C, and the rest will escape to infinity, becoming radially outward asymptotically, like those of a point charge $10\ C - 5\ C = 5\ C$.

Consider finally a case of a continuous charge distribution. Two parallel metallic plates carry uniform charge densities $\pm\sigma$ (measured in C/m^2). This is called a *parallel plate capacitor* and is depicted in the top half of Figure 2.6. At the left is the view looking down at an angle and at the right the view end-on, with the plates coming out of the paper. What do the field lines look like? We know they must start at the positive charges and end at the negative charges. The bottom half shows the deflection of a positively charged particle injected from the left with velocity $\mathbf{v}_0$.

If you imagine the plates to be very large in area, the figure shows the part far from the edges. (At the edges the lines bulge out a bit midway between the top to the bottom plate.) You should not simply accept even the qualitative aspects of the preceding picture. Look very near the positive plate. In the absence of the negative plate, the field lines will be emanating perpendicularly away from it with equal density above and below by symmetry. The same goes for the negative plate, but with the lines flowing into it. If you superpose the two plates, you can see the two plates aid each other in the region between, with both producing downward pointing fields

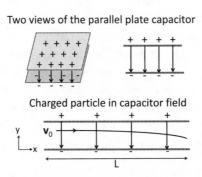

Two views of the parallel plate capacitor

Charged particle in capacitor field

Figure 2.6   The top half shows two views of a parallel plate capacitor and the field inside it. It is uniform except near the edges, where it bulges out (not shown). The bottom shows the trajectory of a positively charged particle shot into it from the left.

there, just like along the line joining the charges in a dipole. But, if you follow the dipole analogy, and consider points just above the top plate, you expect the fields from the two plates to oppose, but with the upper plate winning since it is closer. So some lines must point up just above the upper plate. Yet the figure shows no lines above the upper plate and has all the lines coming straight down, *as if there is a perfect cancellation of the fields due to the two plates, despite the different separations.* The same goes below the lower plate, where there are no field lines. The answer to this mystery will be revealed when we compute the field due to each plate later and find that the field due to an infinite plate of uniform density *does not weaken at all as we move away from the plate!* It is perpendicular to the plate, and it has the same magnitude no matter how far we go, *even though the contribution from the individual charges on the plate fall as* $1/r^2$. Consequently, the plates cancel each other completely outside the plates (above the top plate and below the bottom plate) and aid each other inside. So, the figure is correct only if it represents a finite section of an infinite parallel plate capacitor, or far from the edges of a very large capacitor. The real finite plate problem is far more complicated: doable in principle, but not easy.

Given that the field is limited to the space between the plates, questions still persist. Why is the field uniform between the plates in the infinite capacitor, unchanging as we move up and down or side to side?

First of all, it must be clear that in the infinite capacitor, the field at a given plane parallel to the plates, say at a height $y = 2$ *cm* above the lower one, cannot vary as we move parallel to the plates, say in the x-direction. Every point at some y is like every other point: if we look to the left or right, from any of these points, we see the two plates running to $\pm\infty$.

Here is a more detailed argument, based on cause and effect. Suppose the field varies with the x coordinate, i.e., has a non-trivial profile with some features, some ups and downs in strength. If I slide the plates to the right by 2 *cm*, these features should follow. On the other hand, I can argue that they should not shift since the cause behind the field, namely the infinite, charged plates, look exactly the same before and after I slide them. If the plates look the same after a horizontal shift, so must the field they produce.

Had the plates been finite, this would no longer have been true. There would have been a preferred midpoint and edges where the plates end. If you move this finite system horizontally, it will look different after the shift and so the field need not be x-independent. Indeed, it is not, with bulges at the end.

So the field is constant in x. Why is the field independent of the y coordinate as well? After all, the y dimension is finite and as a result not all y's are equivalent. We *can* tell if we are moving toward or away from either plate. Well, suppose the field got weaker as we approached the middle. The lines must spread out, i.e., the spacing between them must increase. But this is impossible in the infinite case: if you move a line, say second from the left in Figure 2.6, away from its neighbor on the left, to weaken the field to the left, you move it closer to the neighbor on the right, increasing the field between them. Such variations with x are not allowed in the infinite capacitor, as we have seen. So the lines have no choice but to go straight down, preserving their density as y varies. Again, variation in x and y is allowed in a finite capacitor: the lines do get less dense as we move toward the center, and they bulge out at the two ends.

In any event, if the field is uniform, the force will be uniform, just like force of gravity near the surface of the earth. Consequently the particle we shoot in from the left will follow a parabolic path, as depicted in the lower half of the figure. More on this later.

2.5 Field of a dipole

We will now buckle down and calculate the precise value of the electric field due to a dipole. We will write a formula that is good at all points, but evaluate it only at some select places where the calculation is easier. We will examine the field at distances large compared to the separation between the charges. In a later chapter we will find a more efficient way to find the field using the notion of a potential.

Figure 2.7 shows a charge q at $(a,0)$ and a charge $-q$ at $(-a,0)$. Consider the field at a generic point (x,y). (Once we have the field in the xy-plane, we can simply rotate the figure around the x-axis to get

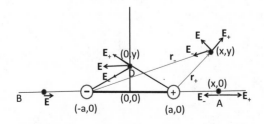

Figure 2.7   Dipole field: $\mathbf{E}_{\pm}$ are due to $\pm q$ located at $(\pm a, 0)$.

the answer in three dimensions. In other words, the cross section on the xy-plane is identical to what we will find in any other planar slice through the x-axis. This point will be fortified soon with symmetry arguments.)

Recall that the field at the point $\mathbf{r}$ due to a single charge q at the origin is

$$E(\mathbf{r}) = \frac{q}{4\pi\,\varepsilon_0}\frac{\mathbf{r}}{r^3}. \tag{2.20}$$

If the charge were not at the origin (as in the application that follows immediately), $\mathbf{r}$ would be the *vector from where the charge is to where we want the field*.

The field due to both $\pm q$ at a generic point (x, y) is the sum of the individual contributions $\mathbf{E}_{\pm}$. These in turn can be evaluated by setting $\mathbf{r} = \mathbf{r}_{\pm}$ in Eqn. 2.20 and adding them as follows:

$$\mathbf{E}_+ = \frac{q}{4\pi\,\varepsilon_0}\frac{\mathbf{r}_+}{r_+^3} = \frac{q}{4\pi\,\varepsilon_0}\frac{\mathbf{i}(x-a)+\mathbf{j}\,y}{((x-a)^2+y^2)^{3/2}} \tag{2.21}$$

$$\mathbf{E}_- = -\frac{q}{4\pi\,\varepsilon_0}\frac{\mathbf{r}_-}{r_-^3} = -\frac{q}{4\pi\,\varepsilon_0}\frac{\mathbf{i}(x+a)+\mathbf{j}\,y}{((x+a)^2+y^2)^{3/2}} \tag{2.22}$$

$$\mathbf{E} = \frac{q}{4\pi\,\varepsilon_0}\left[\frac{\mathbf{i}(x-a)+\mathbf{j}\,y}{((x-a)^2+y^2)^{3/2}} - \frac{\mathbf{i}(x+a)+\mathbf{j}\,y}{((x+a)^2+y^2)^{3/2}}\right]. \tag{2.23}$$

This general formula may be a bit hard to digest. Here are some simpler special cases.

At a generic point on the x-axis ($y = 0$) both $\mathbf{E}_{\pm}$ are horizontal and

$$\mathbf{E} = \frac{q}{4\pi\,\varepsilon_0}\left[\frac{\mathbf{i}(x-a)}{|x-a|^3} - \frac{\mathbf{i}(x+a)}{|x+a|^3}\right]. \tag{2.24}$$

(Remember that $\lim_{y\to 0}[\sqrt{(x\pm a)^2+y^2}]^3 = |x\pm a|^3$ and not $[x\pm a]^3$.) For a point like A with $x > a$, we can drop the absolute value sign and obtain

$$\mathbf{E} = \frac{q}{4\pi\,\varepsilon_0}\mathbf{i}\left(\frac{1}{(x-a)^2} - \frac{1}{(x+a)^2}\right) \tag{2.25}$$

$$= \frac{q}{4\pi\,\varepsilon_0}\mathbf{i}\,\frac{4ax}{(x^2-a^2)^2} \tag{2.26}$$

$$\equiv \frac{\mathbf{p}}{4\pi\,\varepsilon_0}\frac{2x}{(x^2-a^2)^2} \quad \text{where} \tag{2.27}$$

$$\mathbf{p} = 2aq\mathbf{i} \tag{2.28}$$

is called the *dipole moment*. The dipole moment is the product of q and the vector $2a\mathbf{i}$ going from the negative to the positive charge.

For $x \gg a$, the field becomes

$$\mathbf{E}(x \to \infty) \simeq \frac{\mathbf{p}}{2\pi \, \varepsilon_0 \, x^3} = \frac{\mathbf{p}}{2\pi \, \varepsilon_0 \, r^3}, \tag{2.29}$$

because r, the radial distance from the center of the dipole to (x, y) equals x when $y = 0$.

For a point on the axis like E with $x < -a$ you should go back to Eqn. 2.24 and verify that the field is invariant under $x \to -x$ and also points along the positive x-axis.

On the y-axis, at a point D with coordinates $(0, y)$, I leave it to you to show that

$$\mathbf{E} = -\frac{\mathbf{p}}{4\pi \, \varepsilon_0 (y^2 + a^2)^{3/2}}. \tag{2.30}$$

For $y \gg a$, the field becomes

$$\mathbf{E}(y \to \infty) \simeq -\frac{\mathbf{p}}{4\pi \, \varepsilon_0 \, |y|^3} = -\frac{\mathbf{p}}{4\pi \, \varepsilon_0 \, r^3}. \tag{2.31}$$

These results with $\mathbf{E} \propto \mathbf{p}$, when $x \to \infty$ or $y \to \infty$ are to be expected. If we set $a = 0$ in Eqn. 2.23 for the sum of $\mathbf{E}_\pm$, we get $\mathbf{E} \equiv 0$ as we must: the two charges sit on top of each other and fully neutralize each other. The total $\mathbf{E}$ as a function of a vanishes when $a = 0$. The net field is non-zero only because $a \neq 0$ and the non-zero part will start out as the first power of a in a Taylor series (Chapter 16, Volume I). To keep the dimension of the field $\mathbf{E}$ the same, the extra a must really be $\frac{a}{r}$, which is what we find in Eqns. 2.29 and 2.31 since $\mathbf{p} = 2aq\mathbf{i}$ is proportional to a.

Recall that the field of a single charge, which looks like a hedgehog, is isotropic. If I rigidly rotate the distribution of field lines around any axis passing through the origin at any angle, they look the same. We may demand this on the basis of the following symmetry argument. You must agree the charge is the cause and the field is the effect. The effect cannot change if the cause does not. Rotating around the origin leaves the point charge alone: it stays where it is and, being a point, looks the same as well after the rotation. It follows that the resulting field distribution must be unaffected by rotation.

On the other hand, even if the dipole looks like a point as we go far away, $\mathbf{E}$ is not isotropic. The field knows that the dipole near the origin has chosen a direction in space, defined by **p**, unlike a single charge, which does not do that. A generic rotation around an arbitrary axis passing through the origin will change the orientation of the dipole (the cause), and the field (the effect) will change accordingly. On the other hand, a rotation around the axis of the dipole will leave it alone and the **E** configuration it produces should be unaffected by such a rotation. This is why we were satisfied with finding **E** in the xy-plane. The answer in any other plane may be found by a rigid rotation around the x-axis.

2.5.1 Far field of dipole: general case

Far from the dipole, the general formula Eqn. 2.23 simplifies, though it takes some more work to extract the part linear in a. Following the details will enhance your mathematical prowess if you suffer through them. Let us begin with the exact result

$$\mathbf{E} = \frac{q}{4\pi\,\varepsilon_0} \left[\frac{\mathbf{i}(x-a)+\mathbf{j}\,y}{((x-a)^2+y^2)^{3/2}} - \frac{\mathbf{i}(x+a)+\mathbf{j}\,y}{((x+a)^2+y^2)^{3/2}} \right]. \quad (2.32)$$

The answer is some function of a (and of course x and y), which vanishes at $a = 0$. Near this zero, the function will have a Taylor expansion in a. By dimensional analysis, the series has to be in a divided by a length and the only possible candidate is r, the distance from the center of the dipole. We are content to find just the first correction to zero. It will be proportional to a or, equally well, the dipole moment $\mathbf{p} = 2aq\mathbf{i}$.

Eqn. 2.32 has two parts, each with a numerator divided by the denominator, or the numerator times the inverse denominator. We can get the single power of a from either term and the a^0 term from the other. If we get a^1 from the numerator we may set $a = 0$ in the denominator and vice versa.

Consider the contribution from the positive charge

$$\mathbf{E}_+ = \frac{q}{4\pi\,\varepsilon_0} \frac{\mathbf{i}(x-a)+\mathbf{j}\,y}{((x-a)^2+y^2)^{3/2}} \quad (2.33)$$

$$= \frac{q}{4\pi\,\varepsilon_0} \frac{\mathbf{r}-a\mathbf{i}}{((x-a)^2+y^2)^{3/2}} \quad (2.34)$$

$$\simeq \frac{q}{4\pi\,\varepsilon_0} \frac{\mathbf{r}}{(x^2-2ax+y^2)^{3/2}} + \frac{q}{4\pi\,\varepsilon_0} \frac{-a\mathbf{i}}{(x^2+y^2)^{3/2}}. \quad (2.35)$$

Here is some explanation. In the last line, the first term comes from keeping the a^0 term, namely $\mathbf{r}$, in the numerator and keeping up to linear terms in the denominator (and hence dropping the a^2 in the expansion of $[x - a]^2$). The second term comes from keeping the a term in the numerator and setting $a = 0$ in the denominator. The terms kept are then

$$\mathbf{E}_+(\text{to order } a) = \frac{q}{4\pi\,\varepsilon_0} \left(\frac{\mathbf{r}}{(r^2 - 2ax)^{3/2}} - \frac{a\mathbf{i}}{r^3} \right). \tag{2.36}$$

The $\mathbf{E}_-$ terms are obtained by changing $q \to -q$, $a \to -a$:

$$\mathbf{E}_-(\text{to order } a) = -\frac{q}{4\pi\,\varepsilon_0} \left(\frac{\mathbf{r}}{(r^2 + 2ax)^{3/2}} + \frac{a\mathbf{i}}{r^3} \right) \tag{2.37}$$

to give a total of

$\mathbf{E}(\text{to order } a)$

$$= \frac{q}{4\pi\,\varepsilon_0} \left(-\frac{2a\mathbf{i}}{r^3} + \frac{\mathbf{r}}{(r^2 - 2ax)^{3/2}} - \frac{\mathbf{r}}{(r^2 + 2ax)^{3/2}} \right) \tag{2.38}$$

$$= \frac{q}{4\pi\,\varepsilon_0} \left(-\frac{2a\mathbf{i}}{r^3} + \frac{\mathbf{r}}{r^3(1 - \frac{2ax}{r^2})^{3/2}} - \frac{\mathbf{r}}{r^3(1 + \frac{2ax}{r^2})^{3/2}} \right) \tag{2.39}$$

$$= \frac{q}{4\pi\,\varepsilon_0 r^3} \left[-2a\mathbf{i} + \mathbf{r}\left(1 + \frac{3}{2}\frac{2ax}{r^2} \right) - \mathbf{r}\left(1 - \frac{3}{2}\frac{2ax}{r^2} \right) \right] \tag{2.40}$$

$$= \frac{q}{4\pi\,\varepsilon_0 r^3} \left[-2a\mathbf{i} + \mathbf{r}\frac{3}{2}\cdot 2\cdot\frac{2ax}{r^2} \right] \tag{2.41}$$

$$= \frac{1}{4\pi\,\varepsilon_0 r^3} \left[-\mathbf{p} + 3\mathbf{r}\left(\frac{\mathbf{p}\cdot\mathbf{r}}{r^2} \right) \right] \tag{2.42}$$

where I have invoked $\mathbf{p} = 2aq\mathbf{i}$, $\mathbf{p}\cdot\mathbf{r} = 2axq$, and applied $(1 + z)^n = 1 + nz + \ldots$, to obtain

$$\left(1 - \frac{2ax}{r^2} \right)^{-3/2} = 1 + \left(\frac{-3}{2} \right)\cdot\left(\frac{-2ax}{r^2} \right) + \text{higher powers of } a. \tag{2.43}$$

2.6 Response to a field

Having seen how to find the field in a variety of situations using Coulomb's law, let us now consider the response of charges to the field using $\mathbf{F} = q\mathbf{E}$, starting with the parallel plate capacitor with a uniform field $\mathbf{E} = -\mathbf{j}E_0$ in between the plates, as indicated in Figure 2.6. Suppose I shoot a particle of mass m and charge q from the left, with a velocity $\mathbf{v}_0$. What will be its position and velocity as it exits the plates?

The force on the charge is a constant, $\mathbf{F} = -qE_0\mathbf{j}$, just like the force of gravity, which will produce an acceleration

$$\mathbf{a} = -\mathbf{j}\frac{qE_0}{m}. \tag{2.44}$$

The particle will follow a parabolic path given by

$$\mathbf{v}(t) = \mathbf{v}_0 + \mathbf{a}t = \mathbf{v}_0 - \mathbf{j}\frac{qE_0}{m}t \tag{2.45}$$

$$\mathbf{r}(t) = \mathbf{r}_0 + \mathbf{v}_0 t + \frac{1}{2}\mathbf{a}t^2 = \mathbf{r}_0 + \mathbf{v}_0 t - \mathbf{j}\frac{1}{2}\frac{qE_0}{m}t^2. \tag{2.46}$$

To compute its y coordinate when it exits the capacitor, we need to know for how long it "falls" at the rate above. That time is clearly $t^* = L/v_0$ where L is the width of the capacitor. (Even though the capacitor is of finite width, we use the constant $\mathbf{E}$ field of the infinite capacitor as a simplification.) As in the case of gravity, the time to go a certain distance horizontally is determined by the initial horizontal velocity and is unaffected by the acceleration in the vertical direction. So if you set $t = t^*$ in $\mathbf{r}(t)$, you will find out where it will end up.

Here is one way in principle to make pictures on television: shoot electrons from the left into the region between *two* pairs of plates, one as shown (perpendicular to the page) and another pair parallel to the page, with one member of the pair above and one below the page. This will cause motion up and down and also in and out of the page. Place a fluorescent screen at the right, perpendicular to the beam. If you apply the right electric field, the electron will land on the screen and make a little glowing dot just where you want it. By scanning the screen many times a second, and by varying the field appropriately and modulating the intensity of the beam, you create the impression of a steady picture. (Actually, magnetic fields were used to deflect electrons in old cathode ray tubes.)

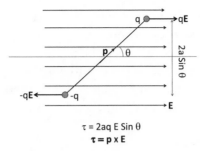

$$\tau = 2aq\, E\, \text{Sin}\, \theta$$
$$\boldsymbol{\tau} = \mathbf{p} \times \mathbf{E}$$

Figure 2.8   The forces and torque $\boldsymbol{\tau}$ on a dipole $\mathbf{p}$ due to a uniform horizontal field $\mathbf{E}$. The torque, computed with respect to the negative charge, has a magnitude $\tau = 2aqE\sin\theta$ and tends to align it with the applied field. The vector $\boldsymbol{\tau} = \mathbf{p} \times \mathbf{E}$ vanishes only when $\mathbf{p}$ and $\mathbf{E}$ are parallel or anti-parallel.

2.6.1 Dipole in a uniform field

What is the force of a uniform electric field on a dipole? Figure 2.8 shows a dipole made of charges $\pm q$ a distance $2a$ apart in a horizontal uniform electric field. It is assumed the charges are mounted at the ends of some rigid structure, like a rod. The force on the two charges is $\pm q\mathbf{E}$ as shown. So the dipole as a whole will not feel any net force, because the two parts are getting pulled by opposite amounts. (If the electric field were not uniform, say it were stronger at the plus charge, the dipole would accelerate to the right.) The forces, which add up to nothing, collaborate in producing a torque. I hope you can see that the torque wants to align the dipole with the field. Recall that when the total force vanishes, the torque may be computed with respect to any point. Choosing it to be the location of $-q$, we find it has a magnitude (see Figure 2.8)

$$\tau = qE \cdot 2a\sin\theta, \tag{2.47}$$

which turns clockwise. As a vector, the torque is given by the cross product

$$\boldsymbol{\tau} = \mathbf{p} \times \mathbf{E}, \tag{2.48}$$

which points into the page. If you mount this dipole so it can swing in the plane of the paper, you could use it as an "electrical compass," which will point along the local electric field. (We assume the rod supporting the charges at its ends has a non-zero moment of inertia I and the support has some friction, so that if it started out non-parallel to **E**, it will quickly align with **E** after some damped oscillations.)

The torque also vanishes when the dipole is anti-parallel to **E**. This is a state of unstable equilibrium: if disturbed, it will not return there but end up parallel to **E**. We can understand this in term of energy.

Recall that a conservative force $F(x)$ and the associated potential $U(x)$ are related as follows:

$$F(x) = -\frac{dU}{dx} \quad \text{which integrates to} \tag{2.49}$$

$$U(x_1) - U(x_2) = \int_{x_1}^{x_2} F(x)\,dx. \tag{2.50}$$

Next recall the SAT analogy: "Torque is to force as angle is to displacement." The torque here is $\tau = -pE\sin\theta$, where the minus sign reflects its tendency to rotate the dipole clockwise, in the direction of *decreasing* θ. So we may now write

$$U(\theta_1) - U(\theta_2) = \int_{\theta_1}^{\theta_2} (-pE\sin\theta)\,d\theta \tag{2.51}$$

$$= pE\cos\theta_2 - pE\cos\theta_1 \tag{2.52}$$

$$U(\theta) = -pE\cos\theta = -\mathbf{p}\cdot\mathbf{E}. \tag{2.53}$$

In going from Eqn. 2.52 to 2.53 we have dropped a possible additive constant in $U(\theta)$.

You see in Figure 2.9 that $U(\theta)$ is an inverted cosine with a minimum at $\theta = 0$, which is a point of stable equilibrium, and a maximum at $\theta = \pi$, which is a point of unstable equilibrium. The points $\pm\pi$ are one and the same. When perturbed about $\theta = 0$, the dipole will execute simple harmonic motion. For small angles, κ, the restoring torque per angular displacement, and ω, the frequency of oscillations, will be (in terms of the moment of inertia I)

$$\tau = -pE\sin\theta \simeq -pE\theta \tag{2.54}$$

$$\kappa = -\frac{\tau}{\theta} = pE \tag{2.55}$$

$$\omega = \sqrt{\frac{pE}{I}}. \tag{2.56}$$

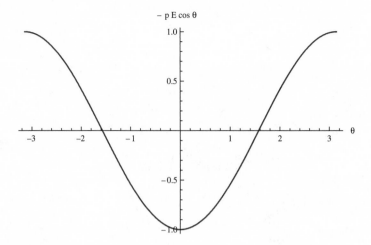

Figure 2.9 The potential energy of a dipole, $U = -pE\cos\theta$ as a function of the angle θ it makes with a field.

CHAPTER 3

Gauss's Law I

In the last chapter we learned that we should think in terms of electric fields and not direct action-at-a-distance between charges according to Coulomb's law. In this parlance, we say charges produce fields as per Coulomb's law, and the fields in turn act on charges as per $\mathbf{F} = q\mathbf{E}$.

The field $\mathbf{E}(\mathbf{r})$ is a condition at a point $\mathbf{r}$, even if there is no charge at that point. This condition is revealed when we place a test charge q there and find a force $q\mathbf{E}(\mathbf{r})$ acting on it. The field due to many charges is the sum of the fields due to each.

Strictly speaking, Coulomb's law is to be applied only in a static situation when the charges do not move, though we do apply it in some situations where they move slowly compared to c, as in circuits. In this chapter, we will assume a static distribution of charges and apply Coulomb's law.

We saw how field lines can depict the state of the electric field: the lines point along the local field, and their areal density (lines per unit area perpendicular to the lines) is proportional to the field magnitude. We could use any number of lines per coulomb, but once we agreed on a convention, say 64 lines per coulomb, we had to stick to it. We looked at the field lines of a dipole as well as that of two equal charges.

We considered the dipole field quantitatively. The answer was expressed in terms of $\mathbf{p} = q(\mathbf{r}_+ - \mathbf{r}_-)$, the dipole moment of charges $\pm q$ located at $\mathbf{r}_{\pm}$.

We found a general expression for the field due to a dipole. We evaluated it exactly along the dipolar axis and perpendicular to the dipolar axis. As for a general direction, we considered the field only for distances $r \gg a$, the distance between the charges. The main point was that the leading term for **E** fell as $1/r^3$.

We studied how charges responded to a field. We saw what happened to a charge shot into the space between plates of a capacitor, where the field was assumed to be uniform and perpendicular to the plates. Finally, we saw that a dipole moment in a field experiences a torque, $\mathbf{p} \times \mathbf{E}$, which tries to line it up with the field. With that torque one can associate a potential energy $U = -\mathbf{p} \cdot \mathbf{E}$.

3.1 Field of an infinite line charge

Here is a standard problem. We have an infinite line of charge parallel to the x-axis, say a charged wire, of which a finite part is shown in Figure 3.1. Somebody has sprinkled it with a continuous density of λ coulombs per meter. If we cut out one meter of this wire, we will find λ coulombs there. We want to compute the electric field everywhere due to this charge distribution using Coulomb's law. (Let us assume λ is positive; if it is negative, we just have to reverse the field everywhere.)

Consider a point $P = (0, a)$ at a distance a from the wire. What can we say about the field there without doing the full calculation?

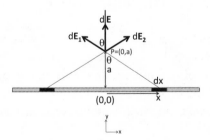

Figure 3.1 The field due to an infinite line charge with linear charge density λ. It is found by adding the contributions from tiny segments of width dx treated as point charges. The figure shows clearly that the fields $d\mathbf{E}_1$ and $d\mathbf{E}_2$
due to segments at x and $-x$ have the same y-components and opposite x-components.

First of all, it must be intuitively clear that the field will be the same at all points at the distance a from the wire. Any x-dependence leads to the following contradiction. Suppose the field had a variation in the x-direction. If I slide the wire to the right by some amount, this pattern will shift by that amount and look different. On the other hand, since the wire looks the same before and after the shift, so must the field it produces. If the cause (the wire) looks the same after a shift in x, so must be the effect, the field it produces.

The field may, however, depend on y and it does.

Next we may argue that the field has to point radially away from the wire; it cannot be tilted to one side or the other since the infinite wire does not distinguish right from left. Here is another way to say it. Suppose the field were tilted to the right. Now rotate the wire by π around an axis perpendicular to it (the y-axis in Figure 3.1) and passing through P. The field lines will rotate as well and end up tilted to the left. But the rotated wire looks the same as the unrotated one and so must the field it produces. The only configuration that is unaffected by this rotation is a field that is everywhere perpendicular to the wire.

The argument fails if the wire is finite. A finite wire has some distinct features and special points like the midpoint and end points. It does not look the same if you slide it parallel to itself and so the field can vary with x. The field lines may tilt toward the left end if the point P is left of center and likewise to the right for points to the right of center. This distribution will still turn into itself under any operation that leaves the wire invariant, such as the above-mentioned rotation by π about its midpoint.

Returning to the infinite wire, let us find how the perpendicular field varies with a.

Look at Figure 3.1. Let us take a segment of wire centered at x and of length dx, which is so small that we can treat it as a point charge. Now the dx as drawn is not a point, but in the end, we're going to make it arbitrarily small. The segment is like a point-charge $q = \lambda dx$ at a distance x from the origin. The infinitesimal electric field it produces at P has a magnitude

$$dE_1 = \frac{\lambda \, dx}{4\pi \, \varepsilon_0 (a^2 + x^2)} \tag{3.1}$$

and points along the vector joining $(x, 0)$ and the point P. We need only keep its y component since the ultimate x component has to be zero, either by our earlier symmetry arguments or by the explicit consideration of the contribution from the similar segment at $-x$. Convince yourself by looking at the figure that when the two contributions are added, the horizontal parts will cancel and the vertical part will be double that due to either segment. Let us therefore double the vertical contribution from the segment on the right but remember to consider only $x \geq 0$. Using

$$\cos\theta = \frac{a}{\sqrt{a^2 + x^2}} \tag{3.2}$$

to project out the vertical part, we find the total vertical field by integration:

$$\mathbf{E}(a) = \mathbf{j} \int_0^\infty \frac{2 \times \lambda \, dx}{4\pi \, \varepsilon_0 (a^2 + x^2)} \frac{a}{\sqrt{a^2 + x^2}}. \tag{3.3}$$

What next? The integral can be done by a clever substitution. What if that trick does not occur to us? It turns out we can go quite far by dimensional analysis. Let us express the coordinate x in terms of a, the only length in the problem, via the dimensionless variable w as

$$x = a \cdot w. \tag{3.4}$$

Then the limits for the integral over $w = \frac{x}{a}$ are 0 and ∞. Since $dx = a \, dw$ we have

$$\mathbf{E}(a) = \mathbf{j} \times 2 \times \frac{a\lambda}{4\pi \, \varepsilon_0} \times \int_0^\infty \frac{a \, dw}{a^3 (1 + w^2)^{3/2}} \tag{3.5}$$

$$= \mathbf{j} \frac{\lambda}{2\pi \, \varepsilon_0 a} \times \int_0^\infty \frac{dw}{(1 + w^2)^{3/2}} \tag{3.6}$$

$$= \mathbf{j} \frac{\lambda}{2\pi \, \varepsilon_0 a} \times N \quad \text{where} \tag{3.7}$$

$$N = \int_0^\infty \frac{dw}{(1 + w^2)^{3/2}} \quad \text{is some number, independent}$$
of all parameters. $\tag{3.8}$

Thus we have the answer up to an overall multiplicative constant N, which is independent of λ and a. Even before we evaluate N we see a surprising thing: the field falls like $1/a$ and not $1/a^2$, even though each piece of the wire makes a contribution that falls like the inverse square of the distance. This is surprising but also inevitable for dimensional reasons. The answer now had to be proportional to λ, which is a charge *per unit length* and not charge. The presence of the prefactor λ robs the denominator of one power of the length, leaving behind one power of the only length in the problem, which is a.

This argument fails if the wire is finite: now we have another length L, the length of the wire, which can bring in factors like L/a (or any function of the dimensionless variable L/a) without messing up the dimensionality of the answer. Indeed, in this case we expect, and find, that if we go to distances much greater than L, the field will be that of a point charge λL.

Let us now evaluate N by making the substitution

$$w = \tan\theta. \tag{3.9}$$

(I call the substitution variable θ rather than some other Greek letter, because in this case it is actually the θ in the figure: $w = \tan\theta$ means $x = a\tan\theta$.) Observe that the change of variable is such that all possible values of w can be obtained by some choice of θ because $\tan\theta$ can go from 0 to ∞. Had we made the substitution $w = \cos\theta$, we could never obtain $w > 1$ (or $x > a$).

Continuing, we find

$$\frac{dw}{d\theta} = \sec^2\theta \tag{3.10}$$

$$dw = \sec^2\theta \; d\theta \tag{3.11}$$

$$N = \int_0^{\pi/2} \frac{\sec^2\theta \; d\theta}{(1+\tan^2\theta)^{3/2}} \tag{3.12}$$

$$= \int_0^{\pi/2} \cos\theta \; d\theta = 1. \tag{3.13}$$

For future use remember that the field due to an infinite linear charge density at a distance a from the line and lying in the xy-plane has a magnitude

$$E(a) = \frac{\lambda}{2\pi\varepsilon_0 a}, \tag{3.14}$$

and points perpendicularly away from the wire.

While Figure 3.1 is two-dimensional, the wire and the field pattern live in three dimensions. What we have in the figure is a slice taken through the xy-plane. The full field configuration will be obtained by rigidly rotating the configuration shown about the x-axis. We can slice

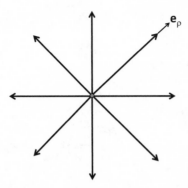

Figure 3.2   The field due to an infinite line charge seen end-on, with the wire perpendicular to the page. The wire and the field distribution it produces are invariant under a rotation of the wire about itself.

that three-dimensional configuration through any plane passing through the x-axis and we will get the same two-dimensional configuration. This is demanded by symmetry or by the cause-effect relationship. If I rotate the wire around the x-axis by some angle, it looks the same. Therefore the field configuration it produces should also be invariant under that rotation. The field pattern in which the lines radiate uniformly and radially away from the line is the unique electrostatic configuration meeting this requirement. (The lines could also point radially inward, but that would correspond to negative $\lambda < 0$.) The end view, with the wire running perpendicular to the page, is shown in Figure 3.2.

   In three dimensions, it is common to denote the distance measured perpendicular to the wire by ρ (and not a). So we should write

$$\mathbf{E} = \mathbf{e}_\rho \frac{\lambda}{2\pi\varepsilon_0\rho} \tag{3.15}$$

where $\mathbf{e}_\rho$ is a unit vector in the direction perpendicular to the wire. (In the xy-plane, $\mathbf{e}_\rho = \pm\mathbf{j}$.)

3.2   Field of an infinite sheet of charge

Imagine an infinite plane with an areal charge density σ depicted in Figure 3.3. This means that if you cut out a tiny part of it, of area dA, it will have a charge $\sigma\, dA$. (By convention, λ stands for charge per unit length, σ for

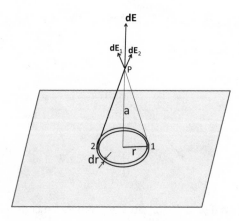

Figure 3.3   The field due to an infinite plane with charge density σ. Shown is one contributing annulus of radius r and thickness dr. It produces a field $d\mathbf{E}$ perpendicular to the plane, as shown by the long dark arrow. There is no parallel part due to cancellations between parts of the annulus that are diametrically opposite. Shown are two such contributions, $d\mathbf{E}_1$ and $d\mathbf{E}_2$, due to the two darkened parts of the annulus. The sum of such vectors due to all parts of the annulus is $d\mathbf{E}$. The integral of $d\mathbf{E}$ over all annuli will give the final $\mathbf{E}$ due to the entire plane.

charge per unit area, and ρ for charge per unit volume.) We want to stand at a point P a distance a from the plane and ask for the field there.

Once again, before jumping into the calculation, let us see what features follow from general considerations.

I think we can agree that the electric field at some point a meters in front of the plane will be independent of the other two coordinates parallel to the plane. Suppose the field varied as we moved parallel to the plane at fixed a, with some ups and downs in field strength. If I move the plane to the right by one inch, the pattern should follow. But the shifted plane looks exactly like the unshifted one. It has to produce exactly the same field, of the same magnitude and direction. This can happen only if the field does not vary under displacements parallel to the plane.

As for the direction, it has to be perpendicular to the plane, again for symmetry reasons. If you tilt it away from the perpendicular, which way would you tilt it? The infinite plane defines no unique direction except the one perpendicular to it. Suppose the field is tilted away from the perpendicular by an angle of 30 degrees, say in the direction of $d\mathbf{E}_1$.

If I now rotate the plane around an axis perpendicular to it and passing through P, the direction of the tilt will rotate as well (ending up parallel to $d\mathbf{E}_2$ after a rotation by π). But the rotated infinite plane looks the same as the unrotated one, and so must be the field it produces. The only field configuration that meets this demand is the one where the field is everywhere normal to the plane.

We can also argue from symmetry that the magnitudes of the field should be the same at two points that are on opposite sides of the plane and at the same distance from it. If the charged sheet lies in the xy-plane we require that $E(z) = E(-z) = E(|z|)$. The charges on the sheet repel a test charge at a given distance from the plane with the same intensity whether the test charge be on one side of the plane or the other. The directions will of course be opposite, pointing away from the plane. Thus we may assert that

$$\mathbf{E}(\mathbf{r}) = \mathbf{k}E(|z|) \quad z > 0 \tag{3.16}$$

$$= -\mathbf{k}E(|z|) \quad z < 0 \tag{3.17}$$

where $\mathbf{k}$ is a unit vector along the z-axis.

While this is intuitively obvious, we could provide the cause-effect argument by demanding that the field configuration should be unaffected if the charged plane is flipped over like a pancake by a rotation around the x-axis by π since the plane looks the same before and after. The configuration written above meets that requirement.

Armed with these anticipations based on general symmetry arguments, we turn to the calculation that will yield results in agreement with our expectations. Our strategy is as follows. We will draw a perpendicular to the plane passing through the point P where we want the field, as shown in Figure 3.3. We will divide the plane into concentric annuli or rings of radius r and width dr, find the contribution $d\mathbf{E}$ from each ring, and integrate them over all rings.

The contribution from a given ring may be readily inferred from Eqn. 1.12 for the force on charge q due to a ring carrying a linear density λ, at a point on its symmetry axis, z meters above it:

$$F_{\perp} = \left[q \times \frac{2\pi r\lambda}{4\pi\varepsilon_0} \right] \cdot \left[\frac{1}{r^2 + z^2} \right] \cdot \left[\frac{z}{\sqrt{z^2 + r^2}} \right]. \tag{3.18}$$

The first factor is the $q_1 q_2/(4\pi \varepsilon_0)$ appropriate to the test charge and the loop, the second reflects the inverse square law, and the third is the cosine factor that projects out the component perpendicular to the plane of the loop, which alone survives when all contributions from the loop are added.

We may import this result after three modifications:

- Drop the test charge q to get the field from the force.
- Set $z = a$.
- Relate σ to λ, the charge per unit length of the annulus. A segment of length 1 along the annulus will have an area $1 \cdot dr$ and contain $1 \cdot \sigma \, dr$ coulombs. Thus the linear charge density in our problem is related to the areal charge density by

$$\lambda = \sigma \, dr. \tag{3.19}$$

The resulting field, at a distance a from the plane, is

$$E_\perp = \frac{2\pi \, r(\lambda = \sigma \, dr)}{4\pi \, \varepsilon_0} \cdot \frac{1}{r^2 + a^2} \cdot \frac{a}{\sqrt{a^2 + r^2}}. \tag{3.20}$$

Since $E_\perp$ in Eqn. 3.20 is the infinitesimal contribution from a ring of infinitesimal width dr, we rewrite it explicitly as an infinitesimal

$$dE_\perp = \frac{2\pi \, r\sigma \, dr}{4\pi \, \varepsilon_0} \cdot \frac{1}{r^2 + a^2} \cdot \frac{a}{\sqrt{a^2 + r^2}} \tag{3.21}$$

and obtain the total field by integration

$$E_\perp = \frac{2\pi \, \sigma}{4\pi \, \varepsilon_0} \int_0^\infty \frac{r \, dr}{r^2 + a^2} \cdot \frac{a}{\sqrt{a^2 + r^2}}. \tag{3.22}$$

Once again we may use scaling to figure out the a-dependence as follows. Setting

$$r = aw \quad \text{which implies} \quad dr = adw, \tag{3.23}$$

we obtain

$$E_\perp = \frac{2\pi\sigma}{4\pi\varepsilon_0} \int_0^\infty \frac{aw\, a dw}{a^2(1^2 + w^2)} \cdot \frac{a}{a\sqrt{1^2 + w^2}} \tag{3.24}$$

$$= \frac{\sigma}{2\varepsilon_0} \int_0^\infty \frac{w dw}{(1 + w^2)^{3/2}} \tag{3.25}$$

$$= \frac{\sigma}{2\varepsilon_0}. \tag{3.26}$$

The integral equals 1, as can be shown by the substitution $z = w^2$.

The preceding result is so important, I will repeat it and suggest you memorize it:

$$\text{Field of infinite plane with charge density } \sigma = \frac{\sigma}{2\varepsilon_0}. \tag{3.27}$$

The most striking aspect of the result is that the field does not decrease as we move away from the plane. It is independent of a, the perpendicular distance to the plane. Since each part of it makes a contribution that falls like $1/(a^2 + r^2)$, and we are increasing the distance to every segment of the plane as we increase a, the field should get weaker, right? And yet that does not happen.

We can understand why this had to be so on dimensional grounds. The field has dimensions of charge over distance squared. (Forget the ubiquitous $4\pi\varepsilon_0$, which is a constant.) For a single charge q the distance in question had to be r, the distance between the charge and the field location. For a line charge, the answer had to be linear (by the superposition principle) in λ, which had units of charge over length, leaving room for just one length a, the distance to the wire, to appear in the denominator. For the plane the inevitable factor linear in σ, which has units of charge over distance squared, has used up all the inverse powers of length, leaving no room for a to appear either in the numerator or denominator. As I mentioned before in connection with the wire, the argument fails if the plane is of a finite extent, say a square of side L. In this case the answer is allowed to have factors like L^2/a^2 and indeed it will: for $a \gg L$, the field will be that of a point charge $q = \sigma L^2$.

To understand the a-independence of $\mathbf{E}$ in pictorial terms, consult Figure 3.3. Let us start at some a and reduce it to get closer to the plane.

We find the contributions from individual segments of each of the rings do indeed go up since $a^2 + r^2$ decreases. However, the contributions, which point along the line joining the segment to the field location, become increasingly parallel to the plane as we approach it. (Look at $d\mathbf{E}_1$ and $d\mathbf{E}_2$ in the figure.) But we have seen that the parallel part gets canceled by symmetry (within each ring) and only the (tiny) perpendicular part survives. So there are two opposing factors as we get close to the plane: the contributions from individual segments of any given ring get bigger, but the useful component that survives the sum over segments, the perpendicular part, gets smaller. So you can give arguments why the field should get weaker and arguments why it should get stronger as a varies. To show that these two tendencies exactly cancel, you have to bite the bullet and do the calculation.

We can now find the field between the plates of a parallel plate capacitor (ignoring edge effects) with $\pm\sigma$. In the region between the plates the fields due to the two plates add to a total of σ/ε_0, pointing from the positive to the negative plate. In the region outside the plates the field vanishes because the two fields cancel, being of equal and opposite strength and independent of distance.

3.3   Spherical charge distribution: Gauss's law

Now we turn to the more difficult case of a spherical charge distribution. Rather than attack it frontally, I will introduce you to a powerful idea called Gauss's law, which will provide a shortcut.

Imagine a solid ball of charge density ρ (measured in C/m^3). We want to find the field due to this ball.

Now, when we did a similar problem in gravitation, we assumed that when you're outside the sphere, the whole sphere acts like a point mass with the entire mass sitting at the center, and that when you are inside (as in our analysis of dark matter), the mass inside the chosen radius acts like a point mass at the center and the mass outside does not contribute.

Since the electrostatic force also obeys the inverse square law, it should not be surprising that we may replace the word "mass" by the word "charge" in the preceding paragraph. But now we want to prove all this, rather than assume it.

This is what took Newton a long time. He knew it was true but he couldn't prove it, because for that, he had to first develop integral calculus. Even today, to find the field due to a sphere using integration

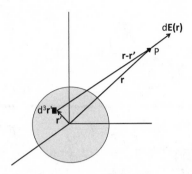

Figure 3.4   The field due to a spherical charge distribution. Each tiny cube $d^3\mathbf{r}'$ located at $\mathbf{r}'$ makes its contribution $d\mathbf{E}(\mathbf{r})$ to $\mathbf{E}(\mathbf{r})$ as per Coulomb's law. These contributions have to be vectorially added to obtain $\mathbf{E}(\mathbf{r})$.

is quite difficult. Think about what you have to do. Look at Figure 3.4. You want the field at point P at location $\mathbf{r}$. You have to divide the sphere into tiny little cubes centered at $\mathbf{r}'$, each carrying charge equal to the density $\rho(\mathbf{r}')$ (which happens to be constant in this case) times the volume of the cube, $d^3\mathbf{r}'$. A typical cube will create a field $d\mathbf{E}(\mathbf{r})$ as shown. You have to integrate the $d\mathbf{E}(\mathbf{r})$'s from every tiny cube in the sphere. But the contribution from each cube will have a different magnitude and direction. Adding all these vectors is a tough problem that we are going to finesse by invoking a very powerful notion called *Gauss's law*. As a prelude, we need to cover some mathematical ideas involving areas and surface integrals.

3.4   Digression on the area vector $d\mathbf{A}$

Imagine I am holding up a tiny little planar area, like a postage stamp, in three dimensions at some location $\mathbf{r}$. I want you to be able to visualize this area. What can I do to specify it besides telling you it is located at $\mathbf{r}$? The first thing I can tell you is how big it is. I say it is dA square meters in size. I then have to tell you in which plane it lies. How do I do that?

Suppose it lies in the xy-plane. Rather than say "lies in xy-plane," I could just as well say it lies perpendicular to the z-axis. *I could then associate a vector $d\mathbf{A}$ with this area, of magnitude dA and direction along the z-axis.* But there are two ways to draw the perpendicular to the xy-plane: up or down the z-axis. To further specify the area, to make it an *oriented or signed area*, I will draw arrows that run around its perimeter in one of

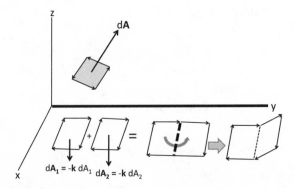

Figure 3.5   The figure shows a generic (shaded) area floating in three dimensions. The area vector $d\mathbf{A}$ is given by the right-hand rule applied to the arrows running around the edges. Also shown are two areas lying in the xy-plane with a common edge. Their sum is an area with the common edge (shown by a dotted line) deleted. If we use that dotted line as a hinge and rotate the second area out of the xy-plane (as indicated), their sum is a non-planar area, bounded by the uncanceled edges.

two possible directions. *The area vector $d\mathbf{A}$ will point along the thumb of our right hand if we curl the fingers around the loop in the sense of the arrows.* This is called the *right-hand rule.* It is illustrated in Figure 3.5 by the two areas in the xy-plane, given by $d\mathbf{A}_1 = -\mathbf{k}dA_1$ and $d\mathbf{A}_2 = -\mathbf{k}dA_2$. (An area without the arrows on its perimeter is like a vector without its head and tail marked.)

The upper part fof the figure shows a generic (shaded) area vector $d\mathbf{A}$, floating in three dimensions. Its direction is determined by the sense of the arrows running around the edges as per the right-hand rule.

Only a planar area can be represented as a vector. All infinitesimal areas can be treated as planar. Finite areas that are non-planar, like a hemisphere or magic carpet, cannot be represented by a single vector: we cannot reconstruct an entire macroscopic surface, with all its undulations, given just a magnitude and a direction.

The use of the right-hand rule in defining areas might remind you of the cross product and indeed there is such an interpretation of areas. Consider an area shaped like a parallelogram, whose adjacent edges are defined by two vectors $\mathbf{B}$ and $\mathbf{C}$ with angle θ between them. Then $\mathbf{A} = \mathbf{B} \times \mathbf{C}$ is the area of the parallelogram, with magnitude $|BC\sin\theta|$ and direction given by the right-hand rule. Infinitesimal areas are bounded by infinitesimal vectors.

Using vectors to describe areas or combining two vectors to get a third by the cross product is possible only in three dimensions where every plane has a unique normal, up to a sign. In four dimensions you cannot have a cross product of two vectors that yields a vector. If you pick two non-planar vectors, the plane they define will have *two* orthogonal directions perpendicular to it.

3.4.1 Composition of areas

Even though infinitesimal areas are given by vectors, the natural rule for combining them is different from vector addition, unless all the areas lie in one plane. I introduce the rule through an analogy, with one fewer dimension; see Figure 3.6.

Suppose we want to construct a curve in two or three dimensions, given any number of tiny vectors. Each vector has two boundary points: its tip and its tail, which are assigned opposite signs. To form the curve, we string these vectors along: the tail of the second vector touches the tip of the first, the tail of the third the tip of the second, and so on to the last one. The resulting curve has only two boundary points: the tail of the first and the tip of the last. All other boundary points have canceled in pairs when we joined them head-to-tail. Of course, the perfectly smooth curve is realized only in the limit of an infinite number of infinitesimal vectors.

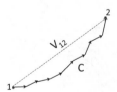

Figure 3.6   A curve C joining points 1 and 2 in the plane, composed of little vectors added tip-to-tail. The tip and tail are the boundaries of each arrow. When two arrows are glued, the touching tip and tail are erased. At the end only the tail of the first vector and the tip of the last vector survive. These are the boundaries of C. The formation of the curve by gluing arrows is not to be confused with vector addition, which would give $\mathbf{V}_{12}$. If the points 1 and 2 were also glued, we would have a closed loop, while the vector sum $\mathbf{V}_{12}$ would vanish.

Do not confuse this *composition of the curve* with the *vector sum*, which would be a straight line going from the tail of the first vector to the tip of the last. Whereas the vector sum remembers only the bottom line, the curve remembers every vector that went into its composition. For example, if the curve is closed, say a circle, the vector sum would simply vanish.

There is a similar rule for combining areas to form two-dimensional surfaces. Consider the two areas dA_1 and dA_2 in Figure 3.5. To combine them, we superpose the right edge of dA_1 and the left edge of dA_2 with their opposing arrows. (This is analogous to placing the tail of one vector on the tip of the previous in forming a curve.) We delete the overlapping parts that carry opposite arrows. *The "sum" of the areas is bounded by the remaining edges.*

Look at the deleted portion shown by a dotted line. If we use that dotted line as a hinge and rotate the second area out of the *xy*-plane, their sum, bounded by the uncanceled edges, is now a *non-planar area*. In this manner, a generic surface in three dimensions may be formed by gluing together little areas or *plaquettes* and deleting the common edges, as illustrated in Figure 3.7. The arrows that used to run around the interior plaquettes have been canceled by the neighboring plaquettes with counter-propagating arrows. What remains are arrows around the perimeter, which run along the boundary of their union or sum.

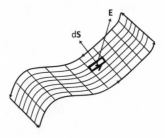

Figure 3.7   A generic surface in three dimensions obtained by gluing together tiny areas or plaquettes. The arrows that used to run around the interior plaquettes have been canceled by the neighboring plaquettes with counter-propagating arrows. What remains are arrows around the perimeter, which define the boundary of the sum. Also shown for later use is one highlighted interior area dS and the electric field vector E at that point. The orientation of this area is indicated by the arrow on one edge.

3.4.2 An application of the area vector

Let us put the concept of the area vector to work. Imagine a tube with a rectangular cross section of height h and width w carrying some fluid moving with velocity $\mathbf{v}$ parallel to its length, as shown in Figure 3.8. What is the *flux* Φ, the volume of fluid flowing past any cross section per second?

To find Φ, we pick as a checkpoint the leftmost area $\mathbf{A}$ in the figure and ask how much fluid goes past it in one second. To this end, at some time $t = 0$ we introduce some tiny beads into the fluid at $\mathbf{A}$. After 1 second, the beads would have moved a distance $v \cdot 1$ and will be resident on the middle area in the figure, which is a shifted duplicate of $\mathbf{A}$. The fluid that has crossed the checkpoint in one second is contained between these two areas. It is a parallelepiped of base $A = wh$ and height $v \cdot 1$ as shown in the figure. Thus

$$\Phi = whv = Av. \tag{3.28}$$

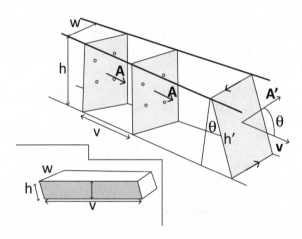

Figure 3.8   A tube of cross-sectional area $A = wh$, carrying a fluid with a velocity $\mathbf{v}$ parallel to $\mathbf{A}$. To monitor the flux (volume flow per second) past the area $\mathbf{A}$ shown at the left, we sprinkle some beads into the fluid at $t = 0$. One second later the beads end up at the middle area. The volume between these two fronts is the flow per second, $\Phi = Av = \mathbf{A} \cdot \mathbf{v}$. The rightmost area $\mathbf{A}'$ is bigger than $\mathbf{A}$ by a factor $1/\cos\theta$ but intercepts the same amount of flux or flow per second. As shown in the text, $\Phi' = \mathbf{A}' \cdot \mathbf{v} = A'v\cos\theta = Av = \mathbf{A} \cdot \mathbf{v} = \Phi$. The inset shows the volume contained between two tilted areas $\mathbf{A}'$ at times $t = 0$ and $t = 1$, separated by $v \cdot 1$ meters.

Because v and A are the magnitudes of the parallel vectors $\mathbf{v}$ and $\mathbf{A}$, we may rewrite the Φ above as their dot product:

$$\Phi = Av = \mathbf{A} \cdot \mathbf{v}. \tag{3.29}$$

Remember that the area, A, if considered as a planar object, lies *perpendicular* to the flow but the *area vector* $\mathbf{A}$, as defined above, is *parallel* to $\mathbf{v}$. So the $\cos\theta$ factor that enters the dot product is simply $\cos 0 = 1$.

Invoking dot product in the present case, when it is just the product of the magnitudes of the parallel vectors $\mathbf{A}$ and $\mathbf{v}$, seems like overkill. But it is introduced to cover a more general case depicted in Figure 3.8. Look at the right-most area $\mathbf{A}'$, which also goes from the ceiling to the floor but with its plane tilted by an angle θ from the vertical. Now

$$A' = wh' = w \cdot \frac{h}{\cos\theta} = \frac{A}{\cos\theta} \tag{3.30}$$

has the same base w as A but a longer side $(h/\cos\theta)$. Let us compute the flux through $\mathbf{A}'$. If we wait one second, the points in $\mathbf{A}'$ will move a distance $v \cdot 1$ downstream and create a replica of $\mathbf{A}'$ there. The flux Φ' is the volume trapped between these two tilted areas. This volume (shown in the inset) is the product of the width w and the area of the parallelogram of base v, side h', and height h. Recall that the area of a parallelogram is base times height. Thus

$$\Phi' = vwh = vwh' \cos\theta = \mathbf{v} \cdot \mathbf{A}', \tag{3.31}$$

which is the same as $\Phi = \mathbf{v} \cdot \mathbf{A} = vwh$. Thus even though $\mathbf{A}'$ is bigger than $\mathbf{A}$ by a factor $1/\cos\theta$, it intercepts the same flux because it is tilted by θ relative to $\mathbf{v}$.

A given area can intercept the greatest flux (say of a fluid) by orienting its plane perpendicular to the flow, or its area vector parallel to the flow. Likewise, it intercepts no fluid at all if it lies in a plane parallel to the flow, or if its area vector is perpendicular to the flow. Most importantly, for intermediate angles, the correct multiplicative factor to use with vA is $\cos\theta$. This appears naturally in the dot product, which therefore seems tailor-made for computing fluxes.

We shall use the term "flux" to denote the dot product of an area vector with any *other vector* $\mathbf{V}$, *even if* $\mathbf{V}$ *is not a velocity. In what follows the vector in question will be* $\mathbf{E}$, *the electric field.*

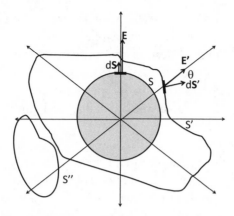

Figure 3.9   The figure shows the two-dimensional cross section of field lines emanating from a charge q. Thus the circle S represents a sphere. It is evident that these lines cross *any* surface enclosing the charge. Two surfaces, a sphere S and a generic one S', are shown. Since the number of lines crossing a surface is proportional to the surface integral of **E**, it means the latter has the same value on any surface surrounding q. The side views of tiny areas dS and dS' on the two surfaces are shown, along with the local value of the field **E** and **E**$'$. Whereas dS and **E** are parallel on the sphere S, dS' and **E**$'$ corresponding to the general case are at an angle θ. The third surface S'' on the lower left-hand corner encloses no charge and has no *net* lines flowing in or out.

3.5   Gauss's law through pictures

Consider a charge q and the field lines coming out of it. Let us assign k lines per coulomb, where k is an arbitrary constant. The following statements should be obvious from Figure 3.9.

- The number of lines passing through a sphere S centered on the charge is independent of its radius r and equals kq, the number emanating from q.
- The same number of lines pass through *any* closed surface such as S' that surrounds the charge.
- If there are several charges $q_i, i = 1 \ldots n$ inside the closed surface S', the number of lines crossing is simply the sum $k\sum_i q_i$. This may not be so obvious, since when many charges are present, the lines assume complicated shapes instead of going straight to infinity. So look at the

field lines due to two positive charges in Figure 2.5. The charges emit kq_1 and kq_2 lines respectively (which happen to be equal in this example). None of these lines can terminate on the other charge, since they are both positive. So all the $kq_1 + kq_2$ lines have to go out of S' and terminate on negative charges outside or escape to infinity. (Again if S' is very convoluted, a line may go in and out of it an odd number of times before finally escaping.)

Suppose next q_2 is negative, i.e., $q_2 = -|q_2|$ with $|q_2| < q_1$. Now $k|q_2|$ lines will terminate on q_2 and the rest, $kq_1 - k|q_2| = k(q_1 + q_2)$, will terminate on negative charges outside S or escape to infinity, after possibly going in and out of S a few times. The argument is readily generalized to any number of charges, of any sign and magnitude. We may assert that if q_i are the charges inside a generic surface S',

$$\text{Lines flowing out of any closed surface } S' = k \sum_i q_i \equiv kq_{enc} \tag{3.32}$$

where q_{enc} is the total charge enclosed in S'.

If you understand Eqn. 3.32 based on the pictures, you understand Gauss's law, for this is what it essentially is, *once we express the number of lines leaving S' in terms of the electric field.*

We will do that in stages. First consider the special case when S' is *a sphere S centered on a single charge q* and consider the areal density of flux lines. These lines cross the sphere perpendicularly, or, if you like, are parallel to the normal to the surface.

$$\text{Lines crossing unit area of } S = \frac{\text{lines coming out}}{\text{area of sphere}} = \frac{kq}{4\pi r^2}. \tag{3.33}$$

But since $E(r)$, the magnitude of the electric field on S, is given by

$$E(r) = \frac{q}{4\pi \varepsilon_0 r^2} \tag{3.34}$$

we may write

$$\text{Lines crossing unit area of } S = \frac{kq}{4\pi r^2} = k\varepsilon_0 E(r). \tag{3.35}$$

Therefore *the electric field is proportional to the lines per unit area,* where the area lies in a plane perpendicular to the lines of **E** or, equivalently, the area vector is parallel to the lines and to **E**.

Consider now a tiny area $d\mathbf{S}$ sitting on the surface of this sphere. I use $d\mathbf{S}$ instead of $d\mathbf{A}$ to signify that this little area is part of a surface S, and I will follow this notation from now on. Letting dS stand for its magnitude,

Lines crossing area $dS = $ (Lines crossing unit area of S)

$$\times dS = k\varepsilon_0 E(r)dS. \qquad (3.36)$$

Let us now re-express product $E(r)dS$ in terms of the corresponding vectors **E** and $d\mathbf{S}$, which are both radial. Thus

$$\mathbf{E} = \mathbf{e}_r E(r) \qquad (3.37)$$

$$d\mathbf{S} = \mathbf{e}_r dS \qquad (3.38)$$

$$\mathbf{E} \cdot d\mathbf{S} = E(r)dS\mathbf{e}_r \cdot \mathbf{e}_r = E(r)dS. \qquad (3.39)$$

This now allows us to reach a very important relationship:

$$\text{Lines crossing } d\mathbf{S} = k\varepsilon_0 E(r)dS = k\varepsilon_0 \mathbf{E} \cdot d\mathbf{S}. \qquad (3.40)$$

Therefore $\mathbf{E} \cdot d\mathbf{S}$, *the electric flux coming out of the area* $d\mathbf{S}$, is proportional to the lines crossing the surface. The proportionality constant is $\varepsilon_0 k$, where ε_0 is a fixed number and k is up to us to choose (once and for all).

If we cover the surface of the sphere with tiny little patches $d\mathbf{S}$ and add the contributions from all of them to the two sides of Eqn. 3.40, make the patches smaller and smaller, and turn the sum into an integral we obtain:

$$\text{Total number of lines crossing } S = k\varepsilon_0 \oint_S \mathbf{E} \cdot d\mathbf{S}. \qquad (3.41)$$

The integral on the right is called the *surface integral of* **E** *over* S. The symbol $\oint$ means the surface is closed.

Since the lines crossing the sphere are independent of the radius, we may now assert that *the surface integral of* **E** *over the sphere* S *is also independent of its radius* r.

Next consider an arbitrary surrounding surface S' surrounding the charge q as shown in Figure 3.9. We know the total number of lines crossing it are again the same, namely kq. How do we express this result in terms of $\mathbf{E}'$? If we cover this surface with patches, the area vectors $d\mathbf{S}'$ will not generally be radial. The number of lines these patches intercept will not be the product $k\varepsilon_0 E'(r)dS'$, but rather $k\varepsilon_0 E'(r)dS'\cos\theta$, where θ is the angle between $\mathbf{E}'$ and $d\mathbf{S}'$. If you think of the lines as the flow of something, from the fluid flow analogy it is evident that a given area will intercept the most lines if its area vector is parallel to $\mathbf{E}'$, and that as it rotates off this direction, their number will diminish by the geometrical factor $\cos\theta$. We have therefore the result that

$$\text{number of lines crossing } d\mathbf{S}' = k\varepsilon_0 \mathbf{E}' \cdot d\mathbf{S}'; \tag{3.42}$$

$$\text{number of lines crossing } S' = k\varepsilon_0 \oint_{S'} \mathbf{E}' \cdot d\mathbf{S}'. \tag{3.43}$$

Since the number of lines crossing a generic surface is independent of its shape as long as it surrounds the charge, we deduce the corresponding fact about the surface integral of the electric field over any generic surface:

$$k\varepsilon_0 \oint_S \mathbf{E} \cdot d\mathbf{S} = \text{lines crossing } S$$

$$= \text{lines emitted by } q = kq \tag{3.44}$$

$$\oint_S \mathbf{E} \cdot d\mathbf{S} = \frac{q}{\varepsilon_0} \tag{3.45}$$

where we have dropped the prime on $\mathbf{E}$ *and* S, *where the latter, from now on, will refer to the general surface, spherical or not. Eqn. 3.45 is Gauss's law for a single charge.*

There is no arbitrary constant k in this relation and there should not be. Whereas the lines we draw to aid our imagination have a density that does depend on k, the electric field at a point is uniquely defined by the charges that produce it or the force it exerts on a test charge. Therefore its integral on a closed surface better not depend on k. The result above is simply a property of the electric field as given by Coulomb's law and does not rely on the notion of field lines. The field lines helped us anticipate the final answer, which can, however, be derived by explicit computation.

As an illustration, consider the field of a point charge and *spherical surface S* centered on it. By direct computation

$$\oint_S \mathbf{E} \cdot d\mathbf{S} = \oint_S \frac{q}{4\pi\varepsilon_0 r^2} \mathbf{e}_r \cdot \mathbf{e}_r dS \tag{3.46}$$

$$= \frac{q}{4\pi \, \varepsilon_0 r^2} \oint dS \qquad (3.47)$$

$$= \frac{q}{4\pi \, \varepsilon_0 r^2} 4\pi \, r^2 = \frac{q}{\varepsilon_0}. \qquad (3.48)$$

The steps leading to Eqn. 3.48 need some explanation. There a surface integral is evaluated by inspection and the answer is simply written down. What happened to the integration? The answer is that the integrand, $E(r) = \frac{q}{4\pi \varepsilon_0 r^2}$, is a constant on the sphere. So $E(r)$ may be pulled out of the integral, like a number 19 can be pulled out. The integral of $E(r)dS$ over the sphere then reduces to the product of this constant $E(r)$ and the area of the sphere. (Here is an analogy. If $f(x) = f_0$, a constant, the definite integral over an interval of length L is the area of a rectangle of height f_0 and base L. More formally, f_0 may be pulled out of the integral and the remaining integral of dx is just L.)

If the surface S is not spherical, it takes more work to show that Eqn. 3.48 still holds by invoking the notion of a solid angle. The pictorial argument in terms of lines spared us that effort.

We now want to extend Gauss's law to many charges $q_i, i = 1 \ldots n$. Now we forget all about lines of force, which can be very complicated. Instead we use superposition of the fields to these charges, each of which obeys Gauss's law. Each charge q_i produces its own $\mathbf{E}_i$ that obeys

$$\oint_S \mathbf{E}_i \cdot d\mathbf{S} = \frac{q_i}{\varepsilon_0}, \qquad (3.49)$$

for *any closed surface S* containing the charge. By summing both sides over i we obtain *Gauss's law in all its generality,*

$$\oint_S \mathbf{E} \cdot d\mathbf{S} = \frac{1}{\varepsilon_0} \sum_i q_i = \frac{1}{\varepsilon_0} \cdot q_{enc} \qquad (3.50)$$

where $\mathbf{E} = \sum_i \mathbf{E}_i$ is the total electric field and q_{enc} is the total charge enclosed by S.

The charges q_i have to be *inside S* to contribute to the surface integral of $\mathbf{E}$, or, equivalently, the lines flowing out of S. Consider for example an empty surface S'' in Figure 3.9 with the charge q lying *outside*. Any field line emanating from the charge that enters the surface will necessarily also exit since there is no charge inside for it to terminate. Lines coming in are counted as negative and those coming out are described as positive, and the positive and negative contributions will cancel precisely. In terms of the electric field, $\mathbf{E} \cdot d\mathbf{S}$ will be negative where lines enter and positive where they leave, and the integral over S'' will be zero.

For future use I repeat the algorithm for computing the surface integral of $\mathbf{E}$ over any surface S, *closed or not*. Consult Figure 3.7.

- Tile the surface with tiny areas or patches $d\mathbf{S}(\mathbf{r}_i)$ located at $\mathbf{r}_i$. For a closed surface the area vectors are defined to point outward.
- On each tiny area compute the flux $d\Phi(\mathbf{r}_i) = \mathbf{E}(\mathbf{r}_i) \cdot d\mathbf{S}(\mathbf{r}_i)$.
- Do the sum $\sum_i d\Phi(\mathbf{r}_i) = \sum_i \mathbf{E}(\mathbf{r}_i) \cdot d\mathbf{S}(\mathbf{r}_i)$.
- Repeat with smaller and smaller patches till the sum converges to some limit. That defines $\int_S \mathbf{E} \cdot d\mathbf{S}$.

In some special cases like the field of a point charge the integral can be done analytically, but in all cases it has a well-defined numerical value, which can be determined as above.

3.5.1 Continuous charge density

Suppose S contains a continuous blob of charge, with charge density $\rho(\mathbf{r})$ instead of discrete charges q_i. To write down Gauss's law we need the total charge enclosed. A tiny cube of size $d^3\mathbf{r} = dx\,dy\,dz$ at $\mathbf{r}$ will enclose $\rho(\mathbf{r})d^3\mathbf{r}$ coulombs, and the enclosed charge will be this quantity integrated over the volume V within the closed surface S.

So the form of Gauss's law we will find most useful is as follows:

$$\oint_{S=\partial V} \mathbf{E}(\mathbf{r}) \cdot d\mathbf{S} = \frac{1}{\varepsilon_0} \int_V \rho(\mathbf{r})d^3\mathbf{r} \tag{3.51}$$

where

$S = \partial V$ means the closed surface S is the boundary

of the volume V. $\qquad\qquad$ (3.52)

In future I will also use

$C = \partial S$, which means the closed loop S is the

boundary of the surface S. $\qquad\qquad$ (3.53)

CHAPTER 4

Gauss's Law II: Applications

In the last chapter we encountered Gauss's law:

$$\oint_{S=\partial V} \mathbf{E} \cdot d\mathbf{S} = \frac{1}{\varepsilon_0} \int_V \rho(\mathbf{r}) d^3\mathbf{r} = \frac{1}{\varepsilon_0} \rho_{enc}. \qquad (4.1)$$

On the left-hand side we have the surface integral of the electric field over a *closed* surface S, which is the boundary of a volume V. On the right-hand side is the total charge enclosed by S divided by ε_0. The charge enclosed is the volume integral of the charge density $\rho(\mathbf{r})$ if continuous, and the sum over point charges q_i if discrete.

The surface S, called the *Gaussian surface*, is a theoretical construct to help our calculations. It may be chosen at will, and for every choice of S, there is a corresponding equality. The Gaussian surface could sometimes coincide with a real surface (say of a conductor).

As for the left-hand side, recall the algorithm for computing the surface integral: divide S into little areas or patches $d\mathbf{S}$, add the contributions $\mathbf{E} \cdot d\mathbf{S}$ from each area (where $\mathbf{E}$ is the electric field on that tiny area), and take the limit of an infinite number of patches of infinitesimal size. Often the only way to do this integral is by numerical means, though occasionally an analytic evaluation may be possible, such as when $\mathbf{E}$ is due to a point charge and S is a surrounding concentric sphere.

As for the right-hand side, the integrals of ρ can be done by inspection in all the cases we will discuss in this chapter. In general you will have to do a multiple integral of the charge density ρ over the volume V.

4.1 Applications of Gauss's law

As a first application, consider the field due to a uniform spherical ball of charge Q and radius R centered on the origin. Since the sphere of charge looks the same if we rotate it around any axis passing through the origin, the field distribution must have this property. The only solution to this requirement is the hedgehog field, with lines fanning out equally in all directions, with the same density at all points of a given r. In terms of the field, the allowed configuration is of the form

$$\mathbf{E}(\mathbf{r}) = \mathbf{e}_r E(r). \tag{4.2}$$

We just need to find $E(r)$ and will do so using Gauss's law.

To find the field outside the charged sphere, we choose as the Gaussian surface S, a sphere of radius $r > R$, as depicted in the top left

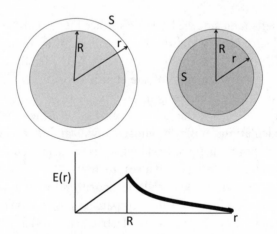

Figure 4.1   The use of Gauss's law to find the field due to a uniform solid ball of charge outside (top left) and inside (top right) its radius R by using a spherical Gaussian surface S of appropriate radius. The graph at the bottom shows $E(r)$, the radial field as a function of r. It rises linearly for $r \leq R$ and thereafter falls as $1/r^2$.

half of Figure 4.1. The calculation proceeds as with a point charge:

$$\oint_S \mathbf{e}_r E(r) \cdot \mathbf{e}_r dS = \oint_S E(r) dS \tag{4.3}$$

$$= E(r) \oint dS$$

since $E(r)$ is a constant all over S (4.4)

$$= E(r) \, 4\pi \, r^2 = \frac{Q}{\varepsilon_0} \quad \text{by Gauss. So} \tag{4.5}$$

$$E(r) = \frac{Q}{4\pi \, \varepsilon_0 r^2} \tag{4.6}$$

$$\mathbf{E}(\mathbf{r}) = \mathbf{e}_r E(r) = \mathbf{e}_r \frac{Q}{4\pi \, \varepsilon_0 r^2}, \tag{4.7}$$

which is the field of a point charge Q at the origin.

Note that *Gauss's law gives just one piece of information about* $\mathbf{E}(\mathbf{r})$: *its integral over a surface S*.

One cannot infer from that a whole function $\mathbf{E}(\mathbf{r})$. For example, if I say that

$$\int_{-1}^{+1} f(x)dx = 14, \tag{4.8}$$

what can you say about $f(x)$? It could be $f(x) = 7$, $f(x) = 7 + \sin x$, etc. But if I said $f(x)$ is a constant f_0 over the region of integration, you could deduce $f_0 = 7$ as follows:

$$\int_{-1}^{+1} f(x)dx = \int_{-1}^{+1} f_0 dx = f_0 \int_{-1}^{+1} dx = f_0 \cdot 2 = 14. \tag{4.9}$$

The moral is that if a function is a constant over a region of integration, its integral equals that constant times the length or area or volume of the integration region.

This is what happened in Eqn. 4.3: the surface integral was $E(r) \times 4\pi \, r^2$. Equating this to Q/ε_0, we obtained Eqn. 4.7.

Gauss's law can be used to deduce the entire field $\mathbf{E}(\mathbf{r})$ only when symmetry arguments can be used to reduce the unknown to just one number, $E(r)$ on the Gaussian surface. Had S been a sphere, but the charge a non-sphere with bumps and lumps here and there, the surface integral of $\mathbf{E}$ would still be known to be Q/ε_0, but one could not use this to find $\mathbf{E}(\mathbf{r})$ anywhere because it would vary over S. Similarly, had the charge been spherical but S not spherical, we would again have a result that was true, but not helpful in finding $\mathbf{E}(\mathbf{r})$ anywhere on S.

Next we want to find the field *inside* the sphere of charge. So we take for the Gaussian surface a sphere of radius $r < R$, as shown in the top right half of Figure 4.1. The calculation proceeds as for $r > R$ but with one change: the charge enclosed is not all of Q but only q_{enc}, the amount enclosed by the sphere of radius r. Since the density is uniform, the ratio of the enclosed charge to the total charge is the ratio of their volumes:

$$\frac{q_{enc}}{Q} = \frac{r^3}{R^3}. \tag{4.10}$$

If you do not like this argument, let me rewrite this result as follows:

$$q_{enc} = \frac{Q}{\frac{4}{3}\pi R^3} \cdot \frac{4}{3}\pi r^3 \tag{4.11}$$

where the first factor is the charge density and the second factor is the volume in question.

The surface integral of the field is the same as before and Gauss's law takes the form

$$E(r) \cdot 4\pi r^2 = \frac{q_{enc}}{\varepsilon_0} = \frac{1}{\varepsilon_0} \frac{Q r^3}{R^3} \tag{4.12}$$

$$E(r) = \frac{Qr}{4\pi \varepsilon_0 R^3} \tag{4.13}$$

$$\mathbf{E}(\mathbf{r}) = \mathbf{e}_r \frac{Qr}{4\pi \varepsilon_0 R^3}. \tag{4.14}$$

Thus the field actually *grows* from zero as we move out, and it reaches a maximum of $Q/(4\pi \varepsilon_0 R^2)$ at the surface. Thereafter, it falls like $1/r^2$. The field, radial in all cases, is as follows for all values of r:

$$E(r) = \frac{Qr}{4\pi \varepsilon_0 R^3} \quad r \leq R \tag{4.15}$$

$$= \frac{Q}{4\pi \varepsilon_0 r^2} \quad r \geq R \tag{4.16}$$

and is depicted in the bottom part of Figure 4.1. The two expressions agree on the surface of the ball $r = R$.

Why does $E(r)$ *grow* (linearly) with r when $r < R$? Because, as r increases, the enclosed charge grows as r^3 and the field it produces, acting as a point charge at the origin, falls as $1/r^2$. Once we go outside the sphere, for $r > R$, the field falls like $1/r^2$, since we do not pick up any extra charge as we increase r, the radius of the Gaussian surface.

These results may be taken over verbatim for gravity, with the understanding that the force is always attractive. Consider in particular the linear force, which points toward the center inside a spherical mass. This linear (restoring) force *implies simple harmonic motion*. If the spherical mass in question is the earth, this has the following interesting consequence. If you drill a very narrow hole passing through the center of the earth (so narrow that the mass you scooped out does not affect the preceding answer for the field) and drop an object into it, it will oscillate back and forth between where you are and the diametrically opposite point on the globe. I invite you to show that $\omega = \sqrt{GM/R_E^3}$. (First write down Gauss's law for gravity.)

4.2   Field inside a shell

Consider a uniformly charged solid sphere of radius R_2 from which a concentric sphere of radius $R_1 < R_2$ has been scooped out. We want the field due to this hollow shell. By Gauss's law, for $r > R_2$ this hollow sphere will act like a point charge centered at the origin. How about inside the hollow region, for $r < R_1$? By applying Gauss's law to a Gaussian surface of radius $r < R_1$, we see that the field inside is zero because the charge enclosed is zero. This result is equally true for the gravitational force.

Let us try to understand the absence of the field inside a hollow shell directly in terms of Coulomb's law. This discussion is optional.

I will only show that the field inside a hollow shell of radius R and *infinitesimal thickness* is zero. I am done, because the original shell of finite thickness $R_2 - R_1$ can be built out of concentric, infinitesimally thin shells of radius ranging from R_1 to R_2, each of which contributes a zero to the total.

Consider then a point P inside such a shell of radius R and infinitesimal thickness, as depicted in Figure 4.2. Assume the shell has a surface charge density σ. (I invite you to show that $\sigma = \rho\,dr$, where ρ is the uniform density of the charged sphere and dr is the thickness of the shell.)

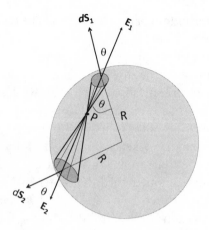

Figure 4.2   The aim is to show that the field at a generic point P inside a hollow shell is zero. The figure shows two oppositely pointing cones of identical opening angle that meet at P and intersect the sphere in two caps, shown as dark ellipses. The same number of field lines emitted by the test charge at P pierce the two caps. This is shown in the text to imply that the charges on these two caps exert equal and opposite forces on a test charge at P. It is possible to cover the entire shell using canceling pairs of cones.

If P is the center of the shell, we can argue by symmetry that the field there has to vanish: a non-zero $\mathbf{E}$ at the center necessarily has to point in some arbitrary direction, violating the rotational symmetry of the problem. But we can see more directly that the field has to be zero because for every tiny patch of charge on the shell pushing a test charge one way, there is a diametrically opposite patch that exerts an equal and opposite force.

But the result is stronger; it says $E(r) = 0$ even for a point off-center, like P in the figure. We would like to show that this too follows from the cancellation of forces exerted by charges in different segments of the shell. To this end consider two cones of the same opening angle pointing away from P in opposite directions and intersecting the shell on two caps. The opening angle of the cones is infinitesimal, as are the planar areas they pierce through, denoted by $d\mathbf{S}_1$ and $d\mathbf{S}_2$.

Instead of showing that (the charges on) the caps exert equal and opposite forces on a unit test charge at P, we will show the unit test charge exerts equal and opposite forces on the (charges on) the two caps. We are then done because action and reaction are equal and opposite in Coulomb's law: if the forces the test charge at P exerts on the caps are equal and opposite, so are the forces the caps exert on the test charge. Since it is possible to surround the point P with such canceling pairs of cones, we know the net force of the shell on the charge at P will be zero.

So imagine a unit test charge placed at P and the lines emanating isotropically from it. *Since the cones have the same opening angle, they contain the same number of field lines and thus the number of lines crossing the two caps is equal.* Now the number of lines crossing the caps is, by Eqn. 3.42, $k\varepsilon_0 \mathbf{E}_1 \cdot d\mathbf{S}_1$ and $k\varepsilon_0 \mathbf{E}_2 \cdot d\mathbf{S}_2$ where $\mathbf{E}_1$ and $\mathbf{E}_2$ are the fields produced at the caps by the test charge at P.

Next we collect some relevant facts.

- The area vectors $d\mathbf{S}_1$ and $d\mathbf{S}_2$ are radial, being parts of a sphere.
- The electric fields $\mathbf{E}_1$ and $\mathbf{E}_2$ point outward along the symmetry axis of the two head-to-head cones.
- The angles between the area vectors and the corresponding field vectors are the same in both patches, and are denoted by θ. This equality follows from the fact that the angles opposite to the indicated θ's lie at the base of an isosceles triangle (whose two equal sides are the radius R and whose base is the chord connecting $d\mathbf{S}_1$ and $d\mathbf{S}_2$).

We put all this together and reason as follows:

$$\text{lines crossing cap } 1 = k\varepsilon_0 d\mathbf{S}_1 \cdot \mathbf{E}_1 = k\varepsilon_0 d\mathbf{S}_2 \cdot \mathbf{E}_2$$

$$= \text{lines crossing cap 2} \tag{4.17}$$

$$dS_1 E_1 \cos\theta = dS_2 E_2 \cos\theta \tag{4.18}$$

$$dS_1 E_1 = dS_2 E_2 \tag{4.19}$$

$$\sigma\, dS_1 E_1 = \sigma\, dS_2 E_2 \tag{4.20}$$

$$dq_1 E_1 = dq_2 E_2 \tag{4.21}$$

where $\sigma\, dS_i = dq_i$ is the charge on cap $i, i = 1$ or 2. The caps will behave as point charges $\sigma\, dS_1$ and $\sigma\, dS_2$ when we take the opening angles of the cones to zero.

Look at Eqn. 4.21. It says $dq_1 E_1$, the magnitude of the force the unit test charge at P exerts on the charges residing in dS_1 through the field E_1 it creates there, is equal to $dq_2 E_2$, the force the unit test charge at

P exerts on the charges residing in dS_2 through the field E_2 it creates there. The two forces of course have opposite directions, pointing away from the test charge. But if the test charge exerts equal and opposite forces on the caps, they in turn must exert equal and opposite forces on the test charge, because in Coulomb's law action and reaction are equal and opposite. (Recall $\mathbf{F}_{12} = -\mathbf{F}_{21}$.)

The argument relating the flux lines intercepted by the two caps to the fields $\mathbf{E}_1$ and $\mathbf{E}_2$ relies on the inverse square law of the electric force. Conversely one of the earliest tests of the inverse square law was the absence of field inside a hollow sphere.

4.3   Field of an infinite charged wire, redux

We have already seen how symmetry demands that the field of an infinite wire with linear charge density λ is constant if we move parallel to the wire at a fixed distance ρ, and points radially away from it:

$$\mathbf{E}(\mathbf{r}) = \mathbf{e}_\rho E(\rho). \tag{4.22}$$

We found that $E(\rho) = \lambda/(2\pi\varepsilon_0\rho)$ by doing an integral along the wire.

Now we will rederive $E(\rho)$ by using Gauss's law. The trick is to find a Gaussian surface on which there is a single unknown, $E(\rho)$. A natural choice is a cylinder of radius ρ coaxial with the wire, as shown in Figure 4.3, since the field is constant in magnitude all over it. However, it is not enough to take just the curvy sides of the cylinder; we need the two flat sides at both ends, since the Gaussian surface has to be closed in order for the law to work, for it to enclose a definite amount of charge.

The radius of the cylinder is clearly ρ since we want $E(\rho)$, but what should be its length L? Since the Gaussian surface is a figment of our imagination and not really wrapped around the wire, we can choose any length we want and then desperately hope that the answer will not depend on this arbitrary L.

Look at Figure 4.3. The charge enclosed within the cylinder is $\lambda \cdot L$, from the very definition of λ as the charge per unit length. So we begin with

$$\oint_S \mathbf{E} \cdot d\mathbf{S} = \frac{\lambda L}{\varepsilon_0}. \tag{4.23}$$

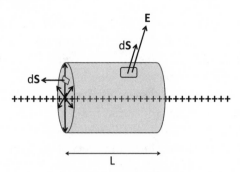

Figure 4.3   By symmetry, the field due to an infinite wire is radial and of constant magnitude at a fixed distance ρ from the wire. The Gaussian surface is a coaxial cylinder of radius ρ and has an arbitrary length L. The charge enclosed is simply λL. The two flat faces make no contribution to the flux since $\mathbf{E}$ and $d\mathbf{S}$ are perpendicular. The curved face, on which the flux density is constant, makes a contribution $E(\rho) \cdot 2\pi\rho L$.

The surface breaks up into three parts: the two flat ends and the curved face parallel to the wire.

We seem to have a problem with the flat faces, since $E(\rho)$ is not a constant on the entire face because different parts of it are at different distances from the wire. On the other hand, we have seen that Gauss's law is useful only when there is just one constant E on the entire surface. Luckily we are saved by the fact that the area vectors $d\mathbf{S}$ and field $\mathbf{E}$ are perpendicular on these two faces: $d\mathbf{S}$ is parallel to the wire while $\mathbf{E}$ is perpendicular to it so the flux through the flat faces is zero. Or if you like, the field lines run parallel to the flat faces and so none cross it.

We are then left with the curved face on which the area vectors $d\mathbf{S}$ are radial and $\mathbf{E}(\rho)$ is a constant and radial. (Remember that for a closed surface, the area vector is defined as positive if it points outward.) So Gauss's law tells us

$$\int_{curved\ face} \mathbf{e}_\rho E(\rho) \cdot \mathbf{e}_\rho\, dS = E(\rho)(2\pi\rho L) = \frac{\lambda L}{\varepsilon_0} \qquad (4.24)$$

$$E(\rho) = \frac{\lambda}{2\pi\,\varepsilon_0\rho} \quad \text{in agreement with the old result.} \qquad (4.25)$$

The arbitrary length L has canceled out, as it must.

We get the answer so easily only because of the high symmetry of the problem. For example, if the wire had been non-uniformly charged, with $\lambda = \lambda(x)$, we could still equate the flux over the cylinder to the charge enclosed (the integral of $\lambda(x)$ over the length L). However, since $\mathbf{E}$ varies in magnitude and direction (not always radial) this will only tell us something about the integral of $\mathbf{E}$ over the surface and not about its value at any one place. On the other hand, if the line charge is replaced by a uniform cylindrical distribution, $\mathbf{E}$ may be found everywhere using Gauss's law and symmetry.

4.4   Field of an infinite plane, redux

Consider an infinite plane, which we take to be the xy-plane. It has a uniform charge density σ.

Recall what the symmetry arguments tell us. The field is independent of x or y (but could depend on z) and must point perpendicularly away from the plane with the same magnitude at z and $-z$. That is, $\mathbf{E}$ must have the form

$$\mathbf{E}(z) = \mathbf{k}E(|z|) \quad z > 0 \tag{4.26}$$

$$= -\mathbf{k}E(|z|) \quad z < 0. \tag{4.27}$$

To find $E(|z|)$ we need a Gaussian surface on whose various parts E is either constant or perpendicular to the area vector. Such a surface is shown in Figure 4.4. It is a cylinder of cross section A, with its symmetry axis parallel to the z-axis and its flat faces at $\pm z$.

The area A is arbitrary and hopefully will drop out of the answer. The charge enclosed is clearly σA, where A is the area of the circle the cylinder encloses as it pierces the plane. In contrast to the infinite wire, this time the curved side of the cylinder makes no contribution to the surface integral since the field is parallel to the curved side and the area vector is normal to it. (The field lines cross the two flat faces but not the curved face.) As for the flat faces, on the upper face we have $\mathbf{k}A \cdot \mathbf{k}E(|z|) = A \cdot E(|z|)$. The same contribution comes from the lower face where both $\mathbf{E}$ and the area have flipped their orientation to yield $(-\mathbf{k}A) \cdot (-\mathbf{k}E(|z|)) = A \cdot E(|z|)$. Gauss's law then tells us

$$2AE(|z|) = \frac{\sigma A}{\varepsilon_0} \tag{4.28}$$

$$E(|z|) = \frac{\sigma}{2\varepsilon_0}. \tag{4.29}$$

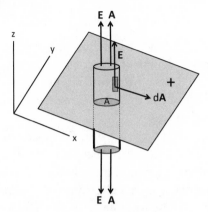

Figure 4.4   Shown is an infinite plane with charge density σ. Symmetry tells us the field is everywhere normal to the plane and constant in magnitude as we move parallel to the plane. The Gaussian surface is a cylinder of area A and height $2z$, symmetrically located with respect to the plane. The charge enclosed is σA. As for the flux, or surface integral of $\mathbf{E}$, the curved side makes no contribution because the area vector and field are perpendicular, while the two flat faces make equal contributions of $E(|z|)A$ each, where $E(|z|)$ is the constant value of the field strength a distance $|z|$ from the charged sheet.

The area has dropped out as it must, and, remarkably, there is no dependence on z, the coordinate perpendicular to the plane.

4.5   Conductors

Consider a chunk of copper, which is a good conductor. In a good conductor not all of the electrons in the atoms are tied to the nuclei, but shared communally. They are free to move around the material but not to leave it. The conductor is like a swimming pool for the electrons: they can swim freely inside but cannot scale the walls at the boundary. If they try that, all the nuclei will exert a force to pull them back. The energy needed to rip an electron out of the material is called the *work function*. There are good and bad conductors, and we will discuss a perfect conductor in which the charges can move freely in response to the smallest field.

We will now make many predictions about conductors, mainly using Gauss's law.

4.5.1 Field inside a perfect conductor is zero

The first property that follows by definition is that in a static situation, the electric field inside a perfect conductor is zero. Had there been a field, the charges would have been moving but we have been assured it is an electrostatic situation. So there can be no field. The no-field rule does not hold in the non-static case.

For example, it fails if there is a field **E** in space and I suddenly insert a chunk of conductor shaped like a rectangular slab into that region, as shown in the top half of Figure 4.5. Initially there *will* be a field inside the conductor. It will start moving the electrons (whose charge is negative) in the opposite direction. These will pile up on the left face as shown, leaving behind positive charges on the other face, due to nuclei whose electrons have drifted away. The pileup will continue until the internally generated field due to the two layers cancels the applied field **E**. For an infinite slab, we

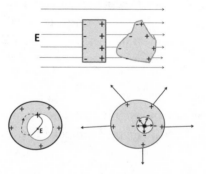

Figure 4.5   Top: Two conductors placed in an external field **E**, which gets screened inside by polarization. In the rectangular slab, the internal field σ/ε_0 due to the charges on the two faces neutralizes **E**. Bottom left: A conductor with a hole in it and some positive charges deposited on it. By Gauss's law, these must be on the outer surface and the charge on the inner surface has to be zero. If two canceling charges reside on the inner surface, they would produce a field **E**, which can do work on a test charge moving from the $+$ to the $-$. The test charge can then be brought back to the $+$ for free inside the conductor along the dotted line. The cycle violates energy conservation. Bottom right: A charge q placed inside the hole. The lines it emits terminate on the inner wall (on the negative charges from the conductor that piled up there) and the lines are re-emitted by the positive charges that are on the outer wall.

know the two faces would produce a field σ/ε_0 in the region between them, opposing the external **E**. For a finite slab there will be some complications near the ends but the field inside the conductor will still end up vanishing.

If a time-dependent electromagnetic field encounters a metal, it will be screened (and reflected) if the incident frequency is below the *plasma frequency* of order $10^{16}Hz$. The reciprocal of this frequency, $\simeq 10^{-16}s$, is roughly the time it takes the disturbed system to settle down to equilibrium.

In the case of the rectangular conducting slab, we are able to anticipate the way the charges in the conductor would rearrange themselves to cancel or screen the external field. What is remarkable is that even if the conductor has a crazy shape, say like a potato, it will find a way to rearrange its charges so as to kill the field in the interior. Even for a simple conductor like the sphere, it takes a lot of work to compute theoretically the final charge distribution that will exactly annul the external field within the conductor. Yet the electrons in a metal are able to figure this out for any shape, almost instantaneously! But you should not be too impressed. They do this rather mindlessly. First they migrate against the applied field (since they have negative charge) and soon the new immigrants start discouraging newer immigrants from joining them, using their Coulomb repulsion. Eventually this repulsion will balance the force of the external field and the migration will stop.

4.5.2 The net charge on a conductor will reside at the surface

Suppose we throw some positive charges on a neutral conductor. They will run as far away from each other as possible. Since they cannot leave the conductor, you may conjecture that they will end up at the surface. This is indeed so and we can prove it using Gauss's law as follows. Take any closed infinitesimal surface anywhere inside the conductor and apply the law. Since the field is zero, its surface integral is zero, and so is the enclosed charge.

Not only is an external field screened by the conductor, but the additional charges we throw in will also produce zero field inside the conductor. They have to, for if this were not so, the mobile charges would move till it is so. If the conductor is a sphere, we know that the charge we dump on to it will spread uniformly over the surface—this being the configuration that produces zero field in the interior. But, amazingly, even if the conductor looks like a potato, the charges will find a way to arrange themselves on the surface so as to produce zero internal field.

4.5.3 A conductor with a hole inside

Suppose we throw some charge on a conductor with a hole inside, as shown in the lower half of Figure 4.5. Will all the charges end up on the outer surface or will there be some on the inner surface? It turns out all charges will be on the outer surface. To prove this, consider a Gaussian surface that tightly encloses the inner surface and lies entirely within the conductor, an infinitesimal distance away from the hole. Since the field on this surface is zero, so will be the charge enclosed.

Could this zero be made of equal numbers of positive and negative charges occupying different parts of the inner surface? Even if it starts out that way, the opposite charges are free to race across the inner surface and neutralize each other.

This would be obviously true were it not for the charges on the outer surface. Could they somehow exert a force on these charges to prevent this reunion? Suppose there were two opposite charges on the inner boundary, as shown in the bottom left of Figure 4.5. The field lines leaving the positive charge and ending on the negative charge have to do so within the cavity. (The lines cannot go into the conductor.) If we release a test charge near the positive charge, it will be accelerated along the field lines till it gets to the negative charge. We could suck up its kinetic energy (for use elsewhere) and bring it back to the positive charge *inside the conductor*. This return trip will cost no energy since there is no field inside the conductor. We could do this cycle *ad infinitum* and extract an infinite amount of energy from nowhere, violating the law of conservation of energy. The only way to avoid this is for the opposite charges to meet and neutralize each other.

Next, suppose we place a charge q (assumed positive) inside the hole, as shown in the lower right of Figure 4.5. Will the world outside know about it? Since no field can enter the conductor, how can it tell the outside world it is there? Yet a Gaussian surface outside the conductor should yield a surface integral corresponding to an enclosed charge q. The answer is shown in the bottom right of the figure. The neutral conductor splits into positive and negative charges $\pm q$, and the positive charges go to the outer surface and the negative ones to the inner one. (To be specific, the electrons will go to the inner surface and leave behind unbalanced protons on the outer surface.) The field lines leaving the q we placed inside the hole will terminate on the $-q$ sitting on the inner surface, while the $+q$ sitting on the outer surface will emit the lines that penetrate our Gaussian surface.

4.5.4 Field on the surface of a conductor

Consider a conductor, not necessarily spherical, on which we have placed some charge that is now sitting on the surface. While no field can enter the conductor, what can we say on the surface? The field cannot have a component parallel to the surface, for this will set charges in motion along the surface, contrary to the assumed static situation. (Motion of charges *along the boundary* is not forbidden by the nuclei; they just won't let them escape outside.) So the field has to be normal to the surface. We will now relate this $E_\perp$ to the *local* charge density σ.

Figure 4.6 shows a (tilted) Gaussian cylinder of infinitesimal height, of base dS, its axis normal to the surface, and situated half inside and half outside the conductor. The charge enclosed is $\sigma\,dS$. The flat face inside

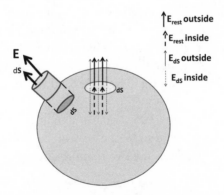

Figure 4.6 The field at the surface of a charged conductor is calculated using a Gaussian cylinder half inside and half outside with its axis normal to the surface. There is non-zero flux only on the flat face outside. There is no field inside and no flux on the curved side outside, which runs parallel to the field. Also shown are the field at a small area dS due to the charges on it (thin arrows, solid outside and dotted inside) and the charges on the rest of the surface (thick arrows, solid outside and dotted inside). The two contributions exactly cancel inside and double up outside. The charge density σ and $\mathbf{E}_\perp$ can vary from point to point.

the conductor does not contribute to the flux as $\mathbf{E}$ is zero inside. As for the curvy side, $\mathbf{E}$ is either zero on it (if it is inside) or parallel to it (if it is outside), and in neither case contributes to the flux. The top face contributes $E_\perp \, dS$. Gauss's law tells us

$$E_\perp \, dS = \frac{\sigma \, dS}{\varepsilon_0} \tag{4.30}$$

$$E_\perp = \frac{\sigma}{\varepsilon_0}, \tag{4.31}$$

a result worth committing to memory.

We understand this result as follows. Divide the charged surface into a very tiny patch dS where we are computing $E_\perp$, and the rest with a hole where the patch is. Arbitrarily close to dS, for distances much smaller than its linear dimensions, the patch will behave like an infinite plane and produce a field $\mathbf{E}_{dS}$ pointing normally out on the outside and normally in on the inside with equal strength $\frac{\sigma}{2\varepsilon_0}$. This discontinuity between inside and outside is familiar from the infinite plane and is due to the charge density that divides the two regions. To this we must add the contribution $\mathbf{E}_{rest}$ from the rest of the surface. This contribution will be continuous across the hole because the charges in the rest of the surface do not reside in the hole to cause any discontinuity. This continuous field must be pointing normally out with strength $\sigma/(2\varepsilon_0)$ to kill the normally inward field due to dS, so that the net field inside the conductor will be zero. However, when we go outside the conductor, the very same field will reinforce and double that due to dS.

In short, the field due to dS switches sign at the surface (due to the surface charge), while that due to the rest of the surface is continuous across the surface. This is why the two reinforce outside and cancel inside.

CHAPTER 5

The Coulomb Potential

There are two parts to electrodynamics: find the field $\mathbf{E}(\mathbf{r})$ produced by all the charges at the location of a charge q and find its response to the field using $\mathbf{F} = q\mathbf{E}$. This is a very complicated problem because each charge is playing a dual role: producing the field others respond to and responding to the field others produce. The fields depend on the past positions of all the particles due to the retardation demanded by relativity.

So far we have been making life tolerable by dealing with static charges. Despite the forces between them, we assume some other force is holding them in fixed positions so we may use Coulomb's law to find $\mathbf{E}$. But there is no fun in finding $\mathbf{E}$ if none of the charges is free to respond to it. So we are going to relax things a little bit: all but one charge will be held fixed and produce a field $\mathbf{E}$ given by Coulomb's law, and the one solitary charge q will be free to respond to this field. As it moves, the force it exerts on the other charges will vary, but that does not matter because they are not free to move in response.

We are going to start with

$$m\frac{d^2\mathbf{r}}{dt^2} = q\mathbf{E}(\mathbf{r}) \tag{5.1}$$

where $\mathbf{E}(\mathbf{r})$ is due to all the fixed charges. Eqn. 5.1 is all we need in principle. Given this equation, as well as the particle's initial position $\mathbf{r}(0)$ and velocity $\mathbf{v}(0)$, we can determine the subsequent fate of the particle analytically in some rare cases and numerically in all cases, given a fast

computer. Using the initial velocity, we find the position a short time
dt later as

$$\mathbf{r}(dt) = \mathbf{r}(0) + \mathbf{v}(0)dt \qquad\qquad (5.2)$$

and given the initial acceleration (decided by the field at its initial position)
we can find the velocity at time dt as

$$\mathbf{v}(dt) = \mathbf{v}(0) + \frac{q\mathbf{E}(\mathbf{r}(0))}{m}dt. \qquad\qquad (5.3)$$

At time dt we can repeat the process and move forward in time in
increments of dt. The errors vanish in the limit $dt \to 0$.

5.1 Conservative forces and potential energy

This is a topic that was covered extensively in Volume I. I present here a
brief review in the interest of continuity.

If a mass m connected to a spring of force constant k is pulled by
some amount A and released, we can find its subsequent position $x(t)$
by solving the differential equation, and from that we can find $v(t)$ by
differentiation. But we found that certain questions can be answered much
more easily, such as "What will be its velocity when it is at $x = x_0$?"
The trick is to invoke the law of conservation of energy, which tells us in
this case

$$\frac{1}{2}mv_1^2 + \frac{1}{2}kx_1^2 = \frac{1}{2}mv_2^2 + \frac{1}{2}kx_2^2 \qquad\qquad (5.4)$$

where the subscripts 1 and 2 refer to two points on the mass's trajectory. If
x_1 and v_1 are the initial position and velocity, we can find v_2 (up to a sign)
at the point x_2 by solving for it in Eqn. 5.4.

More generally we would have

$$K_1 + U_1 = K_2 + U_2 \qquad\qquad (5.5)$$

where $K = \frac{1}{2}mv^2$ is the kinetic energy and $U_1 \equiv U(x_1)$ and $U_2 \equiv U(x_2)$
denote the potential energy that depends on the forces acting on the body.

Eqn. 5.5 is easily derived in $d = 1$ starting with Newton's law. Here is one way.

$$\frac{dK}{dt} = \frac{m}{2}\frac{dv^2}{dt} = \frac{m}{2} \cdot 2v\frac{dv}{dt} = m\frac{dv}{dt} \cdot v \quad (5.6)$$

$$= F(x)\frac{dx}{dt} \quad (5.7)$$

$$\int_{t_1}^{t_2} \frac{dK}{dt}dt = \int_{t_1}^{t_2} F(x)\frac{dx}{dt}dt \quad (5.8)$$

$$\int_1^2 dK = K(t_2) - K(t_1) = \int_{x_1}^{x_2} F(x)dx, \quad (5.9)$$

which is the *work-energy theorem*.

The quantity

$$dW = F(x)dx \quad (5.10)$$

is the *work done by the force* when the body moves by dx and the work-energy theorem relates the work done to dK, the change in the kinetic energy of the body. This theorem relies on just $F = ma$ and is valid for all F, including friction.

Now, the definite integral of any function of one variable may be expressed as

$$\int_{x_1}^{x_2} F(x)dx = G(x_2) - G(x_1) \quad \text{where} \quad (5.11)$$

$$\frac{dG}{dx} = F(x). \quad (5.12)$$

Combining this with Eqn. 5.9 we find

$$K_2 - K_1 = G_2 - G_1 \quad (5.13)$$

$$K_2 - G_2 = K_1 - G_1 \quad (5.14)$$

$$K_2 + U_2 = K_1 + U_1 \quad \text{where} \quad (5.15)$$

$$G = -U. \quad (5.16)$$

The following reciprocal relations between F and U are worth remembering:

$$F = -\frac{dU}{dx} \quad (5.17)$$

$$U(x_2) - U(x_1) = -\int_{x_1}^{x_2} F(x)\,dx. \tag{5.18}$$

They allow us to go from the potential to the force or vice versa.

Where does the derivation fail if there is friction? Eqn. 5.11 does not apply since the force of friction is not just a function of x; it depends on the velocity, being always opposed to it in direction. As long as the particle is moving in one direction, we can pick a sign for the frictional force to find its impact on K using the work-energy theorem, but we cannot derive a law of conservation of energy for motion with changes in direction, such as in a damped oscillation.

Deriving Eqn. 5.5 in two (or higher) dimensions may not be possible, even if there is no friction. Let us recall the problem and its resolution.

We begin with the natural definition of kinetic energy in higher dimensions:

$$K = \frac{1}{2}m|\mathbf{v}|^2 = \frac{1}{2}m\mathbf{v}\cdot\mathbf{v} = \frac{1}{2}m\left(v_x^2 + v_y^2 + v_z^2\right) \tag{5.19}$$

and take its time derivative:

$$\frac{dK}{dt} = m\left(v_x\frac{dv_x}{dt} + v_y\frac{dv_y}{dt} + v_z\frac{dv_z}{dt}\right) \tag{5.20}$$

$$= \mathbf{v}\cdot m\frac{d\mathbf{v}}{dt} \tag{5.21}$$

$$= \mathbf{F}\cdot\mathbf{v} = \mathbf{F}\cdot\frac{d\mathbf{r}}{dt} \tag{5.22}$$

$$dK = \mathbf{F}\cdot d\mathbf{r}. \tag{5.23}$$

So far there is no problem. The change in kinetic energy when the force pushes the body over a vector distance $d\mathbf{r}$ is unambiguous: it is $\mathbf{F}\cdot d\mathbf{r}$.

The trouble comes when we string together little $d\mathbf{r}$'s to make a finite path connecting two points 1 and 2 as shown in Figure 5.1: there are infinitely many possible paths, two of which are shown.

The line integral in the relation

$$K_2 - K_1 = \int_1^2 dK = \int_1^2 \mathbf{F}\cdot d\mathbf{r} \tag{5.24}$$

is generally path-dependent and cannot be written as $U_1 - U_2$.

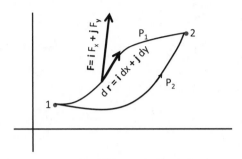

Figure 5.1   Two paths P_1 and P_2 connecting the same end points 1 and 2. The line integral, which is the sum over $\mathbf{F} \cdot d\mathbf{r}$, will generally depend on the path.

If, however, the line integral is of the form

$$\int_1^2 \mathbf{F} \cdot d\mathbf{r} = U_1 - U_2 \tag{5.25}$$

independent of the path, and a function of only the end points, we may write

$$K_2 - K_1 = U_1 - U_2 \tag{5.26}$$

and obtain the law of conservation of energy.

A force for which the line integral is path-independent is called a *conservative force*. One may think such forces are a rarity, but there is a recipe for manufacturing any number of them. Pick *any* function $U(x, y, z)$ and define the force by

$$F_x = -\frac{\partial U}{\partial x} \tag{5.27}$$

$$F_y = -\frac{\partial U}{\partial y} \tag{5.28}$$

$$F_z = -\frac{\partial U}{\partial z} \text{ or, more compactly,} \tag{5.29}$$

$$\mathbf{F} = -\left(\mathbf{i} \frac{\partial U}{\partial x} + \mathbf{j} \frac{\partial U}{\partial y} + \mathbf{k} \frac{\partial U}{\partial x} \right) \equiv -\nabla U \tag{5.30}$$

where ∇U is called the *gradient of U* and pronounced "grad U."

Let us see why such a force is conservative. Since

$$d\mathbf{r} = \mathbf{i}dx + \mathbf{j}dy + \mathbf{k}dz, \tag{5.31}$$

$$\mathbf{F} \cdot d\mathbf{r} = -\nabla U \cdot d\mathbf{r} \tag{5.32}$$

$$= -\left(\frac{\partial U}{\partial x}dx + \frac{\partial U}{\partial y}dy + \frac{\partial U}{\partial z}dz\right) = -dU. \tag{5.33}$$

Thus $\mathbf{F} \cdot d\mathbf{r} = dU$ is the first order change (linear in dx, dy, and dz) in the function U due to changes in x, y, and z. Consequently

$$\int_1^2 \mathbf{F} \cdot d\mathbf{r} = -\int_1^2 dU = U_1 - U_2 \tag{5.34}$$

is the total change in U between the end points. This leads to

$$E_1 \equiv K_1 + U_1 = K_2 + U_2 \equiv E_2. \tag{5.35}$$

Thus the *function U, which generates* $\mathbf{F}$, *is also the potential energy in the formula* $E = K + U$.

In the case of $d = 2$ it is useful to think of U as a height measured above the point (x, y). Since $\mathbf{F} \cdot d\mathbf{r} = -\nabla U \cdot d\mathbf{r} = -dU$ measures (minus) the change in "height" when we move by $d\mathbf{r}$, the line integral is the height difference between points 1 and 2 and is clearly independent of whichever interpolating path we take.

Once again, here are the reciprocal relations between the potential and the force in higher dimensions:

$$\mathbf{F} = -\nabla U \tag{5.36}$$

$$U_2 - U_1 = -\int_1^2 \mathbf{F} \cdot d\mathbf{r}. \tag{5.37}$$

If $\mathbf{F}$ is a conservative force acting on the body and we want to move the body against it (without accelerating it), we need to apply a force $-\mathbf{F}$ that exactly balances $\mathbf{F}$. The right-hand side is the work *we must do* to move it from 1 to 2, and the left-hand side is the gain in potential energy.

This recipe for producing a conservative force is exhaustive: *every* conservative force is the gradient of some U.

Thanks to this we can see if a given force is conservative or not as follows. Consider two dimensions first. If $\mathbf{F}$ is conservative, we know its components have the form

$$F_x = -\frac{\partial U}{\partial x} \tag{5.38}$$

$$F_y = -\frac{\partial U}{\partial y} \tag{5.39}$$

for some U. Consequently

$$\frac{\partial F_x}{\partial y} - \frac{\partial F_y}{\partial x} = -\frac{\partial^2 U}{\partial y \partial x} + \frac{\partial^2 U}{\partial x \partial y} = 0 \tag{5.40}$$

since the order of partial derivatives does not matter.

For example,

$$\mathbf{F} = -\mathbf{i}6xy - \mathbf{j}3x^2 \quad \text{is conservative with } U = 3x^2 y; \tag{5.41}$$

$$\mathbf{F} = -\mathbf{i}6x^2 y - \mathbf{j}3x^2 \quad \text{is not conservative.} \tag{5.42}$$

In three dimensions we have two more equations like 5.40 obtained by the *cyclic permutations* $x \to y, y \to z, z \to x$.

Instead of saying the line integral of a conservative force is path-independent we could say the *line integral of a conservative force over any closed loop is zero*.

Here is the logic. Consider two different paths P_1 and P_2 connecting the same points 1 and 2 in Figure 5.1. Start with what we are given and proceed as follows:

$$\int_1^2 \mathbf{F} \cdot d\mathbf{r} \quad \text{(along path } P_1)$$

$$= \int_1^2 \mathbf{F} \cdot d\mathbf{r} \quad \text{(along path } P_2) \tag{5.43}$$

$$\int_1^2 \mathbf{F} \cdot d\mathbf{r} \quad \text{(along path } P_1)$$

$$- \int_1^2 \mathbf{F} \cdot d\mathbf{r} \quad \text{(along path } P_2) = 0 \tag{5.44}$$

$$\int_1^2 \mathbf{F} \cdot d\mathbf{r} \quad \text{(along path } P_1)$$

$$+ \int_2^1 \mathbf{F} \cdot d\mathbf{r} \quad \text{(along path } P_2) = 0 \tag{5.45}$$

$$\oint \mathbf{F} \cdot d\mathbf{r} = 0. \tag{5.46}$$

The passage from Eqn. 5.44 to Eqn. 5.45 uses the fact that when the end points 1 and 2 are exchanged, the integral changes sign: on the backward path $\mathbf{F}$ is the same at every point, while every $d\mathbf{r}$ is reversed.

Eqn. 5.46 states that the integral over *any* closed loop $1 \to 2 \to 1$ is zero.

5.2 Is the electrostatic field conservative?

You know it must be, given the time I spent reviewing conservative forces. But here is a more substantial piece of reasoning.

We will say a *field* $\mathbf{E}$ is conservative if it has zero line integral around every closed loop. Given this, the *force* $\mathbf{F} = q\mathbf{E}$ it exerts on a charge q will also be conservative.

How am I going to show that in every possible electrostatic field, created by every possible arrangement of static charges, the line integral of $\mathbf{E}$ around every possible loop is zero?

The key step is to use superposition: if I can show that the field due to a point charge is conservative, the field due to many charges, which is the sum of such conservative fields, is also conservative.

Consider for example two conservative fields $\mathbf{E}_1$ and $\mathbf{E}_2$ obeying

$$\oint \mathbf{E}_1 \cdot d\mathbf{r} = 0 \tag{5.47}$$

$$\oint \mathbf{E}_2 \cdot d\mathbf{r} = 0 \tag{5.48}$$

where both integrals are over the same (but arbitrary) loop. Now add the two equations to find

$$\oint \mathbf{E}_1 \cdot d\mathbf{r} + \oint \mathbf{E}_2 \cdot d\mathbf{r} = 0 \tag{5.49}$$

$$\oint (\mathbf{E}_1 + \mathbf{E}_2) \cdot d\mathbf{r} = 0, \tag{5.50}$$

which means that $\mathbf{E}_1 + \mathbf{E}_2$ is also conservative.

In other words, if I add two fields with zero line integral around any closed loop, I get a field that also has zero line integral around any closed loop because the integral of a sum of integrands is the sum of the corresponding integrals.

To show that $\mathbf{E}$ due to a point charge is conservative, I will show it is (minus) the gradient of function V, called the *electrical potential* or simply *potential*:

$$\mathbf{E} = -\nabla V. \tag{5.51}$$

Here is the potential due to a charge q at the origin:

$$V(\mathbf{r}) = \frac{q}{4\pi\,\varepsilon_0 r}. \tag{5.52}$$

Let us see if it does what it should, namely, is

$$-\nabla V = \mathbf{E}(\mathbf{r}) = \mathbf{e}_r \frac{q}{4\pi\,\varepsilon_0 r^2}? \tag{5.53}$$

Consider first the x-component of $-\nabla V$.

$$-\frac{\partial V}{\partial x} = -\frac{q}{4\pi\,\varepsilon_0}\frac{\partial(1/r)}{\partial x} = -\frac{q}{4\pi\,\varepsilon_0}\frac{\partial}{\partial x}\left[\frac{1}{\sqrt{x^2 + y^2 + z^2}}\right] \tag{5.54}$$

$$= \frac{q}{4\pi\,\varepsilon_0}\left[\frac{1}{2}\right]\frac{2x}{(x^2 + y^2 + z^2)^{3/2}} \tag{5.55}$$

$$= \frac{q}{4\pi\,\varepsilon_0}\frac{1}{r^2}\frac{x}{r}. \tag{5.56}$$

It follows that

$$-\nabla V = -\mathbf{i}\frac{\partial V}{\partial x} - \mathbf{j}\frac{\partial V}{\partial y} - \mathbf{k}\frac{\partial V}{\partial z} = \frac{q}{4\pi\,\varepsilon_0 r^2}\frac{\mathbf{i}x + \mathbf{j}y + \mathbf{k}z}{r} \tag{5.57}$$

$$= \frac{q}{4\pi\,\varepsilon_0 r^2}\frac{\mathbf{r}}{r} = \frac{q}{4\pi\,\varepsilon_0 r^2}\mathbf{e}_r = \mathbf{E} \tag{5.58}$$

as desired.

As usual we may add a constant to this potential V without changing $\mathbf{E}$. The present choice makes V vanish at spatial infinity: $V(r = \infty) = 0$.

By construction, the reciprocal relation

$$-\int_1^2 \mathbf{E}\cdot d\mathbf{r} = V_2 - V_1 = \frac{q}{4\pi\,\varepsilon_0 r_2} - \frac{q}{4\pi\,\varepsilon_0 r_1} \tag{5.59}$$

has to follow.

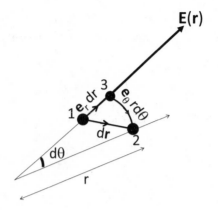

Figure 5.2   The work done by the electric field **E** when the particle moves by a tiny amount $d\mathbf{r}$ is either given by $\mathbf{E} \cdot d\mathbf{r}$ or as a sum of the work done on a radial segment $\mathbf{e}_r dr$ and an angular segment $\mathbf{e}_\theta \, r d\theta$ that connect the same end points. The angular part does not contribute to the work done.

However, to gain practice, let us derive the above relation anyway by setting

$$\mathbf{E} = \mathbf{e}_r \frac{q}{4\pi \varepsilon_0 r^2} \tag{5.60}$$

$$d\mathbf{r} = \mathbf{e}_r dr + \mathbf{e}_\theta \, r d\theta \tag{5.61}$$

where an arbitrary infinitesimal step $d\mathbf{r}$ between $\mathbf{r}_1$ and $\mathbf{r}_2$ is written as the vector sum of a radial part $\mathbf{e}_r dr$ $(1 \to 3)$ and an angular part $\mathbf{e}_\theta \, r d\theta$ $(3 \to 2)$ as shown in Figure 5.2. The field is assumed to be a constant $\mathbf{E}(r)$ over this infinitesimal loop $1 \to 3 \to 2 \to 1$. Using $\mathbf{e}_r \cdot \mathbf{e}_\theta = 0$, we find

$$\mathbf{E} \cdot d\mathbf{r} = \mathbf{e}_r E(r) \cdot (\mathbf{e}_r dr + \mathbf{e}_\theta \, r d\theta) = E(r) dr \tag{5.62}$$

$$= \frac{q}{4\pi \varepsilon_0} \frac{dr}{r^2}. \tag{5.63}$$

If we now glue together such infinitesimal segments $d\mathbf{r}$ to form a finite curve, the integral will be the sum of contributions from each one given above. The result, for arbitrary points 1 and 2, is

$$\int_1^2 \mathbf{E} \cdot d\mathbf{r} = \int_{r_1}^{r_2} \frac{q}{4\pi \varepsilon_0} \frac{dr}{r^2} \tag{5.64}$$

$$= \frac{q}{4\pi \varepsilon_0} \left. \frac{-1}{r} \right|_{r_1}^{r_2} \tag{5.65}$$

$$= \frac{q}{4\pi \varepsilon_0} \left(\frac{1}{r_1} - \frac{1}{r_2} \right) = V_1 - V_2. \tag{5.66}$$

The potential at a point $\mathbf{r}$, due to charges $q_1, q_2, \ldots q_i \ldots q_N$ located at $\mathbf{r}_1, \mathbf{r}_2 \ldots \mathbf{r}_i \ldots \mathbf{r}_N$, is by superposition,

$$V(\mathbf{r}) = \sum_{i=1}^{N} \frac{q_i}{4\pi \varepsilon_0 |\mathbf{r} - \mathbf{r}_i|} \tag{5.67}$$

where $|\mathbf{r} - \mathbf{r}_i|$ is the distance between q_i and where we want the potential. (This generalizes Eqn. 5.52 describing just one charge $q_1 = q$ at the origin $\mathbf{r}_1 = 0$.) The corresponding total electric field $\mathbf{E} = -\nabla V$ is conservative by superposition.

Note that there are no vectors involved in Eqn. 5.67: each charge contributes a scalar and these are simply added to give the total potential. The power of this approach will be demonstrated shortly when we find the field due to a dipole.

For a charge q moving in the field produced by any number of fixed charges, the law of conservation of energy takes the following form in terms of the V in Eqn. 5.67:

$$E_1 \equiv \frac{1}{2} m |\mathbf{v}_1|^2 + q V(\mathbf{r}_1) = \frac{1}{2} m |\mathbf{v}_2|^2 + q V(\mathbf{r}_2) \equiv E_2. \tag{5.68}$$

Some closing remarks on the potential: It is called V and not U because $-\nabla U$ is the *force* $\mathbf{F} = q\mathbf{E}$ while $-\nabla V = \mathbf{E}$, the *field*. Thus the electrical *potential* V is related to the *potential energy* U of a charge q in that field by

$$U = qV. \tag{5.69}$$

In the case of gravitation near the earth where $U = mgh$, the corresponding $V = gh$. Thus V is the potential energy of unit mass in the gravitational case, and V is the potential energy of unit charge in the electrostatic case. (In many advanced courses one uses ϕ to denote the potential instead of V.)

The unit for potential, joules per coulomb, is a *volt*. You should use units in all of your calculations. Without units an answer like 23 is meaningless. You must always use units. I may not always use units but then I have tenure. Once you have tenure you don't have to use units, pay taxes, show up for jury duty, or avoid fire hydrants when parking. Life after tenure resembles that of a deep sea mollusk that permanently attaches itself to a rock when it reaches adulthood and eats its brain for food.

5.3 Path independence through pictures

Let us understand the path independence of the line integral of **E** in visual terms. Figure 5.3 shows two representative paths that go from A to B in the field of a point charge. One path goes radially out from A to 4 and then at fixed r to B in the angular direction. The angular part $4 \to B$ does not contribute since **E** is radial and $d\mathbf{r}$ is tangential. In option $A \to 1 \to 2 \to 3 \to B$, the angular parts $A \to 1$ and $2 \to 3$ do not contribute for the same reason, while the two radial parts $1 \to 2$ and $3 \to B$ together contribute what the radial part $A \to 4$ did in the other option.

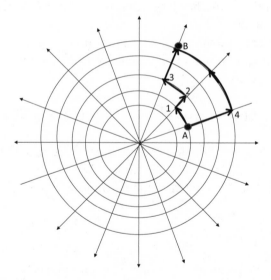

Figure 5.3   The work done in going from A to B by the field of a point charge is path-independent. One path goes radially out from A to 4 and then in the angular direction at fixed r to B. The angular part $4 \to B$ does not contribute since **E** and $d\mathbf{r}$ are orthogonal. In the other path $A \to 1 \to 2 \to 3 \to B$, the angular parts $A \to 1$ and $2 \to 3$ do not likewise contribute, while the radial parts $1 \to 2$ and $3 \to B$ together contribute the same as the radial part $A \to 4$ in the other path.

Why are the radial contributions the same? The path from A to 4 receives contributions of the form $\mathbf{E}(r) \cdot d\mathbf{r}$ from the radial segments $d\mathbf{r} = \mathbf{e}_r dr$ that constitute it. Now look at the figure. *To every segment in this path there is a corresponding radial segment in either $1 \to 2$ (for $r_1 \leq r \leq r_2$) or $3 \to B$ (for $r_3 \leq r \leq r_B$) in which $\mathbf{E}(r) \cdot d\mathbf{r} = E(r)dr$ has the same value.* This is because $\mathbf{E}$ and $d\mathbf{r}$ on $1 \to 2$ and $3 \to B$ are simply rigidly rotated versions of $\mathbf{E}$ and $d\mathbf{r}$ on $A \to 4$ and the dot product is unaffected by the joint rotation of the two vectors.

In general one can draw any path joining A and B made up of radial and angular segments and get the same answer in all of them. The angular segments will not contribute and the sum of the contributions from all the radial parts will equal that of the one-shot move from A to 4.

It seems reasonable that by making the grid finer and finer we can approximate any smooth path by such radial and angular segments. But there are some subtleties. Even though the smooth path and the jagged one made of angular and tangential parts may appear indistinguishable to the naked eye, some properties may be very different. Consider two paths connecting diagonally opposite points on a unit square. A straight path along the diagonal will have a length $\sqrt{2}$ while a staircase path that moves in tiny steps parallel to the sides and closely follows the straight line path will have a length 2. So it is not obvious that the line integral of some vector field $\mathbf{V}(\mathbf{r})$ along the smooth and jagged paths will be equal. Fortunately $\int \mathbf{E} \cdot d\mathbf{r}$ is indeed the same on the continuous path and the jagged approximation made of radial and angular segments, as was shown in discussions accompanying Figure 5.2.

We can also consider paths that leave the plane of the paper or are not monotonic in r while going from A to B. The angular parts (which now lie on a sphere of fixed r) will again make no contribution since $\mathbf{E}$ is radial, and the contribution of the radial parts will add up to the contribution of $A \to 4$.

Once we understand why the field due to one charge is conservative, we may use superposition to infer the same of the field due to many charges. (Pictures will not help in this case because the total $\mathbf{E}$ can be very complicated.)

5.4 Potential and field of a dipole

Recall how we found the field of a dipole by adding the *vector contributions* from $+q$ and $-q$. The fact that the two vectors came with different magnitudes and directions contributed to the complexity. I urge you to go over that derivation in Section 2.4 before proceeding.

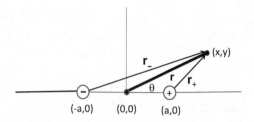

Figure 5.4 The potential at the point (x, y) is simply the sum of the two scalar contributions from $\pm q$ at $(\pm a, 0)$.

We will now do it differently, by *first* computing the potential due to $\pm q$ and *then* taking the gradient. This will prove to be a lot easier because the potential is a scalar, no vector addition is required, and taking derivatives is an act that can not only be done mindlessly, but is better done that way.

Consider Figure 5.4. It is clear that

$$V(x, y) = \frac{q}{4\pi\varepsilon_0}\left[\frac{1}{r_+} - \frac{1}{r_-}\right] \tag{5.70}$$

$$= \frac{q}{4\pi\varepsilon_0}\frac{r_- - r_+}{r_+ r_-} \quad \text{where} \tag{5.71}$$

$$r_+ = \sqrt{(x-a)^2 + y^2} \quad \text{and} \quad r_- = \sqrt{(x+a)^2 + y^2}. \tag{5.72}$$

We want to evaluate this expression for $r \gg a$. When $a = 0$, we have $V \equiv 0$ since the charges are on top of each other and $r_\pm = r$. We want the first non-zero term in the answer when $a > 0$, the term proportional to the first power of a. Anything that goes like a^2 or higher will be dropped.

In Eqn. 5.71 the numerator $r_- - r_+$ contains at least one power of a (since it vanishes when $a = 0$):

$$r_- - r_+ = \sqrt{(x+a)^2 + y^2} - \sqrt{(x-a)^2 + y^2} \tag{5.73}$$

$$\simeq \sqrt{x^2 + 2ax + y^2} - \sqrt{x^2 - 2ax + y^2}$$

(upon dropping the a^2 term) $\tag{5.74}$

$$= \sqrt{r^2 + 2ax} - \sqrt{r^2 - 2ax}, \quad \text{using } x^2 + y^2 = r^2 \tag{5.75}$$

$$= r\left(1 + \frac{2ax}{r^2}\right)^{1/2} - r\left(1 - \frac{2ax}{r^2}\right)^{1/2} \qquad (5.76)$$

$$\simeq r\left[1 + \frac{1}{2}\frac{2ax}{r^2} + \ldots - 1 + \frac{1}{2}\frac{2ax}{r^2} + \ldots\right]$$

$$(\text{using } (1+z)^{1/2} = 1 + \tfrac{1}{2}z + \ldots) \qquad (5.77)$$

$$= \frac{2ax}{r}. \qquad (5.78)$$

Since this expression goes into the numerator of Eqn. 5.71, and it contains one power of a and we want no more, we may evaluate the denominator at $a = 0$, i.e., set $r_\pm = r$ in to obtain

$$V(x,y) = \frac{2aqx}{4\pi\varepsilon_0 r^3} = \frac{px}{4\pi\varepsilon_0 r^3} = \frac{\mathbf{p}\cdot\mathbf{r}}{4\pi\varepsilon_0 r^3} \qquad (5.79)$$

where

$$\mathbf{p} = 2aq\mathbf{i} = p\mathbf{i} \qquad (5.80)$$

is the dipole moment.

If we write $x = r\cos\theta$, where θ is the angle between $\mathbf{r}$ and the x-axis, we see that V falls like $1/r^2$. When we take its gradient to find $\mathbf{E}$, it will fall as $1/r^3$. Here are the details.

$$E_x = -\frac{\partial V}{\partial x} = -\frac{p}{4\pi\varepsilon_0}\frac{\partial}{\partial x}\left[\frac{x}{(x^2+y^2)^{3/2}}\right] \qquad (5.81)$$

$$= -\frac{p}{4\pi\varepsilon_0}\left[\frac{1}{(x^2+y^2)^{3/2}} - \frac{3x}{2}\frac{2x}{(x^2+y^2)^{5/2}}\right] \qquad (5.82)$$

$$= \frac{p}{4\pi\varepsilon_0 r^3}\left(3\frac{x^2}{r^2} - 1\right) = \frac{p}{4\pi\varepsilon_0 r^3}\left(3\cos^2\theta - 1\right). \qquad (5.83)$$

Similarly

$$E_y = -\frac{\partial V}{\partial y} = -\frac{p}{4\pi\varepsilon_0}\frac{\partial}{\partial y}\left[\frac{x}{(x^2+y^2)^{3/2}}\right] \qquad (5.84)$$

$$= \frac{p}{4\pi\varepsilon_0}\frac{3}{2}\frac{2xy}{r^5} \qquad (5.85)$$

$$= \frac{p}{4\pi\varepsilon_0 r^3}3\sin\theta\cos\theta. \qquad (5.86)$$

Before we combine E_x and E_y to form the vector $\mathbf{E}$, let's derive some results we will need. From Figure 5.4 we see that

$$\mathbf{r} = \mathbf{i}\,x + \mathbf{j}\,y \tag{5.87}$$

$$= \mathbf{i}\,r\cos\theta + \mathbf{j}\,r\sin\theta \tag{5.88}$$

$$= r(\mathbf{i}\cos\theta + \mathbf{j}\sin\theta) \equiv r\mathbf{e}_r, \quad \text{which gives us} \tag{5.89}$$

$$\mathbf{e}_r = \mathbf{i}\cos\theta + \mathbf{j}\sin\theta. \tag{5.90}$$

Given that the dipole moment $\mathbf{p} = \mathbf{i}p$, it follows that

$$\mathbf{p}\cdot\mathbf{e}_r = p\cos\theta. \tag{5.91}$$

Armed with Eqns. 5.90 and 5.91 we proceed as follows:

$$\mathbf{E} = \mathbf{i}E_x + \mathbf{j}E_y \tag{5.92}$$

$$= \frac{p}{4\pi\,\varepsilon_0 r^3}\left[3\cos^2\theta\,\mathbf{i} - \mathbf{i} + 3\sin\theta\,\cos\theta\mathbf{j}\,\right] \tag{5.93}$$

$$= \frac{1}{4\pi\,\varepsilon_0 r^3}\left[3p\cos\theta(\mathbf{i}\,\cos\theta + \mathbf{j}\,\sin\theta) - \mathbf{p}\right] \tag{5.94}$$

$$= \frac{1}{4\pi\,\varepsilon_0 r^3}\left[(3\mathbf{p}\cdot\mathbf{e}_r)\mathbf{e}_r - \mathbf{p}\right], \tag{5.95}$$

in agreement with Eqn. 2.42.

CHAPTER 6

Conductors and Capacitors

Let us begin with the highlights from the last chapter. We focused on the idea that the electric field **E** is conservative. This means that its line integral between points 1 and 2 is independent of the path connecting them, or equivalently that its line integral around every closed loop is zero.

A necessary and sufficient condition for this to be true was that **E** be expressible as the gradient of a scalar function:

$$\mathbf{E} = -\nabla V \tag{6.1}$$

where V is called the *potential* and is measured in volts. When we multiply both sides of Eqn. 6.1 by q, we obtain the electric force

$$\mathbf{F} = q\mathbf{E} = -\nabla(qV) = -\nabla U, \tag{6.2}$$

which is also conservative. For a particle moving in an electrostatic field this leads to the law of conservation of energy with

$$U = qV \tag{6.3}$$

as the potential energy:

$$E_1 \equiv K_1 + qV_1 = K_2 + qV_2 \equiv E_2. \tag{6.4}$$

Just as **E** is the force on a unit charge, V is the potential energy of a unit charge. In the gravitational analogy, if $h(x)$ is the height of a mountain at point x, we may factorize the potential energy as

$$U = mgh(x) = m \times gh(x) \tag{6.5}$$

so that $gh(x)$ essentially encodes the altitude of the mountain and $mgh(x)$ the work done to lug a *particular* mass m to that height from sea level.

In electrostatics the voltage V is the electrical height (with respect to some reference) and qV is the work you need to do to drag a charge q to that point from the reference point where $V = 0$.

Given the potential V we obtain the field as a gradient. For example, if in two dimensions

$$V(x, y) = x^2 \sin y \tag{6.6}$$

then

$$\mathbf{E} = -\nabla V = -2x \sin y \mathbf{i} - x^2 \cos y \mathbf{j}. \tag{6.7}$$

The reciprocal relation to $\mathbf{E} = -\nabla V$ is

$$V_2 - V_1 = -\int_1^2 \mathbf{E} \cdot d\mathbf{r} \tag{6.8}$$

where the line integral may be evaluated along any path with end points 1 and 2.

Eqn. 6.8 equates the gain in potential energy to the work you do when you precisely balance the electric force and drag a unit charge from 1 to 2.

To prove that $\mathbf{E}$ is conservative, I just wrote down the potential

$$V(\mathbf{r}) = \frac{q}{4\pi \varepsilon_0 r} \tag{6.9}$$

for charge at the origin and verified that (minus) its gradient gave the field:

$$-\nabla \left[\frac{q}{4\pi \varepsilon_0 r} \right] = \mathbf{e}_r \frac{q}{4\pi \varepsilon_0 r^2}. \tag{6.10}$$

I also showed how to go backward from $\mathbf{E}$ to V integrating $\mathbf{E}$ as per Eqn. 6.8.

For many charges q_i located at $\mathbf{r}_i$, the potential was, by superposition,

$$V(\mathbf{r}) = \sum_i \frac{q_i}{4\pi \varepsilon_0 |\mathbf{r} - \mathbf{r}_i|}. \tag{6.11}$$

Pictorial arguments were given to explain the path independence of the line integral. For a single charge we saw how going from A to B on different paths, made of different radial and angular segments, gave the same answer: the angular parts never contributed (since $\mathbf{E}$ and $d\mathbf{r}$ were orthogonal) while the radial parts always added up to the same number on every path. This was because $E(r)dr$, the work done on a segment of radial extent dr on one path, was also done on the other path as it crossed that range of r. The vectors $\mathbf{E}$ and $d\mathbf{r}$ on one path were the rotated versions of their counterparts on the other path, and the dot products between the field and displacement were unaffected by this rotation and made the same contribution $E(r)dr$.

Finally we saw it was easier to find the field due to many charges by adding their potentials, which were just some scalars, and then taking the gradient, which was a relatively mindless process, in contrast to adding the individual vector contributions to $\mathbf{E}$. This was illustrated by computing $\mathbf{E}$ due to a dipole and reproducing results found earlier.

6.1 Cases where computing V from $\mathbf{E}$ is easier

There are a few cases where it is easier to find $\mathbf{E}$ from V than the other way around. An example is the problem of a hollow spherical shell of radius R with some charge Q spread uniformly on its surface.

To find V directly we could slice the hollow shell into rings whose centers lie on the line joining the origin to the point $\mathbf{r}$ where we want the potential, as shown in Figure 6.1. Since all points on the ring are equidistant from $\mathbf{r}$, its contribution is just the charge on it divided by the distance from points on the ring to $\mathbf{r}$ (ignoring the $4\pi\varepsilon_0$ for now). We then need to integrate over all such rings, the closest one being at a distance $r - R$ (and zero radius) and the farthest one at $r + R$ (also of zero radius). It can be done, of course, but this painful calculation is totally avoidable in this case.

The spherical symmetry of the problem allows us to use Gauss's law to find $\mathbf{E}$ very easily and then integrate it to find V.

For $r > R$, the sphere produces the field of a point charge Q at the origin, while inside the sphere the field vanishes:

$$\mathbf{E}(\mathbf{r}) = \frac{Q}{4\pi\varepsilon_0 r^2}\mathbf{e}_r \quad r > R \tag{6.12}$$

$$= 0 \quad r < R. \tag{6.13}$$

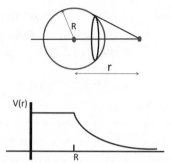

Figure 6.1   To find V due to a spherical shell carrying charge Q we can slice it into rings and integrate contributions from them. Details of this complicated integration are not discussed since there is an easier way using Gauss's law, and the lower half shows the resulting $V(r)$. The entire interior of the sphere is at the same potential as the surface, because $\mathbf{E} = 0$ inside. The surface potential is that of a point charge Q at the origin.

To find V we invoke the formula

$$V(\mathbf{r}_2) - V(\mathbf{r}_1) = -\int_{\mathbf{r}_1}^{\mathbf{r}_2} \mathbf{E}(\mathbf{r}) \cdot d\mathbf{r}. \tag{6.14}$$

For $\mathbf{r}_1$ we choose the point at infinity, for $\mathbf{r}_2$ the coordinate $\mathbf{r}$ of any point outside the sphere where we want the potential, and for the path (which we can choose at will due to path independence) a radial line from ∞ to radius r. The potential $V(\mathbf{r})$ will of course only depend on the radial coordinate r by the spherical symmetry of the charge distribution. We find:

$$V(r) - V(\infty) = -\int_{\infty}^{r} e_r E(r) \cdot e_r dr$$

$$= -\int_{\infty}^{r} E(r) dr = \frac{Q}{4\pi \varepsilon_0 r}. \tag{6.15}$$

In our convention $V(\infty) = 0$ and so we drop it to obtain

$$V(r) = \frac{Q}{4\pi \varepsilon_0 r} \qquad r \geq R. \tag{6.16}$$

To find the potential inside the sphere we must continue the line integral into the sphere. But there is no field inside the sphere! This does not mean $V = 0$ inside, but rather that the line integral receives no further contributions as we go inside. Its value everywhere inside equals $V(R)$, the value at the surface. Figure 6.1 shows a plot of $V(r)$.

6.2 Visualizing V

We have seen how drawing electric field lines gives us a nice way to visualize the salient features of $\mathbf{E(r)}$. Even if we have a formula for it, it helps to draw pictures. We are going to do the same with the potential V.

Consider the simple case of two infinite parallel plates with charge density $\pm\sigma$. We know the field between them is σ/ε_0 pointing from the positive to the negative plate. The upper plate pushes down a unit test charge with force $\sigma/(2\varepsilon_0)$ and the lower one pulls it down equally hard to produce a total of σ/ε_0. In the region outside, that is, above the upper plate and below the lower one, the fields cancel because the fields due to such infinite plates do not diminish with distance.

The lower plate, being a conductor, will be at some fixed potential, because $\mathbf{E} = 0$ in a conductor. We choose this constant potential to be 0. (The usual choice $V(\infty) = 0$ is not so useful in this context, or in electrical circuits.) If we lift a unit test charge upward, against the downward pointing field, the work done is just the constant field E times distance. So the potential at a height y above the negative plate is

$$V(y) = Ey = \frac{\sigma}{\varepsilon_0}y. \tag{6.17}$$

This is just like the gravitational problem where the potential energy of unit mass at a height y above the ground is gy. Figure 6.2 shows a few lines of constant V. These are called *equipotentials*. The figure corresponds to a case when the upper plate is at a potential 4 volts above the lower one. If a 10 coulomb charge falls from the upper to the lower plate, it will gain a kinetic energy of $40J$. If a proton of charge $1.6 \cdot 10^{-19}C$ fell, it would gain a kinetic energy of

$$K = 4V \cdot 1.6 \cdot 10^{-19}C = 4 \cdot (1.6 \cdot 10^{-19})J \equiv 4eV \tag{6.18}$$

where eV stands for *electron volt* and has the value:

$$1eV = 1.6 \cdot 10^{-19}J. \tag{6.19}$$

It is the energy a proton gains if it falls down a voltage difference of one volt. It is still called an electron volt because an electron (which does

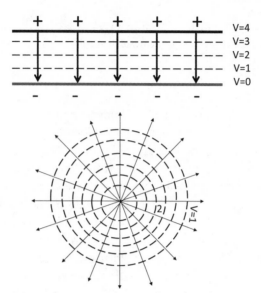

Figure 6.2   A two-dimensional cross section of the equipotentials (dotted lines) due to the uniform electric field (solid arrows) between parallel plates (top) and that of point charge (bottom). Note that the field is always perpendicular to the equipotential surfaces: planes in the first case and concentric spheres in the second.

most of the falling) that "falls" *from the negative to the positive terminal* of a $1.5\,V$ battery will *gain* the same kinetic energy of $1.5\,eV$. This is not crazy: in the figure, an electron released at the lower plate will "fall" toward the upper plate. The analogy between voltage and height in a gravitational field breaks down here because unlike mass, which always falls down along the gravitational field, a charge can go either way depending upon its sign. Had there been objects of negative mass, they would be like helium-filled balloons that have to be tied down to the floor to keep them from rising to the ceiling.

An electron volt is a convenient unit of energy not only when discussing electrons, which do all the charge carrying in our daily life, from lightning to electrical circuits, but also all atomic scale particles whose charges are small multiples of the electronic charge. This choice of unit eliminates the constant use of numbers like 10^{-19}.

For example, the total energy of an electron in the innermost orbit of hydrogen is $-13.6\,eV$. This means that it can be knocked out of the atom if this energy or more is furnished, say by radiation. This removal of the electron is called *ionization*.

6.3 Equipotentials

Coming back to the parallel plates, note that the lines of constant V, the equipotentials, are perpendicular to the lines of **E**. I will now consider one more case where this is again true and then explain why this is always true.

The example is the point charge q at the origin whose field lines radiate isotropically. How about the contours of constant V? Since $V(r) \propto 1/r$, these are spheres of fixed radius. The radial field lines are then perpendicular to the equipotential spheres, a planar cross section of which is shown in Figure 6.2.

Suppose you are asked to bring a coulomb from infinity to the origin, where there is a charge q. The closer you get to the summit, the harder it is to climb Mount Coulomb because q is pushing you away with force that diverges like $1/r^2$. The contours at fixed V tell you how you are doing. Sadly, you will never get to the top, which is at an unattainable $V = \infty$. On the other hand, if you are carrying a coulomb from the negative plate to the positive plate, which is higher by $4V$, the equally spaced equipotentials will mark your steady progress toward the top.

Consider now the electric field of a dipole and its equipotentials. We know they will be mutually orthogonal very close to either charge, where we can ignore the finite field due to the other charge compared to its own divergent $1/r^2$ contribution and where the field lines and equipotentials will resemble what you see in the lower half of Figure 6.2. We could establish their mutual orthogonality everywhere by analyzing the formula for **E** and V. Instead, we will establish this orthogonality once and for all for all of electrostatics.

From the defining relation

$$\mathbf{E} = -\nabla V \quad \text{it follows} \tag{6.20}$$

$$\mathbf{E} \cdot d\mathbf{r} = -\nabla V \cdot d\mathbf{r} = -\frac{\partial V}{\partial x} dx - \frac{\partial V}{\partial y} dy - \frac{\partial V}{\partial z} dz = -dV.$$

$$\tag{6.21}$$

If you are at some point $\mathbf{r}$, this equation tells you how much V will change if you move by an amount $d\mathbf{r}$, which may be in any direction. But some directions will produce more change than others for a given value of $|d\mathbf{r}|$, the length of the step you take. This can be quantified if we rewrite the dot product in its alternate form

$$dV = -\mathbf{E} \cdot d\mathbf{r} = -|\mathbf{E}||d\mathbf{r}|\cos\theta \qquad\qquad (6.22)$$

where θ is the angle between the field and the displacement. Let us keep the step length $|d\mathbf{r}|$ fixed and study the impact of the angle θ relative to $\mathbf{E}$.

If you move in the direction of $\mathbf{E}$, ($\theta = 0$) you experience the biggest drop in V. Thus *the electric field points in the direction of the greatest rate of drop in V*. If V were really a height of a volcanic mountain and you wanted to race to the bottom before it blew up, you should compute the gradient at each point and move against it, or compute the field and move along it. If, however, you were racing to the top to beat the approaching tsunami, you should do the opposite.

But suppose you were very happy at your altitude. You *could* stay where you were, but you could also move perpendicular to the gradient or $\mathbf{E}$ and maintain the altitude: now $\cos\theta = 0 = dV$. In three dimensions the region perpendicular to $\mathbf{E}$ will be a two-dimensional plane. Of course, you can only go an infinitesimal distance along this plane, because the direction of $\mathbf{E}$ could change as you moved and you would have to find the plane orthogonal to the field at the new location. By patching together these little planar areas you will reconstruct the equipotential surface, which will be everywhere perpendicular to $\mathbf{E}$.

In the simplest case of the oppositely charged parallel (infinite) plates, where $\mathbf{E}$ has a constant downward direction, the equipotential surfaces you get in this manner will be planes parallel to the charged plates. In the case of a point charge your little equipotential patches will approximate spheres and become spheres as the patch sizes go to zero. In the dipolar case they will be more complicated surfaces that reduce to spheres near the charges.

6.4 Method of images

The notion of equipotentials can be exploited to solve a class of problem using a trick called the *method of images.*

Consider the following problem depicted in Figure 6.3. A charge q is placed at a distance a to the left of an infinite conducting plane perpendicular to the x-axis. With respect to the origin $(0,0,0)$ shown in the figure, the charge has coordinates $(-a,0,0)$. The plane is grounded, i.e., held at zero potential by the earth, which, given its size, can give or

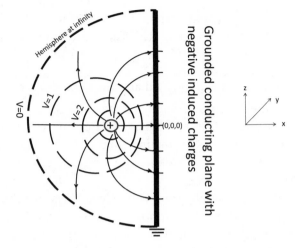

Figure 6.3   The main features of the field due to a point charge q in front of an infinite grounded conducting plane that passes through the origin and is perpendicular to the x-axis. The figure is a cross section in the xz plane. The field lines leave the charge radially and approach the conductor normal to its surface and terminate on the induced negative charges. Also shown is a hemisphere of infinite radius, which along with the plane forms a closed surface S at $V = 0$.

take charges to hold the plane at its own potential, which is taken to be zero. What will be the electric field in all of space?

We can guess some broad features that are indicated in Figure 6.3. Very close to the charge the field lines will be isotropic and radial. If the plane did nothing, the radial field will hit the plane with a component parallel to the plane and pointing away from $(0,0,0)$. But a parallel field at the surface is not allowed in a conductor in the electrostatic situation since charges will move in response to it. Indeed, this is what they will do initially. There will be a current in the direction of this field pointing away from $(0,0,0)$. This flow will lead to an accumulation of unbalanced negative charges until the parallel field due to q is annulled. In reality, the current is not made of positive charges (which do not move) but of electrons, which move against the field of q. They cannot leave the conductor and fall on top of q, so they will instead be concentrated in front of it, with the maximum surface density at $(0,0,0)$.

The same picture emerges if we think in terms of the potential. The initial effect of q is to place different parts of the plane at different potentials: since $V \propto q/r$, points on the plane closer to q will be at a higher potential than those further away. This initial situation will be quickly remedied as electrons flow in (from the ground) to the high potential region to even out the potential to zero everywhere.

In any event, when things settle down to a static configuration, the field lines from q will approach the conductor normal to its surface and terminate on the induced negative charges. These must add up to $-q$ in order to gobble up the lines emanating from $+q$. The field will be zero to the right of the plane since no lines can penetrate a conductor. These features are sketched in Figure 6.3.

What if the plane was not grounded? When $+q$ is brought in front of it, the neutral plane will separate or polarize into charges $\pm q$. The charges $-q$ will place themselves in front of the external $+q$ so as to bring the plane to an equipotential or equivalently to cancel the parallel field. The charge $+q$ will spread itself over the plane to keep it an equipotential V_0. The finite charge q spread over an infinite plane will lead to zero charge density σ and zero field.

Can we go beyond these qualitative aspects and answer some quantitative questions? What exactly will be the final field configuration to the left of the plane, where **E** is non-zero? What will be the distribution of the induced negative charges on the conducting plane? What will be the force of attraction between q and the negative charges in the plane?

It turns out we can answer all these questions *exactly* by employing the following clever trick.

Forget our problem and look at Figure 6.4, which shows the equipotentials of a dipole. Focus on the infinite plane that perpendicularly bisects the line joining the charges. All over this plane $V = 0$ because points on it are equidistant from the two opposite charges and get exactly canceling contributions. *The dipole field to the left of the plane $x < 0$ shares many features with our problem: the field lines emerge in a spherically symmetric manner from q and terminate on the plane orthogonally.* Does the similarity end here, or is the dipole field in the region $x < 0$ the actual answer to our problem of a charge in front of the conducting plane?

The answer is yes, but the reason is quite subtle. It is based on the following *uniqueness theorem*:

The potential V inside a closed surface S is uniquely determined by its values on S and the distribution of the charges inside.

If these are given, there is a unique answer for V.

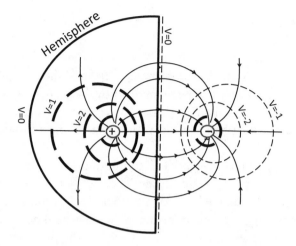

Figure 6.4   Equipotentials in a dipole field. Their shape changes from spherical very close to the charges to the infinite plane at $V = 0$ that is the perpendicular bisector of the line joining the charges. The $V = 0$ plane and the hemisphere at infinite radius form a closed surface S on which $V = 0$, and inside which is q.

Postponing the proof of this uniqueness theorem, let us ask how it is to be applied to our problem. We had a charge q in front of an infinite plane. We need a closed surface S enclosing the charge if we are to invoke the theorem. To this end we glue on to the plane an infinite hemisphere that extends for all $x < 0$. On this closed surface we have $V = 0$ and inside it we have a single charge q at $x = -a$.

The dipole problem also has a closed surface at $V = 0$, namely the infinite equipotential that bisects the dipole moment and the hemisphere of infinite radius that lives in $x \leq 0$. This closed region also contains a charge q at $x = -a$. The dipole potential obeys the laws of electrostatics using which it was constructed.

Since the two problems have the same value of $V = 0$ on the surrounding surface (the infinite plane glued to the hemisphere that extends for $x < 0$) and the same charge distribution inside (charge q at $x = -a$) they must have the same potential and field everywhere inside S.

Thus, to solve the problem of a charge q in front of a grounded infinite plane, we take the dipole field and potential for $x < 0$ and throw away the right half with $x > 0$.

The right halves are different in the two cases: in the given problem there is no field or charge there, while in the dipole problem invoked by the trick, there is a charge $-q$ at $x = a$ and the dipolar field due to both the charges $\pm q$.

The crucial point is that the field in the region of interest, $x < 0$, can be produced in two ways: by the charge q in front of the conducting plane and all the induced negative charges on it, or by the charge q and the charge $-q$ at $x = a$ and no conducting plane. The charge $-q$ is called the *image charge*. The image is a phantom, like your image behind a mirror, and does not exist in the original problem.

But the phantom is good for computing the force of attraction between q and the plane. Here is how. The induced negative charges on the plane attract the charge q through the field they produce in the region $x < 0$. But this is the same field the image charge would have produced in that region. So the charge q will be attracted to the plane with the same force that $-q$ would exert on it:

$$F = \frac{(-q) \times q}{4\pi \varepsilon_0 (2a)^2}.$$
(6.23)

We can calculate the induced density σ on the plane as follows. We first find the normal electric field at any point on the plane by adding the electric field vectors due to q and the image $-q$. Then we recall that the normal electric field at the surface of a conductor equals σ / ε_0. For example, at the point $(0, 0, 0)$

$$\mathbf{E}(0,0,0) = \mathbf{i}\frac{q}{2\pi \varepsilon_0 a^2}$$
(6.24)

with equal contributions from q and the image charge $-q$. The induced charge density is

$$\sigma(0,0,0) = -\frac{q}{2\pi a^2}.$$
(6.25)

You should verify, using an appropriate Gaussian cylinder, that I have my signs right: $\mathbf{E}$ is positive but the area vector is negative on the flat face at $x < 0$.

Upon integrating the induced charge density over the plane, we will find it equals $-q$. This is to be expected since the lines of force that leave q terminate on the plane in one description and on the image charge in the other.

Here is another problem that can be solved by the method of images. Suppose you place a charge q in front of a grounded ($V = 0$) conducting sphere of radius R (rather than the infinite plane) at a distance a from the

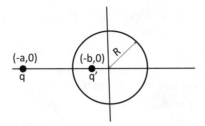

Figure 6.5   The charges q and q' produce an equipotential $V = 0$ in the form of a sphere. The field *outside the sphere* also corresponds to a problem of a charge q placed in front of a grounded conducting sphere of radius R.

center. We know the field will be zero inside the sphere, but what will it be outside? What will be the force of attraction between q and the induced charges on the sphere? What will be the distribution of induced charge on the sphere?

The answer follows from a solved problem depicted in Figure 6.5. We see a charge q at $x = -a$ and a charge $q' = -q\sqrt{b/a}$ at $x = -b$. This pair produces an equipotential $V = 0$ on a sphere of radius $R = \sqrt{ab}$ centered at $x = 0$. Given $R = \sqrt{ab}$ you can also write $q' = -q\sqrt{b/a} = -q\frac{b}{R}$. (I urge you to show that $V = 0$ on the circle $r = R$. Then $V = 0$ on the sphere $r = R$ follows by symmetry.)

To apply the uniqueness theorem we need a closed surface S at the same potential and enclosing the same charge in both problems.

Start with all of space, a sphere of infinite radius, and scoop out a sphere of radius R centered at the origin. This volume has two boundaries: the outer one, a sphere at infinity, and the inner one, a sphere of radius R. That is our S, which encloses the volume of interest in the original problem.

In the original and the image problem the potential $V = 0$ on S. The charge enclosed is $+q$ at $x = -a$ in both cases. So the answer inside S is the same in both cases.

The field *outside the sphere* is due to q and its image charge $q' = -q\sqrt{b/a}$ sitting at $x = -b$. Upon computing the (normal) electric field on the surface of the sphere due to q and q' we may equate it to the surface charge density σ/ε_0. It will integrate to q'. Try to understand why.

Again the method of images gives the correct field only within the surface containing the real charge q. In the rest of the universe, where the image charge is located (inside the sphere of radius R in this case), the situation is different. In the original problem there is no field inside the sphere because it screens the field due to q. In the image problem there are charges and fields in both regions, one containing q and the other q' but no conducting sphere.

Suppose the uncharged conducting sphere is not grounded. It cannot borrow negative charge q' from the ground to realize the $V = 0$ equipotential configuration discussed above. It manages as follows. The sphere, neutral at each point, now separates or polarizes into charges $\pm q'$. The q' (which is negative) will spread itself over the sphere into the σ described in the $V(S) = 0$ problem we just solved, and the $-q'$ (which is positive) will spread itself *uniformly* over the sphere, making its surface an equipotential at $V = -q'/(4\pi \varepsilon_0 R)$. If you send $R \to \infty$, the sphere becomes the infinite plane that we studied earlier and the potential on it becomes $V = -q'/(4\pi \varepsilon_0 \infty) = 0$.

6.4.1 Proof of uniqueness (optional section)

The uniqueness theorem of electrostatics states that given

- a closed surface S,
- the distribution of charges inside S, collectively referred to as q_{in},
- and the value of the potential $V(S)$ on S,

there is only one possible potential V inside S.

First, you will agree that if I specify *all* the charges in the universe, referred to collectively as q_{in} inside S, and q_{out} outside S, you can of course write down a unique V:

$$V(\mathbf{r}) = \sum_{i=in,\ out} \frac{q_i}{4\pi \varepsilon_0 |\mathbf{r}_i - \mathbf{r}|} \tag{6.26}$$

once we choose $V(\infty) = 0$. This has been our approach so far: tell us where *every* charge is and we can write down V *everywhere* using superposition and the choice $V(\infty) = 0$.

But we want something different. We want to pick a part of the universe bounded by a surface S and just want V inside S. The closed surface S could be a mathematical surface, like the Gaussian surface, or real surface, like the boundary of a conductor. We are given q_{in} and nothing about q_{out}. Do we really need to know where every charge q_{out} in the external universe is to find V in our sub-universe? It turns out we do not; all we need is $V(S)$, the value of V on S. *In other words, the specification of $V(S)$ is as restrictive as the specification of all the outside charges q_{out} provided we only want V inside S.*

I repeat: electrostatics allows for only one solution to V inside S given its value on S and q_{in}, the charge distribution inside.

I will now demonstrate this by showing that if V' is another solution that assumes the same value $V(S)$ *on S*, and corresponds to the same q_{in}, then $V = V'$, *inside* all of S.

If $V' \neq V$, there has to be a reason, and it has to be that q_{out} is now different because q_{in} is fixed by assumption. So let q'_{out} be the new distribution. (In Figure 6.6 q'_{out} differs from q_{out} by a third charge q^3_{out}.)

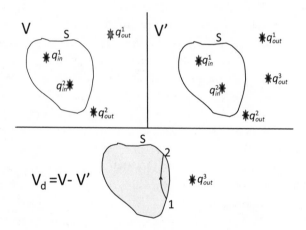

Figure 6.6   Top left: A set of charges producing a potential V with a value $V(S)$ on a closed surface S. Top right: A different set of charges differing only outside S that produce a different potential V', which, however, agrees with V on S. Bottom: The difference of the two sets of charges, non-zero outside S and represented by q^3_{out}, produces a difference potential $V_d = V - V'$, which vanishes on S. If V_d did not vanish inside, it would change from 0 and lead to a field whose lines must leave S (at 1 in the figure) and reenter S (at 2). These lines imply a potential difference in V_d between 1 and 2, which are given to be on the equipotential $V_d = 0$.

Here is what we have:

$$(q_{in}, q_{out}) \text{ produce } V \quad \text{with } V = V(S) \text{ on } S \qquad (6.27)$$
$$(q_{in}, q'_{out}) \text{ produce } V' \quad \text{with } V' = V(S) \text{ on } S. \qquad (6.28)$$

Subtracting the second line from the first, we find

$$(0, q_{out} - q'_{out}) \text{ produce } V - V' = V_d \quad \text{with } V_d = 0 \text{ on } S.$$
$$(6.29)$$

Let me explain the last step. You know that if you add or superpose two sets of charges, you can add or superpose the corresponding potentials and fields. I hope you can see that if you *subtract* one set of charges (q_{in}, q_{out}) from another (q_{in}, q'_{out}), the resultant "difference" potential $V_d = V - V'$ and field will be the corresponding differences. (Instead of subtracting, reverse the second set of charges and add.)

Let us look at the last equation describing $V_d = V - V'$. It vanishes identically on S and is produced by the difference charges $q_{out} - q'_{out}$ that *lie entirely outside S*. (In the figure this is represented by just one charge.) If V_d, which vanishes on S, did not vanish inside S, it would have to change from zero to non-zero as we go in. This change will produce a gradient and a field $\mathbf{E}_d = -\mathbf{\nabla} V_d$. The lines of $\mathbf{E}_d$ cannot begin or end inside S since it is free of charges. So the lines that enter S (at point 1 in the figure) must exit somewhere (point 2) on S. This leads to a contradiction. The line integral of $\mathbf{E}_d$ from 1 to 2 will yield a non-zero potential difference between them, whereas every point on S is supposed be at $V_d = 0$. The only way to avoid the contradiction is for V_d, which vanishes on S, to vanish inside all of S. That is, $V = V'$ inside all of S.

I have only shown you that V is uniquely specified by $V(S)$ and q_{in}, but not how this unique solution is to be found. We are used to getting V given all the q's but not some of the q's and its values on a surrounding surface. This requires more fancy techniques you will learn in advanced courses.

6.4.2 Additional properties of the potential $V(\mathbf{r})$

I will now show you some properties of $V(\mathbf{r})$, partly for their intrinsic value and partly because they provide another route to proving the uniqueness theorem.

Property 1. In a charge-free region $V(\mathbf{r})$ cannot have a maximum or minimum.

Assume to the contrary that there exists a point $\mathbf{r}_0$ at which V is a minimum. This means V increases as we move away from $\mathbf{r}_0$ in *every direction*. This means the gradient ∇V is pointing away from $\mathbf{r}_0$ or that the electric field $\mathbf{E} = -\nabla V$ (the restoring force) is pointing toward $\mathbf{r}_0$ as we approach $\mathbf{r}_0$ in any direction. The surface integral of such an $\mathbf{E}$ over a tiny surface surrounding $\mathbf{r}_0$ will be non-zero (and negative). By Gauss's law, that surface must enclose some negative charge, which violates the assumption the region is charge-free. If $\mathbf{r}_0$ is a maximum, we simply reverse the signs of the field and the enclosed charge in the preceding argument to arrive at a contradiction.

Property 2. If $V(S) = 0$ on a surface S enclosing a charge-free region, $V(\mathbf{r}) \equiv 0$ inside S.

Suppose V had some non-zero values inside S. The largest of these values is a maximum if positive or a minimum if negative, both of which are forbidden by Property 1. So $V \equiv 0$ inside a charge-free S if $V(S) = 0$.

I can now complete the earlier proof of uniqueness in a different way, starting from the point where I showed that the difference potential $V_d = V - V'$ vanishes on S. Because S bounds a charge-free region (the difference charge vanishes inside it), Property 2 implies that $V_d \equiv 0$ inside S.

6.5 Capacitors

Suppose you are willing to do some mechanical work that can be stored and used later. One way is to haul some water up to a tank at some height above the ground, doing work *mgh*. When you are ready to cash in, you allow the water to flow down to the ground along a pipe. The kinetic energy of the water can be used to turn a turbine blade or to run a mill.

Capacitors are the electrical analogs of this process. They provide a way of storing electrical potential energy that can be consumed later.

As a simple example consider two parallel conducting plates of area A a distance d apart. Each plate, being a conductor with no electric field allowed in its interior, is at some fixed potential. If they are initially neutral they will both be at the same potential, which we take to be 0. Now we begin to transfer some charge from the lower plate to the upper plate. As we continue it will become harder and harder to transfer charge because the positive charges in the upper plate will repel the newcomers. More precisely, if Q is the charge on the upper plate (and $-Q$ the charge on the lower plate) the electric field opposing the charge transfer will be (ignoring edge effects due to finite size)

$$E = \frac{\sigma}{\varepsilon_0} = \frac{Q}{A\varepsilon_0}. \tag{6.30}$$

The voltage difference between the plates will be the product of this constant field and the spacing d:

$$V = Ed = \frac{Qd}{A\varepsilon_0}. \tag{6.31}$$

If we define the *capacitance* of the pair of plates by

$$C = \frac{Q}{V}, \tag{6.32}$$

we find

$$C = \frac{\varepsilon_0 A}{d} \tag{6.33}$$

for the *parallel plate capacitor*.

Here is another example. Take two concentric spheres of radii $a < b$ and transfer Q coulombs from the outer to the inner one. Since in the region between the spheres the inner sphere will act like a point charge Q centered at the origin, the potential difference is clearly that a point charge would produce in this region:

$$V = \frac{Q}{4\pi\varepsilon_0}\left[\frac{1}{a} - \frac{1}{b}\right] \equiv \frac{Q}{C} \text{ so that} \tag{6.34}$$

$$C = \frac{4\pi\varepsilon_0 ab}{b - a}. \tag{6.35}$$

Let us put our result to a test. Consider the case when the spacing between the spheres $d = b - a$ is negligible compared to a or b. To a tiny creature of size d, the spheres will appear infinitely large and planar and the formula should reduce to that of the parallel plate capacitor. Indeed it does. Upon setting

$$a = R - \frac{d}{2} \qquad b = R + \frac{d}{2}, \tag{6.36}$$

Eqn. 6.35 reduces to

$$C = \frac{4\pi \varepsilon_0 ab}{b-a} = \frac{4\pi \varepsilon_0 \left(R^2 - \frac{1}{4}d^2\right)}{d} = \frac{\varepsilon_0 A}{d} \qquad (6.37)$$

upon dropping d^2 compared to R^2 in the numerator and setting $4\pi R^2 = A$, the area of the sphere.

More generally, we can build a capacitor out of any two conducting objects. Each will be at some definite potential (being a conductor). If they are initially uncharged we may take their common potential to be $V = 0$. As we transfer charge Q from one to the other, a potential difference proportional to Q will develop and we may define

$$C = \frac{Q}{V} \qquad (6.38)$$

as the capacitance of this pair. For an arbitrary pair of conductors, it may be hard or impossible to compute C analytically.

Let us understand why V has to be proportional to Q (and not, say Q^2). Take some arrangement of charges $\pm Q$ on the two conductors that produces some potential difference V. Suppose you increase the local charge density *at each point* by a factor λ. By superposition, the resultant field $\mathbf{E}$ and potential V will also go up by λ. The fact that $V \to \lambda V$ when $Q \to \lambda Q$ implies V is linear in Q.

Capacitance is measured in coulombs per volt and is referred to as a *farad*, in honor of Michael Faraday (1791–1867). A capacitor with $C = 1F$ can hold one coulomb when the voltage difference between the two conductors inside is one volt. A farad is actually quite a big unit and typically you run into capacitances of order millifarads (mF) or microfarads (μF).

6.6 Energy stored in a capacitor

Suppose we have moved a charge Q' from the negative to positive conductor and the voltage difference is $V = Q'/C$. If we transfer an extra dQ' coulombs against this potential, we have to do work

$$dW = VdQ' = \frac{Q'}{C} dQ'. \qquad (6.39)$$

The total work done when we have transferred a charge Q is

$$W = \int_0^Q \frac{Q'}{C} dQ' = \frac{Q^2}{2C}. \tag{6.40}$$

The work done is the energy stored in the capacitor, denoted by U:

$$U = \frac{Q^2}{2C}. \tag{6.41}$$

Look at Eqn. 6.39. It should remind you of

$$dW = kxdx, \tag{6.42}$$

which is the work we have to do to stretch the spring from x to $x + dx$. Just as it gets progressively harder to increase x because the spring resistance grows linearly with x, it gets progressively harder to transfer dQ' as Q' increases because the electric field opposing the transfer grows linearly with Q'.

The energy stored in the capacitor, Eqn. 6.41, may also be rewritten, using $Q = CV$, as

$$U = \frac{1}{2} CV^2. \tag{6.43}$$

6.7 Energy of a charge distribution

Suppose we want to bring a whole set of charges, $q_1, q_2, \ldots q_N$, which were infinitely separated from each other to a configuration where the q_i are at some finite locations $\mathbf{r}_i$. We take all charges to be positive and if they are not, we know how to put in the minus signs and change the word "repulsion" into "attraction" as needed.

When they are infinitely far they don't even know about each other. They don't feel any force. The question is, how much work do we have to do to bring them to the final configuration? First let's take charge 1. Let us place it at $\mathbf{r}_1$. This takes no work since there are no other charges at a finite distance from it to exert a force on it. Then we bring charge 2 from infinity and put it at $\mathbf{r}_2$. The work done, is by definition, q_2 times the potential at $\mathbf{r}_2$ due to q_1:

$$W = q_2 \frac{q_1}{4\pi \varepsilon_0 |\mathbf{r}_2 - \mathbf{r}_1|}, \tag{6.44}$$

which is also the stored energy:

$$U = \frac{q_1 q_2}{4\pi \varepsilon_0 |\mathbf{r}_2 - \mathbf{r}_1|}. \tag{6.45}$$

This energy will be given back to us if we let q_2 (or q_1) fly off to infinity. To prevent the flying off, we assume the two charges are held in place by an unspecified force.

Then we bring q_3 from infinity to $\mathbf{r}_3$. How much work should we do? It is given by q_3 times the potential at $\mathbf{r}_3$ due to q_1 and q_2. The total stored energy is

$$U = \frac{q_1 q_2}{4\pi \varepsilon_0 |\mathbf{r}_1 - \mathbf{r}_2|} + \frac{q_1 q_3}{4\pi \varepsilon_0 |\mathbf{r}_1 - \mathbf{r}_3|} + \frac{q_2 q_3}{4\pi \varepsilon_0 |\mathbf{r}_2 - \mathbf{r}_3|}. \tag{6.46}$$

The first term is the work done to assemble q_1 and q_2, the second is the work done to drag in q_3 from infinity against the force due to q_1, and the last one is the work done to drag in q_3 from infinity against the force due to q_2.

Notice that the final expression for U does not depend on the order in which the charges were brought in from infinity.

Finally for N such charges the stored energy is

$$U = \frac{1}{2} \sum_{i=1}^{N} \sum_{j=1}^{N} \frac{q_i q_j}{4\pi \varepsilon_0 |\mathbf{r}_j - \mathbf{r}_i|} \quad j \neq i. \tag{6.47}$$

Let us understand this sum. First, it disallows $i = j$, i.e., the self-interaction of charge q_i, the energy needed to assemble charge q_i. We assume charge q_i, say an electron, is given to us by nature. Our job is simply to bring these preexisting charges close to each other from infinity. Next is the factor of $\frac{1}{2}$. We know from the case of $N = 3$ (Eqn. 6.46) that we should count each pair only once. The sum counts each pair twice and then divides by 2. Try this out for small values of N.

Now, let me give another simple example. I want to take a hollow sphere of radius R and uniformly deposit Q coulombs on its surface. How much work must I do? As the first couple of charges come in, they don't run into any opposition. But as the sphere charges up it starts fighting back. At some intermediate stage, when the charge on this sphere is Q' and I want to bring in a charge dQ', how much work do I have to do? When the charge is Q' the potential of the surface of the sphere is $Q'/(4\pi \varepsilon_0 R)$. The whole sphere is at that potential and I'm trying to bring in a tiny more dQ' from infinity and smear it on. The work for that will be

$$dW = \frac{Q'\,dQ'}{4\pi\,\varepsilon_0 R} \quad \text{so that} \tag{6.48}$$

$$W = \frac{1}{4\pi\,\varepsilon_0 R}\int_0^Q Q'\,dQ' = U = \frac{Q^2}{8\pi\,\varepsilon_0 R}. \tag{6.49}$$

If I write the stored energy as $Q^2/2C$, I find

$$C = 4\pi\,\varepsilon_0 R. \tag{6.50}$$

Compare this to the capacitance of two concentric spheres of radii $a < b$:

$$C = \frac{4\pi\,ab\varepsilon_0}{b-a}. \tag{6.51}$$

If you send the outer radius $b \to \infty$ and set the inner radius $a = R$, you will find $C = 4\pi\,\varepsilon_0 R$. This makes sense because when you charge a single sphere of radius R, you are bringing charges from infinity, which is imagined to be an equipotential sphere of infinite radius and $V = 0$.

CHAPTER 7

Circuits and Currents

Toward the end of the last chapter we learned about capacitors. You can make a capacitor out of any two conductors. Just move a charge Q from one to the other. At every stage each conductor will be an equipotential, and there will exist a well-defined potential difference V, which has to be linear in Q by superposition. The capacitance is defined by the relation

$$V = \frac{Q}{C}. \tag{7.1}$$

In general it is not possible to analytically derive a formula for C, though we succeeded in two simple examples fabricated from parallel plates and concentric spheres.

At some intermediate stage when charge Q' has been transferred, the voltage is $V' = Q'/C$ and the work done to transfer an extra dQ' is

$$dW = V' dQ' = \frac{Q'}{C} dQ' \quad \text{by the definition of } V' \tag{7.2}$$

$$W = \int_0^Q \frac{Q'}{C} dQ' = \frac{Q^2}{2C} = \frac{1}{2}CV^2. \tag{7.3}$$

We see that charging a capacitor is like stretching a spring: the opposition grows linearly with the extension x in one case and with the charge Q' in the other. The work you do is stored in the charges on the two plates: they have been separated despite their mutual attraction. They want to

119

recombine but do not have a path connecting the two conductors. When a path is provided, say in the form of a wire, electrons will run from the negative to the positive plate gaining kinetic energy. Along the way they can light up a flashbulb.

7.1 Energy in the electric field

But there is another manifestation of the work done: there is now an electric field between the conductors while there was none to begin with. For example, in the parallel plate capacitor there is a constant field $E = \sigma/\varepsilon_0 = Q/(A\varepsilon_0)$ pointing from the positive to the negative plate. Let us now relate the field to the energy $U = \frac{Q^2}{2C}$ through the following steps:

$$U = \frac{Q^2}{2C} = \frac{\sigma^2 A^2}{2C} \tag{7.4}$$

$$= \frac{\varepsilon_0^2 E^2 A^2}{2C} \tag{7.5}$$

$$= \frac{\varepsilon_0^2 E^2 A^2}{2(\varepsilon_0 A/d)} \quad \text{using } C = \varepsilon_0 A/d \tag{7.6}$$

$$= \frac{\varepsilon_0}{2} E^2 A \cdot d. \tag{7.7}$$

But $A \cdot d$ is the volume between the plates where the field exists (ignoring fringe effects), which gives us the following formula for u_E, the *energy density* or energy per unit volume, due to the electric field:

$$u_E = \frac{1}{2}\varepsilon_0 E^2. \tag{7.8}$$

Although we derived the formula in the context of a simple capacitor, the energy density due to any $\mathbf{E}(\mathbf{r})$ is given by

$$u_E(\mathbf{r}) = \frac{1}{2}\varepsilon_0 E^2(\mathbf{r}) \tag{7.9}$$

no matter how it was created. Even if $\mathbf{E}$ is a time- and space-dependent field produced by a radio station, this formula for the energy density holds at that space-time point. It is like saying that the energy in a spring extended by A is $\frac{1}{2}kA^2$ no matter what agency (human, Yeti) brought about this extension.

Since it takes energy to establish the electric field, it cannot just disappear. The law of conservation of energy will require that you account for it.

7.2 Circuits and conductivity

I'm going to assume you have seen circuits before and I will be brief. Let us begin with the definition of current in a wire. Imagine the wire as a perfect cylinder of cross section A. You pick some cross-sectional area and measure the number of the coulombs that go by per second. That gives the current in amperes, denoted by A. The ampere was originally defined in macroscopic terms by the magnetic effects of currents in wires. At that time we did not know about atoms or electrons.

What is the connection between such a macroscopic electric current and what's going on microscopically? We know electrons carry the current when they move. Now we come to one of the biggest irritants in life. Because the electron charge is defined to be negative, when you draw a picture with the current flowing to the right, electrons are actually moving to the left. We will need to keep an eye on just the direction of the current. We will imagine that there are objects carrying charge $+e$ moving in the direction of the current. At any time you can go back to real life by reversing the velocity and charge of these carriers to find out what the electrons are doing.

Back to the current in the wire: assume there are n carriers per unit volume and each has a charge e. From our earlier discussions of flux we know that in one second the volume that flows past any cross section of area A is Av, where v is the velocity of the carriers. The number of carriers in this volume will be Avn and the charge in this volume will be $Anve$. Thus the current will be

$$I = nevA. \tag{7.10}$$

We may write the current as a product of the area A and the *current density* j, which is the current per unit area:

$$I = jA \quad \text{where} \tag{7.11}$$

$$j = nev. \tag{7.12}$$

As the area vector **A** and current density are parallel (pointing along the wire) we could write $I = jA$ as a dot product of **A** and the current density vector **j**:

$$I = \mathbf{j} \cdot \mathbf{A}. \tag{7.13}$$

(Unfortunately, **j** is also the symbol we use for the unit vector in the y-direction. I will try to keep them from both appearing in the same discussion. Unless stated otherwise, **j** will be the current density vector.)

If the current density is not uniform across an area we should use the surface integral of the current density **j** to find

$$I = \int_A \mathbf{j} \cdot d\mathbf{A} \tag{7.14}$$

for the total current. In our discussions we will assume the current density in wires is uniform. In addition, in steady state, the current I will be assumed to be a constant along the length of the wire: were it not so, there would be a charge buildup at some point that would eventually stop the flow.

Why does current flow in a wire? It is not simply due to electronic motion. While electrons do indeed move very rapidly in solids, with typical speeds of order one million meters per second, this motion is random and varies from electron to electron. The net current due to such motion is zero: for every electron moving very fast one way there is another moving equally fast the opposite way. Over time the electrons may swap momenta but as a population they have this random velocity distribution with zero average. In fact, you could argue that without an external agency that singles out a direction, the average of these velocities has to be zero. The electrons are like a swarm of mosquitoes going nowhere.

Things change if you now apply a field **E** along the wire. The velocity acquires a non-zero average, called the *drift velocity*. The swarm drifts with this average velocity, which translates into a current. You might say, "I thought there was no electric field inside a conductor." Yes, there is no field inside a *perfect* conductor in electro*static* equilibrium. The charges in electrostatic equilibrium are indeed at rest, at the macroscopic level. The equilibrium in a current-carrying wire that we are discussing is *dynamic*. There is a net drift, but the drift has attained a steady value. Instead of the fixed positions the carriers have in electrostatic equilibrium, they have a fixed average drift velocity (and the associated steady current) in this dynamic equilibrium.

How can an electric field produce a steady velocity? Should not the carriers keep accelerating and should not the current grow indefinitely with time? This would be the case in a perfect conductor. What happens in a real conductor like copper is the following, according to classical electrodynamics. (A modification due to quantum effects will be discussed later.) Pick a particular carrier at some time. It experiences a force $F = eE$ and accelerates in response. (I am dropping vector symbols since everything is one-dimensional and along the wire in this discussion.) In addition to its random initial velocity, it now picks up a coherent piece, a *drift velocity*, along the applied E. It then collides with the nuclei in the solid. In such a collision, it typically loses some energy (which appears as resistive heat) and typically loses all memory of its original velocity and emerges from the collision in a totally random direction. It loses whatever drift velocity it had built up. Let us now ignore the random motion (which does not contribute to current) and focus on the drift velocity along the applied **E**. If I look at an assembly of such carriers what will I see? For each carrier, the drift velocity will depend on how long it has been accelerating since its last collision when its (drift) velocity was reset to zero. If it has been t seconds since the last collision, the drift velocity along the field will be

$$v(t) = \frac{eE}{m}t. \tag{7.15}$$

The drift velocity averaged over all carriers will be

$$\bar{v} = \frac{eE}{m}\tau \tag{7.16}$$

where $\tau = \bar{t}$, the *mean collision time*, is the average time since the last collision.

Earlier we wrote a formula $j = nev$ assuming all carriers were moving at one velocity v. We see that the picture is more complicated. Henceforth when we write $j = nev$ it will be with the understanding that v is really $\bar{v}$. Thus the current density in field E will be

$$j = nev = ne\bar{v} = ne \cdot \frac{eE}{m}\tau = \frac{ne^2\tau}{m}E. \tag{7.17}$$

We define the *conductivity* σ of the material as the ratio of the current density to the field that causes it:

$$j = \sigma E. \tag{7.18}$$

Eqn. 7.17 tells us that in our simple model, attributed to Paul Drude (1863–1906),

$$\sigma = \frac{ne^2\tau}{m}. \tag{7.19}$$

Let us study Eqn. 7.17. Does it make sense? We know why the E is there: without it pointing the way and producing a coherent drift velocity, the random motion of electrons will cause no current. That the current is bigger if you have a bigger density of carriers is obvious. The inverse dependence on the mass of the carriers just comes from $a = F/m$. The bigger the τ, the bigger the response, because the carriers can go for a longer time on average before colliding, and therefore they have more time to pick up speed in the direction of the field. The e^2 is interesting. One factor of e comes because the force on the carrier is eE. The second e comes because the current it carries is itself proportional to e. Notice that the current is independent of the sign of e. If you make it negative, the carriers accelerate the other way, but because they have the opposite charge, the current will be the same. This means that you cannot tell the sign of the current carriers by measuring the conductivity.

We can rewrite $j = \sigma E$ in a way that will be more familiar. Consider the situation where the field E is obtained by applying the voltage difference V between the two ends of a wire of length L. Then $V = EL$ by definition. The total current in the wire is the current density times area, $I = jA$. Rather than saying the current density is driven by the electric field, let's say the current is driven by the voltage difference between the two ends of the wire. We now end up with

$$I = jA = \sigma EA = \sigma A \frac{V}{L} = GV \quad \text{where} \tag{7.20}$$

$$G = \frac{\sigma A}{L} \tag{7.21}$$

is the *conductance*. Whereas the conductivity $\sigma = ne^2\tau/m$ depends only on the material (copper versus aluminum), the conductance G depends additionally on the dimensions of the wire. A large conductance could come from material with small conductivity if the wire had a large cross section and a small length.

We are more familiar with *resistance* R than conductance, which appears when we rewrite Eqn. 7.20

$$I = \frac{\sigma A}{L} V$$

as

$$V = IR \quad \text{where} \tag{7.22}$$

$$R = \frac{L}{A\sigma} \equiv \frac{\rho L}{A} \quad \text{and} \tag{7.23}$$

$$\rho = \frac{1}{\sigma} \quad \text{is the *resistivity*.} \tag{7.24}$$

Again, resistivity is a property of the material, and the resistance depends additionally on the dimensions of the wire.

Eqn. 7.22 is the well-known Ohm's law named after Georg Ohm (1789–1854). Resistance is denoted by the symbol R and is measured in ohms, represented by Ω. Thus a 5Ω resistor will allow a current of $2A$ to flow when a voltage $V = 10$ volts is applied.

According to Eqn 7.23, the bigger the resistivity, the bigger the resistance, which is to be expected. In addition, if the wire is made twice as long, say by joining two identical pieces of wire, the resistance will be twice as big. This is true because each resistor will suffer the same voltage drop, and it will take double the voltage to drive the same current. If you double the area, resistance is turned into half its value. This makes sense if you think of the wider wire as two identical wires, glued side by side, each responding to the voltage V across it and carrying its own current.

Two caveats are needed here.

First, I will often refer to ideal conducting wires or leads in a circuit that have no resistance and no voltage drop across them. The current $I = \frac{0}{0}$ seems indeterminate, but only if seen in isolation. The current I through such an ideal lead is decided by the other circuit elements and batteries. In Figure 7.1 it is decided by C and R and the charge on the capacitor. In reality all leads have some resistance and some voltage drop across them, but both these are too small to make a difference and chosen to be zero for simplicity. An ideal lead is like a massless string that transmits a force between two massive objects, one of which is being pulled by a force. Both the force on the string and its mass are zero and its acceleration $a = \frac{0}{0}$ seems indeterminate, but it is not. It is decided by the two non-zero masses and the applied force. Massless strings are also idealizations introduced to simplify the calculation and to focus on the objects of significant mass.

Next, when I said that the resistivity was due to the carriers bumping into the nuclei, I was simplifying things. In a perfect solid, where each nucleus sits at a precise location on a periodic lattice, the electrons do not bump into them at all. (The average time since the last collision τ will be infinite.) This is due to quantum mechanics. The wave theory of electrons (more on this later in the book) allows them to navigate around these nuclei the way a blind person can navigate around a room full of furniture *placed at predictable and fixed locations.* At zero temperature the nuclei sit at well-defined positions on a regular lattice and the conductance is infinite. At non-zero temperature, the nuclei start jiggling around their nominal positions as part of random thermal fluctuations. This unpredictability leads to collisions with the electrons and to resistance. There are other sources of resistance as well, such as impurities, which are foreign atoms that are embedded in the solid.

The conductivity will generally depend on the temperature, the purity of the sample, and the strength of interaction between electrons. Even the mass m is not the mass of the electron in free space; it is modified by the lattice and electron-electron interactions. Computing σ is a big industry that calls for a sophisticated quantum mechanical treatment.

7.3 Circuits

Let us begin with a simple circuit depicted in Figure 7.1. I take a capacitor, charge it up to some amount $Q(0)$, and then connect it to a resistor R via a switch. What happens when I close that switch at time $t = 0$? The positive charges were dying to get over to the negative plate, but they could not traverse the vacuum between plates. But if you give them a path, in the form of a wire, they will go through that and come back to the other plate and neutralize their opposites. The capacitor gets discharged in this process and the voltage between its plates is diminished.

We want to calculate the currents and voltages in this circuit as a function of time. We follow two rules due to Gustav Kirchoff (1824–1887):

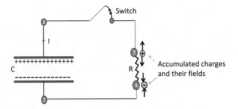

Figure 7.1 The *RC* circuit. If *R* is removed, charges will build up at points 3 and 4. Their fields (shown by arrows) kill the field in the leads and reinforce each other in the gap where *R* is to be connected.

1. If there is a branch in the circuit at some node, currents entering the node must equal the currents leaving it. This is to enforce charge conservation and prohibit charge buildup at the nodes, for this will eventually stop the current.
2. The sum of the changes in voltage as we go around any loop must add up to zero. This is so because the line integral of the electric field on a loop is 0, or equivalently because the potential at any point is like an electrical height: if you add all the changes in height as you go around a loop, you must get zero.

In this circuit there are no branches and just one current $I(t)$.

Next let us add the height changes around the loop, starting at point 1. When we go up through the capacitor to 2, we go up in electrical height by an amount $\frac{Q}{C}$ volts. The leads to the resistor have zero resistance and there is zero voltage drop between 2 and 3. A resistor will not carry current unless there is a voltage applied to it. Since current flows downhill, we drop by RI volts when we come from 3 to 4. There is no further drop as we go back to point 1 along the perfectly conducting leads. Thus we have

$$0 = \frac{Q}{C} - IR. \tag{7.25}$$

A few words are needed on the perfectly conducting leads. We know there can be no field inside them. Yet we want a field inside the resistor to drive the current. How does this field suddenly appear just within *R*? Here is a very simplified explanation. First remove the resistor and let the two ends 3 and 4 connected to it dangle. Some tiny positive charge will initially flow from the positive plate to the tip of the upper wire (point 3) till its own field blocks the arrival of more positive charges. The field due to this accumulated charge balances the field in the leads due to charges in the positive plate. The lead becomes field-free and equipotential. The same happens at the other lead, which will have some tiny negative charge at its tip. Notice that the fields due to charges accumulating at points 3 and 4 aid each other (both pointing down in the figure) where the resistor will be placed. If we now reinstate the resistor, the accumulated charges will drive the current in the resistor.

Back to our circuit equation, 7.25. What is the relation between I and Q? In a time dt the current I, as shown in the figure, *carries away* a charge

$$dQ = -Idt \tag{7.26}$$

from the upper plate. Thus

$$I = -\frac{dQ}{dt}. \tag{7.27}$$

Feeding this into Eqn. 7.25 we obtain a differential equation for Q:

$$R\frac{dQ}{dt} + \frac{Q}{C} = 0. \tag{7.28}$$

Upon integrating both sides of

$$\frac{dQ}{Q} = -\frac{dt}{RC} \tag{7.29}$$

from the initial time of 0 (when the switch was closed) to time t, we find

$$\ln\left[\frac{Q(t)}{Q(0)}\right] = -\frac{t}{RC}, \tag{7.30}$$

which means

$$Q(t) = Q(0)e^{-t/RC}. \tag{7.31}$$

The current is

$$I(t) = -\frac{dQ}{dt} = \frac{Q(0)}{RC}e^{-t/RC} \equiv I(0)e^{-t/RC}. \tag{7.32}$$

The charge on the capacitor starts out as $Q(0)$ and decays exponentially once you close the switch. When will it completely discharge? The answer is "Never!" Why is the capacitor not able to discharge completely? As it drives current through the resistor, it begins to discharge, the voltage across it drops, and it is less able to drive current through the resistor. It is trying to discharge itself, but soon its ability to do that plummets: there is less and less Q on it to drive any more Q away through the resistor. So $Q(t)$ will never hit zero and neither will $I(t)$. But in practice, it is essentially all over after a few times the *time-constant*

$$t_0 = RC. \tag{7.33}$$

The reason is that when $t \gg t_0$ we have e raised to a big negative number, which is negligible. For example, if we set $t = 3RC = 3t_0$ in Eqn. 7.32, we find $e^{-3} \simeq 1/20$. If the time elapsed is large compared to the time-constant t_0, the decay is essentially complete. Here is another way to understand the time-constant t_0. Consider the initial rate of decay of the current as per Eqn. 7.32:

$$\left. \frac{dI}{dt} \right|_0 = -\frac{I(0)}{RC} = -\frac{I(0)}{t_0}. \tag{7.34}$$

We may rewrite this as

$$t_0 \cdot \left. \frac{dI}{dt} \right|_0 = -I(0), \tag{7.35}$$

which means that *if* the current continued to decay at the initial rate, it will reach zero in time t_0. (Of course, this is not what happens—the rate of decay itself drops as the current drops and the current is non-zero for all finite t.)

So capacitors can be pretty dangerous. If you open an old amplifier, even though it's not plugged in, there could be capacitors inside that are charged and the R in the diagram could be you. That's why they always tell you, "Do not take this amplifier into your bathtub."

To operate the flashbulb in your camera, you charge up a capacitor, and when you squeeze the shutter you close the circuit and let it discharge through the bulb. Here you want the time constant to be very small, because after a while people will stop smiling.

Before we closed the switch, we had a fully charged capacitor with energy

$$U(0) = \frac{Q^2(0)}{2C}. \tag{7.36}$$

At $t = \infty$ the capacitor is discharged and there is no current flowing. What happened to the energy? We know it went into heating the resistor, but we would like to see if the initial stored energy precisely matches the loss over time.

If a current I flows through a resistor across which is a voltage V, it means I coulombs are falling down V volts every second for a loss of

$$P = VI = I^2 R. \tag{7.37}$$

(The kinetic energy gained in the fall is transferred via collisions with the nuclei into heat.) Integrating this power loss over all time we get

$$\text{Total dissipation in } R = \int_0^\infty P(t)\,dt \tag{7.38}$$

$$= \int_0^\infty I^2(t)R\,dt \tag{7.39}$$

$$= R\left[\frac{Q(0)}{RC}\right]^2 \int_0^\infty e^{-2t/RC}\,dt \tag{7.40}$$

$$= \frac{Q^2(0)}{2C}, \tag{7.41}$$

which is exactly the initial energy in the capacitor.

7.4 The battery and the EMF $\mathcal{E}$

The trouble with the RC circuit is that after you close the switch, the current is essentially zero after a few time constants. If you want something more long-lasting you need a battery or cell, shown in Figure 7.2. I want to share with you some fine points about batteries in circuits.

Let us begin with what you might know already. Between the positive and negative terminals of the battery there are some chemicals that essentially remove electrons from the positive terminal and deposit them on the negative terminal. Soon this runs into some opposition: the

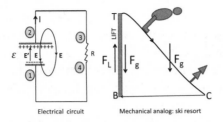

Electrical circuit   Mechanical analog: ski resort

Figure 7.2   Left: The electrical circuit indicating the non-conservative chemical force per unit charge $\mathbf{E}'$ and the electrostatic field $\mathbf{E}$, which are equal and opposite inside the battery. The battery does work $\mathcal{E}q$ on every charge q that goes uphill from the negative to the positive plate. This work is returned by the electrostatic field $\mathbf{E}$ outside the battery when charges flow downhill from the positive to negative terminal via the external circuit. Right: The mechanical analogy in which the non-conservative force $\mathbf{F}_L$ due to the lift does work mgh per cycle. Because it exactly balances the gravitational force $\mathbf{F}_g$, i.e., $\mathbf{F}_L = -\mathbf{F}_g$ *in the lift area*, the work done by $\mathbf{F}_L$ is also the difference in gravitational potential energy.

accumulated charges do not want more of their type to come their way. They set up an electrostatic field $\mathbf{E}$ that opposes the chemical forces. At equilibrium $\mathbf{E}'$, the chemical force per unit charge balances the electrical force per unit charge, $\mathbf{E}$. The potential difference associated with the electric field is the nominal voltage, say 1.5 volts. We are then used to including the battery in the circuit equation as the source of an upward jump of 1.5 volts when we go from the negative to positive terminal.

All this seems familiar but there are some subtle issues that I would like to share with you.

Let us begin with an analogy shown in the right half of the figure. You are coming down a ski slope, from the top of the lift T, to the chalet C. Gravity is pulling you down and speeding you up. If there were no trees that you bump into, you could, in principle, ski right back up the slope to reach the top at zero speed. This is just the law of conservation of kinetic plus potential energy in the gravitational field. But say there are many trees and that you lose all the gained kinetic energy by colliding against them on the way down to C, and then at the same height, to the bottom of the lift B. Once you reach the bottom B, gravity is finished with you. It cannot get you to the top T for the next round. Indeed, gravity, which was with you coming down, will be against you going back up. That has to be so, given that it is a conservative force.

Someone observing you for a whole day will find you delivering energy to the trees every cycle. Something is giving you that energy, or doing that work on you every cycle. That something cannot be gravity since the work done by gravity in a full cycle is zero. It has to be a force with a non-zero line integral over a closed loop. It has to be a non-conservative force.

That force is of course the one due to the ski lift. The lift applies a force $\mathbf{F}_L$ that exactly balances $\mathbf{F}_g$, the force of gravity, as it carries you up from the bottom of the lift B to the top T:

$$\mathbf{F}_g = -\mathbf{F}_L \quad \text{during ascent in lift.} \tag{7.42}$$

The work done by the lift on the upward trip is

$$\int_B^T \mathbf{F}_L \cdot d\mathbf{r} = mgh. \tag{7.43}$$

Non-conservative forces are defined by their *circulation*, which is their line integral over a closed loop. How shall we define the circulation of $\mathbf{F}_L$ that is non-zero only over an open segment $B \to T$ inside the lift? We simply add an extra portion that completes the loop by going back from T to B along *any path* outside the lift, say $T \to C \to B$ in the figure. You can choose a different way to close the loop but it will not matter since $\mathbf{F}_L$ is identically zero outside the segment $B \to T$. Letting $\mathcal{E}$ denote this loop integral, we have the result

$$\mathcal{E} = \oint \mathbf{F}_L \cdot d\mathbf{r} = \int_B^T \mathbf{F}_L \cdot d\mathbf{r} = -\int_B^T \mathbf{F}_g \cdot d\mathbf{r}$$
$$= U(T) - U(B)(= mgh) \tag{7.44}$$

upon invoking

$$-\int_1^2 \mathbf{F}_g \cdot d\mathbf{r} = U_g(2) - U_g(1). \tag{7.45}$$

As $\mathbf{F}_g$ is conservative, potential energy difference $U_g(T) - U_g(B)$ can be traded for kinetic energy along *any* path connecting T to B. You cannot do this inside the lift because the floor keeps you from falling. But you can leave the lift and ski downhill from $T \to C \to B$ during which ride

$$\int_{T \to C \to B} \mathbf{F}_g \cdot d\mathbf{r} = U_g(T) - U_g(B) = \mathcal{E}, \tag{7.46}$$

which nicely relates $\mathcal{E}$, the work done by the lift to increase the gravitational potential energy, to the work done *by* gravity on the skier.

Eqn. 7.44 equates $\mathcal{E}$, the line integral of the *non-conservative* force $\mathbf{F}_L$ of the lift around a closed loop, to the gravitational potential energy difference between the top and bottom of the lift due to the *conservative* force $\mathbf{F}_g$. Such a relation exists because

- $\mathcal{E}$, the integral of $\mathbf{F}_L$ around any closed loop, is simply its integral within the lift, because $\mathbf{F}_L$ is zero everywhere else, and
- $\mathbf{F}_L = -\mathbf{F}_g$ during the climb, so that this integral is also the gravitational potential difference between top and bottom.

Now return to the left half of the figure with the battery. The analogy with the ski lift should help you as you go along.

The electrostatic field $\mathbf{E}$ is set up by the charges deposited at the terminals by the chemicals in the battery and points from $+$ to $-$ inside the battery, just like $\mathbf{F}_g$ but with one trivial difference: $\mathbf{E}$ is the force on a unit charge while $\mathbf{F}_g$ was the force on the skier, not necessarily of unit mass. When the circuit is closed, positive charges can flow from the $+$ to the $-$ terminal through the resistor. (In reality it is the electrons going the other way.) In the resistor they deliver the excess kinetic energy the electric field gives them to the nuclei via collisions (that heat up the resistor) and finally end up at the negative terminal. They cannot go up to the positive terminal using the electrostatic field, which now opposes this motion inside the battery. Here is where the non-conservative chemical force $\mathbf{E}'$ of the battery (the analog of the lift force $\mathbf{F}_L$) comes into play. It lifts the charges against the internal electrostatic field $\mathbf{E}$ and deposits them in the positive terminal. The *electromotive force* or emf is defined as the closed loop integral of $\mathbf{E}'$, which is the work done on a unit charge around a closed loop:

$$\mathcal{E} = \oint \mathbf{E}' \cdot d\mathbf{r}. \qquad (7.47)$$

The loop is composed of the path from the negative to the positive terminal inside the battery and an arbitrary path outside that closes it. It does not matter how we choose this path because the entire contribution to $\oint$ comes from inside the battery on the segment going from the negative to the positive terminal. That is,

$$\mathcal{E} = \oint \mathbf{E}' \cdot d\mathbf{r} = \int_{-}^{+} \mathbf{E}' \cdot d\mathbf{r}. \tag{7.48}$$

Next, because $\mathbf{E}' = -\mathbf{E}$ inside the battery (just like $\mathbf{F}_g = -\mathbf{F}_L$), we deduce

$$\mathcal{E} = -\int_{-}^{+} \mathbf{E} \cdot d\mathbf{r} = V(+) - V(-) = V_{+-} \equiv V$$

the voltage of the battery, like 1.5V. (7.49)

This difference in potential can be converted to kinetic energy on any path going from the $+$ terminal to the $-$ terminal. Any path inside the battery is blocked by the chemical forces (the way the lift keeps the skier at the top from falling to the bottom). But any path outside, provided by the external circuit, is permitted. This is what happens when the circuit is closed.

The main point, which you may not have appreciated in earlier encounters, is that *even though the voltage concept is associated with the conservative electrostatic field* $\mathbf{E}$, *it is numerically equal to the closed loop integral of a non-conservative chemical field* $\mathbf{E}'$. The non-conservative chemical force is needed for the battery to do work cycle after cycle, as charges go around the circuit. The conservative electric field takes energy from the chemical force inside the battery and gives it to the charges in the circuit.

If you do not want to look under the hood, you may simply (and correctly) assume that when you travel across the battery from the negative to the positive terminal, the electrostatic potential goes up by the emf $\mathcal{E}$. Sometimes the voltage across the terminals of the battery is denoted by the more familiar V rather than $\mathcal{E}$ since they are numerically equal.

Let us write an equation for the circuit in Figure 7.2. As we add the changes in voltage starting from point 1, we find it goes up by $V_2 - V_1 = \mathcal{E}$ for $1 \to 2$ and by zero for $2 \to 3$ (perfectly conducting wire with no drop); it drops by IR during $3 \to 4$ and by zero for $4 \to 1$. Thus we have

$$0 = \mathcal{E} - IR \tag{7.50}$$

$$\mathcal{E} = IR, \tag{7.51}$$

which is sometimes written as $V = IR$.

7.5   The RC circuit with a battery

The circuit is shown in Figure 7.3. You have a battery with emf $\mathcal{E}$, a capacitor C, a resistor R, and a switch that is *initially open*. The wire joining the lower plate of the capacitor and the negative terminal of the battery ensures that they are both at the same potential, say $V = 0$. The upper plate of the capacitor is also at $V = 0$ since there is no field between the plates to create a potential difference. The positive terminal of the battery is at $V = \mathcal{E}$ due to the electric field inside. You might think that some of the charges in the negative terminal will flow to the lower plate of the capacitor due to inter-electron repulsion. This does not happen because the negative charges in the negative terminal are bound to the positive charges in the positive terminal along with whom they were created by the chemicals. Flowing to the lower plate of the capacitor would increase their separation from the positive charges and increase the energy. The positive and negative charges in the two terminals would love to reunite inside the battery but are prevented by the chemical forces.

Now let us close the switch. The positive terminal at voltage $\mathcal{E}$ is now connected to the positive plate of the capacitor. Positive and negative charges that wanted to reunite inside the cell but were held back by the chemical forces still cannot reunite, but they can get closer: some positive charges will begin rushing to the upper plate of the capacitor and an equal number of negative charges will rush to the lower plate. Because there is

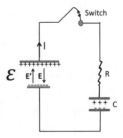

Figure 7.3   The circuit with a battery, resistor, capacitor, and switch. Shown are the non-conservative chemical force per unit charge $\mathbf{E}'$ and the electrostatic field $\mathbf{E}$, which are equal and opposite inside the battery. Before the switch is closed both plates of the capacitor are uncharged and at the same potential. When the switch is closed, equal and opposite charges flow to the plates of the capacitor (as shown), which then begins to oppose the very battery that feeds it.

a resistor in the circuit, the current will be finite. As the current flows, the capacitor will develop a voltage that opposes this current. The current will stop when the opposing voltage exactly balances the battery. Soon we will find out when this happens.

When some positive and negative charges leave the two terminals, the electrostatic force inside the battery momentarily becomes weaker than the chemical force, which immediately deposits opposite charges on the two terminals to bring the voltage difference back to $\mathcal{E}$.

Look at what has happened after the switch was closed. Some positive charges created inside the cell have gone to the positive terminal and continued onward to the positive plate of the capacitor. Some negative charges created inside the cell have gone to the negative terminal of the battery and on to the negative plate of the capacitor. But negative charges flowing from the negative terminal to the negative plate of the capacitor are equivalent to positive charges or current flowing from the negative plate of the capacitor to the negative terminal of the battery. *On the whole it is as if some positive current has flowed around the circuit, even though no charge has flowed across the gap between the plates in the capacitor.*

It is this current we want to describe, qualitatively first and then quantitatively. Initially the current will be $\mathcal{E}/R$ because the battery is the only driving force. But as the current flows, it charges up the capacitor, and if you look at the figure you can see that the capacitor would like to drive a current in the opposite direction from the battery. It bites the hand that feeds it. Eventually we expect that the capacitor will exactly counter the battery and then the current will stop. To know when this will happen we have to do a calculation after first writing down the circuit equation.

Since there are no branches, we just have to deal with just one current $I(t)$.

Starting at a point below the battery, as we move past it, we go up by a voltage $\mathcal{E}$ and then we drop by an amount IR when we cross the resistor and another Q/C when we go from the positive to the negative plate and arrive at the starting point. Setting the sum of all the voltage changes to zero we find

$$0 = \mathcal{E} - IR - \frac{Q}{C}. \tag{7.52}$$

Convince yourself that the current, which is now responsible for *charging* the capacitor (rather than discharging as in the previous example with just a capacitor and resistor), is related to Q by

$$I = +\frac{dQ}{dt}. \tag{7.53}$$

Combining the last two equations we arrive at an equation obeyed by the charge

$$R\frac{dQ}{dt} + \frac{Q}{C} = \mathcal{E}. \tag{7.54}$$

We could solve this equation easily if it were not for the $\mathcal{E}$ on the right. Since it is a constant, we eliminate it as follows. Define $\tilde{Q}$ as follows

$$Q = \tilde{Q} + C\mathcal{E}. \tag{7.55}$$

Eqn. 7.54 now becomes (upon realizing the time derivative of $C\mathcal{E}$ vanishes):

$$R\frac{d\tilde{Q}}{dt} + \frac{\tilde{Q}}{C} + \mathcal{E} = \mathcal{E} \tag{7.56}$$

$$R\frac{d\tilde{Q}}{dt} + \frac{\tilde{Q}}{C} = 0. \tag{7.57}$$

We have already solved this equation before and the answer is

$$\tilde{Q}(t) = \tilde{Q}(0)e^{-t/(RC)}. \tag{7.58}$$

Since the initial charge on the capacitor $Q(0) = 0$, we see from Eqn. 7.55 that

$$\tilde{Q}(0) = -C\mathcal{E} \tag{7.59}$$

and Eqn. 7.55 implies that

$$Q(t) = \tilde{Q} + C\mathcal{E} \tag{7.60}$$

$$= \tilde{Q}(0)e^{-t/(RC)} + C\mathcal{E} \tag{7.61}$$

$$= C\mathcal{E}\left[1 - e^{-t/(RC)}\right] \tag{7.62}$$

$$= Q(\infty)\left[1 - e^{-t/(RC)}\right] \tag{7.63}$$

using the fact that $Q(\infty) = C\mathcal{E}$.

We find from Eqn. 7.62 that the voltage on the capacitor, $Q(t)/C$, always falls short of $\mathcal{E}$ and reaches that value only asymptotically as $t \rightarrow \infty$. Since the capacitor is fighting the very battery that is charging it, it is nourished less and less by it as it approaches the battery in stature. But it cannot ever become its equal.

The current is found by differentiating $Q(t)$:

$$I = \frac{dQ}{dt} = \frac{\mathcal{E}}{R} e^{-t/(RC)}. \tag{7.64}$$

This simple analysis should give you a feeling for how physics works. You develop models of the capacitor, resistor, and battery and write down the circuit (differential) equation that reflects the basic principles like charge conservation and conservative forces. You solve the equations and are stuck with what they predict. The mathematics rules after that point. And whatever it tells you, you rush out to the lab to verify. For example, you may want your capacitor to attain 80 percent of its maximum charge and you may like to know, "How long should I wait?" Simply set $Q(t)/Q(\infty) = .8$ in Eqn. 7.63 and solve for t. (If you want it to hold 100 percent of the maximum charge, that will never happen.)

A final check on energetics. The work done by the battery, W_{Batt}, is the integral of the power P. Since the battery lifts I coulombs per second over a "height" $\mathcal{E}$ volts, the power delivered by it is

$$P(t) = \mathcal{E}I(t), \tag{7.65}$$

and the energy delivered over all time is

$$W_{Batt} = \int_0^\infty \mathcal{E}I(t)dt \tag{7.66}$$

$$= \frac{\mathcal{E}^2}{R} \int_0^\infty e^{-t/(RC)}dt = \mathcal{E}^2 C. \tag{7.67}$$

I leave it to you to verify that this is the sum of the final energy stored in the capacitor and the heat dissipated in the resistor, both contributing equally.

7.6 Miscellaneous circuits

The following is a review of DC circuits. The same rules apply in AC circuits, which we will study later in this course. Look at Figure 7.4.

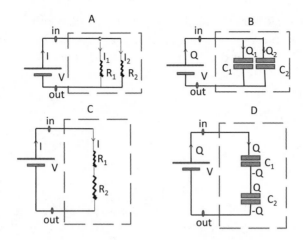

Figure 7.4 A: Adding resistors in parallel. B: Adding capacitors in parallel. C: Adding resistors in series. D: Adding capacitors in series. The circuit elements are enclosed in a black box (dotted line) with just two leads coming out.

Part A shows two resistors R_1 and R_2 in parallel, hidden in a box, shown by dotted lines. Just two terminals marked *in* and *out* are visible. We have to find out what effective resistance resides inside. So we hook the terminals to a battery of known voltage V, measure the current I that flows, and declare that the resistance inside is

$$R = \frac{V}{I}. \tag{7.68}$$

The figure shows the current I entering the box and splitting into two parts I_1 and I_2, which must add up to I. We then reason as follows:

$$I_1 = \frac{V}{R_1} \tag{7.69}$$

$$I_2 = \frac{V}{R_2} \tag{7.70}$$

$$I = I_1 + I_2 = \frac{V}{R_1} + \frac{V}{R_2} \quad \text{(current conservation)} \tag{7.71}$$

$$\frac{I}{V} = \frac{1}{R_1} + \frac{1}{R_2} = \frac{1}{R} \quad \text{(definition of } 1/R \text{)} \tag{7.72}$$

$$R = \frac{R_1 R_2}{R_1 + R_2}. \tag{7.73}$$

This formula says that the final resistance is less than either one. (Check this.) This too makes sense, as a parallel path to either one implies more current and less resistance. You may also check that

$$I_1 = \frac{V}{R_1} = \frac{IR}{R_1} = \frac{R_2}{R_1 + R_2} \cdot I \qquad (7.74)$$

$$I_2 = \frac{R_1}{R_1 + R_2} \cdot I, \qquad (7.75)$$

which states that the current flowing in one branch is proportional to the resistance of the *other*. This makes sense: the greater the opposition to current the other branch offers, the more likely the current is to come your way.

Now turn to part B with two capacitors in parallel. To find out their effective capacitance, we will apply a voltage V, find the charge Q that flows in, and assign a value

$$C = \frac{Q}{V} \qquad (7.76)$$

to the capacitance inside. The charge that flows divides into Q_1 and Q_2 and each capacitor feels the full applied V. Thus

$$Q_1 = C_1 V \qquad (7.77)$$

$$Q_2 = C_2 V \qquad (7.78)$$

$$Q = Q_1 + Q_2 = (C_1 + C_2) V \qquad (7.79)$$

$$\frac{Q}{V} = C_1 + C_2 = C. \qquad (7.80)$$

Thus capacitors in parallel add. You can almost see this from the figure. If you just let the two capacitors touch and become one, the combination has an area equal to the sum of the areas and the same separation between plates. From the formula $C = \varepsilon_0 A / d$ we see $C = C_1 + C_2$. In general if the capacitors are totally different in design, we must return to the more basic notion that capacitance is a measure of how much charge can be held for a given applied voltage and that when connected in parallel, the holding capacity is additive.

Part C shows two resistors in series. Clearly

$$V = IR_1 + IR_2 = I(R_1 + R_2) \tag{7.81}$$

$$R = \frac{V}{I} = R_1 + R_2. \tag{7.82}$$

Finally, part D shows two capacitors in series. The battery sends in $\pm Q$ to the upper plate of C_1 and the lower plate of C_2. If the two plates in the middle do nothing there will be a field between these two plates, with the corresponding energy per unit volume. However, the energy can be reduced if the lower plate of C_1 borrows $-Q$ from the upper plate of C_2. This traps the field lines between the two plates of each capacitor, causing a reduction in energy. Given this arrangement of charges, it is evident that

$$V = \frac{Q}{C_1} + \frac{Q}{C_2} \tag{7.83}$$

$$\frac{V}{Q} = \frac{1}{C} = \frac{1}{C_1} + \frac{1}{C_2} \tag{7.84}$$

$$C = \frac{C_1 C_2}{C_1 + C_2}. \tag{7.85}$$

In summary, capacitances in parallel simply add, just like resistances in series. The inverses of capacitances add in series, just like the inverse of resistances in parallel.

CHAPTER 8

Magnetism I

Every time you think you're done with the laws of physics, somebody does some experiment that doesn't fit what you know, and you have to make up new stuff. That takes us to our next topic: magnetism. Don't believe the myth that magnetism was discovered in Ancient Greece, when parents noticed kids were sticking their art work on the refrigerator using some little black rocks. It is true, however, that magnetic phenomena in lodestone were discovered before the common era and later used to make compass needles.

8.1 Experiments pointing to magnetism

I'm going to give you a string of more modern experiments (depicted in Figure 8.1) that tell you there is something going on that is not described by anything I've described so far in this course, new phenomena that are inexplicable.

Here's the simplest one. There are two parallel wires carrying currents I_1 and I_2 in the same direction. The wires are found to attract each other. This force cannot be electrostatic since the wires are neutral. You can confirm this by placing a test charge next to either wire and finding no response. Next, if you reverse one of the currents the force becomes repulsive. You might guess a new law: parallel (anti-parallel) currents attract (repel). However, it is not going to be easy to find a force vector pointing from one wire to the other by combining the vectors

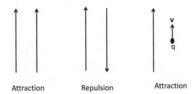

Figure 8.1   Three examples of the magnetic force: parallel currents attracting, anti-parallel currents repelling, and a moving charge $q > 0$ attracted to a current in the same direction. Not shown is the repulsion if the charge moves anti-parallel to the current.

corresponding to the two colinear currents: their dot product will change sign under current reversal but will be a scalar rather than a vector, while the cross product of the colinear current vectors will be zero.

Let us simplify one of the two players in the last discussion and replace one wire by a charge $q > 0$ as shown in Figure 8.1. When the charge q sits next to the wire nothing happens. This is expected since the wire is neutral. The charge then begins to move at speed v parallel to the current. It is now found to be attracted to the wire. It starts bending in toward the wire. That also cannot be due to the electrical force, which doesn't care if the charge is moving or not. And if the charge reverses its velocity and moves anti-parallel to the current, the force becomes repulsive. So, this is one class of phenomena or experiments that eludes description in terms of what I have covered so far.

Consider next the most familiar case: bar magnets. They seem to have a north and a south end, and opposite ends attract and like ends repel just like electric charges. How do you decide which end is north? You can randomly pick one end of a reference magnet as north; if the end of another magnet is attracted to it, that is the south end and, if repelled, the north end. This is how you would decide which charges are positive and which are negative, as a matter of convention. But the words "north" and "south" have an independent connotation (Canada is to the north of the United States) that removes the arbitrariness. If you mount a natural magnet on a pivot so it can swing and form a compass, it will line up in the north-south direction on earth. The end that points to the north (arctic) is the north pole N of the compass needle. This sounds wrong; it should repel the north pole of the earth, instead of being attracted to it. The explanation is that the giant magnet inside the earth, which produces the terrestrial field, is actually upside down—with its magnetic south pole (which will

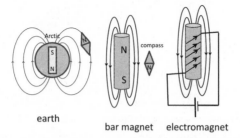

earth

bar magnet   electromagnet

Figure 8.2   The earth, a bar magnet, and an electromagnet. The direction of the field lines is determined at each point by a compass needle. The end marked N is the north pole of the needle. While this end points toward the south pole of the bar magnet, it points toward the geographic north of the earth because the magnetic poles of the earth are aligned opposite to the geographic poles.

attract the north pole of the compass needle) in the arctic and its magnetic north pole in the antarctic, as shown in the leftmost part of Figure 8.2. (This is another nuisance like the minus sign in the electron charge.) To map out the field of a bar magnet, place the compass needle at various places, let it settle down and draw a little vector from its south pole to its north pole, and join the little arrows to define the lines of the magnetic field. You will end up with the familiar picture shown in Figure 8.2. The north end of the bar magnet is where the lines emerge and the south is where they return. If you could go inside the magnet with your needle, you would find the lines that entered the south end continue up the magnet and emerge as the lines leaving the north pole. Magnetic lines form closed loops.

Another baffling experimental fact is that you could make a magnet by driving current through a solenoid as shown in the figure. The compass needle responds to the field of this *electromagnet* as it did to that of the bar magnet. Magnet is reversed if you reverse the current.

All this should be enough to convince you that something beyond electrostatics is at work. Why didn't we need this something before, and why do we need it now? What is new in the phenomena just described to distinguish them from problems we have been studying so far? What feature distinguishes these phenomena from electrostatics?

After some discussion my class was able to zero in on the answer: the charges are now moving. Go back to the charge q that was drawn to the wire carrying current I. The charge in question is moving and so are the charges in the current-carrying wire. Stop the charge or the current and the force goes away. (The bar magnets seem to violate this characterization since nothing is moving. Actually there are circulating atomic currents behind the magnetism. More on this later.)

So, magnetism is caused by moving charges and it is felt by moving charges. We need to figure out how the velocities enter the game in both parts.

Having impressed you with an array of inexplicable, magnetic phenomena, I will now give you the fundamental equations of *magnetostatics*, equations that summarize everything I've described so far. (The "statics" in magnetostatics may seem inappropriate after just saying that magnetism involves moving charges. It refers to the fact that the macroscopic *currents* involved are constant in time.)

There will be two parts to magnetostatics, just as in electrostatics. The first part will specify the force felt by a moving charge in a magnetic field. The second will specify how currents produce a magnetic field $\mathbf{B}$.

The force on a charge is called the *Lorentz force* in honor of Hendrik Lorentz (1853–1928). He did not discover this law but made other profound contributions to electrodynamics. Here it is:

$$\mathbf{F} = q(\mathbf{E} + \mathbf{v} \times \mathbf{B}). \qquad (8.1)$$

The first term in Eqn. 8.1 is the familiar electric force. As mentioned before, this part is unaffected by relativity: it simply relates the $\mathbf{E}$ at some space-time point to the force $\mathbf{F}$ it exerts on a charge at that same point. The fact that there is a delay between cause and effect complicates the *calculation* of $\mathbf{E}$ in terms of the charges that produce it, but not on what it does to the charge q, the latter being a *local* relationship in space-time.

The second term is the magnetic force. It too is unmodified by relativity, with the understanding that the force stands for the rate of change of the correct relativistic momentum $\mathbf{p} = m\mathbf{v}/\sqrt{1 - v^2/c^2}$ and not its low velocity limit $m\mathbf{v}$. You can take the Lorentz force law as the summary of years of experiment.

How will you measure $\mathbf{E}$ and $\mathbf{B}$ at a point given this formula?

We've done it before for $\mathbf{E}$. Take a coulomb and put it at rest where you want $\mathbf{E}$. Find the force on it, and that's $\mathbf{E}$. If you placed 5 coulombs, you divide the force by 5. Finding the electric field is easy because its direction coincides with the acceleration of the charge.

In the magnetic problem, there are lots of vectors involved, as indicated in Figure 8.3.

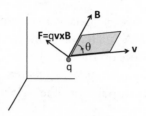

Figure 8.3   The magnetic force on a charge q, moving at velocity $\mathbf{v}$ in a field $\mathbf{B}$, is $\mathbf{F} = q\mathbf{v} \times \mathbf{B}$.

There is $\mathbf{v}$ the velocity of the charge, $\mathbf{B}$ the magnetic field, and finally magnetic force $\mathbf{F}$ given by their cross product. Suppose I ask you, "Which way is B pointing?" You cannot use a compass needle. That is cheating; I want you to use just the Lorentz force law. One option (in principle) is to shoot a few charged particles in different directions and find out how they bend. The ones with velocity exactly parallel to $\mathbf{B}$ won't bend at all since $\mathbf{v} \times \mathbf{B}$ vanishes. Once you have figured that out, you have a plane orthogonal to $\mathbf{B}$ to fire one more particle. The force on it will have a magnitude qvB since $\sin\theta = 1$ in the cross product of perpendicular vectors, and the sense in which the particle bends will tell us along which of the two possible directions normal to the plane $\mathbf{B}$ points.

The unit for the magnetic field is the *tesla*. A one-coulomb charge moving at one meter per second perpendicular to a one-tesla field will experience a force of one newton.

Whenever a force acts on a body, you know $P = \mathbf{v} \cdot \mathbf{F}$ is the power it delivers, the rate at which it does work. If you compute that for the Lorentz force you find

$$P = q(\mathbf{E} + \mathbf{v} \times \mathbf{B}) \cdot \mathbf{v} = q\mathbf{E} \cdot \mathbf{v}. \tag{8.2}$$

The magnetic force makes no contribution because $\mathbf{v} \times \mathbf{B}$ is perpendicular to $\mathbf{v}$ and hence has zero dot product with $\mathbf{v}$:

$$\mathbf{v} \cdot (\mathbf{v} \times \mathbf{B}) = 0. \tag{8.3}$$

The magnetic force is always perpendicular to the velocity of the particle. *That means it never does any work.* So you may say, "Who cares about such a thing?" Electric fields do a lot of work. They speed up particles, they slow them down. By contrast, the kinetic energy of a particle will never change due to the magnetic field. And yet you will see that it is extremely useful as an intermediary in transferring energy, as in a generator or motor.

8.2 Examples of the Lorentz force, the cyclotron

We are now going to do some simple problems to acquaint you with the magnetic force.

In the first problem, depicted in Figure 8.4, I have a beam of particles, all of which have the same mass m and charge $q > 0$, but going from left to right with different speeds.

I want to select out those that have a certain speed. I want a *velocity filter*. Here is how it is done. I take two parallel plates and charge them up so there is an electric field as in Figure 8.4. The particles will then bend downward in this constant downward electric field. Now I introduce a magnetic field **B** going into the page. Throughout this book a vector pointing away from you and into the page is shown by a symbol $\otimes$ and a vector coming toward you from the page is shown by $\odot$ or simply a dot. Now what is $\mathbf{v} \times \mathbf{B}$? It points straight up and has a magnitude qvB, *which varies with v, the particle speed.* Particles with a speed v^* satisfying

$$qE = qv^*B \quad \text{or} \tag{8.4}$$

$$v^* = \frac{E}{B} \tag{8.5}$$

will go undeflected and get out at the right end, while the others will either hit the upper plate $(v > v^*)$ or the lower one $(v < v^*)$ assuming the

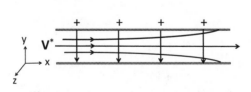

Figure 8.4   A beam of positively charged particles moving along the x-axis with various velocities enters a velocity filter, a region of a crossed electric field **E** (down the y-axis) and magnetic field **B** (into the page, in the $-z$ direction). Those moving at a speed $v^* = E/B$ pass through while faster (slower) ones hit the top (bottom) plate.

plates are long enough. If the particle velocity $v > v^*$, the magnetic force beats the electric force and bends the particle upward. If not, the opposite happens. The device works because the magnetic force cares about the particle velocity, while the electric force does not.

Another standard example of the Lorentz force is shown in the left half of Figure 8.5.

There is a uniform magnetic field going into the page. I shoot a particle with $q > 0$ in the plane of the page as indicated. What will it do? It feels a force $\mathbf{v} \times \mathbf{B}$ and bends to the left. At the new location it again feels a force perpendicular to the instantaneous velocity and bends again. It is like planetary motion. It will go in a circle. It's not speeding up, because the force is always perpendicular to velocity. You don't change the kinetic energy, but you change the direction of motion. If you want to trap charged particles, you put them in a magnetic field. They will not go anywhere, just run around in circles.

These circular orbits have a remarkable property because the magnetic force, unlike gravity, is velocity-dependent. If the orbit has a radius r, non-relativistic Newtonian mechanics (assumed to be valid at

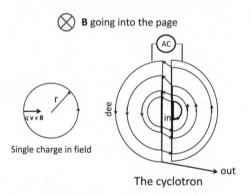

Figure 8.5 Left: A particle of mass m and positive charge q enters a magnetic field perpendicular to the page, which bends it into a counterclockwise circular orbit. A negative charge would orbit clockwise. The frequency ω depends only on q/m and not the orbit radius or velocity. Right: A cyclotron exploits this feature of ω. A particle injected near the center goes along a circular arc within that dee. When it crosses to the other dee it gets a kick due to a voltage drop between the dees. When it comes around back to the first dee, the polarity is reversed and it gets another kick, still falling downhill. At the end it is shot out of the machine.

the velocities in question) tells us to equate the requisite centripetal force to the available magnetic force:

$$\frac{mv^2}{r} = qvB. \tag{8.6}$$

(I didn't write the cross product, because $\mathbf{v}$ is perpendicular to $\mathbf{B}$.) Canceling one power of v from both sides we find

$$\frac{v}{r} = \frac{qB}{m}. \tag{8.7}$$

In a circular orbit the tangential velocity v is related to the angular velocity ω by

$$v = \omega r, \tag{8.8}$$

which then implies

$$\omega = \frac{qB}{m}. \tag{8.9}$$

The striking feature of this result is that the orbital frequency ω, called the *cyclotron frequency*, is independent of the velocity of the particle and the radius of the orbit. It just depends on q/m, the charge-to-mass ratio of the particle in a given magnetic field. This means that if you shoot many such particles into the plane perpendicular to the field at different speeds, they will form orbits of different radii, but all the orbits—big and small, fast and slow—will be traversed in the same time.

Ernest Lawrence (1901–1958), who was on the Yale faculty briefly before going off to Berkeley, made sensational use of this property in devising the *cyclotron*, a particle accelerator. But first consider a simpler accelerator. You take a battery with a voltage V, connect it to two parallel plates, and set up a field between them. You release a proton from the positive plate and it accelerates toward the negative plate, gaining kinetic energy $\frac{1}{2}mv^2 = eV$. When it reaches the negative plate, it finds you have cleverly made a hole that lets it emerge through it with a velocity $v = \sqrt{2eV/m}$. That is your particle accelerator. If you want to accelerate it to higher and higher energies, you can either get batteries with a bigger and bigger voltage or let a series of accelerators like the one described above accelerate the particle in sequence. Indeed this is essentially how the one at the Stanford Linear Accelerator Center (SLAC) works, though it uses a suitable AC voltage that repeatedly keeps kicking the particle over a two-mile stretch.

What Lawrence invented had a different design that uses electric and magnetic fields as follows. Take a closed metallic cylinder with a broad base and a very small height and slice it into two equal halves along a diameter. The halves are called "dees" for obvious reasons. Leave a small gap between the dees as shown in the right half of Figure 8.5. Apply a magnetic field B perpendicular to the plane of the dees. Connect a battery of voltage V to the two dees, thereby placing them at different potentials. A field $\mathbf{E}$ and potential difference V will be created in the space between the dees. Near the center, inject into the positive dee a positively charged particle at some tangential velocity $\mathbf{v}$. It will bend in the magnetic field into a circular orbit, emerge after a half-circuit, and enter the other dee. During the jump, it will gain a kinetic energy qV because of the potential difference between the dees. After another half circle, as it reenters the original dee, it will *lose* the kinetic energy it gained because now the field is opposed to it. What was a downhill journey in the previous jump is now uphill. This is not a good accelerator. Suppose that we very cleverly swap the terminals of the battery just before the second jump so that the particle gets another boost to its kinetic energy of qV. It now travels on a bigger circle due to its increased speed. When it arrives at the next jump we repeat the swap. The particle will keep picking up speed on its spiral path, always going down in potential, like something out of Escher's drawings. Eventually the orbital radius exceeds the dees in size, and the accelerated particle is ejected for the intended collision.

The flaw with this design is that we need to do the terminal swapping very fast. But there is compensating good news hidden in what I said earlier: we need to swap the leads at the same frequency *because ω remains unaffected by the change in speed and radius.* You can probably guess that Lawrence did not swap the polarity of the dees manually or ask his graduate student to do it: he just applied an AC voltage of the desired ω.

Lawrence's first cyclotron was about 5 inches in diameter and could give a proton an energy of $80,000\ eV$. (This is like connecting a battery with $80,000$ volts to our parallel plate accelerator.) Later he used bigger magnets to reach $16,000,000\ eV$'s. His idea was this: you do not need a million-volt battery to impart a million eV of energy to a particle; you just give it a million small kicks of $1\ eV$ each. Eventually a different design was required because the non-relativistic kinematics that went into the preceding derivation no longer applied. The next generation of the accelerators, called *betatrons*, were designed to operate at relativistic energies and will be discussed later.

8.3 Lorentz force on current-carrying wires

The Lorentz formula describes the magnetic force felt by a single charge, like an electron. This is useful in certain contexts, like in designing a cathode ray tube or an accelerator. But often the moving charges are part of a current-carrying wire. Let us derive the force law that applies to macroscopic currents starting from Lorentz's microscopic expression.

Consider a wire of cross section A carrying current I assumed to be uniform along its length. The wire may, however, twist and turn so that the *direction* of the current is variable. We want to find the force on a little segment, which I write as a vector $d\mathbf{l}$. The wire is bathed in a magnetic field $\mathbf{B}(\mathbf{r})$, which may be assumed constant over this tiny segment. There's going to be a force on this segment because there are little charged guys moving in the wire. Each one feels a force $e\mathbf{v} \times \mathbf{B}$. We must add them all up. If this segment has a length dl, how many charges are we talking about? It is the density of carriers, n, times the volume of the segment, which is Adl. Thus the force on the segment is

$$d\mathbf{F} = e \cdot n \cdot A \cdot dl \cdot \mathbf{v} \times \mathbf{B}. \qquad (8.10)$$

Now I'm going to do a little switch here. The force contains the product of dl, the magnitude of the vector $d\mathbf{l}$ and the velocity vector $\mathbf{v}$. Since both $\mathbf{v}$ and $d\mathbf{l}$ point along the wire, we can attach the vector symbol to $d\mathbf{l}$ and

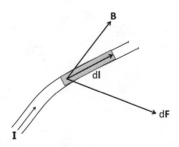

Figure 8.6   The magnetic force $d\mathbf{F}$ on a segment $d\mathbf{l}$ of a wire carrying current I is $Id\mathbf{l} \times \mathbf{B}$. This is simply the sum of the forces on the individual carries inside $d\mathbf{l}$.

replace **v** by its magnitude v:

$$dl\mathbf{v} = v d\mathbf{l}, \tag{8.11}$$

both of which describe a vector of magnitude vdl pointing along the current. It follows that

$$d\mathbf{F} = enAv\, d\mathbf{l} \times \mathbf{B}. \tag{8.12}$$

But $enAv$ is the current I, and so the force on $d\mathbf{l}$ is

$$d\mathbf{F} = I d\mathbf{l} \times \mathbf{B}, \tag{8.13}$$

a result worth remembering.

Here is an illustrative example. There is a uniform magnetic field **B** coming out of the page in the $+z$ direction, and a semicircular wire lying in the plane of the page (the xy-plane), carrying a counterclockwise current I, as shown in the left half of Figure 8.7. Let us find the force on its diameter and semicircular part.

The force on the diameter is easy to figure out since the entire segment points in one direction. It has a magnitude

$$F = B(2R)I \tag{8.14}$$

and points *down* the y-axis.

As for the semicircle, the figure shows a segment $d\mathbf{l}$ at an angle θ from the x-axis. Since **B** and $d\mathbf{l}$ are perpendicular, the force, perpendicular

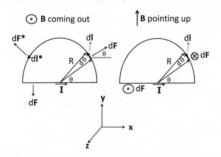

Figure 8.7　Left: Loop in the xy-plane, and the field coming out of the page. Right: Loop and field in the same xy-plane. In both cases the total force on the loop vanishes.

to both, lies in the plane of the page and points radially outward as shown, with a magnitude $dF = IBdl$. The figure also shows another segment dl^* at angle $\pi - \theta$, which feels a force $d\mathbf{F}^*$ with the same y component and opposite x-component. Since only the y-component will survive, we compute only that

$$dF_y = +IBdl\sin\theta. \tag{8.15}$$

Be aware that the $\sin\theta$ above is not the usual factor that gives the dependence of the cross product on the angle between the two vectors. That angle is in fact $\frac{\pi}{2}$ here, because $\mathbf{B}$ (coming out of the page) and $d\mathbf{l}$ (in the plane of the page) are perpendicular. The angle θ here comes from projecting out the y-component of $d\mathbf{F}$. Since $dl = Rd\theta$, the total force on the semicircle, pointing *up* the y-axis is

$$F_y = IBR \int_0^\pi \sin\theta\, d\theta = 2IBR, \tag{8.16}$$

which exactly cancels the downward force on the diameter. So the force on the closed loop is zero.

Suppose $\mathbf{B}$ is now parallel to the page and pointing up the y-axis, as shown in the right half of the figure. The force on the horizontal segment will be of the same magnitude $2IBR$ since the current is still perpendicular to $\mathbf{B}$, but will point out of the page. On the semicircle, $d\mathbf{l}$, the segment located at angle θ from the x-axis and $\mathbf{B}$ are no longer perpendicular but at angle θ. The force of magnitude $dF = IBdl\sin\theta$ will be pushing the segment into the page. This time the $\sin\theta$ *is* the factor that enters the cross product. The integral of dF is once again $2IBR$ and it cancels the force on the diameter.

It can be shown that the force on *any* closed loop in a uniform $\mathbf{B}$ is zero:

$$\mathbf{F} = \oint d\mathbf{F} = I \oint d\mathbf{l} \times \mathbf{B} \tag{8.17}$$

$$= I \left[\oint d\mathbf{l} \right] \times \mathbf{B} \tag{8.18}$$

$$= 0 \tag{8.19}$$

where I have pulled the constant $\mathbf{B}$ out of $\oint$ and used the fact that the vector sum of all the little $d\mathbf{l}$'s forming a *closed* loop vanishes.

8.4   The magnetic dipole

Next consider a rectangular loop of area $A = w \cdot l$ immersed in a uniform field **B** pointing along the z-axis as shown in Figure 8.8. The loop carries a current I in the sense shown $1 \to 2 \to 3 \to 4 \to 1$. Let us choose the arrows running around the area to specify its orientation to be in the same sense as the current. If you curl the fingers of your right hand around the current (don't do this at home) your thumb will point along the area vector **A**.

The field will exert no net force on the loop by symmetry. For every segment $d\mathbf{l}$ pointing one way, there is one in the opposite side of the loop pointing exactly the opposite way, experiencing the opposite force, because **B** is constant.

The field will, however, exert a torque that causes rotation around the axis OO'. The torque is due to the segments 12 and 34, which experience a perpendicular force of magnitude BwI pointing away from the loop. (The other two sides, 23 and 41, which also experience a force pointing out of the loop, do not contribute to any torque.) The "lever arm" for the torque is $l\sin\theta$ where θ is the angle between **A** and **B**. This may be clearer in the side view in the lower left of the figure, looking at the loop along the rotation

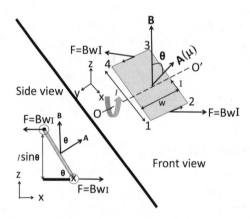

Figure 8.8   A current loop of area $A = l \cdot w$ in a magnetic field **B**. The net force on it is zero and the net torque is $\boldsymbol{\tau} = \boldsymbol{\mu} \times \mathbf{B}$, where the magnetic moment $\boldsymbol{\mu} = I\mathbf{A}$. The torque is due to the forces on segments 1–2 and 3–4, which try to rotate it around the axis OO' in the sense indicated. Forces on 2–3 and 4–1 try to stretch out the loop but not turn it. The inset in the corner shows a side view along the axis OO' to clarify the torque calculus.

axis OO' straight at the edge 41. So the torque on the loop is

$$\tau = BwIl\sin\theta = IAB\sin\theta. \qquad (8.20)$$

The $\sin\theta$ tells us clearly that we are dealing with the cross product of two vectors, **B**, and the *magnetic moment* $\boldsymbol{\mu}$, which is parallel to **A**:

$$\boldsymbol{\mu} = I\mathbf{A}. \qquad (8.21)$$

That is,

$$\boldsymbol{\tau} = \boldsymbol{\mu} \times \mathbf{B}. \qquad (8.22)$$

The torque $\boldsymbol{\tau}$ wants to align the moment $\boldsymbol{\mu}$ with the field **B**. Thus a little loop can be used as a compass: the normal to it will point along **B**.

We call $\boldsymbol{\mu}$ the magnetic moment of the loop because it is analogous to the electric dipole moment. You might remember that an electric dipole **p** in a field **E** experiences a torque

$$\boldsymbol{\tau} = \mathbf{p} \times \mathbf{E}, \qquad (8.23)$$

which also tries to align **p** with **E**. So a current loop looks like a magnetic dipole in a magnetic field. In other words, the loop behaves like a pair of opposite magnetic charges, separated in the direction of **A**. If magnetic charges or monopoles existed, this would be simply a magnetic dipole, aligning itself with the magnetic field. So far we have not found reliable and reproducible evidence of magnetic charges, or *monopoles*, that would produce a radially outgoing or incoming magnetic field. We have only loops that behave like dipoles. Just as the electric moment p is the product of the charge and the separation between them, the magnetic μ is the product of the current and the area.

Besides responding to a magnetic field the way an electric dipole responds to an electric field, the loop also produces a magnetic field that looks like the electric field of the electric dipole at long distances. We will eventually compute this field, but only for a simple case.

Given the torque, we can integrate it to obtain a potential energy

$$U = -\boldsymbol{\mu} \cdot \mathbf{B} = -\mu B\cos\theta. \qquad (8.24)$$

The potential is minimized (maximized) when the moment is parallel (anti-parallel) to the field. Unlike in the case of the electrical dipole, this energy only keeps track of the mechanical work done to turn the loop, but not the electrical work done to keep the current in the loop constant as it turns.

8.5   The DC motor

Now it turns out I can make some money out of this torque. I can build a device. The device I'm going to build is an electric motor. I take two bar magnets and place them north-to-south as shown in the top left of Figure 8.9. In this region of a constant **B**, I place a current loop fed by a battery. The loop is free to rotate about an axis parallel to the leads. If the current is as shown, the loop will rotate till μ aligns itself with **B**. Assuming its motion is damped by some little friction, it will stop in that position. That will be the brief life of my motor. It's going to turn till the moment lines up with the field and that's the end. And if it's already lined up, it won't even do that. This is not going to sell. So what do I have to do?

A very good suggestion that came up in class was to use an AC supply. But what if I just have a DC source? Switching the poles every half cycle got a good laugh but no one thought it was a good idea. The actual solution is very clever. Let us get there in stages.

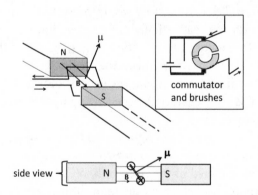

Figure 8.9   Top left: A view of a current loop in the field of two permanent magnets. The torque on it will rotate it till μ and **B** are parallel. If the current is then reversed, it will swing by another 180 degrees and so on. This reversal is done using the commutator shown in the inset (top right). The dotted line is a spring that holds the brushes in place. Bottom: Side view of the motor.

First, instead of switching the poles, I can switch the leads at the battery that feeds the rotating loop. Every time the loop thinks it has found happiness, that it has reached an energy minimum, I say no. I turn the energy minimum to a maximum just when it gets there by reversing the current. So it goes another half turn and I do it again. The motor will now work, but I can never leave this motor and go anywhere, because I have to stick around to switch the leads. It is high maintenance.

Now for the real answer. If you don't know the answer, you will be very impressed, as I was. That is the gap between pure science and applied science. It makes all the difference. The practical guys invented the *commutator*, shown in the insert in Fig. 8.9. The two leads from the spinning loop are connected to two semicircular metallic half-rings, which rotate with the loop and have a tiny space between them. The battery is not hard-wired to the loop. Instead it is connected to the half-rings with two spring-loaded metallic brushes; this allows the half-rings to spin without breaking the contact. Now you can see what happens from the figure. Initially the positive terminal is connected to the bottom half-ring, the negative terminal to the top half-ring, so that the current goes in and comes out as shown. But half a revolution later, the half-rings switch places, and the polarity and current are reversed.

CHAPTER 9

Magnetism II: Biot-Savart Law

We have finished the first part of magnetism, which concerns the magnetic forces and torques *on* moving charges and current-carrying wires. We now turn to the second part, which deals with how the magnetic fields are produced *by* moving charges and currents.

At the microscopic level the magnetic field is produced by moving charges, but the formula for that is quite difficult to calculate, because the charges are moving around producing the magnetic field, and the field at any one location depends on what they were doing at various times in the past. This is just like electric fields, which are also difficult to calculate if charges are moving because relativity forbids instantaneous action-at-a-distance. In electrostatics, we beat the problem by saying, "Look, none of these charges ever moved. They've been there forever." Consequently where they are now is where they were at any time in the past and we could calculate the field. In the magnetic case we cannot stop the charges, for we will then stop the current that produces the field. *Instead we say that the currents are steady, time-independent.* This is a clever way out for now: the charges are moving in the wire but the current, which causes the field, is constant. Electron Joe who is here in the wire now may be replaced by electron Shmoe a little later, but that makes no difference to the current. It is steady. So the magnetic field that is produced will also be time-independent. That is what we mean by magnetostatics.

Do not confuse a steady current with a single particle moving at a steady velocity. That is not a steady current. You see the difference? With a steady current, if you sit at any one point in a wire, the current going past you is always the same. If you have an ammeter that measures the current, its reading will be steady. By contrast, a single charge at constant velocity causes a current only at its location. When it moves, the current goes with it. There is a current only where there is charge. It is like saying that when I go on the freeway at 40 miles an hour, I do not myself constitute steady traffic, because there is no traffic where I am not. Once I pay the toll collector and pass the tollbooth, it is all over for the revenue. On the other hand, with a steady traffic the money will be pouring in steadily.

So the question is, what is the magnetic field produced by a tiny element, a tiny piece of current-carrying wire? The magnitude of current in the wire is I no matter where we slice it, but the wire can twist and turn, and the element in question is represented by a vector $d\mathbf{l}$. It's part of a bigger loop, which feeds it the current and takes it out, but its contribution depends on its orientation encoded in $d\mathbf{l}$. Let us say it is located at $\mathbf{r}'$. I want the field at the point $\mathbf{r}$. Every segment of wire will produce a little magnetic field, $d\mathbf{B}$. The expression for the field is called the *Biot-Savart* law:

$$d\mathbf{B} = \frac{\mu_0 I}{4\pi} \frac{d\mathbf{l} \times (\mathbf{r} - \mathbf{r}')}{|\mathbf{r} - \mathbf{r}'|^3} \equiv \frac{\mu_0 I}{4\pi} \frac{d\mathbf{l} \times \mathbf{e}_{\mathbf{r}-\mathbf{r}'}}{|\mathbf{r} - \mathbf{r}'|^2} \quad \text{where} \quad (9.1)$$

$$\frac{\mu_0}{4\pi} = 10^{-7} \frac{N \cdot s^2}{C^2} \quad \text{in our units.} \quad (9.2)$$

That constant $\frac{\mu_0}{4\pi}$ (like its electric counterpart $\frac{1}{4\pi\varepsilon_0}$) is cooked up so that if I measure the current in amperes and the distances $\mathbf{r}$ and $\mathbf{r}'$ in meters, the field comes out in tesla.

That is one nasty formula, unlike Coulomb's law. Whereas the cause of the electric field is a point charge, the cause of the magnetic field is a vector $d\mathbf{l}$. That in turn is so because the cause of the magnetic field is charge in motion, which introduces its own velocity vector.

Because there is no vector associated with a point charge, the field had to point along the line joining the charge to the point where we want the field, along the separation vector. There is no way any other vector can get into the act. The current element, on the other hand, has got its own direction, in addition to where it is. It describes the way the wire is going at that point. It is the presence of this extra vector $d\mathbf{l}$ that allows the formation of yet another vector by combining it with the separation vector $\mathbf{r} - \mathbf{r}'$ into a cross product. That is how you get these cross products.

9.1 Practice with Biot-Savart: field of a loop

As the first illustration of the law we will be finding the field of a circular loop carrying current I of radius R, centered at the origin, and lying in the xy-plane, as shown in Figure 9.1. We will only consider the field at a point with coordinates $(0, 0, z)$ or position vector $\mathbf{r} = \mathbf{k}z$.

As with a circle of charge, we divide the loop into segments, find the field due to each, and add up the result. First consider the indicated segment $d\mathbf{l}$ located at a point $\mathbf{r}'$ on the y-axis. It is half in and half out of the page (just like the loop itself), with the current going in. We take the cross product of $d\mathbf{l}$ and $(\mathbf{r} - \mathbf{r}')$ and divide by some scalars. The cross product has to (i) lie in the plane of the page since $d\mathbf{l}$ is normal to it and (ii) be normal to $\mathbf{r} - \mathbf{r}'$, the other vector in the cross product. So $d\mathbf{B}$ has to point in the yz-plane. The magnitude of this vector is

$$|d\mathbf{B}| = \frac{\mu_0 I}{4\pi} \frac{dl}{R^2 + z^2}. \qquad (9.3)$$

There is no "$\sin\theta$" factor in the cross product, the vectors in question being perpendicular.

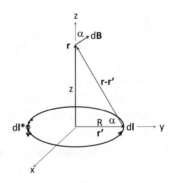

Figure 9.1   Field due to current loop (half in and half out of the page) at a point on its symmetry axis. The vector $d\mathbf{B}$ is the contribution from a segment $d\mathbf{l}$ that is going into the page and is perpendicular to the separation vector $\mathbf{r} - \mathbf{r}'$. We keep only the z-component since the part parallel to the xy-plane will be canceled by the diametrically opposite element $d\mathbf{l}^*$.

Only the component of $d\mathbf{B}$ pointing up the z-axis,

$$dB_z = |d\mathbf{B}|\cos\alpha = |d\mathbf{B}|\frac{R}{\sqrt{R^2 + z^2}}, \tag{9.4}$$

is going to survive, because the diametrically opposite segment $d\mathbf{l}^*$, coming out of the page, will make a contribution with the same z-component and opposite y-component. The final result for the total field is

$$\mathbf{B}(0,0,z) = \mathbf{k}\frac{\mu_0 I}{4\pi}\frac{1}{R^2 + z^2}\frac{R}{\sqrt{R^2 + z^2}}\int dl \tag{9.5}$$

$$= \mathbf{k}\frac{\mu_0 I}{4\pi}\frac{2\pi R^2}{(R^2 + z^2)^{3/2}} \tag{9.6}$$

using $\int dl = 2\pi R$.

At the center of the loop, the origin,

$$\mathbf{B}(0,0,0) = \mathbf{k}\frac{\mu_0 I}{2R}, \tag{9.7}$$

while as $z \to \infty$,

$$\mathbf{B}(0,0,z \to \infty) = \mathbf{k}\frac{\mu_0(\pi R^2 I)}{2\pi z^3} \tag{9.8}$$

$$= \boldsymbol{\mu}\frac{\mu_0}{2\pi z^3} \quad \text{where the magnetic moment} \tag{9.9}$$

$$\boldsymbol{\mu} = I\mathbf{A} = \mathbf{k}\,\pi R^2 I. \tag{9.10}$$

This field is exactly what we found for the electric field of the electric dipole at long distances, apart from the inevitable substitution $\varepsilon_0 \leftrightarrow 1/\mu_0$.

Finding the magnetic field at a point off the symmetry axis is very complicated because we no longer have all the symmetries. The result, which I state without proof, is that far from the dipole the field is exactly like the electric field of the electric dipole, depicted in Figure 9.2.

As claimed earlier, not only do the magnetic dipole and electric dipole experience similar torques in the corresponding fields, the fields they produce are also identical *at long distances*. Things are very different up close. If you go close to an electric dipole, you find two opposite charges. If you go close to a magnetic dipole, you'll find no magnetic charges at the center, just a current loop. So nature gives us magnetic dipoles, but not magnetic monopoles.

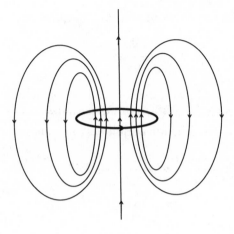

Figure 9.2 Field due to a current loop at generic off-axis points (schematic). It resembles the electric field of an electric dipole until you get close to the origin: rather than running into a pair of oppositely charged monopoles, you run into a current loop.

9.2 Microscopic description of a bar magnet

Now suppose I stack a whole lot of loops coaxially, say by wrapping a wire around a cardboard cylinder many times into a spiral. Given that a single loop produces the dipole field, it is plausible that what we get looks like the field of an electromagnet depicted in Figure 8.2. This field also looks exactly like that of a bar magnet. As far as a compass needle is concerned, it behaves the same way around both.

Consider the following. We can have a magnetic field produced by current-carrying loops. We can also have a permanent magnet with no currents in sight. This magnet is not connected to anything. So we have an option. Either we can say that's a new kind of magnetism produced by God knows what, or we can say, "We believe that everything is coming from electric currents." If we take the second point of view the question is, where are the currents in a bar magnet? *They come from the electrons in the atoms.* Every atom has electrons going around the nucleus, and every moving electron is a current. (This picture of orbits will be modified by quantum mechanics, but not the final result.) Imagine nine electrons in the plane of the paper, all going around their atoms as shown in the left half of Figure 9.3. In the region between the atoms, they go in opposite directions. They cancel. The only thing that doesn't cancel is the current

Edge currents on a
single atomic layer

Surface currents on a
bar magnet
with many layers

Figure 9.3   Left: The uncanceled edge currents of a layer of nine atoms. Right: The surface currents of a permanent magnet made of many such layers.

along the perimeter or edge. Thus a single layer of atoms can produce a current at the edge. It will be permanent since atomic currents are.

Now that is just a two-dimensional current loop due to a single atomic layer, but you can think of a magnetic solid as made of layers of such atoms. At the edge of each layer is its current. So a magnetic material can effectively have a sheet of current on its surface, which will produce a magnetic field. In the case of the cylindrical magnet shown in Figure 9.3 the electronic orbits lie in a plane perpendicular to the length of the cylinder and the edge currents flow along the curved side, producing a field along the axis, as per the right-hand rule.

Even this crude description of magnetism leaves us with some questions. Why isn't everything magnetic? Why not a potato? It has atoms that have electrons, right? Why aren't their orbits lined up to produce a magnetic field? And if they do line up, which direction should they choose for the north-south axis?

First of all, some materials may not become magnets because the electrons they contain move in orbits whose *net* contribution to the magnetic moment of the atom is zero. For example, there could be two electrons orbiting in opposite directions. If the atom as a whole does not have a magnetic moment there is no question of macroscopic magnetism.

Even if every atom has some uncanceled magnetic moment, the moments from *different* atoms may point in random directions, adding up to nothing. The random orientation is a reflection of thermal agitation. Things like to jiggle when you heat them. If you take a bar magnet on your fridge and put it on a hotplate for a while, you will find it becomes less magnetic. And if you heat it above the *Curie temperature* the jiggling will be so intense that magnetism will be destroyed. But if you now cool it below the *Curie temperature*, magnetism will be restored.

A deep question arises at this point. It is clear how a magnet, with its north and south poles pointing aligned in some direction, becomes non-magnetic upon heating. But if as it is cooled below the Curie temperature it magnetizes, *which way will the magnetization point?* (Here we assume the crystal the atoms form does not provide a direction.) How are the little moments to agree on a common direction to point along in the magnetically ordered state? The answer is that they need some help in the form of an *external* field. The presence of such a magnetic field nudges the moments to align with it, because they are dipoles in a field. What happens if you turn off the magnetic field? In some cases the chaos sets in right away and the magnetization disappears. In ferromagnetic materials below the Curie temperature, the dipoles remain aligned in that direction *even after you turn off the external field.* Why? Because when aligned by the external field, the dipoles produce their own magnetic field that is strong enough to keep them aligned even after the external field is removed. It pulls itself up by its own bootstraps. (It is like helping a kid ride a bike by giving an initial push to impart a minimum sustainable velocity and then letting the kid take over.) Thus magnetism is a cooperative effect. It can exist only if the thermal agitation is not too strong to kill the ordering tendency generated by the moments themselves.

9.3 Magnetic field of an infinite wire

Now for a classic problem, the magnetic field of an infinite straight wire carrying current I. Figure 9.4 shows such a wire along the x-axis. We will find the field at the point $\mathbf{r} = (0, a)$ in the xy-plane. As usual we take some segment of length dx located at $\mathbf{r}' = (x, 0)$ and find its contribution. The segment $d\mathbf{l} = \mathbf{i}\,dx$ is along x and the separation vector $\mathbf{r} - \mathbf{r}'$ is as shown and also lies in the xy-plane. Their cross product points out of the page and has magnitude

$$dB = \frac{\mu_0 I}{4\pi \left(x^2 + a^2\right)} dx \sin\theta \tag{9.11}$$

$$= \frac{\mu_0 I}{4\pi \left(x^2 + a^2\right)} dx \frac{a}{\sqrt{a^2 + x^2}} \tag{9.12}$$

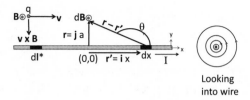

Figure 9.4   Left: Contribution from a segment $d\mathbf{l} = \mathbf{i}dx$, which is at an angle θ relative to the separation vector $\mathbf{r} - \mathbf{r}'$. The resulting $d\mathbf{B}$ comes out of the page. (Vectors coming out of the page are shown by a circle with a dot at the center.) The segment $d\mathbf{l}^*$ at $-x$ makes the same contribution. The field everywhere is found by translations along the wire and rotations around the x-axis. A charge q moving parallel to the wire feels an attractive force. Right: View looking into the x-axis, with the current coming out of the page.

$$B = \frac{\mu_0 I}{4\pi} \int_{-\infty}^{\infty} \frac{adx}{(a^2 + x^2)^{3/2}} \tag{9.13}$$

$$= \frac{\mu_0 I}{2\pi a}. \tag{9.14}$$

(The segment $d\mathbf{l}^*$ at $-x$ makes the same contribution as $d\mathbf{l}$.)

You may fill in the missing step by showing that

$$\int_{-\infty}^{\infty} \frac{adx}{(a^2 + x^2)^{3/2}} = \frac{2}{a} \tag{9.15}$$

upon making the familiar substitution $x = a\tan\theta$. As before, by writing $x = aw$ and changing variables you can show that the x-integral is a constant (a dimensionless integral) times $1/a$. Even more simply, the integral has dimensions of inverse length and the only length in town in a.

Symmetry now allows us to get the field everywhere. First, the field will be the same as we move parallel to the infinite wire. Next, the wire lives in three dimensions and what we see in the figure is cross section in the xy-plane. We can obtain the full configuration by rotating what we see around the x-axis. A view looking into the wire with the current coming out of the page is shown to the right. In terms of $\mathbf{e}_\phi$, a unit vector in the azimuthal direction, we may write

$$\mathbf{B} = \frac{\mu_0 I}{2\pi a} \mathbf{e}_\phi. \tag{9.16}$$

The right-hand rule is at work here: if your fingers curl along with **B**, the thumb points along the current.

Whereas the infinite charged wire seen end-on has electric field lines coming radially out, the infinite wire has magnetic field lines that encircle the wire, closing in on themselves. Both fields fall as $1/a$ even though individual segments make contributions that fall as inverse distance squared.

This formula is going to explain a few phenomena that were mentioned at the outset. Look at Figure 9.4, which shows a charge $q > 0$ moving with velocity **v** parallel to the wire and perpendicular to **B**. It feels a force of magnitude

$$F = qvB = \frac{\mu_0 I q v}{2\pi a} \tag{9.17}$$

toward the wire. Reversing the velocity **v** or the current changes attraction to repulsion.

It follows immediately that if we replace the single charge q by a wire carrying current I' in the same direction, it too will be attracted. We can make life easy by considering both wires to be infinite. However, the force between them will also be infinite. So we define the force per unit length on the second wire. Recalling the force due to **B** on a segment $d\mathbf{l}$ carrying current I',

$$d\mathbf{F} = I' d\mathbf{l} \times \mathbf{B}, \tag{9.18}$$

we see that the attractive force per unit length ($|d\mathbf{l}| = 1$) on the second wire due to the first is of magnitude

$$\text{Force per unit length} = \frac{\mu_0 I I'}{2\pi a}. \tag{9.19}$$

Changing the direction of either current changes attraction to repulsion. We also see that the answer is symmetric between I and I' and hence obeys Newton's third law. It is this formula that was used, long before we knew about atoms and electrons, to define the ampere in macroscopic terms: two parallel wires carrying one ampere each and one meter apart will exert on each other a force per unit length $= 2 \cdot 10^{-7}$ N/m (upon setting $\mu_0 = 4\pi \cdot 10^{-7}$ $N \cdot s^2/C^2$ in Eqn. 9.19).

Earlier we asked "How are we going to construct the vector that gives the force between current-carrying wires that's attractive when they

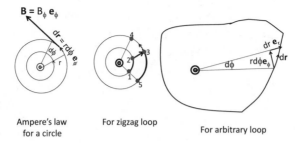

Figure 9.5   Left: Ampère's law for a circle. Middle: Two segments subtending the same angle. Right: Arbitrary Ampèrean loop.

are parallel, and repulsive when they are anti-parallel?" Nothing simple involving the current vectors would work. The dot product was a scalar and the cross product vanished. We see that the correct answer is a rather complicated sequence of *two* cross products. The first cross product is from the Biot-Savart law and yields the **B** due to the first wire as a cross product of every segment $d\mathbf{l}$, and the separation vector. The second is the cross product of this **B** with the current segment $d\mathbf{l}'$ of the second wire. (The same scenario describes the force between a wire and a moving charge if we replace $d\mathbf{l}$ by **v**.) Whereas force of attraction between two charges is simply $q_1 q_2 / r^2$, the force of attraction between two wires, even though it is given by a simple formula, hides an orgy of cross products.

9.4   Ampère's law

Ampère's law is to magnetostatics what Gauss's law was to electrostatics. Recall what we did there. We took the field of a point charge q and computed its surface integral on a sphere centered on it. We found the answer was q/ε_0 independent of the radius of the sphere because the area of the sphere went as r^2 while the field decreased as $1/r^2$. We then went on to show that the surface integral was the same on any closed surface enclosing the charge. Finally, we used superposition to show that the surface integral of **E** on any closed surface was the total charge enclosed divided by ε_0.

Now for Ampère. Consider the field **B** due to an infinite wire carrying current I. Let us see the wire end-on, as shown in the leftmost part of Figure 9.5, with the current coming out at us and the field lines circulating counterclockwise.

At the left third of the figure we have a circular path of radius r encircling the current. Consider the line integral of $\mathbf{B}$ around this loop, called the *circulation*. Both the line segment and field are in the azimuthal direction:

$$\mathbf{B} = \mathbf{e}_\phi \frac{\mu_0 I}{2\pi r} \tag{9.20}$$

$$d\mathbf{r} = \mathbf{e}_\phi r d\phi, \tag{9.21}$$

which makes the line integral very simple:

$$\oint \mathbf{B} \cdot d\mathbf{r} = \int_0^{2\pi} \mathbf{e}_\phi \frac{\mu_0 I}{2\pi r} \cdot \mathbf{e}_\phi r d\phi \tag{9.22}$$

$$= \mu_0 I \frac{1}{2\pi} \int_0^{2\pi} d\phi \tag{9.23}$$

$$= \mu_0 I. \tag{9.24}$$

The line integral or circulation is independent of the radius of the circle. This is analogous to the statement that the surface integral of the electric field is the same for any sphere centered on the charge, independent of its radius.

Next consider loops made of radial and angular segments. Parts of two such loops subtending the same angle at the origin are shown in the middle of the figure. Consider the segment that goes along a circle from $5 \rightarrow 3 \rightarrow 4$. The contribution of this segment is

$$\frac{\mu_0 I}{2\pi} \cdot (\text{angle subtended by the arc } 5 \rightarrow 3 \rightarrow 4).$$

Consider another segment that subtends the same angle but along the path $1 \rightarrow 2 \rightarrow 3 \rightarrow 4$. The angular part $1 \rightarrow 2$ gives

$$\frac{\mu_0 I}{2\pi} \cdot (\text{angle subtended by the arc } 1 \rightarrow 2),$$

the radial part $2 \rightarrow 3$ gives nothing since $\mathbf{B}$ is azimuthal, and finally the angular part $3 \rightarrow 4$ gives

$$\frac{\mu_0 I}{2\pi} \cdot (\text{angle subtended by the arc } 3 \rightarrow 4).$$

The final result is clearly the same for both paths since it depends only on the total angle swept in the journey. Consequently the circulation will be the same on any path that encloses the current and is composed of any number of radial and angular parts.

The rightmost part of the figure considers an *arbitrary* loop and a segment $d\mathbf{r}$ that is neither radial nor angular but a little bit of both:

$$d\mathbf{r} = \mathbf{e}_r dr + \mathbf{e}_\phi r d\phi. \tag{9.25}$$

In the line integral only the angular part survives the dot product with the purely azimuthal magnetic field:

$$\oint \mathbf{B} \cdot d\mathbf{r} = \oint \mathbf{e}_\phi \frac{\mu_0 I}{2\pi r} \cdot \left[\mathbf{e}_\phi r d\phi + dr \mathbf{e}_r \right] \tag{9.26}$$

$$= \mu_0 I \frac{1}{2\pi} \int_0^{2\pi} d\phi \tag{9.27}$$

$$= \mu_0 I. \tag{9.28}$$

Thus the line integral of **B** around any closed path equals μ_0 times the current enclosed.

This is analogous to the statement that the surface integral of the electric field on any surface surrounding the charge equals the enclosed charge divided by ε_0. This is the most general Gauss's law for a *single charge*. From this we can get Gauss's law for any collection of point charges by using superposition for the fields they produce. Let us similarly extend Ampère's law, from a single current to many.

Suppose there are many currents $I_1, \ldots I_N$ enclosed by the contour C lying in the plane of the page, as shown in Figure 9.6. We may superpose the corresponding magnetic fields to obtain a result relating the circulation or the line integral of the total **B** around a closed contour C to the total current enclosed, the celebrated *Ampère's law*:

$$\oint_C \mathbf{B} \cdot d\mathbf{r} = \mu_0 \sum_{j=1}^{N} I_j. \tag{9.29}$$

Remember the convention. If your fingers are curving along the counterclockwise contour, your thumb will stick out of the page, and a current coming out of the page ($\odot$) is counted as positive. A current going into the page ($\oplus$) will be counted as negative. The current in the

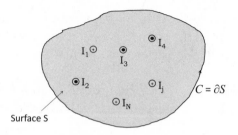

Figure 9.6   Ampère's law for a contour C enclosing many currents, some coming out $\odot$ and some going in $\oplus$. The currents may be written as the surface integrals of the current density $\mathbf{j}$ over the surface $S = \partial C$ bounded by C.

right-hand side of Eqn. 9.29 has to be counted with this sign. For example, if two one-amp currents came out and two went in, the line integral of $\mathbf{B}$ will vanish. A current not enclosed by the contour will not contribute to the line integral. If we repeat the preceding derivation with a contour not encircling the current, the answer will still be proportional to the angle swept out by the contour, but this angle will be zero: as we traverse the loop, the angle will first go up and then go down back to its initial value as we complete the circuit. Draw a figure to convince yourself if needed.

All this is just like Gauss's law, which equates the surface integral of $\mathbf{E}$ over a closed surface S to the total charge enclosed, paying attention to the sign of each charge. A charge not enclosed by S will not contribute: any flux from it that enters the closed surface will also leave it, having nowhere to terminate in the interior.

But there are some differences.

First of all, Eqn. 9.29 describes a situation in which the currents I_j were assumed to be carried by *infinitely long and straight* wires for which the formula giving $\mathbf{B}$ was simple. We want to lift that restriction. Let us first rewrite the currents enclosed by C as the surface integral of the current density $\mathbf{j}$ over the surface S (shaded in the figure). Ampère's law takes the form

$$\oint_C \mathbf{B} \cdot d\mathbf{r} = \mu_0 \sum_{n=1}^{N} I_n = \mu_0 \int_S \mathbf{j} \cdot d\mathbf{S}. \tag{9.30}$$

(The current density will be non-zero only where the wires cross S. The integral of $\mathbf{j}$ over the cross section of wire n will be I_n.)

It seems reasonable that this relation between the line integral of **B** around C and the surface integral of current densities over a surface S bounded by C should depend only on the current densities *on that surface* and not on whether they were carried by infinite wires (as in our derivation) or some other set of wires that crossed S with the same currents but otherwise unrestricted away from S. This reasonable guess is actually correct and can be proved from the Biot-Savart law using somewhat more advanced methods. So by Ampère's law we shall mean Eqn. 9.30 with no restriction on the currents away from S.

Now for the second difference. I proved Gauss's law in three dimensions. The closed surface S lived in $3d$ and enclosed the charge that was the integral of the charge density ρ over the volume enclosed by the S. *This volume was uniquely defined.* On the other hand, the preceding derivation of Ampère's law was done in two dimensions: the contour C lay in a plane (of the page) perpendicular to the current. The contour enclosed a unique (planar) surface S and the current in the right-hand side of Ampère's law penetrated that surface. But wires live in three dimensions. What happens to Ampère's law given that *a contour C in $3d$ can encircle a wire without lying in a plane and given that one can draw an infinite number of surfaces S for which the same contour is the boundary C?* Make sure you follow this. Imagine a closed metallic rim that you dip in some soap solution. The soap film will form some surface with the rim as the boundary. If the loop is not planar, neither will the surface be. You can apply Ampère's law to this case, with the soap film as the surface and the rim as the boundary. Now blow some air into the film. It will bulge out and define a new surface, but the rim will still be its boundary. Figure 9.7 illustrates this point. Will Ampère's law continue to hold with the same rim and the bulging surface? That is the question we address.

In Figure 9.7, the contour C is the boundary of both S and S'. If we integrate **B** over C, will it equal μ_0 times the current crossing both S and S'? The answer is affirmative because the same current I crosses both. So either surface can be used in Ampère's law. (If the current entering S was not the current leaving S', there is either non-conservation of charge or a continuous time-dependent pileup of charge in the volume bounded by S and S', which are glued at C.)

Proving Ampère's law for a non-planar S given its validity on planar S is quite easy and is illustrated in the lower half of Figure 9.7. First take an infinitesimal loop, labeled 1, which we can treat as planar. The line integral of **B** around that loop, the circulation C_1 (in the same sense as

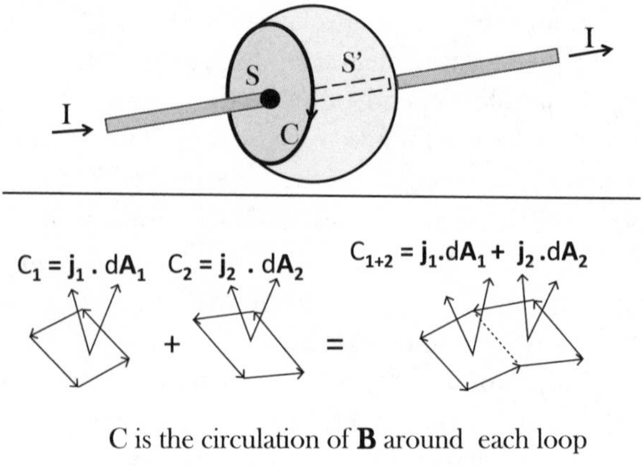

$$C_1 = j_1 \cdot dA_1 \quad C_2 = j_2 \cdot dA_2 \qquad C_{1+2} = j_1 \cdot dA_1 + j_2 \cdot dA_2$$

C is the circulation of **B** around each loop

Figure 9.7   Top: The contour C is the boundary of both S and S'. Since the same current I crosses both, by charge conservation, either can be used in Ampère's law. Bottom: Ampère's law for a composite non-planar surface made by gluing two planar surfaces with a common edge. The circulations of the two loops add, as do the currents crossing them. I have suppressed μ_0 and shown the canceled part of the common edge by a dotted line.

the arrows along its edges specifying its orientation), is equal to (μ_0 times) the current crossing it, $\mathbf{j}_1 \cdot d\mathbf{A}_1$. Now glue to that another planar area, $d\mathbf{A}_2$, with one common edge traversed in the opposite sense, just as when we glued two infinitesimal areas. This defines a larger area with the common edge deleted. Although the loops share an edge, *they need not and do not lie in the same plane*. The circulation of **B** around the combined loop C_{1+2} is the sum of the circulations around each because the common edge cancels. The current crossing the combined area is the sum of the currents crossing each. Proceeding in this manner we can prove Ampère's law for a non-planar boundary of an arbitrary non-planar surface in three dimensions:

$$\oint_{C=\partial S} \mathbf{B} \cdot d\mathbf{r} = \mu_0 I_{enc} = \mu_0 \int_S \mathbf{j} \cdot d\mathbf{S}. \tag{9.31}$$

9.5 Maxwell's equations (static case)

We now break for a mathematical interlude. Given the Lorentz formula for the forces the fields exert on the charges, what we need to conclude the story is a complete set of rules for computing the fields due to any set of static charges and time-independent currents. What we have so far is Gauss's law for electrostatics, derived from Coulomb's law, and Ampère's law for magnetostatics, derived from the Biot-Savart law. Here they are

$$\oint_{S=\partial V} \mathbf{E} \cdot d\mathbf{S} = \frac{q_{enc}}{\varepsilon_0} = \frac{1}{\varepsilon_0} \int_V \rho \, d^3 r \quad \text{Gauss} \tag{9.32}$$

$$\oint_{C=\partial S} \mathbf{B} \cdot d\mathbf{r} = \mu_0 I_{enc} = \mu_0 \int_S \mathbf{j} \cdot d\mathbf{S} \quad \text{Ampère} \tag{9.33}$$

where S is a closed surface that bounds the volume V in Gauss's law and C is the contour that bounds the open surface S in Ampère's law.

The preceding equations specify the surface integral (flux) of $\mathbf{E}$ and the line integral (circulation) of $\mathbf{B}$. What about the surface integral (flux) of $\mathbf{B}$ and the line integral (circulation) of $\mathbf{E}$?

We already know that too. First, because $\mathbf{E}$ is conservative,

$$\oint_C \mathbf{E} \cdot d\mathbf{r} = 0 \quad \text{($\mathbf{E}$ is conservative).} \tag{9.34}$$

Next, given that magnetic lines never start or end (there being no monopoles) it follows that the lines entering any closed surface will have to also leave it. This means there can be no net magnetic flux coming out of a closed surface

$$\oint_{S=\partial V} \mathbf{B} \cdot d\mathbf{S} = 0 \quad \text{(no monopoles).} \tag{9.35}$$

Equations 9.32 to 9.35 are called the integral Maxwell equations *for the static case.* (A more common version, fully equivalent, involves derivatives and emerges when the loops and surfaces become infinitesimal.)

They are the best way to summarize what we have learned so far. This is so because of the mathematical result that a vector field like $\mathbf{E}$ or $\mathbf{B}$ is uniquely determined if it vanishes at infinity and if its circulation around every loop and the integral over every closed surface are specified. This is exactly what the Maxwell equations do in terms of charges and currents, which are assumed to be given. There is also a procedure for finding the fields given this data. We will not discuss this procedure since it calls for a lot more mathematical machinery. We will be content with being able to find the fields in a few problems endowed with a high degree of symmetry.

CHAPTER 10

Ampère II, Faraday, and Lenz

We have just finished learning Ampère's law. We will now put it to work for us by using it to compute the magnetic field in certain situations with a high degree of symmetry. Recall the law:

$$\oint_{C=\partial S} \mathbf{B} \cdot d\mathbf{l} = \mu_0 I_{enc} = \mu_0 \int_S \mathbf{j} \cdot d\mathbf{S} \qquad (10.1)$$

where S is a surface with boundary C, and I_{enc} is the sum of all the currents crossing S, given by the surface integral of $\mathbf{j}$ over S. If your right hand encircles the contour in the sense in which it is traversed, your thumb defines the positive direction for the currents. For a contour traversed counterclockwise in the plane of the page, the positive direction is straight out of the page. Note that C is a specific closed loop but S can be *any* surface with C as its boundary.

The right-hand rule is everywhere and you should master and exploit it. Our ability to use the thumb against the four fingers is what distinguishes us from the lower primates, who just do not get the right-hand rule. There are cave drawings of motors and generators that were doomed to failure since those cave dwellers were curving all five fingers. Then the right-hand rule was invented. It is an invention that matches the wheel and fire in significance and after that there was no stopping us.

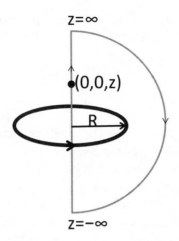

Verifying Ampère's law for known **B**

Figure 10.1   Verification of Ampère's law on an infinite semicircle.

10.1   Field of an infinite wire, redux

Ampère's law is like Gauss's law: it makes a statement about the integral of the field. It is always true, but only on special occasions can you deduce the field everywhere from the knowledge of its integral. Recall that Gauss's law applies to every charge distribution and every surface surrounding it. This will not help us find **E** everywhere. How can you expect to find the integrand given just the integral? You cannot, except in highly symmetric situations. In a problem with spherical or cylindrical symmetry, the integrand was a constant on the entire Gaussian surface, so that the integral was simply this constant times the region of integration. For example, all over a Gaussian sphere of radius r surrounding a spherically symmetric distribution of charge, the field was known to be radial and of constant magnitude $E(r)$. Thus the surface integral was $4\pi\, r^2 E(r)$. Relating this to the charge enclosed in this sphere we could deduce $E(r)$.

So it is with Ampère's law. Like Gauss's law, it is always true, but its efficacy in finding the field relies a lot on symmetry. Here is an example where we *do not* have such symmetry. Consider the field of a ring in the xy-plane, of radius R, centered at the origin and carrying a current I, as shown in Figure 10.1. Though we computed the field only at points $(0,0,z)$

lying on the z-axis, it will suffice for our purposes. Recall that the field points up the z-axis (from the right-hand rule applied to the current) and has magnitude

$$B_z(z) = \frac{\mu_0 I}{2} \frac{R^2}{(R^2 + z^2)^{3/2}}. \tag{10.2}$$

Let us do the line integral of **B** on an infinite semicircular loop with the z-axis as its diameter. Thus the loop begins at $z = -\infty$, goes up the z-axis through the center of the loop to $z = \infty$, and bends around in a huge semicircle, which closes the loop at $z = -\infty$. First consider the integral on the infinite semicircle. We do not know the field off axis in detail, but we do know that the dipolar field falls like $1/r^3$. (This is evidently true on the axis as $z \to \infty$ in the formula above.) An integrand that falls like $1/r^3$, when integrated over a curve whose length grows only as r, vanishes as $r \to \infty$. On the straight path from $-\infty$ to ∞ the contribution is

$$\int_{-\infty}^{\infty} \frac{\mu_0 I}{2} \frac{R^2 \, dz}{(R^2 + z^2)^{3/2}} = \mu_0 I, \tag{10.3}$$

which is perfect, since the current enclosed is indeed I and goes into the page, as required by the clockwise contour.

But the point is that we cannot go backward: we cannot deduce $B_z(z)$ given that the integral on this contour is $\mu_0 I$, because the integrand $B_z(z)$ *varies along the contour.*

Having made this point, I turn to a case where there is enough symmetry to find **B** from Ampère's law: the field of an infinite wire. Look at the wire shown in Figure 10.2, with its current coming out of the page. What can we say without doing a calculation? Any field distribution we end up with has to be invariant under translations along the wire since the current is. It must be invariant under rotations around the axis of the wire since the current is. These are very general statements stemming from translational and rotational symmetry. In the case of the electric field of a charged wire, we also argued the field at any point cannot be tilted to the right or left along the axis of the wire, since if we rotated the wire and field pattern by π around an axis perpendicular to the wire, the line of charge would look the same but the field would have reversed its tilt. This would constitute a change in effect without a change in cause. This argument does not hold for the current-carrying wire: the current distinguishes left from right. So we peek into the underlying Biot-Savart law, the cross product in

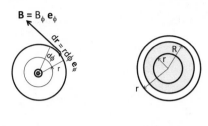

Finding **B** using Ampère
Wire of zero radius

Finding **B** using Ampère
Wire of radius R

Figure 10.2   Left: The use of Ampère's law to find the field of an infinite wire.
The figure shows the view staring into the current, which is coming out of the
page. The Ampèrean contour lies on a plane normal to it. The displayed features
of the field can be deduced by symmetry. Right: Finding **B** for a wire of non-zero
thickness and uniform current density. The dark circle of radius R represents the
current-carrying conductor and the other two circles the Ampèrean loops.

which precludes a component of **B** parallel to $d\mathbf{l}$ segment by segment. So let
us take a slice perpendicular to the wire, with the current coming out of the
page. Since the wire and the current it carries look the same if the wire is
rotated around its *own* axis, there are only two possible configurations with
this property: the lines go in or out radially or the lines go in circles around
the wire. The radial configuration is ruled out for so many reasons, some
of which I give just to give a feeling for such arguments: radiating lines
imply monopoles that do not exist, the cross product in the Biot-Savart
law prevents a radial field segment by segment, and, finally, when I rotate
the wire by π around a perpendicular axis, the current flips sign but the
radial field does not.

So we can be pretty sure the lines encircle the wire, with the
circulation in the sense determined by the right-hand rule, as shown in the
left half of Figure 10.2. This configuration meets the requirement that if
I rotate the current and field configuration by π about an axis perpendicular to the wire, both the direction of current and the sense of circulation
reverse.

The Ampèrean loop of choice is a concentric circle of radius r.
We know that on this loop

$$\mathbf{B} = \mathbf{e}_\phi B(r) \tag{10.4}$$

where $B(r)$ is a constant because r is. The circulation is

$$\oint B_\phi \mathbf{e}_\phi \cdot \mathbf{e}_\phi \, r d\phi = B(r) 2\pi r. \tag{10.5}$$

From Ampère's law

$$2\pi r B(r) = \mu_0 I \tag{10.6}$$

$$B(r) = \frac{\mu_0 I}{2\pi r} \tag{10.7}$$

$$\mathbf{B} = \mathbf{e}_\phi \frac{\mu_0 I}{2\pi r}, \tag{10.8}$$

which is what we got by integrating the Biot-Savart law (Eqn. 9.16).

It is reasonable to object that in the time it took to furnish all the symmetry arguments we could have done the integral in the Biot-Savart law. That is perhaps right, but consider the following variation shown in the right half of Figure 10.2. We replace the infinitely thin wire by one with a circular cross section of radius R. The current I is uniformly distributed across the circular cross section. The current density now is

$$j(r) = \frac{I}{\pi R^2} \quad r \le R \tag{10.9}$$

$$= 0 \quad r > R. \tag{10.10}$$

What is the magnetic field? If we try going directly to the Biot-Savart law we will be looking at a nasty three-dimensional integral due to the non-zero thickness of the wire. But Ampère's law applied to a circular contour allows us to use all the symmetry arguments we invoked for an infinitely thin wire. We find for $r \le R$

$$2\pi r B(r) = \mu_0 I_{enc} = \mu_0 \underbrace{\left[\frac{I}{\pi R^2}\right]}_{\text{current density}} \underbrace{\pi r^2}_{\text{area}}$$

$$= \mu_0 I \frac{r^2}{R^2} \quad r \le R, \text{ whereas,} \tag{10.11}$$

$$2\pi r B(r) = \mu_0 I \quad r > R, \tag{10.12}$$

which means

$$\mathbf{B} = \mathbf{e}_\phi \mu_0 I \frac{r}{2\pi R^2} \quad r \leq R \tag{10.13}$$

$$= \mathbf{e}_\phi \frac{\mu_0 I}{2\pi r} \quad r > R. \tag{10.14}$$

Like the electric field inside a sphere of charge, the field rises linearly with r inside the wire, peaks at $r = R$, and drops off like $1/r$ beyond. The initial rise is due to the fact that the current enclosed by the Ampèrean circle inside the wire grows like r^2, while its influence drops like $1/r$. Outside the wire, increasing the radius of the contour does not lead to any increase in enclosed current, while the field due to it drops like $1/r$.

Similarly, just as the electric field of a spherical charge outside its radius is that of a point charge at the center, the magnetic field outside the radius of a wire carrying a uniform current density is that of a zero-thickness wire carrying all the current.

The analogy continues. Suppose you scooped out a coaxial cylindrical region of radius a from the interior of the wire. It is now hollow for $0 \leq r \leq a$ and carries the current only in the region $a < r \leq R$. Ampère's law and symmetry will tell you that inside the hollow region there will be no magnetic field.

10.2 Field of a solenoid

Imagine a cardboard tube of cross-sectional radius R around which you wrap N turns of a wire carrying a current I, as shown in the left half of Figure 10.3.

We know that the field of a single loop is like that of a tiny magnet with its north-south ends lined up along the dipole moment $\boldsymbol{\mu}$. The lines go up inside the loop, and they return outside the loop and join up below the loop. The solenoid, made of many turns, is like a stack of these dipoles NSNSNS . . . lined up end to end. We should not be surprised that it should create the field of a cylindrical bar magnet. On the plane $P_\perp$ that bisects the solenoid, the field outside will be pointing straight down. As the length of the solenoid approaches infinity, the curved parts near the end will also get pushed to infinity and the field lines outside will be pointing straight down everywhere. In other words, for an infinitely long solenoid, every plane perpendicular to the axis will look like $P_\perp$. (This is like a parallel plate capacitor, whose curved field lines near the edges are banished to infinity as the plates become infinitely large. The electric field lines we will see in the finite part of the universe will be parallel to each other and perpendicular to the plates.) We want to use Ampère's law to find $\mathbf{B}_{in}$ and $\mathbf{B}_{out}$, the field strengths inside and outside the infinite coil, pointing up and down respectively. These fields could depend on the distance from the axis.

You may be tempted to choose a circular Ampèrean loop coaxial with the solenoid. This will, however, give $0 = 0$: no current crosses it (meaning any surface bounded by it) and the field has no azimuthal component. To get to the right loop we must slice the solenoid parallel to its axis, bisecting its cross section, one half of which is shown at the right in Figure 10.3.

First consider $\mathbf{B}_{out}$ and Ampère's law applied to the contour C':

$$\oint_{C'} \mathbf{B} \cdot d\mathbf{r} = \mu_0 I_{enc}. \tag{10.15}$$

The horizontal sides 23 and 41 do not contribute to the line integral because the field and $d\mathbf{r}$ are perpendicular. The oppositely oriented vertical sides 21 and 43 are parallel (anti-parallel) to $\mathbf{B}_{out}$ and contribute $+B_{out}(12) L$ and $-B_{out}(34) L$ respectively. These contributions must cancel each other since no current is enclosed by C'. This means $\mathbf{B}_{out}$ has the same magnitude on both these sides: $B_{out}(12) = B_{out}(34)$. Now let us widen the loop, sending the side 12 off to infinity where $\mathbf{B}_{out}$ must vanish. It follows

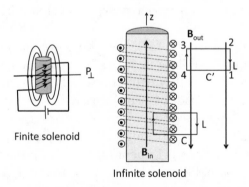

Figure 10.3   Left: A finite solenoid. The field lines go up the solenoid inside and return outside. Right: The cross section of the infinite solenoid. The field is parallel to the solenoid inside and outside. C and C' are two Ampèrean loops.

it must vanish on 34 as well. Since we can place 34 anywhere (outside the solenoid) we conclude $\mathbf{B}_{out} \equiv 0$. (The $\mathbf{B}$ due to an infinite coil will vanish at infinity while $\mathbf{E}$ due to an infinite sheet does not, because the former is infinite in one dimension, the length of the solenoid, while the latter is infinite in two dimensions. This can be verified by working out $\mathbf{B}$ for longer and longer solenoids.)

Next consider Ampère's law on contour C, partly in and partly out of the solenoid as shown. The horizontal sides contribute zero individually. The vertical side outside does not contribute since we have shown that $\mathbf{B}_{out} = 0$. The vertical side inside contributes $B_{in}L$. With the contour traversed as shown, the current enclosed is positive if going into the page. If there are n turns per unit length, Ampère's law tells us

$$B_{in}L = \mu_0 nLI. \tag{10.16}$$

Notice two things. First, the length L cancels out, as it should, since it characterizes a fictitious Ampèrean loop and cannot be present in the answer for the field. Second, the current enclosed does not depend on where the vertical side of the loop is inside the solenoid. It follows that $\mathbf{B}_{in}$ is constant inside the solenoid. So here is the final answer, assuming the axis of the solenoid coincides with the z-axis:

$$\mathbf{B} = \mathbf{k}\,\mu_0 nI \quad \text{inside} \tag{10.17}$$

$$= 0 \quad \text{outside}. \tag{10.18}$$

This is another result worth memorizing.

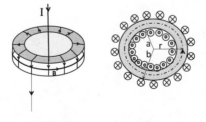

Figure 10.4   Left: Front view of toroid. Right: The mentally sliced-up toroid. The Ampèrean loop shown is the dotted concentric circle of radius $a < r < b$. A loop that is smaller or larger encloses no net current and implies zero field.

The infinitely long solenoid is an idealization in which the return flux is banished to infinity. Any finite solenoid is going to have the return flux as well as complications at the ends. The lines that leave the north pole have to return to the south pole so that they may close in on themselves. This makes it impossible to use Ampère's law to find the field of a finite solenoid. (In practice we use the infinite solenoid result for a finite solenoid as long as we do not go near the ends or too far off axis.)

A toroidal solenoid beats this problem by being finite and yet free of ends. The trick is to bend the linear solenoid we have been discussing into a hula hoop, joining the top and bottom. The result looks like a donut, with the flux trapped inside (where the dough would be in a donut) and closing on itself. Often the core is filled with iron, which encourages the flux to stay inside the donut. Figure 10.4 should give you an idea. I have chosen the cross section of the dough to be rectangular instead of circular for simplicity. To find the field using Ampère's law we need to slice the donut the way we would to butter it. The cross section that emerges is shown in the right half of the figure. (The slice can bisect the donut along the equator shown by the line marked **B**, or lie above or below this. In all cases the cross section will be the same because of the assumed rectangular cross section.) The slice is bounded by two concentric circles of radius $a < b$. The wires sliced (mentally) that are inside the inner circle have currents coming out of the page and those outside the outer circle have the currents going in. The Ampèrean contour is a circle of radius $a < r < b$ shown by the dotted line. The field is azimuthal and has a magnitude $B_\phi(r)$ at radius r. Note that the direction of the field agrees with the direction of current flow as required by the right-hand rule. Ampère's law tells us

$$2\pi r B_\phi(r) = \mu_0 NI \tag{10.19}$$

where N is the total number of turns. Thus

$$B_\phi(r) = \frac{\mu_0 NI}{2\pi r} \quad a < r < b. \tag{10.20}$$

The field is not constant within the donut: it is strongest on the inner rim $r = a$ and gets weaker as we go out to $r = b$.

The field is clearly zero when r is not between a and b because the total current enclosed is either trivially zero ($r < a$) or a zero due to cancellation of opposite currents ($r > b$).

We can subject our result to a test. Imagine the inner and outer radii of the toroid have become astronomical but their difference $b - a$ is finite. In this limit, any finite section of the toroid will look like a straight tube because we cannot detect the curvature of such a large circle. The azimuthal field B_ϕ will become a field along the axis of this tube. The variation of $B_\phi(r)$ within $a < r < b$ can be neglected because the function $1/r$ hardly varies in the interval $a < r < b$ for astronomical a and b and fixed $b - a$. In this limit the field should approach that inside an infinite linear solenoid and indeed it does:

$$B = \mu_0 I \frac{N}{2\pi R} = \mu_0 n I \qquad (10.21)$$

where R can be a or b, it does not matter, and we may take either $N/2\pi a$ or $N/2\pi b$ as being equal to n, the number of turns per unit length.

This wraps up our study of electrostatics and magnetostatics. A complete mathematical characterization of everything we have done is given below:

$$\mathbf{F} = q(\mathbf{E} + \mathbf{v} \times \mathbf{B}) \quad \text{(Lorentz force)} \qquad (10.22)$$

$$\oint_{S=\partial V} \mathbf{E} \cdot d\mathbf{S} = \frac{q_{enc}}{\varepsilon_0} = \frac{1}{\varepsilon_0} \int_V \rho \, d^3 r \quad \text{(Gauss)} \qquad (10.23)$$

$$\oint_{C=\partial S} \mathbf{B} \cdot d\mathbf{r} = \mu_0 I_{enc} = \mu_0 \int_S \mathbf{j} \cdot d\mathbf{S} \quad \text{(Ampère)} \qquad (10.24)$$

$$\oint_C \mathbf{E} \cdot d\mathbf{r} = 0 \quad \text{(\textbf{E} is conservative)} \qquad (10.25)$$

$$\oint_{S=\partial V} \mathbf{B} \cdot d\mathbf{S} = 0 \quad \text{(no monopoles)} \qquad (10.26)$$

where S is a closed surface that bounds the volume V in Eqns. 10.23 and 10.26 and any open surface S bounded by the curve C in Ampère's law (Eqn. 10.24).

The Lorentz force law tells you what the fields do to the charges and currents, and the four Maxwell equations tell you how the fields in turn are determined by the charges and currents.

This would have been the end of the story in a world where charges and currents did not change with time.

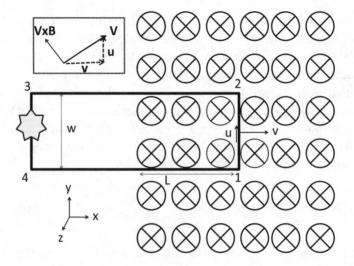

Figure 10.5   The rectangular conducting loop pulled by me to the right at speed
v in a magnetic field **B** going into the page (shown by a circle with a cross). The
carriers move in the wire (counterclockwise) from 1 to 2 with a speed u. The total
velocity of the carriers is $\mathbf{V} = \mathbf{v} + \mathbf{u}$ as shown in the inset. Work is done *by* the
field along **u** and by me along **v**. The lightbulb glows due to the transfer of
mechanical to electrical energy mediated by **B**, which does no net work.

10.3 Faraday and Lenz

But of course they do change with time! And we have to deal with that. I am
aware of the mental load you have to carry as one new idea after another
is introduced. "Drinking out of the fire hose" was an expression that often
came up in class. But we are not too far from the end of our discussion of
electromagnetism and you will enjoy the way the missing pieces fall into
place. It is one of the finest examples of mathematical physics.

I am now going to lead you through some experiments that force us
to change some of these Maxwell equations of the static case to their final
form.

In the first experiment, depicted in Figure 10.5, there is a uniform
magnetic field **B** going into the page to the right of some line, say $x = 0$. It
is zero to the left.

The solenoid or magnet producing the **B** is not shown in the figure so
we can focus on the main item, which is a rectangular loop of wire of width
w in the plane of the page. A part of it of length L lies within the field. A

lightbulb is part of the circuit. When the loop is static the bulb does not glow. Now I begin to drag the loop to the right at some speed v. What do you think happens?

Everyone in class was able to guess that the bulb would now glow, the most common reason being that I would not have drawn it otherwise. Let us see if can we go beyond this type of reasoning that helps you ace the SAT. *Why* is the lightbulb glowing? Is there some new physics?

Whenever you see a lightbulb glow, you're looking for a battery. There is no battery in the circuit. And yet there must be an emf, because every time a charge makes one full trip around the loop, it delivers some energy to the glowing bulb. Who is providing the energy? Who is pushing the charges around this loop? We defined the emf to be the line integral of the force per unit charge pushing the charges around the loop. What might the force be? And why does it kick in only when I move the loop?

The last sentence is usually enough of a clue for the students to figure out that the force on the unit charge introduced to compute $\mathcal{E}$ is the $\mathbf{v} \times \mathbf{B}$ Lorentz force.

When studying electricity we defined the emf to be

$$\mathcal{E} = \oint \mathbf{E} \cdot d\mathbf{r}. \tag{10.27}$$

Now that we know about magnetism we must use a more general definition of the emf as the line integral of the electromagnetic Lorentz force on a unit charge:

$$\mathcal{E} = \oint (\mathbf{E} + \mathbf{v} \times \mathbf{B}) \cdot d\mathbf{l}, \tag{10.28}$$

where $d\mathbf{l}$ is a segment of a *physical* loop moving at velocity $\mathbf{v}$. In our problem there is no $\mathbf{E}$ and the entire emf comes from the $\mathbf{v} \times \mathbf{B}$ term.

If you look at the loop you see that the edge 12 is moving to the right at speed v in the field $\mathbf{B}$. The unit charges in that segment feel the $\mathbf{v} \times \mathbf{B}$ force that points from 1 to 2. The force has a magnitude vB and its contribution to the emf is vBw. The forces on the horizontal sections are perpendicular to the sections and make no contribution to the emf. Finally, there is no force on the segment 34 in the field-free region. Thus the emf, computed in the counterclockwise sense, is

$$\mathcal{E} = \oint (\mathbf{v} \times \mathbf{B}) \cdot d\mathbf{l} \tag{10.29}$$

$$= \int_{12} + \int_{23} + \int_{34} + \int_{41} \qquad (10.30)$$

$$= vBw + 0 + 0 + 0 = vBw. \qquad (10.31)$$

So far so good. We understand this experiment without bringing in any new stuff. There is just one paradox to be dealt with. We proved at the very outset that the magnetic field doesn't do any work. Remember, the original argument for why it doesn't do any work was that $\mathbf{v} \cdot (\mathbf{v} \times \mathbf{B}) = 0$. But here, $\mathbf{v} \times \mathbf{B}$ is along the wire and so is the current. It looks as if there is a magnetic field pushing these charges along segment 12 and doing net work every time a charge goes around the circuit. What is happening?

The answer has many parts.

First of all, the actual velocity of the charges in the wire is not just the loop velocity v along the x-axis, but also the velocity u *along* the wire due to the current in the loop. The total velocity (shown in Figure 10.5) is thus

$$\mathbf{V} = \mathbf{i}v + \mathbf{j}u = \mathbf{v} + \mathbf{u}, \qquad (10.32)$$

and the total magnetic force per unit charge is

$$\mathbf{V} \times \mathbf{B} = -\mathbf{i}Bu + \mathbf{j}Bv, \qquad (10.33)$$

and the power delivered vanishes:

$$\mathbf{V} \cdot (\mathbf{V} \times \mathbf{B}) = -vuB + uvB = 0. \qquad (10.34)$$

I want to explain the two canceling pieces.

If I want to pull the loop at a steady speed v I have to balance the leftward component $(-Bu)$ $\mathbf{i}$ of $\mathbf{V} \times \mathbf{B}$. This requires I provide power $P = Buv$. How does this power get transmitted to the bulb?

For this we consider the component $+Bv$ $\mathbf{j}$ pointing up the y-axis.

It does not accelerate the charges up the wire in the y-direction *because it is precisely balanced by an internal electrostatic field* E_c which arises as follows. Imagine there is no bulb and we have an open circuit with a gap between points 3 and 4. As I begin to drag the loop, the magnetic force Bv up the wire will initially pile up positive charges at end 3 and leave an equal and opposite negative charge at 4. These are the charges which produce the electrostatic field E_c. Some of its field lines will point straight down the gap from 3 to 4 and others will enter the wire at 3 and return to 4. The charge pileup will continue till E_c inside the wire balances Bv

in the segment 12. (So what is disallowed in a perfect conductor is not a net field but a net force. Here the electric field arises to cancel the magnetic force along the wire.) Now imagine inserting the bulb and allowing current to flow. The built-up charges will flow down the filament from 3 to 4, converting their potential energy to heat. This flow will initially weaken E_c to below Bv, which will promptly pump in more charges to restore the balance. It is this electrostatic field E_c against which the y-component Bv pumps the charges, doing work at a rate $P = Bvu$. Thus the power (per unit charge) expended by me in pulling the loop is exactly equal to the work done against the electric field E_c, in charging the points 3 and 4 and keeping the bulb glowing.

Though the magnetic field does no work, I need it to push the charges against E_c. I cannot grab them and force them through the bulb with my bare hands. It is the **B** field which converts the force I exert to the right to the upward force on the charges inside, against E_c. It converts macroscopic power provided by me as I drag the loop, to the microscopic power delivered to the charges which in turn deposit it inside the bulb. It takes macroscopic mechanical power from me and turns it into microscopic power provided to the charges.

Here is an equivalent way to check the balance of energy. The power delivered to the bulb is $P_{res} = \mathcal{E}I$. Now, we know that once the loop carries a current, I will have to work against the force $Id\mathbf{l} \times \mathbf{B}$ on each piece of wire $d\mathbf{l}$ carrying current I. The force on segment 12 is $F = IBw$ to the left. The power I supply dragging the loop at speed v against this force is $P_{me} = IBwv$. But since $Bvw = \mathcal{E}$ (force on a unit charge times the distance over which it acts), the power I supply is also $P_{me} = \mathcal{E}I$.

In short, the loop is a generator. If I want to light a bulb, one option is to set up a magnetic field perpendicular to the ground, connect the lightbulb to a metallic rectangle, grab it, and keep running. As long as I keep running, the lightbulb will keep glowing. But there is a problem with this besides having to run non-stop. When the trailing segment 34 crosses into the field, the current will stop. The clockwise contribution to the emf from that segment will oppose the counterclockwise contribution from 12. The line integral of the force on a unit charge will be zero.

Now we fully understand the forces and energies involved in this experiment in terms of the $\mathbf{v} \times \mathbf{B}$ force. There seems to be no need to monkey with the Maxwell equations I wrote down earlier. But there is, and it becomes apparent when I introduce the reasonable assumption that the principle of relativity applies to the laws of electromagnetism. Here is how.

Return to the loop I was dragging and running with to light up the bulb. Let us go to the frame where the loop and I are at rest. I am free to assume I am at rest and the magnet creating the field is moving to the left. Indeed it could be that I was always at rest and I hired some guys to carry the magnet and run the other way. I still expect my lightbulb to glow. Lots of things are relative, but whether a lightbulb glows or not is not relative. A glowing lightbulb is a glowing lightbulb in any frame of reference. The power it consumes may vary, but the fact that it glows is undeniable.

So how am I, in the loop rest frame, supposed to understand the glowing of the lightbulb? It is true someone is now moving the magnet, and $\mathbf{B}$ is time-dependent: if at some time $t = 0$ it was non-zero to the right of the line $x = 0$, then at time t it is non-zero to the right of $x = -vt$. The field has changed from zero to non-zero in the region $-vt < x < 0$. But this cannot produce any $\mathbf{v} \times \mathbf{B}$ force because the loop is at rest and $\mathbf{v} \equiv 0$.

So what force could be pushing the charges around the loop? If we believe the Lorentz force $\mathbf{F} = q(\mathbf{E} + \mathbf{v} \times \mathbf{B})$ is all there is, we are left with just the electric force, now that $\mathbf{v} \times \mathbf{B}$ is dead. In this case, we can deduce that if the principle of relativity applies to electromagnetism, *there must be an electric field in the frame of reference where the loop is fixed and the magnet is moving.* Not only that, it must be an electric field whose line integral around the loop is non-zero: charges in the loop are going round and round doing work every cycle, lighting up the bulb. The corresponding emf must be due to this electric field.

All the electric fields we have studied till now were produced by static charges, determined by Coulomb's law, and conservative. Now we find that without the help of any uncompensated charges, we can get an electric field with non-zero circulation in a changing magnetic field.

Can we say any more about this electric field $\mathbf{E}$ besides the fact that it has a non-zero circulation?

We can, if we apply relativity to a simpler related problem. Go back to the magnet frame, and replace the loop moving at velocity $\mathbf{v}$ with a unit charge traveling with the same velocity $\mathbf{v}$ in the plane of the paper. It will experience a force $\mathbf{v} \times \mathbf{B}$ (along the y-axis) and begin to accelerate along y in response. Now we go to the frame at velocity $\mathbf{v}$ in which the particle is *instantaneously* at rest. Let us work in the low velocity (Newtonian) limit, when acceleration and force are invariant when we change inertial frames. The particle should experience the same

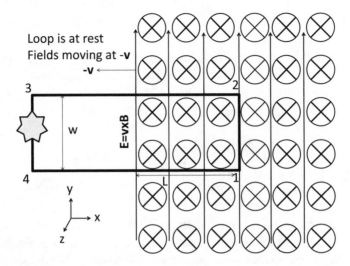

Figure 10.6 The situation in the loop frame. The pattern of **E** (arrows along y) and **B** (into page along $-z$) moves to the left at $-\mathbf{v}$, along with the magnet producing the **B** field (not shown). The electric field $\mathbf{E} = \mathbf{v} \times \mathbf{B}$ produces an emf in the loop because it makes a non-zero contribution only on segment 12.

acceleration or force in its rest frame. If this acceleration is due to an electric field, it must be given by

$$\mathbf{E} = \mathbf{v} \times \mathbf{B}. \tag{10.35}$$

(The exact formula, which we will not derive here, agrees with this in the limit of small velocities and differs by terms of order v^2/c^2 and higher. We do not need the fully relativistic treatment to understand glowing lightbulbs.) Since the loop rest frame is also the particle rest frame, there must be a field $\mathbf{E} = \mathbf{v} \times \mathbf{B}$ in the loop rest frame. This is shown in Figure 10.6. (Not shown in the figure is the moving magnet that produces the **B** field.)

If the loop is partly in and partly out, the $\mathcal{E}$ due to this **E** is just $Ew = vBw$ coming from segment 12. There is nothing from 34 as $\mathbf{B} = 0$ and $\mathbf{E} = 0$ there and finally **E** is perpendicular to the other two segments.

To summarize, the emf in the loop can be understood in two ways in the two frames: as the line integral of the $\mathbf{v} \times \mathbf{B}$ force in the lab or magnet frame or of the electric field $\mathbf{E} = \mathbf{v} \times \mathbf{B}$ in the loop frame.

The new physics is that *a changing magnetic field can produce an electric field with nonzero circulation.*

Now, it turns out there is a master formula that describes the glowing lightbulb not only in these two cases (loop fixed or magnet fixed) but everything in between, where both the loop and field could be changing with time and the emf could be due to both **E** and **B**. It is called *Faraday's law* and it states

$$\mathcal{E} = \oint_C (\mathbf{E} + \mathbf{v} \times \mathbf{B}) \cdot d\mathbf{l} = -\frac{d\Phi}{dt}. \tag{10.36}$$

On the left we have $\mathcal{E}$, defined as the line integral of full electromagnetic Lorentz force on a unit charge on a loop C. *The loop C is a real flexible loop, a conductor carrying charges. It moves and* **v** *is the velocity of the segment d*l. On the right Φ is the magnetic flux penetrating any surface S bounded by the loop C.

The minus sign, associated with *Heinrich Lenz* (1804–1865), gives the sense in which $\mathcal{E}$ will drive a current in the loop: the emf will try to fight the *change* in flux. For example, if the flux is increasing, it will drive a current in the loop, the field due to which will oppose the flux. If the flux is decreasing, it will drive a current that produces a flux in the same sense, trying to prop it up at its old value. Hence what the emf fights is not flux itself, but the *change* in flux.

Lenz's minus sign often takes us to the final answer much faster than all the cross products.

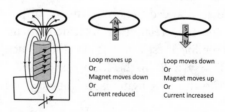

Loop moves up
Or
Magnet moves down
Or
Current reduced

Loop moves down
Or
Magnet moves up
Or
Current increased

Figure 10.7   The current induced in the loop depends only on the rate of change of flux through it, and not on whether the change is due to the moving loop, the moving electromagnet, or the changing current in the magnet. The arrows in the loop show the orientation of the magnetic moment of the loop due to the induced current.

Before applying Faraday's law Eqn. 10.36 to the general case of a flexible loop moving in a space-time–dependent **B**, let us consider an illustrative example that made a profound impression on Einstein, who refers to it in his relativity paper. The leftmost part of Figure 10.7 shows a loop of wire near the north pole of an electromagnet. If we move the loop up, away from the magnet, the flux through it decreases and the current due to $\mathcal{E}$ must flow as shown to fight this decrease. This is also what we would get from computing $\mathbf{v} \times \mathbf{B}$ for the carriers in the moving loop. The same $\mathcal{E}$ arises if the magnet is moved down or the current through it reduced, for they both reduce the flux through the loop. But now $\mathcal{E}$ is attributed to a non-conservative electric field produced by the changing magnetic field.

The opposite $\mathcal{E}$ arises if the loop is moved toward the magnet or the current in it increased. Of course, if the loop and the magnet move, the emf, which still depends only on the rate of change of flux, will be due to both **E** and $\mathbf{v} \times \mathbf{B}$ forces.

This tendency of the loop to oppose change can also be understood in terms of attraction and repulsion of magnetic poles, as shown in Figure 10.7. If you are trying to bring the loop and magnet closer (rightmost part of figure), the magnetic moment $\boldsymbol{\mu}$ in the loop induced by $\mathcal{E}$ will have its north end pointing toward the north end of the electromagnet (so the poles repel). The opposite happens (poles attract) if you are trying to increase the separation (middle part of figure). In both cases the force between the loop and the magnet opposes you.

Faraday's law and Lenz's minus sign explain all cases in one stroke: the generated $\mathcal{E}$ opposes the change of flux.

Let us now return to the loop and the lightbulb and see how Faraday's law explains the emf $\mathcal{E}$ in the lab frame and the loop frame.

First let's do the easy part, when there's a fixed magnetic field and I'm dragging the loop. What is the flux penetrating this loop? It is just the product of the constant magnetic field, the width of the loop and L, the length of the loop *that is in the field*:

$$\Phi = BwL. \tag{10.37}$$

Now let's take minus the time derivative of both sides. On the left is $\mathcal{E}$. On the right B is not changing, w is not changing, but L is changing.

The rate of change of L is v, the speed of the loop. This means

$$\mathcal{E} = -\frac{d\Phi}{dt} = -Bvw. \qquad (10.38)$$

Previously we had seen that the $\mathcal{E}$ due to $\mathbf{v} \times \mathbf{B}$ had a magnitude Bvw in segment 12 and was pushing the charges counterclockwise. The minus sign in Eqn. 10.38 says exactly that in Lenz's convention. The magnetic flux was going into the page. As the loop moved to the right, the flux penetrating the loop into the page increased. Therefore the current produced by $\mathcal{E}$ had to flow in such a way as to reduce the flux going into the page. To produce flux coming out of the page the current had to flow counterclockwise.

If I dragged the loop to the left, the enclosed flux would decrease and the current generated by $\mathcal{E}$ should try to prop it up, and so it will flow clockwise. This agrees with the direction of the $\mathbf{v} \times \mathbf{B}$ force in segment 12.

Finally, when the loop is entirely inside the field, there's going to be no more emf, because the flux through it is not changing. We have already seen this in terms of $\mathbf{v} \times \mathbf{B}$: when the loop is fully in, the contributions to $\mathcal{E}$'s in the segments 12 and 34 due to $\mathbf{v} \times \mathbf{B}$ are equal and opposite.

So far there is nothing in Faraday's law that we could not deduce from just the $\mathbf{v} \times \mathbf{B}$ force. Is there any new content, and if so, when do we encounter it?

We encounter it if we go to the loop frame. Faraday's law tells us that since the flux through the loop is changing (now because the magnet is moving the other way) there will be an emf. It is, however, due to an electric field with non-zero circulation. The law only specifies that the circulation has to equal $-(d\Phi/dt)$, but not what $\mathbf{E}$ is. But in the simple loop experiment we were able to invoke arguments based on relativity to show that $\mathbf{E} = \mathbf{v} \times \mathbf{B}$ and points up in the segment 12, which is in the magnetic field, and is zero in 34, which is outside the field. It is perpendicular to the other two sides. The emf due to this electric field comes from just the segment 12 and equals vBw.

So the new stuff in Faraday's law is the fact that a changing magnetic field implies an electric field of specified circulation. Let us try to extract the precise connection between these two, starting with the definition of $\mathcal{E}$ as the circulation of the electromagnetic Lorentz force on a unit charge

$$\mathcal{E} = \oint_{C=\partial S} (\mathbf{E} + \mathbf{v} \times \mathbf{B}) \cdot d\mathbf{l} = -\frac{d\Phi}{dt} = -\frac{d}{dt} \int_{S(t)} \mathbf{B} \cdot d\mathbf{S}, \quad (10.39)$$

where C is a loop in space around which $\mathcal{E}$ is to be computed, and Φ is the flux penetrating S, which is any surface bounded by C. The loop is a real piece of wire, and $\mathbf{v}$ is the velocity of a segment $d\mathbf{l}$. Thus $\mathbf{v} \times \mathbf{B}$ refers to the magnetic force experienced by the charges in a segment $d\mathbf{l}$ of the wire that have inherited its instantaneous velocity $\mathbf{v}$.

In the right-hand side the rate of change of flux receives contribution from both the changing magnetic field and the changing loop and surface S it bounds. I will show later in this chapter that these two contributions can be nicely separated into two parts that can be identified with the $\mathbf{E}$ and $\mathbf{v} \times \mathbf{B}$ contributions to $\mathcal{E}$ on the left-hand side. Since this derivation is quite tricky, I will first extract the relation between the circulation of $\mathbf{E}$ and the changing magnetic field by a shortcut, leaving the complicated derivation as an option at the end.

The relation between the circulation of $\mathbf{E}$ and the changing $\mathbf{B}$ field is deduced by first considering a loop that is not in motion. This is surely allowed since the answer holds for any state of motion of the loop. Now there is no $\mathbf{v}$ at play and the contour C is fixed. It need not even be associated with any real conductor. It is simply a closed loop in space used to compute the circulation of $\mathbf{E}$. We highlight this by writing a segment of C as $d\mathbf{r}$ instead of $d\mathbf{l}$. We find in this case

$$\oint_{C=\partial S} \mathbf{E} \cdot d\mathbf{r} = -\frac{d\Phi}{dt} = -\frac{d}{dt} \int_S \mathbf{B} \cdot d\mathbf{S}. \qquad (10.40)$$

The derivative d/dt in front of the integral has, in the general case, two parts: one due to the changing C or S, and the other due to the changing $\mathbf{B}$. But now that C is assumed to be fixed, we may take the derivative inside the integral where it can act on $\mathbf{B}$ to give us

$$\oint_{C=\partial S} \mathbf{E} \cdot d\mathbf{r} = \int_S \left[-\frac{\partial \mathbf{B}}{\partial t} \right] \cdot d\mathbf{S}.$$

$$\text{(A final Maxwell equation!)} \qquad (10.41)$$

The partial derivative signifies that we are only computing the rate of change of $\mathbf{B}$ with respect to time and not the spatial coordinates within S. This relation between the fields $\mathbf{E}$ and $\mathbf{B}$, which has no reference to any conductors and how they may be moving, is one of the final four Maxwell equations. It replaces

$$\oint \mathbf{E} \cdot d\mathbf{r} = 0,$$

which we had written down before to express the conserving nature of **E** in electrostatics. *The lesson we have just learned is that in the presence of a time-dependent **B**, the electric field has a non-zero circulation given by Eqn. 10.41.*

Let us see what we have so far. We started with

$$\oint_{C=\partial S} (\mathbf{E} + \mathbf{v} \times \mathbf{B}) \cdot d\mathbf{l} = -\frac{d\Phi}{dt} = -\frac{d}{dt} \int_{S(t)} \mathbf{B} \cdot d\mathbf{S}. \qquad (10.42)$$

In the right-hand side the time derivative generates two terms: one from the time-dependence of **B** and one from the time-dependence of S (because the loop is moving). In other words

$$\oint_{C=\partial S} (\mathbf{E} + \mathbf{v} \times \mathbf{B}) \cdot d\mathbf{l} = \int_{S \text{ fixed}} \left[-\frac{\partial \mathbf{B}}{\partial t} \right] \cdot d\mathbf{S}$$

$$-\text{rate of change of } \Phi \text{ due to changing } S(t). \qquad (10.43)$$

We have just seen that

$$\oint_{C=\partial S} \mathbf{E} \cdot d\mathbf{r} = \int_{S} \left[-\frac{\partial \mathbf{B}}{\partial t} \right] \cdot d\mathbf{S}. \qquad (10.44)$$

It must then be true that the second terms match on both sides:

$$\oint_{C=\partial S} \mathbf{v} \times \mathbf{B} \cdot d\mathbf{l} = -\text{rate of change of } \Phi \text{ due to}$$

$$\text{changing } S(t). \qquad (10.45)$$

If you want to know how this is demonstrated, you must read the next optional section where I discuss the case of a changing loop in a changing field. But in case you skip it, here at least is a brief sketch. Look at Figure 10.8, which depicts a simple case that is somewhat easy to visualize. It shows a circular loop C_1 at time t evolving into a circular loop C_2 at time $t + dt$. The obvious surface to use for computing the flux at $t + dt$ is the planar shaded circular area S_2. But we are free to use any other surface with the same boundary C_2. Let us use S_2', which is just S_1 plus ΔS, the (cylindrical) area swept out by the moving loop. *The advantage is that the contribution to $-\frac{d\Phi}{dt}$ from the changing surface is the contribution from ΔS.* From the figure we see that a portion $d\mathbf{l}$ of the loop moving at velocity **v** sweeps out an area $\mathbf{v}dt \times d\mathbf{l}$ and makes a contribution $-\mathbf{B} \cdot (\mathbf{v}dt \times d\mathbf{l}) = \mathbf{v} \times \mathbf{B} \cdot d\mathbf{l}\, dt$ to $-d\Phi$. The sum of these contributions around the loop gives

$$-\frac{d\Phi}{dt}\bigg|_{\text{due to } \Delta S} = \oint_{C=\partial S} (\mathbf{v} \times \mathbf{B}) \cdot d\mathbf{l}, \qquad (10.46)$$

which precisely matches the $\mathbf{v} \times \mathbf{B}$ term in $\mathcal{E}$. It should be evident that the result holds even if the initial and final loops are not circular and the velocity $\mathbf{v}$ varies with $d\mathbf{l}$. There are a lot of minus signs and orientation of areas to watch out for. All this is described in the next section.

10.4 Optional digression on Faraday's law

Let us return to Faraday's law

$$\mathcal{E} = \oint_{C=\partial S} (\mathbf{E} + \mathbf{v} \times \mathbf{B}) \cdot d\mathbf{l} = -\frac{d\Phi}{dt} = -\frac{d}{dt} \int_{S(t)} \mathbf{B} \cdot d\mathbf{S}. \quad (10.47)$$

We have used it in bits and pieces. We have extracted from it the Maxwell equation relating the circulation of $\mathbf{E}$ and the changing magnetic flux. We have explained the emf of the loop-generator in two situations:

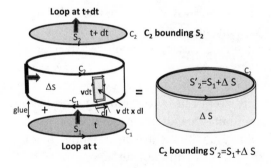

Figure 10.8   The conducting loop C_1 at time t bounds S_1, the lower face of the cylinder. It moves to C_2 at $t + dt$ and bounds S_2, which is the top face of the cylinder. We trade S_2 for $S'_2 = S_1 + \Delta S$, where ΔS is the curved side of the cylinder. This is allowed because the boundary is still C_2: in the sum the common edges C_1 and $-C_1$ (traversed in opposite directions in S_1 and ΔS) get erased, leaving behind C_2. A tiny rectangular part of ΔS is the cross product $\mathbf{v}dt \times d\mathbf{l}$, where $\mathbf{v}dt$ is the vector distance traversed by the segment $d\mathbf{l}$ in time dt. The area vector points inward and the addition of these areas gives ΔS.

- The emf is the integral of **E** and $-d\Phi/dt$ is due to the changing **B**.
- The emf is the integral of $\mathbf{v} \times \mathbf{B}$ and $-d\Phi/dt$ is due to the motion of the loop in a static **B**.

But the remarkable power of Faraday's law is its ability to describe the most general situation, wherein $-d\Phi/dt$ corresponds to a flexible loop moving in a space-time–dependent magnetic field, and the emf is the line integral of both electric and magnetic forces. Let us pursue this feature further.

Let the moving loop bound a surface S_1 at time t and S_2 at $t + dt$, as shown in Figure 10.8. The change in flux is

$$d\Phi = \int_{S_2} \mathbf{B}(t + dt) \cdot d\mathbf{S} - \int_{S_1} \mathbf{B}(t) \cdot d\mathbf{S}. \qquad (10.48)$$

In general S_1 and S_2 can have any shape. However, for the visually and artistically challenged like myself, I limit the discussion to a simple case. (The arguments are good for the general case.) Imagine a closed hollow cylinder made of two flat faces and a curved side. The lower face is our S_1 and its circumference C_1 is the loop of wire at time t. The area vector associated with S_1 points up by the right-hand rule. This also means **B** pointing up contributes positive flux. The upper face of the cylinder is S_2, and its circumference C_2 is where the loop has ended up at $t + dt$. The area vector for S_2 also points up.

In this simple case, the loop has moved straight up the curved face of the cylinder between times t and $t + dt$, with every segment $d\mathbf{l}$ moving with the same velocity **v**. The flat faces can be of any size, but the curved face, ΔS, which is swept out by the moving loop in time dt, should be thought of as an infinitesimal of first order in dt.

It is natural to evaluate the flux penetrating C_2 at time $t + dt$ on the upper face S_2, since it is the simplest surface bounded by C_2. However, the corresponding integral, the first term in Eqn. 10.48, has two effects in one: it is an integral of the field *at a later time on a later surface*. To deal with these two changes one at a time, we will trade S_2 for another surface S_2' with *the same boundary* C_2. We are *allowed* to do this because the flux is going to be the same for any surface with the same boundary. And we *want* to do this because we want to separate the effects of the moving loop and the changing field. What choice of S_2' will do the trick? Imagine that S_2 is a rubber sheet stretched across the circular rim C_2 of a cylindrical drum. Now slowly deform S_2 (blow air into it from above) till it becomes the rest of the cylinder: the curved side, which we call ΔS, and the flat bottom S_1. This is the surface S_2'. Its boundary is still the rim C_2.

It is intuitively clear that S_2' is just S_1 plus the curved face ΔS (which is why we call it ΔS). But let us verify that the two areas have been added as per the rules for gluing areas (deleting oppositely oriented edges that overlap). Consult Figure 10.8.

First observe that ΔS, the surface swept out by the moving loop in the time dt, is itself made of tiny rectangular areas swept out by each segment. Consider a segment $d\mathbf{l}$ of C_1 that moves at velocity $\mathbf{v}$. The area it sweeps out in time dt has a magnitude $|\mathbf{v}|\, dt |d\mathbf{l}|$. The area vector is given by the cross product

$$d\mathbf{S} = \mathbf{v}dt \times d\mathbf{l} \tag{10.49}$$

and points *into* the cylinder. This orientation is to be expected. Originally S_1 and S_2 had area vectors pointing up, by the right-hand rule and by the convention for counting upward flux as positive. If we began with S_2 littered with little upward pointing arrows defining the orientation of the smaller areas it is composed of, and deformed it continuously to the shape S_2', the arrows in the curved side ΔS would end up pointing inward, while the arrows on S_1 would point up.

(In the general case $d\mathbf{l}$ and $\mathbf{v}$ need not be perpendicular and $\mathbf{v}$ need not be the same for all segments. The cross product continues to give the correct area of the *parallelogram* $|\mathbf{v}|dt|d\mathbf{l}|\sin\theta$ swept out.)

When such rectangular areas are glued together to form ΔS, the oppositely oriented vertical edges of neighbors will cancel while the top edges will form C_2 and the bottom edges will form $-C_1$ (which is just C_1 running backward) respectively. The surface ΔS thus has two edges, a lower one $-C_1$, and an upper one, which is C_2. When ΔS is next glued on to S_1 to form S_2' as shown in the figure, the overlapping edges C_1 of S_1 and $-C_1$ of ΔS will get erased and the other edge of ΔS, namely C_2, will become the boundary of S_2'. Since C_2 is also the boundary of S_2, we may swap S_2 for S_2'.

So we *can* trade S_2 for S_2'. I have argued we *should*, because it will sort out the separate contributions from the changing field and changing loop. This will now be shown.

We begin with

$$d\Phi = \int_{S_2} \mathbf{B}(t+dt) \cdot d\mathbf{S} - \int_{S_1} \mathbf{B}(t) \cdot d\mathbf{S} \tag{10.50}$$

$$= \int_{S_2'} \mathbf{B}(t+dt) \cdot d\mathbf{S} - \int_{S_1} \mathbf{B}(t) \cdot d\mathbf{S}$$

because $\partial S_2' = \partial S_2 = C_2$ \hfill (10.51)

$$= \int_{S_1} \mathbf{B}(t+dt) \cdot d\mathbf{S} + \int_{\Delta S} \mathbf{B}(t+dt) \cdot d\mathbf{S}$$

$$- \int_{S_1} \mathbf{B}(t) \cdot d\mathbf{S}$$ \hfill (10.52)

$$= \int_{S_1} \mathbf{B}(t+dt) \cdot d\mathbf{S} + \oint_{C_1} \mathbf{B}(t+dt) \cdot (\mathbf{v}dt \times d\mathbf{l})$$

$$- \int_{S_1} \mathbf{B}(t) \cdot d\mathbf{S}$$ \hfill (10.53)

upon using $d\mathbf{S} = \mathbf{v}dt \times d\mathbf{l}$ in the middle term on the right-hand side of Eqn. 10.52 to arrive at the last equation.

Let us now group the first and third terms, which involve the same surface S_1 but the field at two different times:

$$d\Phi = \int_{S_1} (\mathbf{B}(t+dt) - \mathbf{B}(t)) \cdot d\mathbf{S}$$

$$+ dt \oint_{C_1 = \partial S_1} \mathbf{B}(t+dt) \cdot (\mathbf{v} \times d\mathbf{l})$$ \hfill (10.54)

$$= \int_{S_1} dt \left(\frac{\partial \mathbf{B}}{\partial t} \right) \cdot d\mathbf{S} + dt \oint_{C_1 = \partial S_1} \mathbf{B}(t) \cdot (\mathbf{v} \times d\mathbf{l}).$$ \hfill (10.55)

I changed $\mathbf{B}(t + dt)$ to $\mathbf{B}(t)$ in the second integral since the difference between them is of order dt and there is already a dt in front (from the size of ΔS). Dividing both sides by dt and taking the limit $dt \to 0$ we obtain

$$-\frac{d\Phi}{dt} = -\int_S \left(\frac{\partial \mathbf{B}}{\partial t} \right) \cdot d\mathbf{S}$$

$$- \oint_{C_1 = \partial S} \mathbf{B}(t) \cdot (\mathbf{v} \times d\mathbf{l})$$ \hfill (10.56)

$$= -\int_S \frac{\partial \mathbf{B}}{\partial t} \cdot d\mathbf{S} + \oint_C \mathbf{v} \times (\mathbf{B}(t) \cdot d\mathbf{l})$$

because \hfill (10.57)

$$\mathbf{B}(t) \cdot (\mathbf{v} \times d\mathbf{l}) = -(\mathbf{v} \times \mathbf{B}) \cdot d\mathbf{l}. \tag{10.58}$$

I have removed the subscripts 1 and 2 on S and C for there is only one of each in the limit $dt \to 0$.

So we have in the end

$$\mathcal{E} = \oint_{C=\partial S} (\mathbf{E} + \mathbf{v} \times \mathbf{B}) \cdot d\mathbf{l} = -\int_S \frac{\partial \mathbf{B}}{\partial t} \cdot d\mathbf{S}$$

$$+ \int_C (\mathbf{v} \times \mathbf{B}(t)) \cdot d\mathbf{l}. \tag{10.59}$$

Amazingly, the magnetic parts, which depend on the loop's motion, perfectly match on both sides and can be canceled, leaving us with the Maxwell equation

$$\oint_{C=\partial S} \mathbf{E} \cdot d\mathbf{r} = -\int_S \left(\frac{\partial \mathbf{B}}{\partial t} \right) \cdot d\mathbf{S}. \tag{10.60}$$

There is now no reference to the velocity of the segments $d\mathbf{l}$ of any real loop. What we have instead is a relation between the circulation of $\mathbf{E}$ around some contour C and the rate of change of magnetic flux through a surface bounded by C. To emphasize this I have denoted a segment of this imaginary contour by $d\mathbf{r}$.

Let me make a subtle point about the derivation. The correct magnetic force on the charges in the wire used for computing the emf is really $\mathbf{V} \times \mathbf{B}$ with $\mathbf{V} = \mathbf{v} + \mathbf{u}$, where $\mathbf{v}$ is the velocity of the wire segment $d\mathbf{l}$ and $\mathbf{u}$ is the velocity of the carriers *along the wire* attributed to the current they carry. (This is just like the two parts of the velocity of charges on the leading edge 12 of the rectangular loop being dragged in a magnetic field.) However, in computing the emf we find this extra piece in $\mathbf{V}$ does not matter:

$$\mathcal{E} = \oint (\mathbf{E} + (\mathbf{v} + \mathbf{u}) \times \mathbf{B}) \cdot d\mathbf{l} \tag{10.61}$$

$$= \oint (\mathbf{E} + \mathbf{v} \times \mathbf{B}) \cdot d\mathbf{l}. \tag{10.62}$$

I could set $(\mathbf{u} \times \mathbf{B}) \cdot d\mathbf{l} = 0$ because both the velocity $\mathbf{u}$ and the segment $d\mathbf{l}$ are parallel to the wire.

CHAPTER 11

More Faraday

We have seen that Faraday's law implies that a changing magnetic field will lead to an electric field with a non-zero circulation, as specified by the Maxwell equation:

$$\oint_{C=\partial S} \mathbf{E} \cdot d\mathbf{r} = -\int_S \frac{\partial \mathbf{B}}{\partial t} \cdot d\mathbf{S}. \tag{11.1}$$

As in the case of Gauss's and Ampère's laws, we cannot deduce the induced electric field given just its circulation. However, if the problem has enough symmetry we can. We begin with an example.

11.1 Betatron

The *betatron* was invented to circumvent the problem with the cyclotron at relativistic energies. Recall the operation of the cyclotron. It had two semicircular dees whose diameters were lined up with a tiny space between them. A perpendicular magnetic field penetrated the dees and bent the charge injected. The path of a charge injected into the first dee got bent into a semicircle. As it jumped to the other dee, a downhill voltage was applied across the gap to give it a kick. It then went around the second dee at a higher speed and bigger radius. When it jumped back to the first dee, it got yet another *downhill* kick, because by this time the polarity of the dees had been reversed. After many such downhill kicks, it was ejected from the machine at a high velocity. It was possible to arrange the reversal

200

of the polarities of the dees despite the changing speed and radius because of the following remarkable feature of the kinematics.

Newton's law in the radial direction implies that in a circular orbit

$$\frac{mv^2}{r} = qvB \tag{11.2}$$

$$\frac{v}{r} = \omega = \frac{qB}{m}, \tag{11.3}$$

which means the *frequency of the orbit remains fixed even as the particle speeds up and the orbit size increases.* Thus the requisite alternating voltage between the dees could be provided by simply connecting them to any source of AC voltage of that frequency.

At high velocities the preceding Newtonian kinematics becomes inapplicable. The correct equation is still

$$\mathbf{F} = \frac{d\mathbf{p}}{dt} = q\mathbf{v} \times \mathbf{B} \tag{11.4}$$

but the momentum is not $\mathbf{p} = m\mathbf{v}$ but

$$\mathbf{p} = \frac{m\mathbf{v}}{\sqrt{1 - v^2/c^2}}. \tag{11.5}$$

With this new v dependence of momentum, ω is no longer independent of r.

The betatron does not rely on the constancy of ω or an electrostatic potential to accelerate the particle. It has a totally different design in which a space-time–dependent magnetic field produces a circulating electric field that accelerates the particle. The same magnetic field also bends the particle into a circular orbit of fixed radius. Here are the details.

First a kinematic result. Consider a particle of relativistic momentum $\mathbf{p}$ defined in Eqn. 11.5. Imagine it going around in a circle and also picking up speed. The change in $\mathbf{p}$ has two parts, as shown in the right half of Figure 11.1. Ignore the tangential part due to increase in magnitude dp (which will be produced by a tangential force) and focus on the centripetal part due to changing direction. From the figure it is clear that the change of momentum in the radial direction is

$$dp_r = pd\theta. \tag{11.6}$$

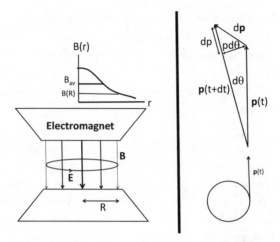

Figure 11.1   Left: The electromagnet produces a field $B(r, t)$ pointing down. Its profile at any typical time is shown as $B(r)$, with an average B_{av}. (The thickness of the downward arrows also indicates the decay of the field with r.) As B grows with time an azimuthal Faraday field $E(R, t)$ accelerates the particle. At each instant the $\mathbf{v} \times \mathbf{B}$ force due to $B(R, t)$ is adjusted to provide the requisite centripetal force to keep it orbiting in a circle of radius R. Right: Top view of the orbit. At time t the particle is moving tangentially at $\theta = 0$ and at $t + dt$ it has acquired some increase in magnitude dp and a change in direction by $d\theta$, and a consequent change in radial momentum $dp_r = p\,d\theta$.

This means the rate of change of momentum in the radial direction is

$$\frac{dp_r}{dt} = p\frac{d\theta}{dt} = p\omega. \tag{11.7}$$

This result, based on geometry and vectors, is true whether $\mathbf{p} = m\mathbf{v}$ as in non-relativistic mechanics or given by Eqn. 11.5. In the non-relativistic case this leads to the familiar result for the centripetal force:

$$p\frac{d\theta}{dt} = p\omega = mv\omega = \frac{mv^2}{r}. \tag{11.8}$$

Now for the betatron. Figure 11.1 shows an electromagnet producing a downward field $B(r, t)$. At some typical time, it has a profile in r as indicated by the graph $B(r)$. The field grows steadily with time, starting from 0 so that, as time goes by, the only change in the profile is a uniform (same at all r) rescaling of the function $B(r)$. On a circle of radius R centered on the symmetry axis of the magnet, this field produces an azimuthal electric field $E(R, t)$ obeying Faraday's law

$$2\pi\, RE(R, t) = \frac{d\Phi(r < R, t)}{dt}, \tag{11.9}$$

where $\Phi(r < R, t)$ is the flux enclosed within the circle of radius R at time t. (Lenz's minus sign is implicit in the direction of $\mathbf{E}$ shown in the figure.)

   We shall assume and ensure as we go along that the particle orbits at the fixed radius $r = R$ even as its speed changes.

   Let us define an average r-independent field $B_{av}(t)$ that will produce the same flux inside $r < R$ as the actual field:

$$\Phi(r < R) = \int_{r<R} \mathbf{B}(r, t) \cdot d\mathbf{S} \equiv \pi R^2 B_{av}(t). \tag{11.10}$$

The electric field may now be related to B_{av}:

$$2\pi\, RE(R, t) = \frac{d\Phi(r < R)}{dt} = \pi R^2 \frac{dB_{av}(t)}{dt} \tag{11.11}$$

$$E(R, t) = \frac{1}{2}R\frac{dB_{av}(t)}{dt}. \tag{11.12}$$

   This azimuthal electric field will change the magnitude of momentum p as follows:

$$\frac{dp}{dt} = qE(R, t) = \frac{q}{2}R\frac{dB_{av}(t)}{dt}. \tag{11.13}$$

Integrating this over time assuming $p(0) = B_{av}(0) = 0$, we obtain

$$p(t) = \frac{q}{2}RB_{av}(t). \tag{11.14}$$

This will be the magnitude of the momentum of the particle at time t.

   Meanwhile the same magnetic field $B(R, t)$ is also required to provide the requisite centripetal force to keep the particle orbiting at $r = R$ despite its growing momentum. We have seen (Eqn. 11.7) that the rate of change of radial momentum is $\frac{dp_r}{dt} = p\omega$. We equate this to the available centripetal force $q\mathbf{v} \times \mathbf{B}$:

$$p\omega = qvB(R, t) = q\omega RB(R, t) \qquad (11.15)$$

using $v = \omega R$. Canceling ω we find

$$p(t) = qRB(R, t). \qquad (11.16)$$

Although all of $B(r, t)$ inside $r < R$ contributes to the changing flux (that in turn generates $\mathbf{E}(R, t)$), only the field at the orbit $B(R, t)$ applies the centripetal $\mathbf{v} \times \mathbf{B}$ force.

Look at Eqns. 11.14 and 11.16. The first tells you the momentum $p(t)$ the particle has acquired in time t due to the acceleration produced by E. The second tells you what value of $p(t)$ the $\mathbf{v} \times \mathbf{B}$ force at R can handle, i.e., manage to bend into a circle. Equating the two expressions to satisfy the assumption of a circular orbit of radius R, we find the condition for operation:

$$\frac{q}{2}RB_{av}(t) = qRB(R, t) \qquad (11.17)$$

$$B_{av}(t) = 2B(R, t). \qquad (11.18)$$

For the betatron to work *the average field within $r < R$ should be double the field at $r = R$ at every instant*. This is what I have tried to convey in the figure by plotting $B(r)$ and B_{av} at one time. If, however, all we do is crank up the current in the electromagnet and *uniformly* raise the profile of $B(r)$ (by the same factor for all r), the condition $B_{av}(t) = 2B(R, t)$ will hold at all times if it holds initially.

The magnetic field is playing a dual role. By its time-dependence inside $r < R$, it is producing the circulating electric field E (which then accelerates the particle) and, through its $\mathbf{v} \times \mathbf{B}$ force at the orbital radius R, it is keeping it in a circle even as the magnitude of p grows.

The betatron beats the relativistic kinematics but it too eventually runs into problems because charged particles emit radiation when accelerating, and the loss invalidates the preceding analysis.

11.2 Generators

Now for a practical topic: a power generator. Remember, I told you that one way to light up a bulb is to take the conducting loop and keep running, making sure the loop is partly in and partly out of the perpendicular magnetic field. Another option was to sit still with the loop and pay someone to run the other way with the magnet. These options for changing the flux are good material for jokes about how many Yalies it takes to light a bulb, but not practical. Here is a better way. Look at the top half of Figure 11.2, which shows a generator from an angle. There is a loop, taken for convenience to be a square of side a. It is free to spin about the axis as shown by the big curved arrow. It is immersed in a constant magnetic field $\mathbf{B}$ produced by a permanent magnet. The loop's area vector $\mathbf{A}$ is perpendicular to the plane of the loop and is at an angle θ relative to $\mathbf{B}$. The flux penetrating this area is

$$\Phi(\theta) = \mathbf{A} \cdot \mathbf{B} = AB\cos\theta. \qquad (11.19)$$

There are two leads coming out. First assume open-circuit conditions, in which the leads are not connected to anything. Ignore the arrows

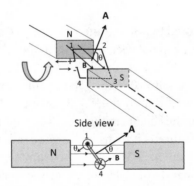

Figure 11.2   The generator. The square loop of side a is in the field of a permanent magnet. When it is rotated, an emf will appear, which is equal to the integral of the $\mathbf{v} \times \mathbf{B}$ force or the rate of change of flux. In the open circuit, this emf will cause charges to pile up at the leads as shown, until their internal electrostatic field balances the emf. When the circuit is closed a current will flow and do work on a bulb, for instance, and work will have to be done by an outside agent to turn the loop.

near the ends. Suppose I begin to turn the loop in the sense of the curved arrow with angular frequency ω. That is

$$\theta(t) = \omega t. \qquad (11.20)$$

There is going to be an emf. As before, we can compute it in two ways. We can integrate the $\mathbf{v} \times \mathbf{B}$ force on each of the four sides of the loop or look at the rate of change of flux penetrating it.

In the first approach, we note that in the sides 23 and 41 the vector $\mathbf{v} \times \mathbf{B}$ is perpendicular to the segments $d\mathbf{l}$ and hence does not contribute to the emf. As for section 12, it is better to see the side view in the lower half of the figure. The segment 12 is rotating counterclockwise at a speed $\omega \cdot a/2$ and the force on a unit charge is $(a/2)\omega B \sin\theta$ pointing from 2 to 1 and its line integral is $(a/2)\omega B a \sin\omega t$. The opposite side 34 makes an equal contribution (in the same sense) for a total of

$$\mathcal{E} = \omega Ba^2 \sin\theta = \omega BA \sin\omega t. \qquad (11.21)$$

It is a lot easier to find $\mathcal{E}$ by differentiating Eqn. 11.19

$$\mathcal{E} = -\frac{d\Phi}{dt} = AB\omega \sin\omega t. \qquad (11.22)$$

Now, $\mathcal{E}$ is supposed to be computed around a *closed* loop and we have an open circuit with a gap between the leads in the edge 41. In the limit of an infinitesimal gap, which I assume here, this makes no difference. Or if you like, you may set $\mathbf{v} \times \mathbf{B} = 0$ in the gap in computing the emf.

What will this emf due to the $\mathbf{v} \times \mathbf{B}$ force do? It will try to drive a current that will fight the change in flux. As shown in the figure, the loop is going to intercept less flux as it turns in the sense indicated. So the current would begin to flow from $4 \rightarrow 3 \rightarrow 2 \rightarrow 1$ to counter it. (You should check this using the right-hand rule.) However, in the open-circuit condition, the current cannot flow around the gap between the leads. So $\pm$ charges will pile up at the open leads as shown, until the electric field they create inside the conductor balances the $\mathbf{v} \times \mathbf{B}$ force. The electric force due to the built-up charges therefore has a line integral equal in magnitude to $\mathcal{E}$ inside the generator. But, being conservative, it must have the same integral on any path joining the terminals outside the generator. This means that in the outside world there will be a path-independent electrostatic voltage difference between the terminals equal to $\mathcal{E}$.

This is exactly what happened in the battery. There a non-conservative chemical force was piling up positive and negative charges in the two terminals and this went on till the Coulomb field set up by these charges in the opposite direction exactly balanced it. The conservative electrostatic force that balanced the non-conservative chemical force inside the battery had to have the opposite line integral inside the battery, equal in magnitude to $\mathcal{E}$. But being a conservative force, it had to have the *same line integral on any path joining the terminals but lying outside* the battery as well. So in the world outside the battery, there was a potential difference $V = \mathcal{E}$ between the terminals waiting to be used to light up a bulb or drive a motor.

There is a subtle issue arising from the fact that the emf in the generator is time-dependent (varies as $\sin \omega t$). The electric field required to balance it is therefore not really static. However, as long as ω is not too big, the retardation effects will be small, and the electric field due to the built-up charges can continue to balance the changing $\mathbf{v} \times \mathbf{B}$ force at every instant. We can continue to use the ideas from electrostatics including that of a potential and voltage.

Back to the battery. Once the battery is connected to a device, the accumulated charges begin to flow downhill from the plus to the minus terminal through the device. This will momentarily weaken the electrostatic field inside the battery, and the chemical forces will briefly win, replenish the terminals, and quickly restore the balance inside. This response will be quick enough for the outside world to get a steady voltage between the terminals if the current drawn is below some limit.

Likewise, once the generator is connected to a device that draws current, the charge buildup at the leads of the loop will momentarily decrease and will not fully balance the $\mathbf{v} \times \mathbf{B}$ force. The uncompensated part of the $\mathbf{v} \times \mathbf{B}$ force will cause some charge accumulation till the Coulomb and $\mathbf{v} \times \mathbf{B}$ forces are rendered equal and opposite. Usually this will happen so quickly that we will not see the momentary voltage drop unless we draw too much current. But if we do, we will see the lights dim for a brief period.

We now face a paradox previously encountered in our discussion of a conducting loop being dragged in a perpendicular $\mathbf{B}$ field. In both cases we have an electric field inside a perfect conductor. Is this not forbidden by its very definition? The answer, as before, is that the real constraint in a perfect conductor that prevents the unlimited acceleration of its free charges is that of zero *net force* and not zero electric field. Thus the moment the $\mathbf{v} \times \mathbf{B}$ force appears due to the rotation of the loop, a compensating electrostatic force generated by the charge buildup is not only allowed but required.

In the open-circuit configuration, there is a voltage available between the terminals, which could be connected to the power outlets in your home, waiting to be used. It does not, however, cost you till you plug in a device and draw current. It costs no energy to turn the loop because there are no currents in any of its four segments to experience the $I d\mathbf{l} \times \mathbf{B}$ force.

This changes when we connect the leads to a resistor R and current begins to flow. The power consumed by the resistor is

$$P_{res} = \mathcal{E}I, \tag{11.23}$$

where $I = \mathcal{E}/R$. Who is paying for this? I am, assuming I am turning the loop. This requires energy because the current-carrying loop experiences a torque opposing the rotation. The energy is of mechanical origin, provided by me turning the crank (or the turbine blades rotated by running water). The mechanical power supplied is torque times angular velocity for rotations. The torque has a magnitude

$$\tau = |\boldsymbol{\mu} \times \mathbf{B}| = AIB\sin\theta \tag{11.24}$$

and the power supplied by me is

$$P_{me} = \omega AIB\sin\theta = \mathcal{E}I = P_{res} \tag{11.25}$$

since $\mathcal{E} = \omega AB\sin\theta$ as per Eqn. 11.22.

The turbines in the real world have a sizable mass and moment of inertia, as well as friction. It takes some power to keep them spinning even without a current load placed by consumers. The minute you plug a toaster into the socket, you start drawing current, and that current flows right through the loop in the generator, making it that much harder to rotate. That's when the steam turbines really get to work. That's what you pay for.

11.3 Inductance

Consider the setup depicted in Figure 11.3. I wrap some turns of wire around a cardboard tube and connect this *primary* solenoid to some alternating voltage. Focus on the situation at *one instant* when the primary

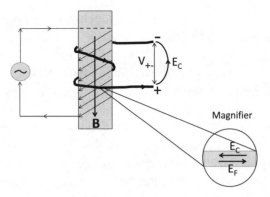

Figure 11.3 *At this instant* the current in the primary coil produces a **B** flux that points down and links with the secondary. If this current increases, $\mathcal{E}$ and the Faraday field $\mathbf{E}_F$ in the secondary will fight it by driving a current that will cause the buildup of $\pm$ charges as shown in the open-circuit configuration. These will set up a Coulomb field $\mathbf{E}_C = -\mathbf{E}_F$ since there can be no net field in a conductor. The magnified view of a piece of the secondary shows this. The conservative field $\mathbf{E}_C$ will, however, have the same line integral between the $+$ and $-$ terminals *outside the coil* as it did inside, and this will translate into a difference in voltage $V(+) - V(-) \equiv V_{+-} = \mathcal{E}$.

current is flowing as shown. There is going to be some magnetic flux going through it, pointing down the solenoid. There is a second wire, the *secondary*, wrapped around the primary a few times and with its ends dangling as leads. What will I find at the leads of the secondary solenoid at this instant?

If the current through the primary changes, so does the flux inside it. That means the flux through the secondary also changes since both coils wrap around the same flux. Let the primary current be increasing at this instant. This means the downward **B** is increasing. There is going to be an emf $\mathcal{E}(t)$ in the secondary to counter this increase. This time the emf is not due to the $\mathbf{v} \times \mathbf{B}$ force, but the induced electric field $\mathbf{E}_F$ as mandated by Faraday's law:

$$\mathcal{E} = \oint \mathbf{E}_F \cdot d\mathbf{r}. \tag{11.26}$$

The direction of $\mathbf{E}_F$ will be as indicated in the magnifier, in order that it may drive a current that will oppose the increase in flux.

For computing $\mathcal{E}$ integrate $\mathbf{E}_F$ counterclockwise along the following loop $\mathcal{L}$: Begin at the end of the lead marked $-$, and then move leftward into the top of the secondary, through it till you emerge at the lower end, then rightward to the point marked $+$, and finally back to point $-$ via the curve marked E_C. (This curve does not correspond to a physical wire; it is a path used for computing $\mathcal{E}$.)

The field $\mathbf{E}_F$ will not succeed in driving the current the way it wants to because of the open circuit. It will, however, pile up charges, leaving the top lead with a net negative charge (as the current flows away from it) and the bottom with a net positive charge (as the current flows into it). These piled up charges will very quickly set up an electrostatic or Coulomb field $\mathbf{E}_C$ that exactly balances the Faraday field $\mathbf{E}_F$ as indicated in the magnified view in the figure. Just as in the battery, we have a conservative field balancing a non-conservative one *inside* the solenoid. This implies a voltage V_{+-} between the $+$ and $-$ leads that equals $\mathcal{E}$, as shown below. However, unlike in the battery, the emf and voltage V_{+-} are time-dependent.

Here is an equivalent demonstration that a voltage difference $V = \mathcal{E}$ will appear between the leads of the secondary in the open-circuit configuration shown.

$$\mathcal{E} = \oint \mathbf{E}_F \cdot d\mathbf{r}$$

around loop $\mathcal{L}$ which includes secondary   (11.27)

$$= \int_-^+ \mathbf{E}_F \cdot d\mathbf{r}$$

just inside secondary, as $\mathbf{E}_F = 0$ outside   (11.28)

$$= \int_-^+ (-\mathbf{E}_C) \cdot d\mathbf{r}$$

inside secondary, as $\mathbf{E}_F = -\mathbf{E}_C$ there   (11.29)

$$= V(+) - V(-) \equiv V_{+-}.$$ (11.30)

Once the secondary circuit is closed and current begins to flow through some device, these $\pm$ charges will begin to migrate to the opposite terminals and disappear. However, as long as the alternating current does not change too fast, there will always be enough electric charges to ensure that the total electric field inside the coil (the sum of $\mathbf{E}_C$ and $\mathbf{E}_F$) continues to vanish and that the voltage difference $V_{+-} = \mathcal{E}$ appears between the leads of the secondary.

Once again we really should not be using Coulomb's law or electrostatics for this problem, since they are applicable only for fixed electric charges. But as long as the retardation effects are negligible, we can continue to use the twin notions of an electrostatic Coulomb force that can instantaneously neutralize a time-dependent Faraday force $\mathbf{E}_F$, and the corresponding potential V_{+-}.

I have devoted considerable time to show you how you may use the notion of a potential difference between the terminals of a battery, generator, and solenoid *in the world outside*, despite the presence of non-conservative forces inside. There is, however, one difference between the solenoid and the other two. The $\mathbf{v} \times \mathbf{B}$ of the generator and the chemical forces of the battery do not preclude the existence of a conservative electrostatic field and its associated potential $V = \mathcal{E}$ between the terminals. But the Faraday field $\mathbf{E}_F$ is a different matter. The time-dependent flux of the solenoid may not be confined to the solenoid—it can leak to the sides and indeed *has* to leave the solenoid during its return from the north end to the south. If this flux penetrates a circuit, we cannot define a path-independent potential in its presence because $\oint \mathbf{E} \cdot d\mathbf{r} \neq 0$. So we must either hope this flux leakage is negligible or find a way to keep it out of the circuit. An excellent way is to wrap the primary and secondary coils around a toroidal iron core. Now almost all of its flux will be trapped in the iron core and not venture into the vacuum outside (due to some energetics that we cannot discuss here).

The bottom line is that with the preceding caveats, when the secondary solenoid is part of a circuit, you may demand that the sum of all the voltage changes is zero if you go around a loop that includes the secondary, with a jump $V_{+-} \equiv V(+) - V(-) = \mathcal{E}$ as we cross the secondary.

11.4 Mutual inductance

Let us relate the emf in the *secondary* coil to the alternating current in the *primary*. We normally write

$$\mathcal{E} = -\frac{d\Phi}{dt} \tag{11.31}$$

for a loop enclosing flux Φ. The emf in the secondary is actually

$$\mathcal{E}_2 = -N_2 \frac{d\Phi}{dt} \tag{11.32}$$

where N_2 is the number of turns in the secondary. The reason for the factor N_2 is that the field $\mathbf{E}_F$ is to be integrated from one end of the solenoid to the other to find $\mathcal{E}_2$ and *each turn* contributes $-d\Phi/dt$. Equivalently, each turn is like a little battery with $\mathcal{E} = -d\Phi/dt$ and N_2 of these have essentially been hooked up in series. So the relevant quantity here is Φ_2, the *flux linked to the coil* 2:

$$\Phi_2 = N_2 \Phi \tag{11.33}$$

where Φ is the flux crossing each turn, the flux running through the length of the primary solenoid. Thus

$$\mathcal{E}_2 = -\frac{d\Phi_2}{dt}. \tag{11.34}$$

Let us calculate Φ_2. The magnetic field inside the primary is

$$B = \mu_0 n_1 I_1, \tag{11.35}$$

where $n_1 = N_1/l$ is the turns per unit length of the primary and I_1 is the current through it. By construction, all the flux inside the primary is linked to every turn in the secondary. The magnetic flux linking with the secondary coil of cross section A_2 is

$$\Phi_2 = N_2 \cdot (\mu_0 n_1 I_1) \cdot A_2 \equiv M_{21} I_1 \tag{11.36}$$

where I have defined the quantity

$$M_{21} = N_2 \mu_0 n_1 A_2 \tag{11.37}$$

called the *mutual inductance* of solenoids 1 and 2. The mutual inductance Φ_2/I_1 is the flux linking with solenoid 2 due to unit current in solenoid 1. That Φ_2 is linearly proportional to I_1 is to be expected based on the superposition principle. If you double the current in the primary you double the field it produces because you can think of the doubled current as the sum of two identical currents flowing in the same wire, each producing its own field. (This also follows from the Biot-Savart law.)

Putting all this together

$$\mathcal{E}_2 = -\frac{d\Phi_2}{dt} \tag{11.38}$$

$$= -M_{21}\frac{dI_1}{dt}. \tag{11.39}$$

Consider the relation

$$M_{21} = M_{12}, \tag{11.40}$$

which claims that the flux linking with solenoid 2 due to unit current in solenoid 1 is the same as the flux linking solenoid 1 due to unit current in solenoid 1. It is not obvious because according to Figure 11.3, all the flux produced by 1 also penetrates 2, but the opposite is not true. The result would be more obvious if both solenoids were wound around the same toroid, for then the flux due to either runs through the same toroidal core. The result, however, is valid even in the non-obvious cases.

In general, we can define and measure the mutual inductance $M_{12} = M_{21} = M$ of any two loops (not necessarily wound around the same core) by driving unit current in either loop and finding how much of its flux links with the other. Mutual inductance can be very important in designing circuits. It can be useful when intentionally coupling two loops. However, at other times, the circuit may have two closed loops in proximity that were not meant to be coupled, but end up experiencing the unwanted emf's due to a changing current in the other.

Inductance is measured in henrys (H) in honor of Joseph Henry (1797–1878).

Consider two coils with N_1 and N_2 turns wrapped around the same donut-shaped core with an alternating current flowing in the primary. *Since the same field penetrates both,* the ratio of the flux linkage is simply in the ratio of the number of turns and this carries over to the ratio of the emf's upon taking the time derivative of the flux:

$$\frac{\mathcal{E}_1}{\mathcal{E}_2} = \frac{N_1}{N_2}. \tag{11.41}$$

We are evidently talking about a *transformer* here. You apply an AC voltage to the primary and a proportional AC voltage appears in the secondary. It could be higher or lower, depending on the ratio N_2/N_1—it could be a step-up or step-down transformer. You can also decide to drive the current through the secondary to get a voltage on the primary with the reciprocal ratio of voltages. Although you can step up or step down the voltage, you cannot create energy this way. You also cannot step up or down DC voltages using this principle.

11.5 Self-inductance

Now we turn to a very important circuit element, the *inductor*. It is a single solenoid and it can be part of a circuit carrying a current $I(t)$. We know that when the current goes through a resistor there is a voltage drop $V_{in} - V_{out} = IR$, between where the current comes in and goes out. What will be the corresponding voltage drop for an inductor?

The wire in the solenoid is a perfect conductor, and therefore it takes no voltage at all to drive a steady current through it. But when the current through the inductor is changing, the drop across it will be non-zero by the Faraday effect, due to the emf generated in the solenoid by its *own* changing current.

Time-dependent currents rise naturally in AC circuits and also in a transient process like the one depicted in Figure 11.4, which we will initially focus on.

Figure 11.4 shows a battery of terminal voltage V_0 connected in series to an inductor L and resistor R via a switch S. When S is closed, the current that begins to flow will produce a magnetic flux in the coil. An emf $\mathcal{E}$ will be generated in the coil to oppose this growth. The emf is the rate of change of Φ_{sel}, flux linking with the coil *due to its own current*. By the superposition principle, the field and flux have to be linear in the current. So we may define the *self-inductance* denoted by L

$$\Phi_{sel} = LI \tag{11.42}$$

as the constant of proportionality. Postponing for a while the computation of L, we proceed to find $\mathcal{E}$ in terms of it:

$$\mathcal{E} = -\frac{d\Phi_{sel}}{dt} = -L\frac{dI}{dt}. \tag{11.43}$$

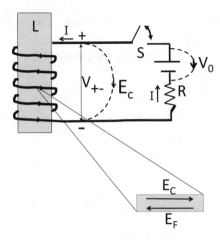

Figure 11.4   An LR circuit. When the current flows, the drop across the inductor (as we follow the current) is LdI/dt. Inside the inductor, the induced field $\mathbf{E}_F$ is neutralized by the Coulomb field $\mathbf{E}_C$ due to built-up charges, as shown in the magnified view of a tiny segment of the coil. The emf $\mathcal{E}$ receives a non-zero contribution to the line integral of $\mathbf{E}_F$ only inside the coil from the $-$ to the $+$ terminal. This in turn equals the line integral of $\mathbf{E}_C$ from $+$ to $-$ inside, which is also the integral outside the coil because $\mathbf{E}_C$ is conservative. This leads to $V(+) - V(-) \equiv V_{+-} = \mathcal{E}$. The dotted lines do not correspond to a physical wire.

We will ignore the minus sign and use it instead as a guiding principle when we consider specific situations and want to know which way a voltage, field, or current will be directed.

As before if we enclose the inductor in a black box, that is to say, we assume its changing flux is somehow confined to its interior and does not link with the rest of the circuit, we can ask what voltage we will measure between the leads. This is going to be a familiar discussion, and let us do it one last time with feeling, but with some variations to relieve the monotony.

Suppose the current is coming in to L as shown in Figure 11.4, and is trying to increase. The Faraday field $\mathbf{E}_F$ will try to push charges in a way that opposes the increasing current, causing the $+$ and $-$ charges to pile up as shown. So it's the same story again. There can be no net field inside the coil, which is a perfect conductor. The Faraday field $\mathbf{E}_F$ is canceled by a Coulomb field $\mathbf{E}_C$. The two will have equal and opposite line integrals inside the coil. So we may equate $\mathcal{E}$, line integral of $\mathbf{E}_F$ from − terminal to the + terminal, to the voltage difference between the external leads. Just to reinforce various concepts, I present the previous arguments in a string of equations:

$$\mathcal{E} = \oint \mathbf{E}_F \cdot d\mathbf{r} \quad \text{(on any loop that includes the coil)} \quad (11.44)$$

$$= \int_{-}^{+} \mathbf{E}_F \cdot d\mathbf{r} \quad \text{(inside coil, as } \mathbf{E}_F = 0 \text{ outside)} \quad (11.45)$$

$$= -\int_{-}^{+} \mathbf{E}_C \cdot d\mathbf{r} \quad \text{(as } \mathbf{E}_F = -\mathbf{E}_C \text{ inside coil)} \quad (11.46)$$

$$= \int_{+}^{-} \mathbf{E}_C \cdot d\mathbf{r} = V(+) - V(-) \equiv V_{+-}$$
$$\text{(by definition of voltage).} \quad (11.47)$$

This means that for people thinking outside the box (that confines the flux), there is a path-independent potential difference

$$V(+) - V(-) = \mathcal{E} = L\frac{dI}{dt} \quad (11.48)$$

between the two ends of the inductor. Once again we assume that the notion of a potential may be extended from the truly static situation to the present one where it is time-dependent.

The implications for circuit theory is that if we follow the current in a circuit, there will be a voltage drop of IR when we pass the resistor and a *drop* of LdI/dt across the inductor, between the end where the current enters and the end where it leaves. If the current is increasing, this really will be a drop. But if the current is *decreasing* (and still flowing in the indicated direction) the drop will actually be *negative*.

Thus unlike in the resistor, the voltage "drop" across the inductor need not be a drop in the direction of the current. It is decided by its rate of change. The arrows in circuit diagrams generally show only the direction of I but not its rate of change. So LdI/dt can have either sign.

Let us return to the *LR* circuit of Figure 11.4 and ask what happens when the switch S is closed. Let us impose the condition of zero voltage change around a closed loop. The loop to use is the following: Start from the positive terminal of the battery, go along the connecting wire to the +

terminal of the inductance, *jump to the − terminal along the dotted curve marked E_C, bypassing the interior of the solenoid with its nasty* $\mathbf{E}_F$ *for a drop* $L dI/dt$, go through the resistor for a drop of IR and onward to the negative terminal of the battery, and go around the battery *against* the dotted curve to the positive terminal for a *gain* of V_0. The sum of all these changes must be 0, or the magnitude of the gain in voltage must equal the magnitude of the drop:

$$L\frac{dI}{dt} + IR = V_0. \tag{11.49}$$

So this is the equation to solve. Since the next chapter is all about solving this and many such equations, let us wrap up the discussion with the calculation of L, defined as

$$L = \frac{\text{flux linking with inductor}}{\text{current producing flux}}. \tag{11.50}$$

The flux linking with itself is the product of the number of turns times the value of $B = \mu_0 nI$ times the cross-sectional area A:

$$\Phi_{sel} = N\mu_0 nIA \quad \text{which means} \tag{11.51}$$

$$L = \mu_0 nNA = \mu_0 \frac{N^2}{l} A \tag{11.52}$$

where l is the length of the solenoid.

11.6 Energy in the magnetic field

How much energy is stored in an inductor carrying current I? This is a meaningful question because when you begin to drive a current through an inductor, you are doing some work. The changing current is opposing you with a voltage $L dI/dt$, and you're ramming it down in spite of that opposition. The power needed is

$$P = VI = L\frac{dI}{dt}I = \frac{1}{2}L\frac{d[I^2]}{dt}. \tag{11.53}$$

Upon integrating both sides from $t = 0$ to $t = t$, and assuming $I(0) = 0$, we find the stored energy is

$$U = \frac{1}{2}LI^2. \tag{11.54}$$

So it takes some energy to build up a current in the inductor just like it takes some energy to charge up a capacitor.

Feeding in the explicit expression for L shown in Eqn. 11.52,

$$U = \frac{1}{2}\mu_0\frac{N^2}{l}AI^2 \tag{11.55}$$

$$= \frac{1}{2}\left[\frac{\mu_0 NI}{l}\right]^2\frac{1}{\mu_0}Al \tag{11.56}$$

$$= \frac{B^2}{2\mu_0} \times Al. \tag{11.57}$$

Since Al is the volume over which the field $B = \mu_0 nI$ exists, the magnetic energy per unit volume is

$$u_B = \frac{B^2}{2\mu_0}. \tag{11.58}$$

For this discussion it is better to consider a toroidal solenoid whose flux is very well confined. The final formula for u_B is exact and can be derived in many other ways.

Recall that the energy density in the electric field is

$$u_E = \frac{1}{2}\varepsilon_0 E^2. \tag{11.59}$$

So u_E and u_B are given by very similar formulas. Both are quadratic in the fields and even the constants behave similarly: μ_0, which is normally upstairs in every formula, comes downstairs here, and ε_0, which is always downstairs in every formula, comes upstairs here.

So let me summarize what you should remember from all of this. The circuit element called an inductor is just a coil of wire that's wrapped around some core. When you change the current through the inductor, it's going to fight it. It's not like a resistor. A resistor fights any current.

An inductor fights only a change in current. All this is summarized in the circuit equation

$$V_0 = L\frac{dI}{dt} + RI. \tag{11.60}$$

Even without solving this equation we can say some things based on what we know. For example, the current in the circuit infinitesimally after the switch is closed must be 0. Why not something else, say .2 A? A current that jumps from zero to something non-zero in zero time would have an infinite derivative. This is not allowed since LdI/dt cannot ever exceed V_0. So the current in the inductor will never jump. On the other hand, if you connect a battery to a resistor the current can immediately assume the value $I = V/R$.

These restrictions follow from energy considerations. The current in the inductor implies a stored energy of $\frac{1}{2}LI^2$. If the current jumps instantaneously, so does the stored energy, implying infinite power in or out, which is impossible. On the other hand, a resistor stores no energy and the current through it can jump when a switch is opened or closed.

CHAPTER 12

AC Circuits

By AC I mean "not DC." The currents and voltages may not be oscillatory in each case, but in all cases they will be varying with time. The circuits could contain resistors, inductors, and capacitors.

12.1 Review of inductors

Let me start by reviewing inductors before returning to circuits containing them.

An inductor is very different from the resistor in circuit theory both in its energetics and its mathematical treatment. When you connect a resistor to some voltage $V(t)$, the current is determined by

$$I(t)R = V(t), \tag{12.1}$$

which is an *algebraic equation*. This means you can use elementary algebra to solve for the current: simply divide both sides by R and obtain

$$I(t) = \frac{V(t)}{R}. \tag{12.2}$$

You can make the network more complicated—add a few more resistors, connect some in series and others in parallel, and so forth. No matter what you do, you can always combine them by the usual rules to find the current leaving the battery. If you follow that current and you run into a branch, there are simple rules to tell you in what ratio the current will split among the branches. You do not need any calculus to deal with this problem.

When you bring in inductors, things are different. If you have a current going through an inductor, there will necessarily be a voltage drop

$$V = L\frac{dI}{dt} \tag{12.3}$$

in the direction of the current. The "drop" could be negative *if the current is decreasing.* The first difference you notice is that the relation between voltage and current is not an algebraic equation, but a differential equation. In due course I will tell you how to solve the differential equations.

The second difference between the inductor and resistor is that when a current flows through a resistor, whatever energy you provide is gone in the form of heat. It is dissipated. The lightbulb glows and that's the end. With an inductor, when you begin to drive a current, you are building a magnetic field inside the inductor and there's an energy associated with the magnetic field. That stored energy will be given back to you later on. So it's like a capacitor. It takes work to charge a capacitor, because you've got to take charges from one plate and keep on piling them in the other plate, despite the opposition you get. But then if you connect the plates to a bulb and squeeze the trigger in your camera, the discharging capacitor gives back the energy you put into it.

Let us start with a simple problem, depicted in Figure 12.1. I apply a fixed voltage V_0 to a resistor R and inductor L connected in series through an open switch S. Ignore for now the part in dotted lines with the large resistor R'. Or imagine $R' = \infty$ so that no current goes there.

When I close the switch how big a current will begin to flow? The circuit equation is

$$L\frac{dI}{dt} + RI = V_0. \tag{12.4}$$

Because the inductor is a resistance-free wire, you may think a current $I = V_0/R$ will start flowing immediately, but we have seen that that is wrong. Instead the current will start to climb continuously from zero.

What is the function $I(t)$ that describes the current? Let us begin with some basic deductions.

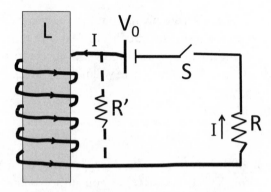

Figure 12.1   The LR circuit coupled to a battery via a switch. The dotted part of the circuit connected to a very large resistor R' can be ignored for now and will be referred to later.

As the current starts climbing up, the resistor uses up a voltage RI and only the balance $V_0 - RI$ is available to sustain dI/dt. As the current increases, the propensity to increase decreases. We expect that after a very long time, it will settle down to some value. We can find it by setting $dI/dt = 0$ in the circuit equation Eqn. 12.4:

$$I(\infty) = \frac{V_0}{R}. \tag{12.5}$$

I call this current $I(\infty)$ because the current will be seen to reach this value only at $t = \infty$. This is reminiscent of a battery trying to charge a capacitor through a resistor. Initially all of V_0 was available for driving the current through R, but as the capacitor starts charging up, it begins fighting the battery. The current gets smaller and smaller but never quite stops since the capacitor can never equal the battery in its opposition. A similar thing happens when a capacitor discharges through a resistor. It never gets fully drained for any $t < \infty$ because as it drains, it has less and less voltage left to discharge through the resistor. In the present case of the LR circuit, as the current grows, it becomes its own enemy due to the increasing drop across R.

However, the current *can reach any fraction* of $I(\infty)$, say .95, in a finite time. To find the time t^* when this happens, we need to buckle down and solve for $I(t)$ starting with

$$L\frac{dI}{dt} + RI = V_0. \tag{12.6}$$

But for the V_0 on the right, we could solve this easily. So we eliminate it as follows. Let us write the current as a sum of the asymptotic value $I(\infty) = V_0/R$ and the rest, denoted by $\tilde{I}$:

$$I(t) = I(\infty) + \tilde{I}(t). \tag{12.7}$$

If we substitute this into Eqn. 12.6, we find (noting that $I(\infty)$ has zero time derivative),

$$L\frac{dI}{dt} + RI = V_0 \tag{12.8}$$

$$L\frac{dI(\infty)}{dt} + L\frac{d\tilde{I}}{dt} + RI(\infty) + R\tilde{I} = V_0 \tag{12.9}$$

$$0 + L\frac{d\tilde{I}}{dt} + V_0 + R\tilde{I} = V_0$$

$$\text{since } RI(\infty) = V_0 \tag{12.10}$$

$$L\frac{d\tilde{I}}{dt} + R\tilde{I} = 0, \tag{12.11}$$

which can be solved by inspection:

$$\tilde{I}(t) = I_0 e^{-tR/L} \equiv I_0 e^{-t/\tau} \quad \text{where} \tag{12.12}$$

$$\tau = \frac{L}{R} \tag{12.13}$$

is the time-constant for the LR circuit and I_0 is arbitrary, as in all linear equations.

To find I_0 we impose the initial condition that the full current vanishes at $t = 0$:

$$0 = I(0) = I(\infty) + \tilde{I}(0) \tag{12.14}$$

$$= \frac{V_0}{R} + I_0 e^{-0}, \quad \text{which means} \tag{12.15}$$

$$I_0 = -\frac{V_0}{R}. \tag{12.16}$$

Armed with this result let us reconstruct the full current $I(t)$:

$$I(t) = I(\infty) + \tilde{I}(t) \tag{12.17}$$

$$= \frac{V_0}{R} + I_0 e^{-t/\tau} \tag{12.18}$$

$$= \frac{V_0}{R}\left[1 - e^{-t/\tau}\right] \equiv I(\infty)\left[1 - e^{-t/\tau}\right]. \tag{12.19}$$

This result again illustrates the interplay between theory and experiment. We study things experimentally, define and measure some physical variables like L, C, R, and I, write down some equations governing them, and solve the equations. Then we get a very precise prediction for what will happen under some given conditions, which we run off to verify experimentally. In the present instance, we don't have to guess at what time t^* the current will come to 95 percent of its maximum value. It is the solution to

$$.95 = \frac{I(t^*)}{I(\infty)} = 1 - e^{-t^*/\tau} \tag{12.20}$$

and has a value $\simeq 3\tau$.

As with the capacitor, the time-constant gives us a natural unit of time appropriate to this problem. We know the current will never reach $I(\infty)$ but we also know that if we wait a long time, it will get really close. It is τ that tells us what "long time" means—it means many times τ.

Let us say we have waited till $t = 1000\tau$. Now we open the switch. What will happen? Normally when you try to reduce the current, the inductor will fight back by driving its own current to prop up the current. But now it is going to be very frustrated because, with the switch open, it cannot drive any current! Also, how is it supposed to get rid of its magnetic energy all of a sudden? The answer is that when you open the switch, the continuing current will begin to pile up charges of opposite types at the two terminals of the switch. The plus charges will be at the terminal where the current was headed before interruption and the minus at the other. Usually this will lead to very high electric fields and cause a spark to jump the gap. The spark is the current carried by air molecules that have been ionized—separated into positive and negative parts—by the strong field.

So it can be very dangerous to interrupt the current in a solenoid. Do you know how people tackle this problem? They connect a large resistor R' in parallel with L as shown in dotted lines in Fig. 12.1. When the switch is in the closed position, R' plays hardly any role; when the current comes to the node where the inductor and R' are in parallel, it takes one look at the huge R' and says, "I'm going the other way." But when you throw the switch open, the current is suddenly all for going through R'. It knows it has no other choice. You have given the inductor a path through R' to discharge its energy, and it will take that path even if R' is large. The current will continue to flow through L in the same direction as before and then return counterclockwise through R' back to L. The resistor will eventually burn up the stored magnetic energy. Let us compute the rate at which that happens, starting with the circuit containing just L and R':

$$-L\frac{dI}{dt} - R'I = 0. \tag{12.21}$$

Let me go over the derivation just to hammer home the question of signs. As we go counterclockwise (the assumed direction of the current) starting at a point below the resistor, we drop by $R'I$ when we get to the upper end of R', and then drop another LdI/dt on crossing the terminals of L. The equation sets the sum of these "drops" to zero. (The "drop" across L will end up being a rise because $dI/dt < 0$.)

Solving this very familiar equation we find the current decays exponentially

$$I(t) = I_0 e^{-R't/L}. \tag{12.22}$$

The time-constant L/R' gives you an idea of how long you have to wait before the inductor is essentially (but never fully) discharged.

Now for the energy check. In the beginning the inductor had $U = \frac{1}{2}LI_0^2$. This better equal the time-integral of the power $P = I^2 R'$ dissipated in the resistor:

$$\text{Loss} = \int_0^\infty I^2(t) R' \, dt \tag{12.23}$$

$$= I_0^2 R' \int_0^\infty e^{-2R't/L} dt \tag{12.24}$$

$$= I_0^2 R' \left. \frac{-e^{-2R't/L}}{2R'/L} \right|_0^\infty \qquad\qquad (12.25)$$

$$= \frac{1}{2} L I_0^2. \qquad\qquad (12.26)$$

12.2 The *LC* circuit

Now I'm going to describe a slightly more complicated circuit with an L and a C hooked up as shown in Figure 12.2.

Assume that at $t = 0$, the capacitor is charged as shown and there is no current. The + charges will find their way around L to the other plate and neutralize the − charges there, and eventually the capacitor will discharge. Had you connected C to a resistor, the story would have ended with the discharge of the capacitor. But when it discharges through L, it's not the end of the story. Why is that? The inductor would be carrying a current by then, and it cannot suddenly stop carrying that current. It is in fact not allowed to, by energy conservation. So it's going to keep driving the current for a while till the current is zero. The inductor has no energy now (since $I = 0$) and is ready to quit, but the capacitor is fully charged and we are almost back to where we started with one difference: the capacitor is charged the opposite way. So you wait another half cycle and you are really back to the beginning and the oscillations go on forever. The figure

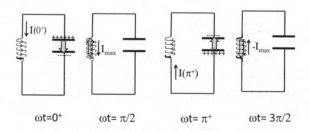

$\omega t=0^+$ $\qquad$ $\omega t= \pi/2$ $\qquad$ $\omega t= \pi^+$ $\qquad$ $\omega t= 3\pi/2$

Figure 12.2   The LC circuit at various times. The electric field in the capacitor and the magnetic field in the inductor oscillate with frequency $\omega = 1/\sqrt{LC}$. The energy alternates between being entirely magnetic in L due to the current and entirely electric in C due to the built-up charge. When the current is at a maximum the charge on the capacitor is zero and vice versa. The electric and magnetic fields in the capacitor and inductance are shown by fat arrows.

shows a few intermediate configurations and where the energy is stored. The frequency of this oscillation will shortly be shown to be $\omega = 1/\sqrt{LC}$.

We can make these heuristic arguments precise by solving the equation

$$-L\frac{dI}{dt} + \frac{Q}{C} = 0. \tag{12.27}$$

As we go counterclockwise around the loop, there is a drop LdI/dt at the inductor and a gain Q/C across the capacitor for the direction of current shown at time 0^+. Since Q is the charge on the $+$ plate, I reduces it if it is flowing as shown. So

$$I = -\frac{dQ}{dt}. \tag{12.28}$$

So the equation for Q is

$$L\frac{d^2Q}{dt^2} + \frac{Q}{C} = 0. \tag{12.29}$$

Now, we have seen exactly this equation before, right? Recall the equation for a mass coupled to a spring

$$m\frac{d^2x}{dt^2} + kx = 0. \tag{12.30}$$

Mathematically, the two equations have essentially the same solution except for a change in symbols. One may involve electric charges and the other may involve masses. You don't care. The equation

$$[\text{cow}]\frac{d^2\text{dog}}{dt^2} + [\text{elephant}]\text{dog} = 0, \tag{12.31}$$

where *dog* is a function of time, has exactly the same solution. What does it matter what you call the unknown variables? Once you assure me that *cow* and *elephant* are time-independent, just as m, k, L, and C are, I can tell you the dog will oscillate at a frequency

$$\omega = \sqrt{\frac{\text{elephant}}{\text{cow}}}.$$

Since the solution to $x(t)$ was

$$x(t) = A\cos(\omega t - \phi) \qquad \omega = \sqrt{\frac{k}{m}} \tag{12.32}$$

where A is the amplitude and ϕ is the phase, the answer for Q is

$$Q(t) = A\cos(\omega_0 t - \phi) \tag{12.33}$$

$$I = -\frac{dQ}{dt} = A\omega_0 \sin(\omega_0 t - \phi) \quad \text{where} \tag{12.34}$$

$$\omega_0 = \sqrt{\frac{1}{LC}}. \tag{12.35}$$

I have set $I = -dQ/dt$ because a positive current in the sense shown depletes the capacitor, and I denote the frequency of oscillations by ω_0, since another frequency ω will appear shortly.

Let us also choose $\phi = 0$, since ϕ is simply a nuisance when we have only one oscillator. (A non-zero ϕ here means that the oscillator does not reach its maximum when $t = 0$. In that case let us reset the clock to coincide with the maximum. There will be no complaints since no one else is using the clock. This would not be true if there were two oscillators, since there can be a fight over who gets to reach the maximum at $t = 0$. Barring coincidences, only one [the winner] can have its maximum at $t = 0$, and the loser must use a non-zero ϕ.)

Figure 12.2 shows the flow of energy between all electric in C and all magnetic in L. When the current is maximum the charge is zero, and vice versa.

We see that the charge does indeed oscillate as anticipated by heuristic arguments. But we know much more having solved the equation. We know that the frequency of oscillations is $\sqrt{\frac{1}{LC}}$. We know that the time it takes to complete a cycle is independent of the amount of initial charge on the capacitor. The analogy with the mechanical oscillator is complete. For example, starting with the capacitor charged to one coulomb and zero initial current is equivalent to pulling the mass by 1 meter and releasing it from rest. Table 12.1 shows a complete dictionary.

Thanks to this table, if you know that an inductor cannot instantaneously change its current, you may infer that the mass cannot instan-

Table 12.1   Mechanical
and electrical equivalents

Mechanical	Electrical
x	Q
v	I
m	L
k	$1/C$
$\frac{1}{2}kx^2$	$\frac{1}{2}Q^2/C$
$\frac{1}{2}mv^2$	$\frac{1}{2}LI^2$

taneously change its velocity. It will be very instructive for you to explore
this analogy further.

12.2.1   Driven LC circuit

Next, we connect L and C in series to an alternating voltage
$V(t) = V_0 \cos \omega t$ as shown in Figure 12.3. The circuit equation is

$$L\frac{d^2Q}{dt^2} + \frac{Q}{C} = V_0 \cos \omega t. \tag{12.36}$$

This ω is not the natural frequency of oscillation, ω_0. It is some externally
given frequency, like 60 Hz from your wall outlet. What happens now? We
have to again guess the solution. We want a function $Q(t)$ such that when
we take two derivatives and add that second derivative to some multiple of
$Q(t)$, we get some constant times a cosine. It is evidently a cosine. So let us
assume a solution of the form

$$Q(t) = Q_0 \cos \omega t \tag{12.37}$$

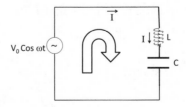

Figure 12.3   The driven LC circuit.

and stick it into the equation. We find

$$\left(-\omega^2 L + \frac{1}{C}\right) Q_0 \cos\omega t = V_0 \cos\omega t. \tag{12.38}$$

Since $\cos\omega t$ is not identically zero, we may cancel it and find that our solution works if the prefactor Q_0 is given by

$$Q_0 = \frac{V_0}{-\omega^2 L + 1/C}. \tag{12.39}$$

So that finally

$$Q(t) = Q_0 \cos\omega t = \frac{V_0}{-\omega^2 L + 1/C} \cos\omega t. \tag{12.40}$$

Actually we may modify the answer as follows:

$$Q(t) = \frac{V_0}{-\omega^2 L + 1/C} \cos\omega t + A\cos(\omega_0 t - \varphi), \tag{12.41}$$

where the extra term is the solution to the case $V_0 = 0$, Eqn. 12.33. You should verify that adding it does not invalidate Eqn. 12.38. For now I choose $A = 0$ to simplify the discussion and promise to address the extra term in depth in the next chapter.

The thing that catches our eye in Eqn. 12.39 is that when

$$\omega^2 = \frac{1}{LC} = \omega_0^2, \tag{12.42}$$

that is, when the driving frequency equals the natural frequency, we have a *resonance* with a diverging amplitude Q_0. You'd better not drive this circuit at the resonant frequency. That's also true of a mechanical oscillator.

Notice that in the LC circuit the voltage goes as $\cos\omega t$, while the current (with $A = 0$) goes as $\sin\omega t$:

$$I(t) = \frac{dQ}{dt} = -Q_0\omega\sin\omega t. \tag{12.43}$$

That's something I want you to think about. The current is not in step with the voltage, whereas in a resistor circuit, the current follows the voltage.

It has the same profile as the voltage, simply divided by R. But here, V is a cosine, and I is a sine. When one guy is at a maximum, the other is at a zero. They are out of phase by 90 degrees.

That means a current as a function of time is not equal to the voltage as a function of time divided by any time-independent quantity, as it used to be in a purely resistive circuit. You cannot divide $\cos\omega t$ by any time-independent quantity and turn it into $\sin\omega t$. It looks like you have to say goodbye to Ohm's law in AC circuits. But there is a way to get some kind of Ohm's law even here, and we will derive it shortly.

12.3   The LCR circuit

We are going to solve for the current in the LCR circuit driven by a cosine voltage, shown in Figure 12.4. The circuit equation is

$$L\frac{dI}{dt} + RI + \frac{Q(t)}{C} = V_0\cos\omega t \qquad (12.44)$$

where $Q(t)$ is the integral of $I(t)$. The equation thus involves the current, its derivative, and its integral.

12.3.1   Review of complex numbers

Solving this equation is going to require complex numbers, which are crucial here and in many other situations. For example, we rely heavily on imaginary numbers when we itemize our tax deductions. I'm assuming you have seen complex numbers in some course or in Volume I, which treats them in great detail. Just to be safe, I'll give you a lightning review.

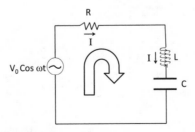

Figure 12.4   The LCR circuit driven by a cosine voltage. For the direction of current shown, note that Q increases with time.

I will only tell you the essentials, but having done so I'm going to assume that you can use them freely and that I can invoke them as often as needed. It is up to you to get prepared for this, based on your past training and the following review of complex numbers.

- A complex number z is written in terms of two real numbers x and y and

$$i = \sqrt{-1} \tag{12.45}$$

as

$$z = x + iy. \tag{12.46}$$

and visualized as a point (x, y) in the xy-plane, as in Figure 12.5. This is the *Cartesian form* of the complex number. All you need to know henceforth is that $i^2 = -1$.

- The *complex conjugate* of z is

$$z^* = x - iy. \tag{12.47}$$

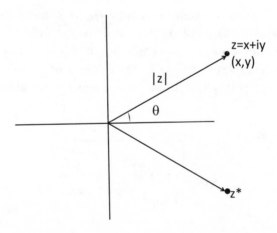

Figure 12.5   The complex plane where $z = x + iy$ is the Cartesian form of z represented by (x, y). The polar form is represented by $|z|$ and $\theta = tan^{-1}(y/x)$. The conjugate z^* has the opposite imaginary part.

We call x and y the *real and imaginary parts* of z. Thus z and z^* have the same real parts and opposite imaginary parts.

- The real and imaginary parts of z may be found as follows:

$$Re[z] = \frac{z + z^*}{2} \qquad (12.48)$$

$$Im[z] = \frac{z - z^*}{2i}. \qquad (12.49)$$

This works for the real and imaginary parts of any function of z. Take, for example,

$$Re\big[f(z)\big] = \frac{f(z) + f^*(z)}{2}. \qquad (12.50)$$

In finding $f^*(z)$ you must complex conjugate not only z but also any complex constants that enter. For example, if

$$f(z) = (3 + 4i)z + 9iz^2 \qquad (12.51)$$

then

$$f^*(z) = (3 - 4i)z^* - 9iz^{*2}. \qquad (12.52)$$

- Two complex numbers are equal if and only if their real and imaginary parts are equal.
- The sum of two complex numbers is

$$\begin{aligned} z_1 + z_2 &= (x_1 + iy_1) + (x_2 + iy_2) \\ &= (x_1 + x_2) + i(y_1 + y_2), \end{aligned} \qquad (12.53)$$

which is just like vector addition. The novelty with complex numbers is that we can also multiply them.

- Their product is

$$\begin{aligned} z_1 z_2 &= (x_1 + iy_1)(x_2 + iy_2) \\ &= (x_1 x_2 - y_1 y_2) + i(x_1 y_2 + y_1 x_2). \end{aligned} \qquad (12.54)$$

- The *modulus* or *absolute value* of the complex number is

$$|z| = \sqrt{zz^*} = \sqrt{x^2 + y^2} \qquad (12.55)$$

and is simply the length of the line joining the origin to (x, y).

- The *phase* (see Fig 12.5) is the angle between the position vector and the real or x axis:

$$\theta = \tan^{-1} \frac{y}{x}. \tag{12.56}$$

- To divide z_1 by z_2, we bring in the modulus of z_2 as follows:

$$\frac{z_1}{z_2} = \frac{z_1 z_2^*}{z_2 z_2^*} = \frac{z_1 z_2^*}{|z_2|^2}. \tag{12.57}$$

We are done, since we can evaluate the product in the numerator and divide the real and imaginary parts by the real number $|z_2|^2$.

- Euler's formula (proved in Volume I) is

$$e^{i\theta} = \cos\theta + i\sin\theta. \tag{12.58}$$

Using $\cos(-\theta) = \cos\theta$ and $\sin(-\theta) = -\sin\theta$

$$e^{-i\theta} = \cos\theta - i\sin\theta. \tag{12.59}$$

You could also obtain this by complex conjugating both sides of Eqn. 12.58, assuming, as we do, that θ is real and only i has to be conjugated to $-i$.

- Thanks to Euler we may write z in *polar form*

$$z = x + iy = |z|\cos\theta + i|z|\sin\theta = |z|e^{i\theta} \tag{12.60}$$

$$z^* = x - iy = |z|\cos\theta - i|z|\sin\theta = |z|e^{-i\theta} \tag{12.61}$$

$$zz^* = |z|e^{i\theta}|z|e^{-i\theta} = |z|^2 \tag{12.62}$$

using $e^{i\theta}e^{-i\theta} = e^0 = 1$.

- To multiply two complex numbers is easy in the polar form:

$$z_1 z_2 = |z_1|e^{i\theta_1}|z_2|e^{i\theta_2} = |z_1||z_2|e^{i(\theta_1+\theta_2)}. \tag{12.63}$$

Thus to multiply one complex number by the second, rescale the modulus of the first by the modulus of the second and rotate it by the phase of the second. Notice and remember that the modulus of a product is the product of the moduli $|z_1 z_2| = |z_1| \cdot |z_2|$.

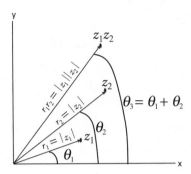

Figure 12.6   The multiplication of z_1 by z_2 rescales z_1 by $|z_2|$ and rotates it by θ_2. Thus $\theta_3 = \theta_1 + \theta_2$.

- Division is equally easy (unlike in the Cartesian case):

$$\frac{z_1}{z_2} = \frac{|z_1| e^{i\theta_1}}{|z_2| e^{i\theta_2}} = \frac{|z_1|}{|z_2|} e^{i(\theta_1 - \theta_2)}. \tag{12.64}$$

Thus complex multiplication and division accomplish two things—rescaling and rotation—in one shot. This is the key feature we will use, and it is illustrated in Figure 12.6.

- Any equation among complex numbers implies another in which both sides are complex conjugated. This is done by complex conjugating all numbers in each side. That is, the real parts are left alone and the imaginary parts are reversed. The reason this works is that if two complex numbers are equal, their real and imaginary parts must separately match. You cannot borrow from the real part and add it to the imaginary part. They are apples and oranges. So if the real and imaginary parts match on two sides of a complex relation, they will match if the imaginary parts are reversed on both sides.

    There is an analogy with vectors in two dimensions. Two vectors are equal only if their components along x and y are separately equal. Thus a vector equation in two dimensions is really two equations, one for the coefficient of $\mathbf{i}$ and the other for the coefficient of $\mathbf{j}$ on the two sides. If two equal vectors are reflected on the x-axis (i.e., their y-components are reversed), the reflected vectors will be equal.

12.3.2 Solving the LCR equation

Now we will use all this machinery to solve the LCR circuit equation:

$$L\frac{dI}{dt} + RI + \frac{1}{C}\int^t I(t')dt' = V_0 \cos\omega t. \tag{12.65}$$

Consider the capacitor term

$$\frac{1}{C}\int^t I(t')dt' = \frac{Q(t)}{C}. \tag{12.66}$$

The indefinite integral might bother you since it leaves the charge on a capacitor uncertain by an amount that depends on the lower limit. You will see that this uncertainty will not prevent us from solving for the current $I(t)$, because $I(t)$ is the derivative of $Q(t)$.

Guessing the answer will be hard. You are trying to find a function $I(t)$ such that when you differentiate it and add to it some multiple of itself and then add some multiple of its integral, you get something proportional to $\cos\omega t$. Neither a pure $\sin\omega t$ nor $\cos\omega t$ can do it.

But you can guess the answer if $V(t) = V_0 e^{\alpha t}$. In this case you can guess that the current will itself be some multiple $I_0 e^{\alpha t}$ of $e^{\alpha t}$ as well. This guess will work because $e^{\alpha t}$ will remain $e^{\alpha t}$ whether you integrate it, differentiate it, or leave it alone. So you can cancel out this time-dependent factor in all the terms in the equation and get a time-independent relation relating I_0 to the voltage amplitude V_0 and the circuit parameters R, L, and C. Unfortunately, no one is interested in this voltage, because it's growing exponentially fast, or, if you put a minus sign in the exponent, it's dying exponentially.

To solve the problem with a $\cos\omega t$ voltage we are going to use a trick based on the superposition principle for linear equations.

Consider the following two equations:

$$L\frac{dI_c}{dt} + RI_c + \frac{1}{C}\int^t I_c(t')dt' = V_0 \cos\omega t; \tag{12.67}$$

$$L\frac{dI_s}{dt} + RI_s + \frac{1}{C}\int^t I_s(t')dt' = V_0 \sin\omega t. \tag{12.68}$$

Thus I_c and I_s are currents driven by the cosine voltage $V_0 \cos\omega t$ and sine voltage $V_0 \sin\omega t$ respectively. We do not know what they are at this point.

Now multiply both sides of Eqn. 12.68 by i and add it to the first to obtain

$$L\frac{d(I_c + iI_s)}{dt} + R(I_c + iI_s) + \frac{1}{C}\int^t (I_c + iI_s)(t')dt'$$
$$= V_0(\cos\omega t + i\sin\omega t) \tag{12.69}$$

$$L\frac{dI_e}{dt} + RI_e + \frac{1}{C}\int^t I_e(t')dt' = V_0 e^{i\omega t} \quad \text{where} \tag{12.70}$$

$$I_e = I_c + iI_s. \tag{12.71}$$

To arrive at the first equation I have simply used the fact that the sum of two derivatives (or integrals) is the derivative (or integral) of the sum. In the second equation I have introduced a complex exponential current

$$I_e = I_c + iI_s, \tag{12.72}$$

which is the response to a complex exponential voltage $V_e = V_0 e^{i\omega t}$.

You may wonder where this is going. Why am I bringing in a complex voltage, when no one asked me to and when I could not even solve the problem with the real cosine potential? Here is the reason.

- I can easily find the current I_e that flows in response to the complex exponential voltage $V_e = V_0 e^{i\omega t}$ thanks to the nice properties of the exponential function.
- The current I really want, namely I_c that flows in response to $V_0 \cos\omega t$, is the real part of I_e.

Look at

$$L\frac{dI_e}{dt} + RI_e + \frac{1}{C}\int^t I_e(t')dt' = V_0 e^{i\omega t}. \tag{12.73}$$

Let us take one time derivative of both sides to eliminate the indefinite integral:

$$L\frac{d^2 I_e}{dt^2} + R\frac{dI_e}{dt} + \frac{1}{C}I_e(t) = i\omega V_0 e^{i\omega t}. \tag{12.74}$$

We can now guess the form of the solution I_e: it is also a complex exponential

$$I_e = I_0 e^{i\omega t} \tag{12.75}$$

where the constant I_0 could itself be complex. This guess is going to work because all three terms on the left—the derivatives and the function—will be the same exponential. Substituting this assumed form into Eqn. 12.74 we find

$$L\frac{d^2 I_0 e^{i\omega t}}{dt^2} + R\frac{dI_0 e^{i\omega t}}{dt} + \frac{1}{C}I_0 e^{i\omega t} = i\omega V_0 e^{i\omega t} \tag{12.76}$$

$$((i\omega)^2 L + i\omega R + \frac{1}{C})I_0 e^{i\omega t} = i\omega V_0 e^{i\omega t}. \tag{12.77}$$

Upon canceling $i\omega e^{i\omega t}$ from both sides we find

$$ZI_0 = V_0 \quad \text{where} \tag{12.78}$$

$$Z = \left(i\omega L + R + \frac{1}{i\omega C}\right) \tag{12.79}$$

is called the *impedance*.

The same relation between I_0 to V_0 and Z is obtained if we start with

$$L\frac{dI_e}{dt} + RI_e + \frac{1}{C}\int^t I_e(t')dt' = V_0 e^{i\omega t} \tag{12.80}$$

and, rather than differentiating it with respect to t as we did, evaluate the indefinite integral as follows:

$$\frac{1}{C}\int^t I_e(t')dt' = \frac{1}{C}\int^t I_0 e^{i\omega t'}dt' = \frac{I_0 e^{i\omega t}}{i\omega C}. \tag{12.81}$$

Thus we simply drop the time-independent contribution from the lower limit. This procedure, which gives the same answer as before, makes the calculations easier in circuit theory because it allows us to assign to the capacitor a contribution $1/(i\omega C)$ to Z. I will resort to it in the future.

The impedance Z has the same units as resistance. For example, if $R = 100\Omega, C = 100\mu F, L = .1H, \omega = 100\pi$,

$$Z = \left[100 + 10\pi i - \frac{100}{\pi}i\right]\Omega. \tag{12.82}$$

Notice the magic of the complex exponential: it has turned an equation involving integrals and derivatives into an algebraic one, Eqn. 12.78 for I_0, which may be found by simply dividing both sides by Z to obtain

$$I_0 = \frac{V_0}{Z} = \frac{V_0}{\left(i\omega L + R + \frac{1}{i\omega C}\right)}. \tag{12.83}$$

12.3.3 Visualizing Z

Let us visualize Z in the complex plane, as in Figure 12.7. It has a real part R and an imaginary part $(\omega L - 1/(\omega C))$.

The magnitude of Z is

$$|Z| = \sqrt{R^2 + \left(\omega L - \frac{1}{\omega C}\right)^2} \tag{12.84}$$

and its phase is

$$\phi = \tan^{-1}\left[\frac{\omega L - \frac{1}{\omega C}}{R}\right]. \tag{12.85}$$

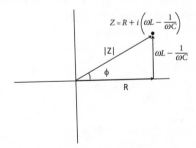

Figure 12.7   The impedance in polar and Cartesian forms. At resonance $Z = R$ and $\phi = 0$. The minimum of $|Z|$ occurs at $\omega = \omega_0 = \sqrt{1/LC}$.

So we may write, in future,

$$Z = |Z|e^{i\phi}. \tag{12.86}$$

For later use remember that

$$\phi > 0 \quad \omega L > 1/(\omega C) \tag{12.87}$$

$$\phi < 0 \quad \omega L < 1/(\omega C) \tag{12.88}$$

$$\phi = 0 \quad \omega L = 1/(\omega C) \quad \text{or when } \omega = \omega_0. \tag{12.89}$$

The current that flows in response to the complex exponential voltage is

$$I_e(t) = I_0 e^{i\omega t} = \frac{V_0}{Z} e^{i\omega t} \tag{12.90}$$

$$= \frac{V_0 e^{i\omega t}}{|Z|e^{i\phi}} \tag{12.91}$$

$$= \frac{V_0}{|Z|} e^{i(\omega t - \phi)}. \tag{12.92}$$

To find $I_c(t)$, the solution to the cosine voltage that is the real part of the exponential voltage, we simply take the real part of $I_e(t)$ and obtain

$$I_c = Re[I_e] = \frac{V_0}{|Z|} \cos(\omega t - \phi) \equiv I_{c0} \cos(\omega t - \phi). \tag{12.93}$$

The amplitude of the physical current, $I_{c0} = \frac{V_0}{|Z|}$, is related to $I_0 = \frac{V}{Z}$, the amplitude of the complex current, as follows:

$$I_0 = I_{c0} e^{-i\phi}. \tag{12.94}$$

Let us write out I_c explicitly so we may analyze it later in some depth:

$$I_c(t) = \frac{V_0}{\sqrt{R^2 + (\omega L - \frac{1}{\omega C})^2}} \cos(\omega t - \phi) \tag{12.95}$$

$$\tan\phi = \frac{\omega L - \frac{1}{\omega C}}{R}. \tag{12.96}$$

12.4 Complex form of Ohm's law

Let us begin with the fact that when the driving voltage is a (complex) exponential, so is the current, and its amplitude I_0 obeys an algebraic equation

$$V_0 = I_0 \left(R + i\omega L + \frac{1}{i\omega C} \right) = I_0 Z, \tag{12.97}$$

which is solved by simply dividing both sides by Z:

$$I_0 = \frac{V_0}{Z}. \tag{12.98}$$

This is as easy as Ohm's law, except for the fact that Z, which plays the role of R in DC circuits, is complex. We can replace the original AC circuit in Figure 12.4 by a DC-like circuit shown in Figure 12.8 where the exponential $e^{i\omega t}$ is removed from the voltage and the current. Only their amplitudes V_0 and I_0 appear and the circuit elements are replaced by their contributions $Z_R = R, Z_L = i\omega L$, and $Z_C = 1/(i\omega C)$ to the impedance. The voltage equation is

$$V_0 = V_R + V_L + V_C \tag{12.99}$$

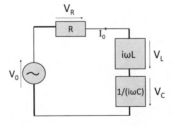

Figure 12.8   The representation of the LCR circuit in DC-like terms where each circuit element is replaced by its impedance. The common factor $e^{i\omega t}$ has been removed from the currents and voltages, whose amplitudes V_0 and I_0 alone appear. The amplitudes of the complex voltage drops across R, L, and C are denoted by $V_R(= RI_0)$, $V_L(= i\omega L I_0)$, and $V_C = \frac{1}{i\omega C} I_0$. The physical time-dependent counterparts are obtained by reinstating the $e^{i\omega t}$ factor and taking the real part.

$$= Z_R I_0 + Z_L I_0 + Z_C I_0 \tag{12.100}$$

$$= R I_0 + (i\omega L) I_0 + \frac{1}{i\omega C} I_0. \tag{12.101}$$

Once we have solved for the complex current amplitude I_0, the real, physical, time-dependent current can be obtained upon reinstating the $e^{i\omega t}$ and taking the real part:

$$I_c(t) = Re\left[I_0 e^{i\omega t} \right]. \tag{12.102}$$

If the physical current I_c is the real part of $I_e e^{i\omega t}$, what is the real, physical, time-dependent voltage $V(t)$ across any circuit element? The easiest case is the resistor. The drop across it is, from first principles,

$$V_R(t) = R I_c(t) = R \cdot Re\,[I_e] = Re\,[R I_e(t)]$$
$$= Re\left[R I_0 e^{i\omega t} \right] \equiv Re\left[V_R e^{i\omega t} \right]. \tag{12.103}$$

Thus, to get $V_R(t)$, the physical voltage, we must multiply the complex time-independent V_R by $e^{i\omega t}$ and take the real part. Only because R was real could we interchange the two operations of taking the real part and multiplying by R.

Next consider the inductor. The real, physical, time-dependent voltage drop across it is

$$V_L(t) = L \frac{dI_c(t)}{dt} = L \frac{dRe\,[I_e(t)]}{dt} \tag{12.104}$$

$$= Re\left[L \frac{d(I_0 e^{i\omega t})}{dt} \right] \tag{12.105}$$

$$= Re\left[(i\omega L) I_0 e^{i\omega t} \right] \tag{12.106}$$

$$= Re\left[Z_L I_0 e^{i\omega t} \right] = Re\left[V_L e^{i\omega t} \right]. \tag{12.107}$$

Again, to get $V_L(t)$, the physical voltage drop across the inductor, we must begin with the complex time-independent amplitude V_L, multiply by $e^{i\omega t}$, and take the real part. Only because L was real could we interchange taking the real part of the current and multiplying it by L.

Likewise the real, physical, time-dependent voltage drop across the capacitor is

$$V_C(t) = \frac{1}{C} \int^t I_c(t')dt' = \frac{1}{C} Re\left[\int^t I_0 e^{i\omega t'} dt' \right]$$

$$= Re\left[\frac{1}{i\omega C} I_0 e^{i\omega t} \right] = Re\left[Z_C I_0 e^{i\omega t} \right]. \tag{12.108}$$

Only because $1/C$ was real could we interchange taking the real part of the current and multiplying it by $1/C$.

CHAPTER 13

LCR Circuits and Displacement Current

The last chapter concluded with an expression for the current in an LCR circuit driven by a cosine potential. The circuit equation was

$$L\frac{dI}{dt} + RI + \frac{1}{C}\int^t I(t')dt' = V_0\cos\omega t. \tag{13.1}$$

This was a differential equation. We managed to turn it into an algebraic equation by following a strategy that I now restate in slightly different language.

We decided to solve instead a different problem where $V(t)$ was a complex exponential and $I_e(t)$ the corresponding current:

$$L\frac{dI_e}{dt} + RI_e + \frac{1}{C}\int^t I_e(t')dt' = V_0 e^{i\omega t}. \tag{13.2}$$

Why? Because, if we could somehow solve this problem, the answer to our original problem would be the real part:

$$I(t) = ReI_e(t). \tag{13.3}$$

This was due to superposition. The voltage $V_0 e^{i\omega t}$ is the sum of a real and pure imaginary voltage

$$V_0 e^{i\omega t} = V_0\cos\omega t + iV_0\sin\omega t, \tag{13.4}$$

244

which must therefore produce the sum of two currents, one real and one purely imaginary. The complex current I_e flowing in response to the exponential can always be written as a sum of its real and imaginary parts:

$$I_e(t) = I_c(t) + iI_s(t).$$ (13.5)

Because R, L and $1/C$ are real, a real voltage $V_0 \cos \omega t$ can only produce a real current, which must therefore be I_c, where the subscript c stands for "cosine." (The purely imaginary part $iV_0 \sin \omega t$ produces the purely imaginary current iI_s. It is the answer to a problem with an oscillating sine voltage.)

The answer to our problem is then the real part of the current produced by $V_0 e^{i\omega t}$.

This modified problem with the complex exponential is very easy to solve by guessing, due to the wonderful property of the exponential that it remains the same whether you leave it alone, integrate it, or differentiate it. So we can readily guess the form of the solution I_e: it is also a complex exponential of the same frequency:

$$I_e = I_0 e^{i\omega t}.$$ (13.6)

Substitution into the circuit equation gives, upon canceling $e^{i\omega t}$ everywhere,

$$ZI_0 = V_0 \quad \text{where}$$ (13.7)

$$Z = \left(i\omega L + R + \frac{1}{i\omega C} \right).$$ (13.8)

Eqn. 13.7 is the algebraic equation analogous to $IR = V$ for a purely resistive circuit. It is solved by dividing by the impedance Z:

$$I_0 = \frac{V_0}{Z}.$$ (13.9)

The time-dependent current produced by the exponential voltage $V_0 e^{i\omega t}$ is

$$I_e(t) = I_0 e^{i\omega t} = \frac{V_0}{Z} e^{i\omega t} = \frac{V_0}{|Z| e^{i\phi}} e^{i\omega t}$$ (13.10)

$$= \frac{V_0}{|Z|} e^{i(\omega t - \phi)}. \tag{13.11}$$

The current produced by the physical cosine voltage, $V_0 \cos \omega t = Re[V_0 e^{i\omega t}]$, is given by the real part of I_e:

$$I_c = \frac{V_0}{|Z|} \cos(\omega t - \phi) = \frac{V_0}{\sqrt{R^2 + (\omega L - \frac{1}{\omega C})^2}} \cos(\omega t - \phi)$$

$$\equiv I_{c0} \cos(\omega t - \phi) \tag{13.12}$$

$$\tan \phi = \frac{\omega L - \frac{1}{\omega C}}{R} \tag{13.13}$$

The complex amplitude of the complex current $I_0 = \frac{V_0}{Z}$ and the amplitude of the real current $I_{c0} = \frac{V_0}{|Z|}$ are related as follows:

$$I_0 = \frac{V_0}{Z} = \frac{V_0}{|Z| e^{i\phi}} = I_{c0} e^{-i\phi}. \tag{13.14}$$

Since in the end the current was real, you could say, "I don't want to deal with complex numbers." You could take an undetermined mixture of $\cos \omega t$ and $\sin \omega t$, put it into the equation, and, after a lot of manipulation, find the same answer. But the beauty of the complex numbers is that the formulas relating current and voltage come out in one package and are as easy to use as Ohm's law. I will later describe more complicated circuits where an approach with just real numbers will be intractable.

13.1 Analysis of LCR results

Let us resume our analysis of the salient features of

$$I_c(t) = \frac{V_0}{\sqrt{R^2 + (\omega L - \frac{1}{\omega C})^2}} \cos(\omega t - \phi)$$

$$\equiv I_{c0} \cos(\omega t - \phi). \tag{13.15}$$

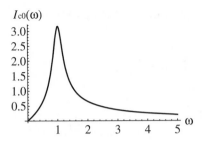

Figure 13.1   The amplitude of the current as a function of frequency ω in units of ω_0 for a typical circuit. The maximum of I_{c0} occurs at $\omega = \omega_0 = \sqrt{1/LC}$.

- Consider first the amplitude of the current $I_{c0}(t)$. Look at

$$I_{c0} = \frac{V_0}{|Z|} = \frac{V_0}{\sqrt{R^2 + (\omega L - \frac{1}{\omega C})^2}}. \tag{13.16}$$

Unlike in a resistive circuit, the size of the current is frequency-dependent. Figure 13.1 is the plot of $I_{c0}(\omega)$ as a function of ω for a typical circuit. As ω varies, so does I_{c0}. When $\omega \to 0$, you've got a $\sqrt{(1/\omega C)^2}$ in the denominator. That's going to beat everything and we find

$$I_{c0}(\omega) \Longrightarrow V_0\, \omega C \text{ as } \omega \to 0. \tag{13.17}$$

The current therefore starts out as 0 at $\omega = 0$. This corresponds to the fact that if the voltage had been a DC source instead of an AC source—that's what $\omega = 0$ means—the capacitor would charge till its voltage equaled V_0 and then the current would stop. That would be the final answer.

As ω increases, I_{c0} will initially grow linearly. It will eventually have to come down because at very large ω, the ωL term in $|Z|$ will dominate and

$$I_{c0}(\omega) \Longrightarrow \frac{V_0}{\omega L} \text{ as } \omega \to \infty. \tag{13.18}$$

Thus I_{c0} will fall like $1/\omega$ at very large frequencies. In between these two extremes, it will reach a maximum. If you're trying to get the maximum current, you want to minimize the denominator $|Z|$. Recall that

$$|Z| = \sqrt{R^2 + \left(\omega L - \frac{1}{\omega C}\right)^2}. \tag{13.19}$$

There's nothing you can do about the R^2 inside the square root. But you can play ωL and $1/\omega C$ against each other and find a frequency when they cancel each other:

$$\omega L = \frac{1}{\omega C} \quad \text{or} \tag{13.20}$$

$$\omega^2 = \frac{1}{LC} = \omega_0^2. \tag{13.21}$$

This happens when the driving frequency is the natural frequency. At that *resonant frequency,* the current amplitude will be simply

$$I_0 = \frac{V_0}{R}. \tag{13.22}$$

It's as if L and R were not there. They have neutralized each other. However, off resonance they do turn on and they are responsible for the sharp resonant peak. Do you know where that comes into play in your daily life?

The answer I had in mind was the radio. Now younger people are always carrying some recorded medium. But if you listen to radio, like in the old days, you run into the following problem. Every room is full of radio signals. Everyone wants your attention. All the radio stations are sending signals right now, and you want to pick just one station that you like. So what happens if your favorite station sends that information at a certain ω_f? If that's all you want, you go to the store and buy an LCR circuit with L and C chosen so that $\omega_0 = \omega_f$. You will get a huge response when you get the signal from that station. Now say there are other stations with different frequencies. You may not want to listen to them, but you may have to listen to some of them, if their frequency is anywhere in the resonant peak. Your radio's response to that station will not be 0. It will be a lot smaller than at the peak but not 0 and you can hear it in the background. If R is very, very small, this response function will be very large at resonance but also very narrow, and you can keep the stations from interfering by assigning them non-overlapping frequencies, differing by at least the width of each peak.

What if you changed your mind and wanted to listen to some other station? What should you do? Buy one radio for this station, one radio for that station, and so on? You know the answer: you fiddle with the dial. What do you think it does? It changes the capacitance. How do you think that is done? Now, don't rush out and smash open your radio. You will see nothing that makes any sense. But in the old days, when all the parts were big, you could look inside and see a *variable capacitor*. How do you vary the capacitance? Recall that for the parallel plate capacitor of plate area A and separation d

$$C = \varepsilon_0 \frac{A}{d}. \tag{13.23}$$

So one option is to change the surface area to vary C, but how does turning the dial do that? The actual geometry is a little different but the idea is this. Suppose the two plates of the capacitor did not fully overlap. Then the effective area A in the formula is not the full area A of each plate but a smaller amount depending on the overlap. Turning the dial changes the overlap. (In practice there are several overlapping plates and they are semicircular.) That will give you a range of resonance frequencies, and that's the range you can hear.

- Next consider the phase of the current:

$$\phi = \tan^{-1} \left[\frac{\omega L - 1/(\omega C)}{R} \right]. \tag{13.24}$$

At small ω, the capacitor term dominates, $\tan \phi$ is negative, and so is ϕ. The current, which goes as $\cos(\omega t - \phi)$, then leads the voltage.

At large ω the inductive term ωL dominates and ϕ is positive and the current lags behind the voltage. Finally, at $\omega = \omega_0$ the phase $\phi = 0$ and the current is in step with the voltage.

Consider the case when the voltage is $\cos \omega t$ and the current lags as $\cos(\omega t - \phi)$. You cannot turn the first cosine into the second upon dividing by any real time-independent function. You cannot get the current from the voltage by dividing by something like resistance. It seems like a farewell to Ohm's law. Yet within complex numbers you can turn $\exp(i\omega t)$ into $\exp i(\omega t - \phi)$ when you divide by $e^{i\phi}$. This possibility in the world of complex numbers, of rescaling and rotating the phase of a complex number in one stroke, by dividing by another complex number, is exactly what the doctor ordered for turning the voltage amplitude V_0 into the current amplitude $I_0 = V_0/Z$. The doctor in question was Dr. Charles Steinmetz (1865–1923), a mathematician and engineer who worked for General Electric and invented this approach to AC circuits using complex numbers.

- The instantaneous power delivered by the source is, from first principles,

$$P(t) = V(t)I(t) = V_0 \cos \omega t \times I_{c0} \cos(\omega t - \phi). \qquad (13.25)$$

(Although $V(t)$ and $I(t)$ are the real parts of the respective complex exponentials, the power $P(t)$ is not the real part of the product of these complex exponentials because the product of real parts is not the real part of the product. More on this later.)

For now, notice $P(t)$ oscillates with time. The oscillations reflect the fact that L and C are either acquiring energy or giving it back. So let us average $P(t)$ over a full cycle using some trig identities:

$$\cos \omega t \times \cos(\omega t - \phi)$$

$$= \cos^2 \omega t \cos \phi + \cos \omega t \sin \omega t \sin \phi \qquad (13.26)$$

$$= \frac{1 + \cos 2\omega t}{2} \cos \phi + \frac{\sin 2\omega t}{2} \sin \phi. \qquad (13.27)$$

The periodic functions all average to zero over a full cycle and we are left with $\frac{1}{2} \cos \phi$. Thus the average power is

$$P_{av} = \frac{1}{2} V_0 I_{c0} \cos \phi \qquad (13.28)$$

where $\cos \phi$ is called the *power factor*.

13.1.1   Transients and the complementary solution

Let me alert you to a problem. The solution I wrote down,

$$I(t) = \frac{V_0}{\sqrt{R^2 + (\omega L - \frac{1}{\omega C})^2}} \cos(\omega t - \phi)$$

$$\equiv I_{c0} \cos(\omega t - \phi), \tag{13.29}$$

has no free parameters in it. You tell me the time, and I tell you the current. Whatever the voltage is, you take that, shift the phase by ϕ, and divide by $|Z|$. But you know that a second order equation in time must have two free parameters. These must correspond to the charge on the capacitor and the current at some time, say $t = 0$. (These are the electrical analogs of the initial position and velocity of the oscillator.) Where are those free parameters going to come from? I will give you a clue and let you ruminate a bit. The clue is this:

$$V_0 \cos \omega t = V_0 \cos \omega t + 0. \tag{13.30}$$

If you still don't have it, here is another clue: superposition.

Anyway, here is the answer. We have seen many times that $V_1 + V_2$ drives a current $I_1 + I_2$ in obvious notation. It follows that $V_0 \cos \omega t + 0$ drives a current $I(t) + I_{com}(t)$ where I_{com} is the current flowing when no voltage is applied. It is called the *complementary solution* You might say, "There is obviously zero current if there is zero voltage," but I have to remind you that you can have current without an external voltage if there is stored energy to begin with. This is like saying that a mass-spring system can oscillate without any driving force if someone had initially stretched the spring and let it go or given the mass a kick imparting to it some kinetic energy. In the electrical case someone could have charged a capacitor and then connected it to R and L or thrown open the switch on an inductor carrying current with stored magnetic energy.

So let us look at

$$L\frac{d^2 I_{com}}{dt^2} + R\frac{dI_{com}}{dt} + \frac{I_{com}}{C} = 0. \tag{13.31}$$

We try an exponential solution

$$I_{com}(t) = Ae^{\alpha t} \tag{13.32}$$

and find the constraint

$$(\alpha^2 L + \alpha R + \frac{1}{C})A = 0. \tag{13.33}$$

Since $A \neq 0$, we get a solution only if α is a root of

$$\alpha^2 + \alpha \frac{R}{L} + \frac{1}{LC} = 0. \tag{13.34}$$

The roots, assuming

$$\frac{1}{LC} = \omega_0^2 > \frac{R^2}{4L^2}, \tag{13.35}$$

are given by

$$\alpha_\pm = -\frac{R}{2L} \pm \sqrt{\frac{R^2}{4L^2} - \omega_0^2} \tag{13.36}$$

$$= -\frac{R}{2L} \pm i\sqrt{\omega_0^2 - \frac{R^2}{4L^2}} \tag{13.37}$$

$$\equiv -\frac{R}{2L} \pm i\omega'. \tag{13.38}$$

The general solution is a sum of the two solutions with arbitrary coefficients:

$$I_{com} = A_+ \exp\left[-\frac{Rt}{2L} + i\omega't\right] + A_- \exp\left[-\frac{Rt}{2L} - i\omega't\right]. \tag{13.39}$$

If this solution is to be real we need $A_\pm$ to be complex conjugates (this will ensure $I_{com} = I_{com}^*$)

$$A_\pm = Ae^{\pm i\chi} \tag{13.40}$$

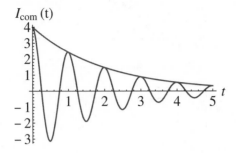

Figure 13.2   The decay of the transient current or complementary solution I_{com}.

where χ is arbitrary and A is some real positive number. This leads to

$$I_{com} = 2Ae^{-Rt/2L}\cos\left[\omega' t + \chi\right] \tag{13.41}$$

illustrated in Figure 13.2. Consequently the complete answer to the driven AC circuit is

$$I(t) = \frac{V_0}{\sqrt{R^2 + (\omega L - \frac{1}{\omega C})^2}}\cos(\omega t - \phi)$$

$$+ 2Ae^{-Rt/2L}\cos\left[\omega' t + \chi\right]. \tag{13.42}$$

The constants A and χ may be chosen to match the initial conditions.

However, the complementary function is a transient: it dies exponentially and if we are only interested in the long-term, we may ignore it. A transient could burn your circuit, but it doesn't matter after a long time, *if you survive the early stages*. It's a lot like this course.

13.2 Power of the complex numbers

I will now explain why it does not help to go back to real numbers to do AC circuit theory. Recall that the answer to the simple LCR circuit was a real cosine, namely $\cos(\omega t - \phi)$, which you could arrive at by substituting a linear combination of $\cos\omega t$ and $\sin\omega t$, and using the equation to determine the coefficients. You may be tempted to avoid complex numbers for this reason. But consider a more complicated circuit illustrated in the upper half of Figure 13.3.

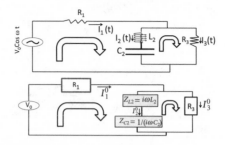

Figure 13.3   A complicated circuit where complex numbers are indispensable. At the top are the actual circuit elements and time-dependent currents and driving voltage. At the bottom is the DC-like description using complex impedances and the voltage amplitude V_0 and current amplitude I_1^0, I_2^0, and I_3^0. The two loops used for the voltage equation are shown by fat arrows.

The circuit has a resistance R_1 connected in series to a parallel circuit, in which one leg has an L_2 and a C_2 in series and the other leg a resistor R_3. The driving voltage is $V_0 \cos \omega t$. The currents flowing are labeled I_1, I_2, and I_3.

Our job is to find these oscillatory currents in magnitude and in phase. Recall the fundamental equations for a circuit. At every branch the incoming current should be equal to the outgoing current:

$$I_1(t) = I_2(t) + I_3(t). \tag{13.43}$$

This means there are only two independent currents, which will be determined by two voltage equations. We can take these currents to be I_2 and I_3. Once we solve for them, I_1 will be given by their sum.

Next we have voltage equations demanding that the sum of the voltage drops be zero in two *independent* loops. Loop 1 includes the source $V(t)$, and elements R_1, L_2, and C_2. The smaller loop 2 includes C_2, L_2, and R_3. Both are traversed in the sense shown.

$$V_0 \cos \omega t = R_1 I_1 + L_2 \frac{dI_2}{dt} + \frac{1}{C_2} \int^t I_2(t') dt' \tag{13.44}$$

$$0 = -L_2 \frac{dI_2}{dt} - \frac{1}{C_2} \int^t I_2(t') dt' + R_3 I_3. \tag{13.45}$$

In the second equation the drops across L_2 and C_2 come with a minus sign because the loop is traversed opposite to the direction of the current I_2 flowing through them. By comparison the loop 1 is traversed in the same sense as I_1 and I_2.

What if you picked a third (outer) loop 3 that included just R_1 and R_3? The corresponding equation

$$V_0 \cos \omega t = R_1 I_1 + R_3 I_3 \tag{13.46}$$

can be obtained as a linear combination of the other two equations. In this case the linear combination is simply the sum. As expected, you can have only two independent equations to determine two currents.

We have a complicated situation here, and it gets even messier with more loops. You have the derivative of this current coupled to the integral of that current and so on. How are we going to solve the equations? Trying to guess can very quickly become intractable.

But if we use complex numbers, we can reduce the problem to something that looks like a DC circuit.

First we replace the given voltage by $V_0 e^{i\omega t} = V_0 \cos \omega t + i V_0 \sin \omega t$. Since all the equations are linear relations between the voltages and currents, with real coefficients R, L and $1/C$, the real part of the voltage can only produce the real part of the current. So we will simply take the real part of the currents at the end.

We assume the complex currents are of the form

$$I_1(t) = I_1^0 e^{i\omega t} \tag{13.47}$$

$$I_2(t) = I_2^0 e^{i\omega t} \tag{13.48}$$

$$I_3(t) = I_3^0 e^{i\omega t}. \tag{13.49}$$

(Previously, when there was only one current in the picture, I used a subscript 0 to denote the current amplitude I_0. Now 0 has become a superscript, the subscript $[1, 2, \text{or } 3]$ being used to distinguish the different currents.) Substituting into the three circuit equations and canceling the common $e^{i\omega t}$ we arrive at

$$I_1^0 = I_2^0 + I_3^0 \tag{13.50}$$

$$V_0 = R_1 I_1^0 + (i\omega L_2) I_2^0 + \frac{1}{i\omega C_2} I_2^0 \tag{13.51}$$

$$0 = -(i\omega L_2)I_2^0 - \frac{1}{i\omega C_2}I_2^0 + R_3 I_3^0. \tag{13.52}$$

These are three linear time-independent equations for three unknowns I_1^0, I_2^0, and I_3^0. Apart from the fact that the coefficients are complex, this situation is no worse than a purely resistive circuit. The lower half of the figure shows how we can replace each element by the corresponding impedance:

$$R \to Z_R = R \tag{13.53}$$

$$L \to Z_L = i\omega L \tag{13.54}$$

$$C \to Z_C = \frac{1}{i\omega C} \tag{13.55}$$

No matter how complicated the circuit, we can keep doing this. We can combine impedances in series by just adding them and impedances in parallel by adding reciprocals and then inverting.

For example, let us consider the total impedance seen by the voltage source. First we compute Z_p, the impedance of the parallel branch:

$$\frac{1}{Z_p} = \frac{1}{R_3} + \frac{1}{i\omega L_2 + 1/(i\omega C_2)}. \tag{13.56}$$

The total impedance seen by V_0 is

$$Z_T = R_1 + Z_p \tag{13.57}$$

and the current flowing out of it is

$$I_1^0 = \frac{V_0}{Z_T}. \tag{13.58}$$

The actual physical current is the real part of $I_1^0 e^{i\omega t}$:

$$I_1(t) = Re\left[\frac{V_0 e^{i\omega t}}{Z_T}\right]. \tag{13.59}$$

The time-dependent voltage drop across various elements is computed from the currents as described at the end of the last chapter.

For example, $V_{R_3}(t)$, the drop across R_3, is

$$V_{R_3}(t) = Re\left[R_3 I_3^0 e^{i\omega t}\right]. \tag{13.60}$$

The voltage drop across L_2 is the real part of the complex voltage drop across it, which in turn is the product of the complex current and the complex impedance:

$$V_{L_2}(t) = Re\left[(i\omega L_2)I_2^0 e^{i\omega t}\right]. \tag{13.61}$$

Suppose I knew I_1 and wanted to know how it divides into I_2 and I_3 at the node. I proceed exactly as in resistive circuits and assign the current to each branch in proportion to the impedance of the *other* branch:

$$I_2^0 = I_1^0 \cdot \frac{R_3}{R_3 + i\omega L_2 + 1/(i\omega C_2)} \tag{13.62}$$

$$I_3^0 = I_1^0 \cdot \frac{i\omega L_2 + 1/(i\omega C_2)}{R_3 + i\omega L_2 + 1/(i\omega C_2)}. \tag{13.63}$$

The time-dependent current $I_2(t)$ will be

$$I_2(t) = Re\left[I_2^0 e^{i\omega t}\right]. \tag{13.64}$$

The recipe of taking the real part at the end fails when we consider power, because the power is *quadratic* in the complex quantities. Let us see what goes wrong.

The power delivered by the voltage source is, from first principles, just the product of the instantaneous voltage and current:

$$P(t) = V_0 \cos\omega t \times I_1(t) \tag{13.65}$$

$$= Re\left[V_0 e^{i\omega t}\right] \times Re\left[\frac{V_0 e^{i\omega t}}{Z_T}\right] \tag{13.66}$$

$$\neq Re\left[V_0 e^{i\omega t} \frac{V_0 e^{i\omega t}}{Z_T}\right]. \tag{13.67}$$

The point is that the product of the real parts of two complex numbers is not the real part of their product:

$$Re[z_1 z_2] = Re\left[(x_1 + iy_1)(x_2 + iy_2)\right] \tag{13.68}$$

$$= x_1 x_2 - y_1 y_2 \tag{13.69}$$

$$\neq Re[z_1]\, Re[z_2] = x_1 x_2. \tag{13.70}$$

This power averaged over a cycle has already been derived (see Eqn. 13.28):

$$P_{av} = \frac{1}{2}|V_0||I_0|\cos\phi. \tag{13.71}$$

We can rewrite the preceding formula as follows:

$$P_{av} = \frac{1}{2}|I_0|^2|Z|\cos\phi. \tag{13.72}$$

In a simple LCR circuit, $P_{av} = \frac{1}{2}|I_0|^2 R$ since $|Z|\cos\phi$ is just R, the real part of Z. Except for the factor $\frac{1}{2}$ that comes from time-averaging, this is the familiar expression for the power consumed by the resistor. The L and C sometimes consume power and sometimes give it back, with zero average over a cycle. The factor of $\frac{1}{2}$ can be eliminated by defining a *root-mean-square or RMS* voltage and current in terms of which P_{av} takes a form identical to that in DC circuits:

$$V_{RMS} = \frac{|V_0|}{\sqrt{2}} \tag{13.73}$$

$$I_{RMS} = \frac{|I_0|}{\sqrt{2}} \tag{13.74}$$

$$P_{av} = I_{RMS}^2 R = V_{RMS} I_{RMS} \cos\phi. \tag{13.75}$$

When we say the voltage in our homes is 110 V we are talking RMS voltage. The maximum magnitude of voltage during a cycle is $\sqrt{2} \cdot 110$.

13.3 Displacement current

We are approaching the finish line for electromagnetic theory. We need to do one last bit of fiddling with the Maxwell equations. So far all the changes to the equations were mandated by new experiments, as when we produced a time-dependent magnetic field by moving a magnet near a loop of wire. But now we are going to consider a change mandated by pure thought. It is due to Maxwell. Look at a part of a circuit shown in Figure 13.4. We don't know where it begins or ends, and we don't care. There's an alternating current flowing in the circuit. (Of course no charge is crossing the gap in the capacitor; it is just sloshing back and forth first in one direction and then in the other.)

Now here is the problem or paradox that Maxwell noticed and resolved. Look at Ampère's law

$$\oint_{C=\partial S} \mathbf{B} \cdot d\mathbf{r} = \mu_0 I_{enc} = \mu_0 \int_S \mathbf{j} \cdot d\mathbf{S} \tag{13.76}$$

where C is a loop to the left of the capacitor and I_{enc} is the current passing through *any* surface with that loop as the boundary.

First we consider a flat surface S bounded by C (assumed to be planar for convenience) with the current $I(t)$ piercing it. Now we say, "That's not the only surface with that loop C as the boundary. We should be able to draw any surface with the loop as the boundary." So we take another surface S'. We are still okay, because whatever current passes this S also passes through S' so as to avert charge buildup, charge violation, or both. The law is still good. So giddy with success, we say, why not S''? That's our new surface. It goes all the way around one of the plates of the capacitor.

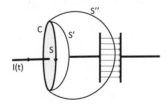

Figure 13.4   Part of an AC circuit. When we apply Ampère's law to a loop C we run into a problem when the surface it bounds goes from S or S' to S'' since no current $I(t)$ passes S''. However, an electric flux does cross S'' and it makes the same contribution for S'' as the current $I(t)$ did for S or S'.

Now we have a problem because there is no current passing through S''. If I draw an even bigger surface that fully encloses the capacitor and goes all the way around the other plate, we are again okay because now the same current $I(t)$ passes through it. It is this surface S'' for which things don't work.

So what would you do if you were Maxwell? You would realize you have to modify the Ampère equation. Sometimes people modify equations based on real experiments, and sometimes based on thought experiments. Einstein loved doing these thought experiments, which are called *gedanken* experiments. You don't really do the experiment, but you say, "If I did this, what will happen?" If that leads to a problem you have to modify your theory.

So we have to add something to the right-hand side of Ampère's equation. That something should not make any contribution on S or S', but on S'' it should make exactly the same contribution the physical current $I(t)$ made on S or S'. There are many ways to find that something depending on how much math you know. Here is an appropriate one.

We all agree that in the region between the plates, we have no current. But we do have something else between the plates we don't have in the wire. You know what that is?

It is the electric field. I am going to relate the current in the wire that penetrates S to this electric field that penetrates S''. Here we go:

$$\mu_0 I_{enc} = \mu_0 \frac{dQ}{dt} \tag{13.77}$$

$$= \mu_0 A \frac{d\sigma}{dt}$$

$$\text{where } A \text{ is the area of the capacitor plates} \tag{13.78}$$

$$= \mu_0 \varepsilon_0 A \frac{dE}{dt} \quad \text{because } E = \frac{\sigma}{\varepsilon_0} \tag{13.79}$$

$$= \mu_0 \varepsilon_0 \int_{S''} \left(\frac{\partial \mathbf{E}}{\partial t} \right) \cdot d\mathbf{S}. \tag{13.80}$$

The term

$$I_d = \varepsilon_0 \int_{S''} \left(\frac{\partial \mathbf{E}}{\partial t} \right) \cdot d\mathbf{S} \tag{13.81}$$

is called the *displacement current* and its density is

$$\mathbf{j}_d = \varepsilon_0 \frac{\partial \mathbf{E}}{\partial t}. \tag{13.82}$$

You might think that equality Eqn. 13.80 allows me to *replace* $\mu_0 I_{enc}$ by $\mu_0 I_d$. Instead I am going to *add* it. Am I double counting? Let us look at the modified Maxwell equation

$$\oint_{C=\partial S} \mathbf{B} \cdot d\mathbf{r} = \mu_0 I_{enc} + \mu_0 \varepsilon_0 \int_{S''} \left(\frac{\partial \mathbf{E}}{\partial t} \right) \cdot d\mathbf{S}$$

$$= \mu_0 \int \left(\mathbf{j} + \varepsilon_0 \frac{\partial \mathbf{E}}{\partial t} \right) \cdot d\mathbf{S} \tag{13.83}$$

and see how this procedure solves the problem.

If you took a surface like S that slices through the wire (a perfect conductor), there is no electric field. Only the $\mu_0 I_{enc}$ term contributes. If you employed S'', which goes between the plates, there is no I there, but there's the rate of change of electric flux and *its contribution is numerically equal to that of I_{enc}*. Thus the circulation of **B** around the contour C is independent of which surface is used.

There was no double counting when we added the contributions from $\mathbf{j}$ and $\mathbf{j}_d$ because only one or the other was non-zero on the surfaces considered.

Here are our final equations for electromagnetism:

$$\mathbf{F} = q(\mathbf{E} + \mathbf{v} \times \mathbf{B}) \qquad \text{(Lorentz force)} \tag{13.84}$$

$$\oint_{S=\partial V} \mathbf{E} \cdot d\mathbf{S} = \frac{1}{\varepsilon_0} \int_V \rho \, d^3 r \qquad \text{(Gauss)} \tag{13.85}$$

$$\oint_{C=\partial S} \mathbf{B} \cdot d\mathbf{r} = \mu_0 \int_S \left(\mathbf{j} + \varepsilon_0 \frac{\partial \mathbf{E}}{\partial t} \right) \cdot d\mathbf{S} \quad \text{(Ampère+Maxwell)} \tag{13.86}$$

$$\oint_{C=\partial S} \mathbf{E} \cdot d\mathbf{r} = \int_S \left(-\frac{\partial \mathbf{B}}{\partial t} \right) \cdot d\mathbf{S} \qquad \text{(Faraday)} \tag{13.87}$$

$$\oint_{S=\partial V} \mathbf{B} \cdot d\mathbf{S} = 0 \qquad \text{(no monopoles)} \tag{13.88}$$

where S is a closed surface that bounds the volume V in Eqns. 13.85 and 13.88 and any open surface S bounded by the curve C in 13.86 and 13.87.

Consider the symmetry of the equations for **E** and **B**. The line integral of the electric field is proportional to the rate of change of magnetic flux. The line integral of the magnetic field is proportional to the rate of change of electric flux but there is another Ampèrean term. The surface integral of **E** is given by the charge enclosed while there is no such right-hand side for **B** since there are no monopoles. However, in free space or vacuum, where $\rho = \mathbf{j} = 0$, the equations become symmetric.

In the next chapter we will find that these equations admit electromagnetic waves as a solution. These waves consist of non-zero electric and magnetic fields arbitrarily far from any ρ or $\mathbf{j}$. One would not expect them from Coulomb's law or the Biot-Savart law. They are possible because of the term Maxwell added.

This is a very important day in your life, because now you finally know all of electromagnetism. It is completely described by Eqns. 13.84 to 13.88. No one knows any more, at least in classical theory. You don't have to pack your head with all kinds of results. You can derive everything I have taught so far given the Lorentz force law and the four Maxwell equations (and an IQ of 600).

CHAPTER 14

Electromagnetic Waves

Now we're going to solve Maxwell's equations and deduce the existence of electromagnetic waves. No matter how many times I talk about it, I remain awestruck. Here again are the Maxwell equations:

$$\oint_{S=\partial V} \mathbf{E} \cdot d\mathbf{S} = \frac{1}{\varepsilon_0} \int_V \rho \, d^3 r \quad \text{(Gauss)} \tag{14.1}$$

$$\oint_{S=\partial V} \mathbf{B} \cdot d\mathbf{S} = 0 \qquad \text{(No monopoles)} \tag{14.2}$$

$$\oint_{C=\partial S} \mathbf{E} \cdot d\mathbf{r} = \int_S \left(-\frac{\partial \mathbf{B}}{\partial t} \right) \cdot d\mathbf{S} \quad \text{(Faraday)} \tag{14.3}$$

$$\oint_{C=\partial S} \mathbf{B} \cdot d\mathbf{r} = \mu_0 \int_S \left(\mathbf{j} + \varepsilon_0 \frac{\partial \mathbf{E}}{\partial t} \right) \cdot d\mathbf{S} \quad \text{(Ampère+Maxwell)}$$
$$\tag{14.4}$$

where S is a closed surface that bounds the volume V in Eqns. 14.1 and 14.2 and any open surface S bounded by the curve C in 14.3 and 14.4.

The first one tells you charges emit or absorb electric field lines or flux depending on their magnitude and sign. So the net amount of charge in the volume controls the net flux coming out. The second one tells you that if you integrate $\mathbf{B}$ over any surface, that is, if you count the net number of field lines coming out, you're going to get 0. That is true because field lines begin and end with charges and there are no magnetic charges or

263

monopoles. Magnetic field lines have neither a beginning nor an end. They close in on themselves. So if you pick any surface, whatever goes in has to come out. The third equation, in the static case, used to say the electric field was conservative. But then we found that a changing magnetic field can sustain a non-conservative electric field. The last one says a changing electric field can produce a magnetic field. In addition, a current can also produce a magnetic field as per Ampère.

For the purpose of studying waves, I'm going to focus on free space, where $\rho = \mathbf{j} = 0$. They could be non-zero arbitrarily far away. In the static case this would mean no $\mathbf{E}$ or $\mathbf{B}$ as per Coulomb or Biot-Savart because both fields drop off like $1/r^2$ or faster as we move away from the charges and currents. But now we will find they can survive on their own, untethered from charges and currents. The reason electromagnetic waves can survive in a vacuum far from all charges and currents is that once you've got $\mathbf{E}$ and $\mathbf{B}$ fields somewhere, they cannot just disappear due to the energy they contain. It's like the LC circuit. If your capacitor is charged to begin with and contains electric field energy, as it discharges it sets up a current in the inductor with stored magnetic field energy. The current does not stop when the capacitor is discharged; it keeps going till it charges the capacitor the opposite way. The current keeps going back and forth. The LC circuit is an example with just one degree of freedom $Q(t)$, the charge in the capacitor (or the current in the circuit, which is the derivative). By contrast, in electrodynamics $\mathbf{E}(x, y, z)$ and $\mathbf{B}(x, y, z)$ are the corresponding variables, with *one vector each for each point in space.*

Without charges and currents the Maxwell equations become very symmetric between $\mathbf{E}$ and $\mathbf{B}$. Neither has a surface integral. The line integral of one guy is proportional to the rate of change of flux of the other.

Deriving the wave equation from Maxwell's equations is a dramatic moment in physics that I am eager to share with you. But if you cannot recognize the wave equation when it miraculously emerges (to the sound of trumpets) you are not going to get the thrill. So I will remind you of some facts, covered in depth in Volume I.

The wave equation in one dimension for a variable $\psi(x, t)$ is

$$\frac{\partial^2 \psi}{\partial x^2} = \frac{1}{v^2} \frac{\partial^2 \psi}{\partial t^2} \tag{14.5}$$

where v is the velocity of the wave.

I would like to show that **E** and **B** obey such a *differential* equation starting with Maxwell's equations. But in the form displayed above, the Maxwell equations involve *integrals* over arbitrary loops, surfaces, and volumes. What we need is a version of the Maxwell equations that involves only derivatives. It is these differential equations that one usually means by Maxwell's equations. The (differential) Maxwell equations follow upon applying the integral version to arbitrary but *infinitesimal* loops, surfaces, and volumes. It is then quite easy to manipulate the differential version to arrive at the wave equation.

First I will derive the differential versions of the Maxwell equations for a *restricted class* of **E** and **B** that depend only on y and t, and have only one component each: **E** along z and **B** along x:

$$\mathbf{E} = \mathbf{k}E_z(y, t) \tag{14.6}$$

$$\mathbf{B} = \mathbf{i}B_x(y, t). \tag{14.7}$$

The resulting pair of equations, obtained by imposing Maxwell's equations on these restricted fields, are simple, but I will show how they lead to the wave equation for **E** and **B**. Though not all its solutions have definite wavelength or frequency (they could describe just localized blips moving at speed c), I will present sinusoidal solutions of definite wavelength and frequency. I will derive a formula for the energy in the electromagnetic waves and discuss their origin.

This will be followed by two optional topics.

The first concerns the derivation of the Maxwell equations for *arbitrary* **E** *and* **B** and in the presence of non-zero ρ and **j**. I express them toward the end in the language of vector calculus. This option is for those who want to see Maxwell's equations in their most general and compact form, having come this far. For completeness I show that when applied to the restricted **E** and **B** of Eqns. 14.6 and 14.7 and $\rho = \mathbf{j} = 0$, we end up with the same pair of equations as in the simpler treatment.

Next I ask if the fields that obey the Maxwell equations on infinitesimal loops, surfaces, and volumes will do so on macroscopic ones. That is, is the passage from the macroscopic to the microscopic reversible? The answer is affirmative. You can either take my word for it, or follow the demonstration of the following fact:

If Maxwell's equations are obeyed on arbitrary infinitesimal loops, surfaces, and volumes, they will be obeyed on all macroscopic ones.

14.1 The wave equation

There are many, many waves: water waves, elastic waves, sound waves, and so on. I am going to discuss waves on a string.

Imagine a string that's been clamped at two ends ($x = 0$ and $x = L$ in Figure 14.1). The thin horizontal line is the x-axis and that is the string's position in static equilibrium. Each point on the string is labeled by the value of x that it will have when the string is in the equilibrium position. The displacement of the string at the point labeled x at time t is denoted $\psi(x, t)$, and is our new dynamical variable. It is the one for which we would like to write the equations of motion.

The string is under some tension T because you have hung some weights at the ends or tightened it with some screws, as in a violin. Without the tension, none of what follows would work, as you will see. The other essential parameter is μ, the mass per unit length. To find it you put the string on a weighing scale, you find the mass, and you divide by the length. For example, if you have a ten-meter string and it weighs one-hundredth of a kilogram, then the mass per unit length is $\mu = 10^{-3}$ kilograms per meter.

Now, I pull or pluck this string in some way, given by the solid curve $\psi(x, 0)$ in the figure, and I want to know what the whole string will do. Compare this to the mass and spring system, oscillating in the y-direction. There you pull the mass out to some new location $y(0)$, let it go, and want to know $y(t)$. There was just one degree of freedom, the location of the mass, $y(t)$. The answer was $y(t) = y(0) \cos \omega t$. Here, at *every point x between* 0 *and* L, I have some segment of the string. I emphasize that

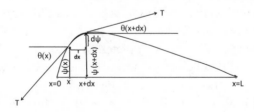

Figure 14.1   The string at some generic time, say $t = 0$. It is under tension T, has mass μ per unit length, and is fixed at $x = 0$ and $x = L$. The highlighted segment has a width dx, with the same tension T pulling the two ends but at slightly different angles. The displacement ψ and angles θ are exaggerated for clarity. The derivation is valid only when all these are very small.

x here is not a dynamical variable, but a label for the dynamical variable $\psi(x)$. Which gives the displacement of each segment from equilibrium. I displace all those infinite degrees of freedom to $\psi(x,0)$ at time 0, and I let them go. I want to know $\psi(x,t)$. For this we need to find the equation satisfied by $\psi(x,t)$.

What principle will decide the behavior of this string? Newton's law is the answer. There are no new laws that I'm going to invoke. I'm not going to say, "Well, we studied masses and springs before; today it's time to study strings and here is the new law of motion." There's only one law of motion. That's $F = ma$. My whole purpose is to show you that this law really does control everything; that's why it's a super law.

The string is a long, extended, and complicated object. I isolate a tiny segment of length dx highlighted in the figure. I am going to calculate the total force on it and equate it to its mass times acceleration. Gravity is not necessary for vibrations, and we will neglect its effect.

The figure shows the forces at the ends of the little segment. Both equal the tension T, which doesn't change from point to point in magnitude. But the angle at which the tension acts is not necessarily the same. It is tangents to the string, and the direction of the tangent (measured from the horizontal) is changing from $\theta(x)$ to $\theta(x+dx)$. The string is curving in general; therefore, the tangents to the string at two ends of the tiny bit are not quite the same and there is generally a net force on the bit.

So, I'm going to find the vertical component of the two forces and take the difference. The upward force at $x+dx$ will be $T\sin(\theta(x+dx))$ and the downward force on the left side will be $-T\sin(\theta(x))$ yielding a total of $T[\sin(\theta(x+dx)) - \sin(\theta(x))]$. That's going to be mass times acceleration. The mass of this little segment is the mass per unit length μ times the length of the segment, which is dx. Now, what is the acceleration in the language of calculus? No, it is not $\frac{d^2x}{dt^2}$ but $\frac{\partial^2\psi(x,t)}{\partial t^2}$ because $\psi(x,t)$ is the vertical coordinate of the string bit. What's jumping up and down is ψ, so the acceleration is its second derivative, and I use the partial derivative because $\psi(x,t)$ can vary with x and t. So $F = ma$ becomes

$$T[\sin(\theta(x+dx)) - \sin(\theta(x))] = \mu\,dx\frac{\partial^2\psi(x,t)}{\partial t^2}. \qquad (14.8)$$

Now, come to the left-hand side and assume the angles involved are very small, i.e., that the string does not deviate too much from being horizontal. If you remember the series

$$\sin\theta = \theta - \frac{\theta^3}{3!} + \dots \tag{14.9}$$

$$\cos\theta = 1 - \frac{\theta^2}{2!} + \dots \tag{14.10}$$

$$\tan\theta = \frac{\theta - \frac{\theta^3}{3!} + \dots}{1 - \frac{\theta^2}{2!} + \dots} = \left(\theta - \frac{\theta^3}{3!} + \dots\right)\left(1 + \frac{\theta^2}{2!} + \dots\right)$$

$$= \theta + \dots \tag{14.11}$$

and keep only terms up to order θ, you may then approximate as follows:

$$\sin\theta \simeq \theta \simeq \tan\theta = \frac{\partial\psi}{\partial x}. \tag{14.12}$$

Eqn. 14.8 becomes

$$T\left[\frac{\partial\psi}{\partial x}\bigg|_{x+dx} - \frac{\partial\psi}{\partial x}\bigg|_{x}\right] = \mu\, dx \frac{\partial^2\psi(x,t)}{\partial t^2}. \tag{14.13}$$

Dividing both sides by T and dx and letting $dx \to 0$, we finally obtain the *wave equation*

$$\frac{\partial^2\psi(x,t)}{\partial x^2} = \frac{\mu}{T}\frac{\partial^2\psi(x,t)}{\partial t^2}. \tag{14.14}$$

This is a *partial differential equation*. It is usually rewritten as

$$\frac{\partial^2\psi(x,t)}{\partial x^2} = \frac{1}{v^2}\frac{\partial^2\psi(x,t)}{\partial t^2} \tag{14.15}$$

$$v = \sqrt{\frac{T}{\mu}}. \tag{14.16}$$

In summary, when you pull a string up, it comes down because the tensions at the two ends of the string bit have vertical components that don't quite cancel. So, the net force depends on the rate of change of $\sin\theta \simeq \tan\theta = \frac{\partial\psi(x,t)}{\partial x}$, i.e, the rate of change of the rate of change, and that's why you get $\frac{\partial^2\psi(x,t)}{\partial x^2}$ on the left-hand side. The second time derivative on the right is just the acceleration of the string bit.

You should verify that v has dimensions of velocity. It will turn out to be the velocity of waves on the string. If you pluck the string and make a little bump and let it go, the bump will move at speed v. One way to deduce this is to consider the nature of the solutions to this equation. What functions do you think will enter? Based on the single oscillator you might be thinking sines and cosines. Such solutions exist, but the class of solutions is much bigger than that. I'm going to write down for you the most general solution to the wave equation: ψ *can be any function you want of $x - vt$*. I don't care what function it is. So if w stands for combination $x - vt$, then

$$\psi(x, t) = f(w) \tag{14.17}$$

where $f(w)$ is whatever function you want. If f depends on x and t only through this combination $x - vt$, it will satisfy the wave equation. To see this, use the chain rule: if $w = x - vt$ then $f = f(w)$ and

$$\frac{\partial f}{\partial x} = \frac{df(w)}{dw} \cdot \frac{\partial w}{\partial x} = \frac{df(w)}{dw} \cdot 1 \tag{14.18}$$

$$\frac{\partial^2 f}{\partial x^2} = \frac{d^2 f(w)}{dw^2} \cdot 1^2 \tag{14.19}$$

$$\frac{\partial f}{\partial t} = \frac{df(w)}{dw} \cdot \frac{\partial w}{\partial t} = \frac{df(w)}{dw} \cdot (-v) \tag{14.20}$$

$$\frac{\partial^2 f}{\partial t^2} = \frac{d^2 f(w)}{dw^2} \cdot (-v)^2 \quad \text{so that finally} \tag{14.21}$$

$$\frac{1}{v^2} \frac{\partial^2 f}{\partial t^2} = \frac{d^2 f(w)}{dw^2} = \frac{\partial^2 f}{\partial x^2}. \tag{14.22}$$

By the same logic $f(x + vt)$ also satisfies the wave equation.

What does it mean for $\psi(x, t)$ to be a function of just $x - vt$? It means that if you change x and change t in such a way that $x - vt$ does not change, the function does not change.

Consider the bell-shaped function

$$\psi(x, t) = A e^{-(x - vt)^2 / x_0^2} \tag{14.23}$$

where x_0 is some constant. It obeys the wave equation even though it does not readily come to your mind when you think of a wave, the way the sine or cosine does. At $t = 0$, the bell-shaped curve is peaked at $x = 0$, where the exponential is largest. At a later time it is peaked at $x = vt$ because that is where the exponential is largest. Hence the peak moves at velocity v. What is true for the peak is also true for any other point, say where $x - vt = 6.5$. It too moves at speed v. The entire curve moves to the right without any distortion at speed v.

This is true for any function $f(x - vt)$, which just slides to the right at speed v: if you increase t by dt and x by vdt, you find f retains its value because $f(x + vdt - v(t + dt)) = f(x - vt)$.

The most general solution to the wave equation is any function you like of $x - vt$ plus any function you like of $x + vt$. The first will describe waves going to the right, and the second will describe waves going to the left. You can superpose them because the wave equation is linear.

14.2 Restricted Maxwell equations in vacuum

As mentioned at the outset, in order to derive the wave equation we need to extract the differential version of the Maxwell equations from the integral one. We consider the fields in vacuum with $\rho = \mathbf{j} = 0$. This is now the first track, in which I derive the restrictions imposed by Maxwell on the restricted class of functions described in Eqns. 14.6 and 14.7 and repeated below:

$$\mathbf{E}(x, y, z, t) = \mathbf{k}E_z(y, t) \tag{14.24}$$

$$\mathbf{B}(x, y, z, t) = \mathbf{i}B_x(y, t). \tag{14.25}$$

The equations fall into two classes: those that involve infinitesimal cubes and those that involve infinitesimal loops.

14.2.1 Maxwell equations involving infinitesimal cubes

The equations of interest in the vacuum are

$$\oint_{S=\partial V} \mathbf{E} \cdot d\mathbf{S} = 0 \qquad \text{(Gauss in vacuum)} \tag{14.26}$$

$$\oint_{S=\partial V} \mathbf{B} \cdot d\mathbf{S} = 0 \qquad \text{(No monopoles)} \tag{14.27}$$

Consider first **E**, assumed to be of the form

$$\mathbf{E} = \mathbf{k}E_z(y, t). \tag{14.28}$$

What conditions do the integral Maxwell equations impose on this function?

The infinitesimal volume we use will be a cube of sides dx, dy, and dz, with its faces parallel to the principal planes and centered at some generic point, as shown in Figure 14.2.

We must add $\mathbf{E} \cdot d\mathbf{S}$ from every face and get 0. That's the condition imposed by Maxwell's equation. There are six faces on this cube. The figure focuses on faces 1, 2, and 3 that we can see, and not $-1, -2$, and -3 on the opposite side that we cannot fully see. Let's look at face 1 and ask what we get for the surface integral of **E**. Clearly $\mathbf{E} \cdot d\mathbf{S}$ is non-zero because $d\mathbf{S}_1$ is parallel to $\mathbf{E}_1$. But on surface -1, **E** is the same since it does not vary with z, but $d\mathbf{S}_{-1}$ points down, in the direction of the outward normal. The same electric field is sitting on the opposite faces of the cube but the area vectors $d\mathbf{S}$ are opposite. The net contribution of these two opposite faces to the surface integral is therefore zero.

We can restate these words in terms of flux lines. The flux is non-zero on 1 and -1 because the lines are perpendicular to the face, but their net contribution is zero because the lines entering one face leave the opposite face with the same density.

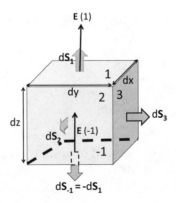

Figure 14.2   The infinitesimal cube on the surface of which **E** is integrated. The three visible faces are labeled 1, 2, and 3 and the ones opposite to them are labeled $-1, -2$, and -3. Only **E** on faces 1 and -1 is shown to avoid clutter.

Then there are other faces, like 3 and -3. Neither makes any contribution to the surface integral because the area vector and field are perpendicular or, if you like, the field lines run parallel to the faces and no flux penetrates them. The same thing goes for 2 and -2. So we get a net surface integral or flux of 0 in the end, either because the field is orthogonal to the area vector, or, if it's parallel, it has the same value on opposite faces with opposite area vectors.

So the surface integral of E vanishes on this tiny cube, given just the assumed functional form. If you repeat the calculation for **B** you encounter pretty much the same logic, except that the lines of **B** run along x. They are parallel to the faces $1, -1, 3, -3$. The only faces they penetrate are 2 and -2. They do not individually vanish but cancel each other: the field is the same on the two faces and the area vectors are opposite.

So our assumed solutions

$$\mathbf{E}(x,y,z,t) = \mathbf{k}E_z(y,t) \tag{14.29}$$

$$\mathbf{B}(x,y,z,t) = \mathbf{i}B_x(y,t) \tag{14.30}$$

identically satisfy the Maxwell equations 14.26 and 14.27 for surface integrals over arbitrary infinitesimal cubes. So no constraint on E_z or B_x emerges by imposing these integral Maxwell equations on them.

14.2.2 Maxwell equations involving infinitesimal loops

Now for the other two Maxwell equations, involving line integrals:

$$\oint_{C=\partial S} \mathbf{E} \cdot d\mathbf{r} = -\int_S \frac{\partial \mathbf{B}}{\partial t} \cdot d\mathbf{S} \tag{14.31}$$

$$\oint_{C=\partial S} \mathbf{B} \cdot d\mathbf{r} = \mu_0 \varepsilon_0 \int_S \frac{\partial \mathbf{E}}{\partial t} \cdot d\mathbf{S}. \tag{14.32}$$

Whereas there is only one kind of infinitesimal cube, there are really three kinds of infinitesimal loops, lying in the principal (xy, yz, and zx) planes. Equations derived from one such loop cannot be derived from the other two, i.e., the loops generate independent equations. On the other hand it can be shown that equations coming from additional loops, lying in arbitrary planes, can be deduced from those coming from the principal planes.

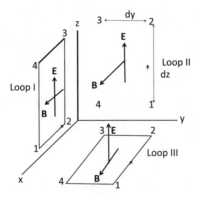

Figure 14.3   Loops in the three principal planes.

Consider loop I in Figure 14.3 first. The area vector is pointing in the positive y direction as per the right-hand rule. We have to now look at the line integral of $\mathbf{E}$ around the loop and demand it equals $-d\Phi_B/dt$. The edges 12 and 34 do not contribute since $\mathbf{E}$ is perpendicular to them. The non-zero contributions from edges 23 and 41 cancel since $\mathbf{E}$ is the same on both (as it is x-independent) but the segments are traversed in opposite directions in the line integral. Therefore $\oint \mathbf{E} \cdot d\mathbf{r}$ around this tiny loop is 0. There better not be any magnetic flux coming out of this loop. This is indeed so: $\mathbf{B}$ runs parallel to the loop and its flux does not penetrate the loop. Equivalently, the area vector points in the $+y$ direction that is normal to $\mathbf{B}$.

Next we demand that the line integral of $\mathbf{B}$ around loop I equals $\mu_0\varepsilon_0$ times the rate of change of electric flux. The magnetic field is perpendicular to the edges 23 and 41, which therefore make no contributions. It is parallel to 34 and anti-parallel to 12, and being z-independent, has the same value on both edges. So the contributions from these two edges cancel. There better not be any electric flux coming out of this loop. This is indeed so because $\mathbf{E}$ runs parallel to the loop, or equivalently, the area vector that points in the $+y$ direction is normal to $\mathbf{E}$.

So far we have obtained no restrictions at all on the fields: all equations reduce to $0 = 0$. But we still have loops in the other two planes. Consider loop II whose area vector is along the $+x$ axis. Let us impose the condition on the circulation of $\mathbf{E}$:

$$\oint \mathbf{E} \cdot d\mathbf{r} = -\frac{\partial \Phi_B}{\partial t}. \tag{14.33}$$

The right-hand side is easily calculated:

$$-\frac{\partial \Phi_B}{\partial t} = -\frac{\partial B_x}{\partial t} dy dz. \tag{14.34}$$

In the left-hand side of Eqn. 14.33 we get nothing from the edges 41 and 23 since they are orthogonal to $\mathbf{E}$. The edges 12 and 34, which are oppositely oriented, make the contribution

$$\oint \mathbf{E} \cdot d\mathbf{r} = E_z(12) dz - E_z(34) dz. \tag{14.35}$$

Even though the segments are oppositely directed their contributions do not cancel since E_z need not be the same on both sides. To first order in dy

$$E_z(12) - E_z(34) = \frac{\partial E_z}{\partial y} dy, \tag{14.36}$$

which when substituted into Eqn. 14.35 gives

$$\oint \mathbf{E} \cdot d\mathbf{r} = E_z(12) dz - E_z(34) dz = \frac{\partial E_z}{\partial y} dy dz. \tag{14.37}$$

Equating this to minus the rate of change of Φ_B given by Eqn. 14.34 we find

$$\frac{\partial E_z}{\partial y} dy dz = -\frac{\partial B_x}{\partial t} dy dz \tag{14.38}$$

$$\frac{\partial E_z}{\partial y} = -\frac{\partial B_x}{\partial t}. \tag{14.39}$$

At last we have a condition on the functions $E_z(y, t)$ and $B_x(y, t)$.

The other equation

$$\oint \mathbf{B} \cdot d\mathbf{r} = \mu_0 \varepsilon_0 \frac{\partial \Phi_E}{\partial t} \tag{14.40}$$

reduces to $0 = 0$. The left-hand side is zero, because $\mathbf{B}$ is perpendicular to the plane of the loop and makes no contribution to the line integral on any of the edges. The right-hand side vanishes because the lines of $\mathbf{E}$ run parallel to the plane of the loop and do not pierce it, or, if you like, the area vector (along $\mathbf{i}$) and $\mathbf{E}$ (along $\mathbf{k}$) are orthogonal.

We can get one more non-trivial condition by considering loop III. I will simply give the result since the steps are quite similar:

$$-\frac{\partial B_x}{\partial y} = \mu_0 \varepsilon_0 \frac{\partial E_z}{\partial t}. \tag{14.41}$$

I urge you to fill in the steps.

14.3 The wave!

Let us begin with the pair of equations our restricted fields must obey to satisfy Maxwell's equation:

$$\frac{\partial E_z}{\partial y} = -\frac{\partial B_x}{\partial t} \tag{14.42}$$

$$-\frac{\partial B_x}{\partial y} = \mu_0 \varepsilon_0 \frac{\partial E_z}{\partial t}. \tag{14.43}$$

This simple pair is enough to deduce the existence of electromagnetic waves.

Take the partial y-derivative of the first equation and add it to the partial t-derivative of the second equation to obtain

$$\frac{\partial^2 E_z}{\partial y^2} - \frac{\partial^2 B_x}{\partial t \partial y} = -\frac{\partial^2 B_x}{\partial y \partial t} + \mu_0 \varepsilon_0 \frac{\partial^2 E_z}{\partial t^2} \quad \text{which implies} \tag{14.44}$$

$$\frac{\partial^2 E_z}{\partial y^2} = \mu_0 \varepsilon_0 \frac{\partial^2 E_z}{\partial t^2} \quad \text{because} \tag{14.45}$$

$$-\frac{\partial^2 B_x}{\partial t \partial y} = -\frac{\partial^2 B_x}{\partial y \partial t}. \tag{14.46}$$

We recognize Eqn. 14.45 as the wave equation for E_z.

By adding $\mu_0 \varepsilon_0$ times the t-partial derivative of the first to the y-partial derivative of the second equation we obtain the wave equation for B_x:

$$\frac{\partial^2 B_x}{\partial y^2} = \mu_0 \varepsilon_0 \frac{\partial^2 B_x}{\partial t^2}. \tag{14.47}$$

This is the first dramatic moment: to discover that Maxwell's equations (including the term Maxwell added) imply electromagnetic waves. What is oscillating now is not some string or medium. It is just the electric and magnetic fields varying in vacuum.

The second dramatic moment follows if we compute the velocity v. Since $1/v^2$ multiplies the second time derivative in the wave equation, we infer that

$$v = \frac{1}{\sqrt{\mu_0 \varepsilon_0}}. \tag{14.48}$$

Now remember that

$$\frac{1}{4\pi \varepsilon_0} = 9 \cdot 10^9 \tag{14.49}$$

$$\frac{\mu_0}{4\pi} = 10^{-7}, \tag{14.50}$$

which means

$$v = \frac{1}{\sqrt{\mu_0 \varepsilon_0}} = \frac{1}{\sqrt{(\mu_0/4\pi)(4\pi \varepsilon_0)}} = \sqrt{9 \cdot 10^{16}}$$

$$= 3 \cdot 10^8 m/s \equiv c, \tag{14.51}$$

which was immediately recognized as the velocity of light. From this Maxwell conjectured that light was an electromagnetic wave. Now $v = c$ doesn't mean that electromagnetic waves are the same as light. For example, we now know that gravity waves also travel at the speed of light. But Maxwell had conjectured correctly. It was clearly demonstrated not long after by Heinrich Hertz (1857–1894) that sparks created in one circuit were able to generate currents in an antenna placed several feet away. By forming standing waves Hertz confirmed the wave velocity was c.

So we now have a new understanding of what light is. It is simply made of electromagnetic waves traveling at speed c. It consists of varying electric and magnetic fields. What we have seen is an example of a simple wave, but one can show in general that if you took the most general $\mathbf{E}$ and $\mathbf{B}$ you would get the following wave equation in vacuum

$$\frac{\partial^2 \Phi}{\partial x^2} + \frac{\partial^2 \Phi}{\partial y^2} + \frac{\partial^2 \Phi}{\partial z^2} = \frac{1}{c^2} \frac{\partial^2 \Phi}{\partial t^2} \tag{14.52}$$

where Φ is any component of $\mathbf{E}$ or $\mathbf{B}$.

Think about how wonderful all this is. You do experiments with charges, with currents, and you describe the phenomena as best as you can.

You measure ε_0 from electrostatics and μ_0 from magnetostatics, throw in Maxwell's displacement current for consistency, and out comes the wave, which turns out to be a description of light! It doesn't get any better than that.

14.4 Sinusoidal solution to the wave equation

As mentioned before, the solutions to wave equations just have to move at speed c; they do not have to be periodic in time or space, i.e., to have a frequency or wavelength. They could represent a single localized pulse that moves at speed c.

But there are periodic solutions. Here is a simple example from our restricted family:

$$\mathbf{E}(y, t) = \mathbf{k}E_0 \sin(\omega t - ky)$$

$$\text{that is } E_z = E_0 \sin(\omega t - ky) \tag{14.53}$$

$$\mathbf{B}(y, t) = \mathbf{i}B_0 \sin(\omega t - ky)$$

$$\text{that is } B_x = B_0 \sin(\omega t - ky) \tag{14.54}$$

where the amplitudes E_0 and B_0 are free parameters, as are the angular frequency ω and *wave number* k, which are related to the more familiar time period T and wavelength λ as follows:

$$\omega = \frac{2\pi}{T} \tag{14.55}$$

$$k = \frac{2\pi}{\lambda}. \tag{14.56}$$

This is confirmed when we write the oscillating function in terms of T and λ:

$$\sin(\omega t - ky) = \sin\left(\frac{2\pi t}{T} - \frac{2\pi y}{\lambda}\right). \tag{14.57}$$

Changing t by T or y by λ changes the argument of the sine by 2π, which remains unaffected. Equations 14.53–14.54 describe *plane waves*: $\mathbf{E}$ and $\mathbf{B}$ have the same values on a plane perpendicular to the y-axis.

Applying the Maxwell equations

$$\frac{\partial E_z}{\partial y} = -\frac{\partial B_x}{\partial t} \tag{14.58}$$

$$\frac{\partial B_x}{\partial y} = -\mu_0 \varepsilon_0 \frac{\partial E_z}{\partial t} = -\frac{1}{c^2} \frac{\partial E_z}{\partial t} \tag{14.59}$$

to the sinusoidal functions above, we find the following constraints:

$$-kE_0 \cos(\omega t - ky) = -\omega B_0 \cos(\omega t - ky) \tag{14.60}$$

$$-kB_0 \cos(\omega t - ky) = -\frac{1}{c^2} \omega E_0 \cos(\omega t - ky). \tag{14.61}$$

Upon canceling $\cos(\omega t - ky)$ from both sides we are left with

$$kE_0 = \omega B_0 \tag{14.62}$$

$$kB_0 = \frac{1}{c^2} \omega E_0. \tag{14.63}$$

Equating the quotient of the left-hand sides to the quotient of the right-hand sides we find

$$\frac{E_0}{B_0} = c^2 \frac{B_0}{E_0} \quad \text{or,} \tag{14.64}$$

$$|E_0| = c|B_0|, \tag{14.65}$$

which tells us the E field is bigger than B by a factor c in this plane wave.

Equating the product of the left-hand sides of Eqn. 14.62 and Eqn. 14.63 to the product of the right-hand sides, we find

$$k^2 = \frac{\omega^2}{c^2}, \tag{14.66}$$

which is a result we could also get if we substituted the sine waves into the wave equation. (Remember, once the two Maxwell equations are satisfied, the wave equation, which results from combining them, is automatically satisfied and will yield no additional constraints.) There are two solutions to this equation

$$k = \pm\frac{\omega}{c}. \tag{14.67}$$

The frequency ω is traditionally treated as positive and the two choices of k correspond to the two directions of propagation. Indeed we find that if we set $k = \omega/c$ in the sine wave it becomes a function of $y - ct$:

$$\sin(\omega t - ky) = \sin\left[\frac{\omega}{c}(ct - y)\right], \tag{14.68}$$

which is a right-moving wave. The other choice $k = -\omega/c$ will yield a function of $y + ct$, describing a left-moving wave.

Another way to write $\omega = kc$ is

$$\frac{2\pi}{T} = \frac{2\pi}{\lambda}c, \quad \text{which implies} \tag{14.69}$$

$$c = \frac{\lambda}{T} = \lambda f. \tag{14.70}$$

This says the source pushes out f cycles per second, each of length λ so that the wave front advances by λf meters per second, which is by definition the wave velocity.

To summarize, the plane waves have two free parameters E_0 and ω, while B_0 and k are related to them by the Maxwell equations.

Here is what we have so far for describing a wave going along $+y$:

$$\mathbf{E} = \mathbf{k}E_0 \sin(\omega t - ky) \tag{14.71}$$

$$\mathbf{B} = \mathbf{i}B_0 \sin(\omega t - ky). \tag{14.72}$$

Observe that the vector $\mathbf{E} \times \mathbf{B}$ points along $+y$, the direction of propagation.

Suppose I want a wave going the opposite way. A reasonable guess is

$$\mathbf{E} = \mathbf{k}E_0 \sin(\omega t + ky) \tag{14.73}$$

$$\mathbf{B} = \mathbf{i}B_0 \sin(\omega t + ky). \tag{14.74}$$

Since it is of the form $f(y + ct)$, the wave is certainly moving along $-y$ and it will satisfy the wave equation. But it will not satisfy all the Maxwell equations. The wave equation is obtained by combining two of the Maxwell equations, and satisfying it does not mean satisfying the two that led to it. Can you see what is wrong with the "solution" above? It is that $\mathbf{E} \times \mathbf{B}$ does not point along $-y$; it still points along $+y$. So we have to reverse **B** to obtain the correct answer

$$\mathbf{E} = \mathbf{k}E_0 \sin(\omega t + ky) \qquad\qquad (14.75)$$

$$\mathbf{B} = -\mathbf{i}B_0 \sin(\omega t + ky). \qquad\qquad (14.76)$$

This is the wave depicted in Figure 14.4.

I know this is the correct answer for another reason. Take the wave going along the $-y$ axis, as shown in Figure 14.4. Rotate the entire configuration by 180 degrees around the z axis. Can you do that in your head? Rotate the whole pattern and it's now going the opposite way, and in the process you can see this **B** will change sign, and point along $+x$ in the first half wavelength. Now, one of the principles of natural laws is that if something is a solution, the rotated thing is also a solution. This is true because space in itself does not have a preference for one direction over another. However, you must rotate all the things that matter. For example, if you have a grandfather clock and you rotate only the clock so it lies on its

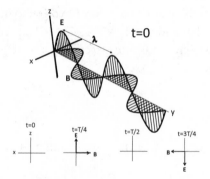

Figure 14.4   A plane wave with $E_z = E_0 \sin(\omega t + ky)$ and $B_x = -B_0 \sin(\omega t + ky)$ at $t = 0$ moving in the negative y-direction or along $\mathbf{E} \times \mathbf{B}$. The wave pattern describes the condition on points lying on the y axis at $t = 0$. This field on the axis is the field on the entire xz-plane normal to that point. The wave is polarized along $+z$. With time, the pattern will move past the origin at speed c in the $-y$ direction. The four insets in the bottom show what is happening at the origin $(x = y = z = 0)$ at various times. At $t = 0$, both fields vanish. The maxima at $y = \lambda/4$ reach the origin at time $t = \lambda/4c = T/4$. At $t = T/2$, the fields vanish again. At $t = 3T/4$, the maximally negative fields reach the origin. At $t = T$ both fields again vanish.

side it won't work, because the clock is very sensitive to the earth. But if you rotate the earth and the clock, the clock will run as before. In fact, that's happening all the time as the earth spins and goes around the sun. What are the relevant things that should be rotated along with the electromagnetic wave? Since it lives in the vacuum, there is nothing else to rotate—we are allowed to rotate just the pattern around any axis and expect the result to be a possible solution.

As another example suppose we rotate the pattern in Figure 14.4 around the y-axis, the axis of propagation. Both $\mathbf{E}$ (i.e., the polarization) and $\mathbf{B}$ will rotate in the same plane, remaining perpendicular to each other. By the time $\mathbf{E}$ in the first half wavelength is rotated to point along $+x$, $\mathbf{B}$ will point along $+z$.

Figure 14.4 shows the electric and magnetic fields at one instant in time (which we choose to be $t = 0$), for a wave moving in the $-y$ direction. The vector $\mathbf{E}$ always lies in the xz-plane and its direction (along $+z$ in the figure) is called the *polarization* of the wave. The field $\mathbf{B}$ also lies in the xz-plane and is parallel to the x-axis. Remember that the vectors in the figure describe the fields *at* points on the y-axis. However, because it is a plane wave, they have the same values on the entire xz-plane passing through that point.

As times goes by, the wave pattern shown moves past the origin at speed c in the $-y$-direction. The four insets in the bottom show what happens at the origin ($x = y = z = 0$) as time goes by. At $t = 0$, the fields vanish at the origin, as shown in the figure. The maxima in $\mathbf{E}$ and $\mathbf{B}$ at $y = \lambda/4$ get to the origin with a delay $t = \lambda/(4c) = T/4$. At $t = T/2$, both fields again vanish. At $t = 3T/4$, they are maximal again but reversed in sign. At $t = T$, the cycle is complete and $\mathbf{E}$ and $\mathbf{B}$ vanish.

The wave is traveling in a direction perpendicular to $\mathbf{E}$ and $\mathbf{B}$, along $\mathbf{E} \times \mathbf{B}$. If it hits an electron the oscillating electric field will make the electron move up and down. The electron will also feel a $\mathbf{v} \times \mathbf{B}$ force. But because $B_0 = E_0/c$, the ratio of magnetic to electric force will be v/c. For electrons in circuits $v/c \ll 1$. So when a radio wave sets the electrons in your antenna in motion, the electric force dominates. But in astrophysical situations, where particles travel at velocities close to c, the two forces can become comparable.

The electromagnetic wave is said to be *transverse*. This just means the oscillation is in the plane perpendicular to the direction of propagation. If I wiggle a taut string tied to a wall, the wiggle will move toward the wall, with the displacement in a plane normal to propagation. So that too is a transverse wave. On the other hand, sound waves are longitudinal: the air molecules set into motion (by my diaphragm) when I speak move back and forth along the direction of the wave.

The perfectly polarized plane wave is hard to find. The light from the bulbs in our homes is a chaotic mixture of different polarizations, frequencies, and phases. Plane waves are also an idealization. The light from the bulb or any point source goes out spherically. But far from the center, when this sphere has a huge radius, the wave may appear planar over small areas.

Let us understand how Polaroid glasses work. The polarizers in the glasses have a preferred direction, called the *polarization axis*. They will allow light to pass completely if it is polarized along this axis and block it completely if it is polarized in the perpendicular direction. For intermediate angles θ, the component of **E** parallel to the polarization axis will be allowed to pass, and the part perpendicular will be blocked. If the stuff coming in is randomly polarized, you can cut down about 50 percent of the light if you use the polarized lens, no matter which way it is oriented. However, when light reflects off a shiny horizontal surface (like a lake) it tends to come to your eyes polarized horizontally. Therefore your lenses should be polarized vertically to be most effective in cutting the glare.

Imagine looking at a light source through two superposed polarized lenses. No matter what kind of light enters the first lens, it will emerge polarized along its polarization axis. As the axis of the second lens is rotated, the amount of light transmitted to your eyes will change and go to zero when the two axes are perpendicular. No light can make it through both.

The light that you and I see has a very limited range of the possible wavelengths, roughly between 400 to 700 nanometers. On the shorter side are ultraviolet light and X-rays; on the longer side are infrared light and radio waves. They are all electromagnetic waves differing only in ω or $\lambda = 2\pi c/\omega$. The prefixes "ultra" and "infra" therefore refer to frequency. Nature designed our eyes to respond only to a range of ω's, probably because those are the frequencies emitted by our most common enemies. If you had different enemies you would have different eyesight. Maybe if you had a lot of enemies you would have eyes all over your head, like some insects. Since Nature gave us just two, I assume we must be pretty safe.

14.5 Energy in the electromagnetic wave

When there is an electromagnetic wave in any region, there is stored energy there. By studying capacitors and inductors we have deduced that the energy per unit volume is

$$u_E = \frac{\varepsilon_0 |\mathbf{E}|^2}{2} \tag{14.77}$$

$$u_B = \frac{|\mathbf{B}|^2}{2\mu_0}. \tag{14.78}$$

You may wonder if this formula also works for fields generated by some radio station. It does, because the formulas above are local. They only care about what the field is at any point and not its origin. For example, it doesn't matter if $\mathbf{E}$ is produced by static charges or by a changing magnetic field. It is like saying that the kinetic energy of a soccer ball is $\frac{1}{2}mv^2$, whether it got that velocity by rolling down a hill or by a kick from you.

Imagine then a wave entering a field-free region. That region now has energy brought in by the wave. For a sinusoidal plane wave the energy densities are

$$u_E = \frac{\varepsilon_0 E_0^2}{2} \sin^2(\omega t - ky) \tag{14.79}$$

$$u_B = \frac{B_0^2}{2\mu_0} \sin^2(\omega t - ky). \tag{14.80}$$

The relation

$$B_0 = \frac{E_0}{c} \tag{14.81}$$

implies that the magnetic force $\mathbf{v} \times \mathbf{B}$ is weaker than the electric force by a factor v/c. You may expect that the magnetic energy density is also smaller. But the energy densities are actually equal:

$$u_B = \frac{B_0^2}{2\mu_0} \sin^2(\omega t - ky) \tag{14.82}$$

$$= \frac{E_0^2}{2\mu_0 c^2} \sin^2(\omega t - ky) \tag{14.83}$$

$$= \frac{\varepsilon_0 E_0^2}{2} \sin^2(\omega t - ky) = u_E \tag{14.84}$$

because $c^2 = 1/(\mu_0\varepsilon_0)$. The total energy density is

$$u_T = u_E + u_B = 2u_E = \varepsilon_0 E_0^2 \sin^2(\omega t - ky). \tag{14.85}$$

This energy density is time-dependent and space-dependent. You can sit at one place and ask, "What is the energy density averaged over a full cycle?" Because the average value of $\sin^2\theta$ over a full cycle is $\frac{1}{2}$, the average energy density is

$$\bar{u}_T = \frac{\varepsilon_0 E_0^2}{2}. \tag{14.86}$$

You will get the same answer if you average over a wavelength at a fixed time. The reason is that given the periodicity in space and time, anything that happens at any one place over a full period will happen at one time over a full wavelength.

What is the *intensity I*, defined to be the watts per meter squared brought in by the wave? If I take a frame one meter by one meter and hold it perpendicular to the wave, I is the number of joules crossing it per second. That's easily calculated from the energy density using familiar reasoning. The wave that passes my one-square-meter frame in one second occupies a volume $1 \cdot c \; m^3$ and therefore contains an energy $u_T c$. Thus

$$I = u_T c. \tag{14.87}$$

For the wave in question, this becomes (upon averaging over a cycle)

$$\bar{I} = c \frac{\varepsilon_0 E_0^2}{2} \tag{14.88}$$

$$= c \frac{\varepsilon_0 E_0 B_0 c}{2} \tag{14.89}$$

$$= \frac{E_0 B_0}{2\mu_0}. \tag{14.90}$$

The *Poynting vector* (not a typo, but in honor of John Poynting [1852–1914])

$$\mathbf{S} = \frac{\mathbf{E} \times \mathbf{B}}{2\mu_0} \tag{14.91}$$

gives not only the direction of propagation, but also $\bar{I}$, the average intensity of the wave.

At the surface of the earth sunlight brings in roughly $I = 1000 W/m^2$. That is pretty amazing: over the entire surface of the earth facing the sun, it is pumping in 1,000 joules per square meter every second! The sun is 93 million miles away emitting power in all directions and we lie on a sphere of radius 93 million miles and still 1000 W/m^2 is our share. You can imagine the prodigious output from the sun. It is interesting to estimate the electric field that comes with sunlight given the energy density. It is roughly 1,000 volts per meter. This means that if that field were uniform, there would be a potential difference of 1,000 volts across one meter. However, the field is not uniform and varies randomly in space and time.

14.6 Origin of electromagnetic waves

Where are these electromagnetic fields coming from? The answer is that they are produced by charges and currents. But did I not say you don't need the currents or the charges, that these waves can exist in free space, arbitrarily far from both? So which is right? The answer is this. Static charges and currents produce fields that die away as $1/r^2$. However, time-dependent charges and currents can radiate electromagnetic waves. *Waves are produced by accelerating charges.* We will not derive this profound fact in this course. Oscillating charges are a special case of accelerating charges. If charges travel at uniform velocity as they do in a straight wire they don't produce electromagnetic waves. Suppose you took a capacitor and connected it to an AC source. The charges and currents will go back and forth. The electric field between the plates will be time-dependent. When you have a time-dependent electric field you'll have a magnetic field going around it because the line integral of **B** will have to be proportional to the rate of change of electric flux. And that induced **B** will itself be time-dependent and will produce an electric field going around *it*. So basically these fields will wind around each other whenever they're dependent on time, and they can then free themselves from the capacitor and take off, the way a soap bubble floats away from the ring it is initially attached to. All you need are two plates and an AC source to make electromagnetic waves. You'll make them at the frequency of the source, so you may not be able to see them. Neither will your dog, but some gadget will be able to pick them up. In the radio station there is a circuit with an oscillating current that emits the waves. The waves reach your radio and set the electrons in the antenna in motion, assuming the circuit is tuned properly.

Thus the waves and their sources are really like you and your parents. At some point you are free from your parents; you are able to manage on your own, but you had parents somewhere, sometime, right? The electromagnetic waves can go on their own, but they are not produced on their own. They are produced by time-varying charges and currents.

14.7 Maxwell equations—the general case (optional)

We now derive the general differential Maxwell equations following from infinitesimal volumes and surfaces, *for arbitrary* **E** *and* **B** and in the presence of ρ and **j**. (The fields are not the restricted ones from Eqns. 14.6 and 14.7.) These are what one means by *the* Maxwell equations.

14.7.1 *Maxwell equations involving infinitesimal cubes*

We will first extract the differential Maxwell equation contained in

$$\oint_{S=\partial V} \mathbf{E} \cdot d\mathbf{S} = \int_V \frac{\rho(\mathbf{r})d^3 r}{\varepsilon_0} \qquad (14.92)$$

by applying it to an infinitesimal cube in Figure 14.5. The three visible faces are labeled 1, 2, and 3 and the ones opposite to them are labeled -1, -2,

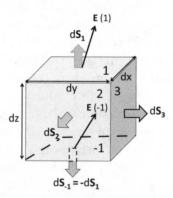

Figure 14.5   The infinitesimal cube on the surface of which **E** is integrated. The three visible faces are labeled 1, 2, and 3 and the ones opposite to them are labeled -1, -2, and -3. Only **E** on faces 1 and -1 is shown to avoid clutter.

and -3. The flux coming out of faces 1 and -1, with areas

$$dS_{\pm 1} = \pm \mathbf{k} dxdy \qquad (14.93)$$

are entirely due to E_z, which pierces them perpendicularly. (The other two components run parallel to these faces and do not contribute to flux.)

Their net contribution is

$$\Phi_E(\text{due to 1 and} -1) = [E_z(1) - E_z(-1)]\, dxdy \qquad (14.94)$$

$$= \frac{\partial E_z}{\partial z} dz \cdot dxdy. \qquad (14.95)$$

The other two pairs of opposite faces make similar contributions for a total of

$$\Phi_E(\text{cube}) = \left(\frac{\partial E_z}{\partial z} + \frac{\partial E_y}{\partial y} + \frac{\partial E_x}{\partial x} \right) dxdydz. \qquad (14.96)$$

According to the integral Maxwell equation this equals

$$\frac{q_{enc}}{\varepsilon_0} = \frac{\rho(x,y,z)dxdydz}{\varepsilon_0}. \qquad (14.97)$$

Upon canceling $dxdydz$ we arrive at one of the Maxwell equations in its final form:

$$\left(\frac{\partial E_x}{\partial x} + \frac{\partial E_y}{\partial y} + \frac{\partial E_z}{\partial z} \right) = \frac{\rho(x,y,z)}{\varepsilon_0}. \qquad (14.98)$$

Since there are no magnetic charges the corresponding equation for $\mathbf{B}$ is

$$\left(\frac{\partial B_x}{\partial x} + \frac{\partial B_y}{\partial y} + \frac{\partial B_z}{\partial z} \right) = 0. \qquad (14.99)$$

To summarize, a non-zero flux out of the cube results from the imperfect cancellation of contributions from opposite faces with oppositely pointing normals, which is why it is determined by the variation of each component of $\mathbf{E}$ and $\mathbf{B}$ along its own direction.

Now for a subtle point. Why don't we consider the variation of $\mathbf{E}$ and $\mathbf{B}$ within a face (taking it to be a constant in computing the flux) but do consider variations between opposite faces? Consider for definiteness the integral of E_z on faces 1 and -1. We are trying to match the surface integral to the enclosed charge, which is proportional to the volume $dxdydz$. The area of the faces 1 and -1 uses up a $dxdy$, leaving us with just a dz that comes from considering the variation *between* the two faces 1 and -1 separated by dz.

14.7.2 Maxwell equations involving infinitesimal loops

Let us extract the Maxwell equation coming from

$$\oint_L \mathbf{E} \cdot d\mathbf{r} = -\left[\frac{\partial \mathbf{B}}{\partial t}\right] \cdot d\mathbf{S} = -\frac{\partial \Phi_B}{\partial t}\bigg|_S \qquad (14.100)$$

on a loop depicted in Figure 14.6, lying in the xy-plane with area vector

$$d\mathbf{S} = \mathbf{k}\,dxdy. \qquad (14.101)$$

The area vector points in the positive z direction as per the right-hand rule, coming out of the page. We have to now look at the line integral of $\mathbf{E}$ around the loop and demand it equals $-\partial \Phi_B/\partial t$.

The right-hand side of Eqn. 14.100 is easily calculated:

$$-\left[\frac{\partial \mathbf{B}}{\partial t}\right] \cdot d\mathbf{S} = -\frac{\partial B_z}{\partial t}\,dxdy. \qquad (14.102)$$

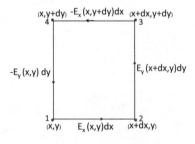

Figure 14.6   Line integral of $\mathbf{E}$ around an infinitesimal loop L in the xy-plane, with contribution from each side shown. Only variations of E_x along y and E_y along x matter, for only they make contributions of order $dxdy$ that matches the rate of change of flux $B_z\,dxdy$.

In the left-hand side, the edges 12 and 34, which are oppositely oriented, make the net contribution

$$\int \mathbf{E} \cdot d\mathbf{r} \text{ (edges 12 and 34)} = E_x(12)dx - E_x(34)dx. \quad (14.103)$$

Even though the segments are oppositely directed their contributions do not cancel since E_x need not be the same on both sides. To first order in dy

$$E_x(12) - E_x(34) = -\frac{\partial E_x}{\partial y}dy \quad (14.104)$$

where the minus sign is present because the 12 has a smaller y coordinate than 34.

When substituted into Eqn. 14.103 the preceding result gives

$$\int \mathbf{E} \cdot d\mathbf{r} \text{ (edges 12 and 34)} = -\frac{\partial E_x}{\partial y}dydx. \quad (14.105)$$

The edges 23 and 41, which are also oppositely oriented, make the contribution

$$\int \mathbf{E} \cdot d\mathbf{r} \text{ (edges 23 and 41)} = \frac{\partial E_y}{\partial x}dxdy. \quad (14.106)$$

Adding the contributions from the four edges gives

$$\oint \mathbf{E} \cdot d\mathbf{S} = \left(\frac{\partial E_y}{\partial x} - \frac{\partial E_x}{\partial y}\right)dxdy. \quad (14.107)$$

Equating this to minus the rate of change of Φ_B (Eqn. 14.102) we find

$$\left(\frac{\partial E_y}{\partial x} - \frac{\partial E_x}{\partial y}\right) = -\frac{\partial B_z}{\partial t}. \quad (14.108)$$

Once again we neglected certain variations, like that of E_x within an edge parallel to x as we integrate along it. We do this because we are trying to get a result proportional to $dxdy$, and a dx is used up in the line integral along the edge, leaving behind a dy for variations across opposite edges separated by dy.

In short, the non-zero circulation around a loop comes from the imperfect cancellation between opposite sides that are traversed in opposite directions, and hence upon the variation of the components of $\mathbf{E}$ and $\mathbf{B}$ in the other orthogonal directions.

By considering loops in the yz and zx planes we will obtain two more such equations with the cyclic permutation of the indices: $x \to y \to z \to x$. Here is the complete set:

$$\left(\frac{\partial E_y}{\partial x} - \frac{\partial E_x}{\partial y} \right) = -\frac{\partial B_z}{\partial t} \tag{14.109}$$

$$\left(\frac{\partial E_z}{\partial y} - \frac{\partial E_y}{\partial z} \right) = -\frac{\partial B_x}{\partial t} \tag{14.110}$$

$$\left(\frac{\partial E_x}{\partial z} - \frac{\partial E_z}{\partial x} \right) = -\frac{\partial B_y}{\partial t}. \tag{14.111}$$

It can be shown that no new *independent* equations emerge by considering loops not in the principal planes.

The other Maxwell equation

$$\oint_L \mathbf{B} \cdot d\mathbf{r} = \mu_0 \mathbf{j} \cdot d\mathbf{S} + \mu_0 \varepsilon_0 \frac{\partial \Phi_E}{\partial t} \tag{14.112}$$

gives three such equations with roles of **E** and **B** reversed and the additional contribution from the current density **j**:

$$\left(\frac{\partial B_y}{\partial x} - \frac{\partial B_x}{\partial y} \right) = \mu_0 \varepsilon_0 \frac{\partial E_z}{\partial t} + \mu_0 j_z \tag{14.113}$$

$$\left(\frac{\partial B_z}{\partial y} - \frac{\partial B_y}{\partial z} \right) = \mu_0 \varepsilon_0 \frac{\partial E_x}{\partial t} + \mu_0 j_x \tag{14.114}$$

$$\left(\frac{\partial B_x}{\partial z} - \frac{\partial B_z}{\partial x} \right) = \mu_0 \varepsilon_0 \frac{\partial E_y}{\partial t} + \mu_0 j_y. \tag{14.115}$$

Together with the two equations relating the derivatives of **E** and **B** to electric and magnetic charges,

$$\left(\frac{\partial E_x}{\partial x} + \frac{\partial E_y}{\partial y} + \frac{\partial E_z}{\partial z} \right) = \frac{\rho(x, y, z)}{\varepsilon_0}; \tag{14.116}$$

$$\left(\frac{\partial B_x}{\partial x} + \frac{\partial B_y}{\partial y} + \frac{\partial B_z}{\partial z} \right) = 0, \tag{14.117}$$

we have a total of eight Maxwell equations.

These equations can be displayed more compactly by introducing the entity

$$\nabla = \mathbf{i}\frac{\partial}{\partial x} + \mathbf{j}\frac{\partial}{\partial y} + \mathbf{k}\frac{\partial}{\partial z}. \tag{14.118}$$

It is not an ordinary vector, because its components are not numbers. It is called a *differential operator*; it is an entity waiting to act on functions to its right. When it does, it will yield numbers, namely the derivatives of the functions.

We are already familiar with one example, the gradient:

$$\nabla V = \mathbf{i}\frac{\partial V}{\partial x} + \mathbf{j}\frac{\partial V}{\partial y} + \mathbf{k}\frac{\partial V}{\partial z}, \tag{14.119}$$

which is a numerical vector for a given function V. Since V is a scalar ∇V is a *vector field*, described by an independent vector at every point in space.

For now treat ∇ as a vector with which we can form dot and cross products with ordinary vectors like $\mathbf{E}$ and $\mathbf{B}$ with one restriction: ∇ must always be to the left of the fields so it may differentiate them.

In this spirit consider the dot product of ∇ with a vector field, say $\mathbf{E}$:

$$\nabla \cdot \mathbf{E} = \left(\mathbf{i}\frac{\partial}{\partial x} + \mathbf{j}\frac{\partial}{\partial y} + \mathbf{k}\frac{\partial}{\partial z}\right) \cdot \left(\mathbf{i}E_x + \mathbf{j}E_y + \mathbf{k}E_z\right) \tag{14.120}$$

$$= \frac{\partial E_x}{\partial x} + \frac{\partial E_y}{\partial y} + \frac{\partial E_z}{\partial z}. \tag{14.121}$$

The expression $\nabla \cdot \mathbf{E}$ is pronounced "divergence of E" or "div E" where "div"rhymes with "give." In this notation we may rewrite Eqns. 14.98 and 14.99 compactly as

$$\nabla \cdot \mathbf{E} = \frac{\rho}{\varepsilon_0} \tag{14.122}$$

$$\nabla \cdot \mathbf{B} = 0. \tag{14.123}$$

Thus the divergence of the electric field is proportional to the charge density and the divergence of the magnetic field is zero, reflecting the absence of monopoles. Since ρ is a scalar so must be $\nabla \cdot \mathbf{E}$, and by extension $\nabla \cdot \mathbf{B}$.

Next consider the cross product of ∇ with $\mathbf{E}$:

$$\nabla \times \mathbf{E} = \mathbf{i}\left(\frac{\partial E_z}{\partial y} - \frac{\partial E_y}{\partial z}\right) + \mathbf{j}\left(\frac{\partial E_x}{\partial z} - \frac{\partial E_z}{\partial x}\right)$$
$$+ \mathbf{k}\left(\frac{\partial E_y}{\partial x} - \frac{\partial E_x}{\partial y}\right). \tag{14.124}$$

The expression $\nabla \times \mathbf{E}$ is pronounced "curl E." This notation allows us to write the other six Maxwell equations 14.109 to 14.111 and 14.113 to 14.115 compactly as

$$\nabla \times \mathbf{E} = -\frac{\partial \mathbf{B}}{\partial t} \tag{14.125}$$

$$\nabla \times \mathbf{B} = \mu_0 \mathbf{j} + \mu_0 \varepsilon_0 \frac{\partial \mathbf{E}}{\partial t}. \tag{14.126}$$

As the right-hand sides of the last two equations are vectors, so must be the left-hand sides, $\nabla \times \mathbf{E}$ and $\nabla \times \mathbf{B}$.

There is just one tricky issue:

$$\nabla \times \mathbf{E} \neq -\mathbf{E} \times \nabla, \tag{14.127}$$

for the left-hand side is a numerical and the right-hand side is still waiting to differentiate something.

I am now ready to state all of classical electrodynamics. These are encoded in the final Maxwell equations in differential and integral form (labeled **I–IV**) and the Lorentz force law:

$$\mathbf{I} \quad \nabla \cdot \mathbf{E} = \frac{\rho}{\varepsilon_0} \leftrightarrow \oint_{S=\partial V} \mathbf{E} \cdot d\mathbf{S} = \frac{1}{\varepsilon_0} \int_V \rho \, d^3\mathbf{r} \tag{14.128}$$

$$\mathbf{II} \quad \nabla \cdot \mathbf{B} = 0 \quad \leftrightarrow \oint_{S=\partial V} \mathbf{B} \cdot d\mathbf{S} = 0. \tag{14.129}$$

$$\mathbf{III} \quad \nabla \times \mathbf{E} = -\frac{\partial \mathbf{B}}{\partial t} \quad \leftrightarrow \oint_{C=\partial S} \mathbf{E} \cdot d\mathbf{r}$$
$$= \int_S \left(-\frac{\partial \mathbf{B}}{\partial t}\right) \cdot d\mathbf{S} \tag{14.130}$$

IV $\quad \mathbf{\nabla} \times \mathbf{B} = \mu_0 \mathbf{j} + \mu_0 \varepsilon_0 \dfrac{\partial \mathbf{E}}{\partial t}$

$$\leftrightarrow \oint_{C=\partial S} \mathbf{B} \cdot d\mathbf{r} = \mu_0 \int_S \left(\mathbf{j} + \varepsilon_0 \frac{\partial \mathbf{E}}{\partial t} \right) \cdot d\mathbf{S} \qquad (14.131)$$

$$\mathbf{F} = q(\mathbf{E} + \mathbf{v} \times \mathbf{B}) \quad \text{Lorentz force.} \qquad (14.132)$$

14.7.3 Consequences for the restricted **E** and **B**

How do these general Maxwell equations constrain the restricted functions

$$\mathbf{E} = \mathbf{k}E_z(y, t) \qquad (14.133)$$

$$\mathbf{B} = \mathbf{i}B_x(y, t) \qquad (14.134)$$

in vacuum, when $\rho = j = 0$? Not surprisingly, the constraints will coincide with the pair of equations we obtained earlier when we derived the Maxwell equations considering only the restricted functions. For completeness, I show how that comes about.

Consider the Maxwell equations *I* and *II* in vacuum:

$$\mathbf{\nabla} \cdot \mathbf{E} \equiv \left(\frac{\partial E_x}{\partial x} + \frac{\partial E_y}{\partial y} + \frac{\partial E_z}{\partial z} \right) = 0 \qquad (14.135)$$

$$\mathbf{\nabla} \cdot \mathbf{B} \equiv \left(\frac{\partial B_x}{\partial x} + \frac{\partial B_y}{\partial y} + \frac{\partial B_z}{\partial z} \right) = 0. \qquad (14.136)$$

They are identically satisfied by the assumed functions in Eqns. 14.6 and 14.7: the only non-zero electric component E_z has no z-derivative and the only non-zero magnetic component B_x has no x-derivative. No constraint on E_z or B_x emerges by imposing the Maxwell equations.

Now for the other two Maxwell equations (in vacuum):

$$\mathbf{\nabla} \times \mathbf{E} = -\frac{\partial \mathbf{B}}{\partial t} \qquad (14.137)$$

$$\mathbf{\nabla} \times \mathbf{B} = \mu_0 \varepsilon_0 \frac{\partial \mathbf{E}}{\partial t}. \qquad (14.138)$$

In the first equation for $\mathbf{\nabla} \times \mathbf{E}$, since **B** in the right-hand side has only an x component, we consider the same component for the curl on the left-hand side as well:

$$\frac{\partial E_z}{\partial y} - \frac{\partial E_y}{\partial z} = -\frac{\partial B_x}{\partial t}. \tag{14.139}$$

This tells us (since $E_y = 0$)

$$\frac{\partial E_z}{\partial y} = -\frac{\partial B_x}{\partial t}. \tag{14.140}$$

You may check that the equations for the other two components reduce to $0 = 0$.

The second equation for $\nabla \times \mathbf{B}$, given the non-zero components of $\mathbf{E}$ and $\mathbf{B}$, leads to just one non-trivial constraint:

$$-\frac{\partial B_x}{\partial y} = \mu_0 \varepsilon_0 \frac{\partial E_z}{\partial t}. \tag{14.141}$$

The fields E_z and B_x we have introduced have to satisfy *just the following two conditions* to obey all the Maxwell equations:

$$\frac{\partial E_z}{\partial y} = -\frac{\partial B_x}{\partial t} \tag{14.142}$$

$$-\frac{\partial B_x}{\partial y} = \mu_0 \varepsilon_0 \frac{\partial E_z}{\partial t}. \tag{14.143}$$

These are just the pair we found on the easier track.

14.8 From microscopic to macroscopic (optional)

We have gone from the integral to the differential version of the Maxwell equations. Can we go the other way, or is there loss of information in taking the infinitesimal limit? Yes, we *can*, just as we can reconstruct a function given its derivative. Using elementary theorems of vector calculus, one can show that the differential Maxwell equations in the left half of Eqns. 14.128 to 14.131 imply the corresponding integral Maxwell equations to their right. In this section I will show you the arguments that lie at the heart of these theorems. Since the differential Maxwell equations simply encode the content of integral Maxwell equations applied to infinitesimal loops, surfaces, and volumes, I just have to prove that *if the (integral) Maxwell's equations are satisfied for every infinitesimal loop, surface, and volume, then they will be satisfied for all macroscopic ones.*

14.8.1 Maxwell equations involving cubes

We begin with the equations relating the surface integrals of **E** and **B** to the enclosed charges. Consider first **E**.

By assumption

$$\oint \mathbf{E} \cdot d\mathbf{S} = \frac{q_{enc}}{\varepsilon_0} \tag{14.144}$$

is valid in every infinitesimal volume V. I'm going to take V to be a cube of sides dx, dy, and dz and S to be its surface as indicated in Figure 14.7. The cube contains charge q_{enc}. The surface is made of the six faces of the cube as shown in the upper left part of Figure 14.7. The area vectors $d\mathbf{S}_i$ $i = \pm 1, \pm 2, \pm 3$ for each face point along the *outward* normal and the field on face i is $\mathbf{E}_i$. By definition, the surface integral of **E** is the sum of $\mathbf{E}_i \cdot d\mathbf{S}_i$ over the six faces. For the reason explained earlier, the value of $\mathbf{E}_i$ is taken to be constant over each face but variations between opposite faces are kept track of.

All this goes for the second cube V' that encloses charge q'_{enc}.

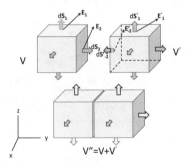

Figure 14.7   Two cubes with outward pointing area vectors at the top are glued to form the solid shown in the bottom. The common faces with oppositely pointing area vectors are deleted in the process. The visible faces are numbered 1, 2, and 3 and the ones opposite to them are numbered -1, -2, and -3.

Start with the given fact that Maxwell's equation holds for each cube:

$$\sum_{i=\pm1,\pm2,\pm3} \mathbf{E}_i \cdot d\mathbf{S}_i = \frac{1}{\varepsilon_0} q_{enc} \qquad (14.145)$$

$$\sum_{i=\pm1,\pm2,\pm3} \mathbf{E}'_i \cdot d\mathbf{S}'_i = \frac{1}{\varepsilon_0} q'_{enc}. \qquad (14.146)$$

By adding the two equations we get

$$\sum_{i=\pm1,\pm2,\pm3} \mathbf{E}_i \cdot d\mathbf{S}_i + \sum_{i=\pm1,\pm2,\pm3} \mathbf{E}'_i \cdot d\mathbf{S}'_i$$
$$= \frac{1}{\varepsilon_0} q_{enc} + \frac{1}{\varepsilon_0} q'_{enc}. \qquad (14.147)$$

Suppose we now glue the two cubes together as shown in the lower half of the figure to form the rectangular solid V''. The right-hand side of the previous equation is the charge enclosed in V''. If the Maxwell equation holds for it, *the left-hand side must be the surface integral of* $\mathbf{E}$ *over the surface of* V''. This must be true despite the fact that the V'' has only 10 of the 12 faces belonging to the two cubes V and V'. Two faces, one from each cube, were lost in the gluing process. Fortunately, their absence does not matter because their contributions cancel each other:

$$\mathbf{E}_2 \cdot d\mathbf{S}_2 = -\mathbf{E}'_{-2} \cdot d\mathbf{S}'_{-2}. \qquad (14.148)$$

This is true because the *same field* is integrated on both faces when they coalesce

$$\mathbf{E}_2 = \mathbf{E}'_{-2}, \qquad (14.149)$$

while the area vectors (pointing outward from the two cubes) are *equal and opposite*:

$$d\mathbf{S}_2 = -d\mathbf{S}'_{-2}. \qquad (14.150)$$

This allows us to rewrite

$$\sum_{i=\pm1,\pm2,\pm3} \mathbf{E}_i \cdot d\mathbf{S}_i + \sum_{i=\pm1,\pm2,\pm3} \mathbf{E}'_i \cdot d\mathbf{S}'_i$$
$$= \frac{1}{\varepsilon_0} \left[q_{enc} + q'_{enc} \right] \quad \text{as} \qquad (14.151)$$

$$\sum_{i=1}^{10} \mathbf{E}_i'' \cdot d\mathbf{S}_i'' = \frac{1}{\varepsilon_0} q_{enc}'' \tag{14.152}$$

where the sum is over the 10 faces of S'', which encloses the volume V'' and charge q_{enc}''. This is precisely Maxwell's equation for V''.

Evidently we can go on to approximate arbitrarily complicated macroscopic volumes by gluing infinitesimal cubes in this manner, and Maxwell's equation will be valid for all of them if it is valid in the little cubes used to form them. The reason will be the same as when two cubes were glued: the charge enclosed in the final volume will be the sum of the charges in the infinitesimal cubes that were glued to form it, and the surface integral on the final volume will be the sum of the surface integrals on the constituent infinitesimal cubes because the internal faces shared by cubes (with opposing area vectors) will make canceling contributions.

The argument works verbatim for $\mathbf{B}$ upon making the substitution $\mathbf{E} \to \mathbf{B}$ and $q_{enc} \equiv 0$.

14.8.2   Maxwell equations involving loops

We begin with the Maxwell equation relating the line integrals of $\mathbf{E}$ to the rate of change of the flux of $\mathbf{B}$. Similar arguments hold if the roles are exchanged and the contribution of the current $\mathbf{j}$ is included with that of the changing electric flux.

Figure 14.8 shows two infinitesimal loops L_1 and L_2 around whose boundary $\mathbf{E}$ is integrated in the sense of the arrows circulating around them. We are given that Maxwell's equations hold for L_1 and L_2

$$\oint_{L_1} \mathbf{E}_1 \cdot d\mathbf{r} = -\left[\frac{\partial \mathbf{B}_1}{\partial t}\right] \cdot d\mathbf{S}_1 = -\left.\frac{\partial \Phi_B}{\partial t}\right|_{L_1} \tag{14.153}$$

$$\oint_{L_2} \mathbf{E}_2 \cdot d\mathbf{r} = -\left[\frac{\partial \mathbf{B}_2}{\partial t}\right] \cdot d\mathbf{S}_2 = -\left.\frac{\partial \Phi_B}{\partial t}\right|_{L_2} \tag{14.154}$$

where $\mathbf{E}_1$ and $\mathbf{E}_2$ are the values of the electric field on the loops 1 and 2 and $\mathbf{B}_1$ and $\mathbf{B}_2$ are the values of the magnetic field on the infinitesimal surfaces or *plaquettes* enclosed by loops 1 and 2.

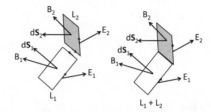

Figure 14.8   Two oriented planar areas or plaquettes are glued to form the non-planar area shown in the bottom. The common edges with oppositely running arrows are deleted.

Suppose we glue the two loops along a common edge to form the loop $L_1 + L_2$. Adding the two previous equations we find

$$\oint_{L_1} \mathbf{E}_1 \cdot d\mathbf{r} + \oint_{L_2} \mathbf{E}_2 \cdot d\mathbf{r} = -\left.\frac{\partial \Phi_B}{\partial t}\right|_{L_1} - \left.\frac{\partial \Phi_B}{\partial t}\right|_{L_2}$$

$$= -\left.\frac{\partial \Phi_B}{\partial t}\right|_{L_1+L_2} \qquad (14.155)$$

because the flux penetrating the composite loop $L_1 + L_2$ is the sum of the fluxes penetrating the two plaquettes. If this is to be the Maxwell equation for $L_1 + L_2$, the left-hand side must equal the line integral of $\mathbf{E}$ around the perimeter of $L_1 + L_2$. Now the perimeter of $L_1 + L_2$ has two edges missing compared to its constituents, the edge from L_1 and the edge from L_2 that were glued. Remarkably these missing edges do not matter because their contributions would have canceled anyway: the same $\mathbf{E}$ lives on the edges when they coalesce but is integrated in opposite directions in the two loops. It follows that if the Maxwell equation held for the smaller loops it would hold for the composite one. We can approximate arbitrarily large and complex loops bounding arbitrary surfaces by gluing smaller ones as above.

Next consider the corresponding Maxwell equation for the magnetic field:

$$\oint \mathbf{B} \cdot d\mathbf{r} = \mu_0 \int \left(\mathbf{j} + \varepsilon_0 \frac{\partial \mathbf{E}}{\partial t}\right) \cdot d\mathbf{S}. \qquad (14.156)$$

Except for the current term $\mathbf{j}$, we are simply exchanging the roles of the $\mathbf{E}$ and $\mathbf{B}$ fields. The analysis goes through with the current density $\mathbf{j}$ because

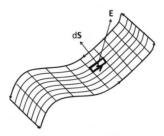

Figure 14.9   A non-planar loop bounding a non-planar surface in three dimensions. It is made by gluing little plaquettes bounded by oppositely oriented edges. The line integral of any field around the perimeter of the macroscopic surface is the sum of the line integrals around each little plaquette that forms it. The contributions from internal edges cancel in pairs leaving behind only the contributions from the edges around the big surface. The flux or current crossing the big surface is the sum of the fluxes or currents crossing the little plaquettes. Consequently, if Maxwell's equation holds on the constituent plaquettes, it holds on the macroscopic surface.

when we glue two loops, the sum of the currents through them is the current through the composite loop, just as it was for the flux.

In three dimensions there is a complication with loops that we do not have with cubes. Consider the macroscopic surface in Figure 14.9. *If Maxwell's equation holds on every plaquette used to tile it, it will hold on the surface because the flux crossing the surface is the sum of the fluxes crossing each plaquette, and the line integral of any field around the boundary is the sum of the line integrals around the constituent plaquettes.* The fact that the plaquettes in question do not lie in the principal planes, whereas the ones for which we established Maxwell's equation did, is not a problem because we may approximate any surface using plaquettes in the principal planes.

CHAPTER 15

Electromagnetism and Relativity

There are many aspects of the interplay between relativity and electromagnetism. I will reluctantly limit myself to two topics.

The first is a more or less mandatory exercise, which shows that if you knew about Coulomb's law for electrostatics and believed in relativity, you could deduce the existence of the magnetic force, $\mathbf{v} \times \mathbf{B}$.

The second is a more formal topic: we ask what modifications, if any, are needed if Maxwell's equations are to conform to the principle of relativity, namely that the equations have the same form for all inertial observers. The answer is that no changes are needed: *electrodynamics, as we have discussed it so far, complies with this principle.* However, electrodynamics does require some cosmetic changes in which certain three-vectors are grouped with scalars to form four-vectors. Once these four-vectors are identified and the equations expressed in terms of them, many separate (vector and scalar) equations collapse into single four-dimensional equations, the way the conservation of energy and momentum of Newtonian mechanics became the conservation of the energy-momentum four-vector P in relativistic mechanics. Most importantly, the four-dimensional equations have the same form for all inertial observers, i.e., invariant under Lorentz transformations.

There is one big difference between mechanics and electromagnetism: while the expressions for energy and momentum had to be first modified (e.g., $p = mv \rightarrow mv/\sqrt{1 - v^2/c^2}$) before the momentum

four-vector could be assembled, no such changes are needed in electrodynamics. For example, the Lorentz force law does not receive any corrections
to higher orders in v/c. The force is still $\mathbf{F} = q(\mathbf{E} + \mathbf{v} \times \mathbf{B})$, though
now it constitutes three components of a four-force determined by the
four-velocity and fields. The fourth component relates the power to the
fields and four-velocity.

15.1 Magnetism from Coulomb's law and relativity

Relativity unifies electricity and magnetism by showing that they are
not independent unrelated phenomena, which is how it seemed in the
static case. Their dynamics got *coupled* when they varied with time, as
in Faraday's law. What I mean by unification is that $\mathbf{E}$ and $\mathbf{B}$ mix with
each other as we change frames, the way x and t do under Lorentz
transformations. I will now demonstrate one amusing result that illustrates
this: given the Coulomb force, you can deduce the velocity-dependent
magnetic force and even derive its magnitude *if you believe in special
relativity*.

Consider a positive charge q moving parallel to the current in an
infinite wire. We know the charge will be attracted to the wire. Pretend
we only know Coulomb's law and are unfamiliar with magnetism. We
cannot explain this attraction, which seems to be due to its velocity v. So
we decide to go to the inertial frame that moves at the same velocity as the
charge. In that frame the charge has come to rest. No new physics beyond
electrostatics must be involved in explaining the behavior of the charge.

You will agree that if the charge was attracted to the wire in the
original frame, it will be attracted in the new frame as well. The transverse
coordinate is not affected by motion parallel to the wire. But now we are
worse off: we see a neutral wire attracting a *static* charge. How come? It
is true the wire as a whole is now moving backward at velocity $-v$ but it
is still neutral! It is also no use saying the wire has Lorentz contracted: a
contracted but neutral wire still cannot attract a static charge.

The resolution, which is not really obvious, is shown in Figure 15.1.

Replace the neutral wire in its rest frame (lab frame) by two
oppositely charged rods, with charge density of $\pm\lambda_0$ coulombs per unit
length. The rods are initially at rest. There is no current and no charge on
the wire. The test charge, at rest, is unaware of these rods. Now slide the
positively charged rod to the right with some speed V and the negatively
charged rod to the left with the same speed, giving it a velocity $-V$. Their

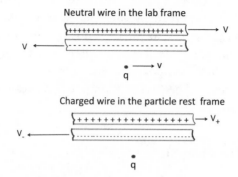

Figure 15.1   Top: The situation in lab frame. A neutral wire modeled by two infinitely long rods with equal and opposite charge densities in their rest frames, moving with equal and opposite velocities (and suffering equal Lorentz contractions), producing a current to the right. A charge $q > 0$ moving to the right is attracted to it. Bottom: The situation in the charge's rest frame. The positive rod appears slower and it is less Lorentz contracted; the opposite is true for the negative rod. The wire appears negatively charged, and the charge is electrostatically attracted to it.

currents add because the left-moving negative charges also constitute a current to the right. Due to length contraction, the charge densities of the rods go up, but by the same amount, to

$$\lambda_\pm = \pm \frac{\lambda_0}{\sqrt{1 - V^2/c^2}}, \tag{15.1}$$

keeping the wire still neutral. A static charge will still feel no force toward this neutral wire.

What is the current, the number of coulombs crossing any checkpoint? It is the charge per unit length times the length of rod that slides past the checkpoint in one second, namely V:

$$I_+ = \lambda_+ \cdot V = \frac{\lambda_0}{\sqrt{1 - V^2/c^2}} V \tag{15.2}$$

$$I_- = \lambda_- \cdot (-V) = \frac{\lambda_0}{\sqrt{1 - V^2/c^2}} V \tag{15.3}$$

$$I_{\text{total}} \equiv I = \frac{2\lambda_0}{\sqrt{1 - V^2/c^2}} V. \qquad (15.4)$$

To make my point without too much algebra *I am going to consider small velocities and drop terms of higher order than* V^2/c^2. The current then becomes

$$I = 2\lambda_0 V \qquad (15.5)$$

as expanding the denominator of Eqn. 15.4 using the binomial theorem will produce terms of order V^3/c^3 or higher.

Now give the charge q some positive velocity v. It is found to be attracted to the wire by virtue of its motion. We do not understand the physics behind this. So we move to a frame traveling at v to bring the charge to a halt. In this rest frame of the charge, the positively charged rod has slowed down from V to V_+, its Lorentz contraction has gone down, and so has its charge density. (In the limit of low velocities, $V_+ = V - v$.) The negatively charged rod has sped up (to $V_- = V + v$ in the small velocity limit) and its charge density has gone up due to increased Lorentz contraction. In the rest frame of the test charge, *the wire is negatively charged!* No wonder the charge is attracted to it.

To find the magnitude of this attraction, I will introduce another simplification: I will assume the particle moves at speed V also. In its rest frame the positive rod is at rest with density

$$\lambda_+ = \lambda_0 \qquad (15.6)$$

and the negative rod, moving at velocity $-2V$ (with corrections of order V^3/c^3) has density

$$\lambda_- = \frac{\lambda_0}{\sqrt{1 - (4V^2/c^2)}} \simeq \lambda_0 \left(1 + \frac{2V^2}{c^2} + \dots \right), \qquad (15.7)$$

leading to a net charge density on the two rods of

$$\lambda_{\text{net}} = +\lambda_0 - \lambda_0 \left(1 + \frac{2V^2}{c^2} \right) = -\frac{2V^2 \lambda_0}{c^2}. \qquad (15.8)$$

This charged wire exerts the familiar attractive electric force on a charge q at distance r,

$$F = qE = q \cdot \frac{\lambda_{\text{net}}}{2\pi \varepsilon_0 r} \tag{15.9}$$

$$= -q\frac{2V^2\lambda_0}{c^2} \frac{1}{2\pi \varepsilon_0 r} \tag{15.10}$$

$$= -qV\frac{I}{2\pi \varepsilon_0 c^2 r} \quad \text{using } I = 2V\lambda_0, \tag{15.11}$$

where the minus sign means the force is toward the wire. In the low velocity Newtonian limit, force and acceleration are the same in the lab frame. We therefore expect that in the lab frame *a charge q moving at velocity V a distance r from the wire will experience an attractive force*

$$F_{\text{attractive}} = qV\frac{I}{2\pi \varepsilon_0 c^2 r} \equiv qV\frac{\mu_0 I}{2\pi r} \tag{15.12}$$

where I have introduced a new constant $\mu_0 = 1/(\varepsilon_0 c^2)$, at least in this discussion, where magnetism was never explicitly introduced but c was there from the beginning in all the relativistic formulas.

We of course encountered this force along a different path. In our initial study of magnetism μ_0 was introduced in the Biot-Savart law as an independent constant describing a new phenomenon called magnetism. The force between the moving charge and current-carrying wire came from calculating **B** using the Biot-Savart law for an infinite wire and invoking the Lorentz force. The relation $\mu_0 \varepsilon_0 = 1/c^2$ emerged only in the last chapter.

You may think the previous demonstration may become invalid if the assumption $v = V$ that I made for convenience was relaxed. It does not and you are free to give the charge a different velocity $v \neq V$ and work to the same order to obtain Eqn. 15.12.

Meanwhile consider this: the final answer expresses the force in terms of the current I with no separate reference to the V of the rods or their charge density. Therefore we expect that if we could change the velocity of the rod without changing the current, the answer must be the same. This of course we can do: the same I can be produced by rods with a smaller linear density and larger velocity or larger density and smaller velocity.

To summarize, given just electrostatics, the existence of the velocity-dependent magnetic force can be deduced if we believe in Einstein's relativity. Conversely, electromagnetic theory, which includes magnetism in just the right way mandated by relativity, is already consistent with special relativity.

15.2   Relativistic invariance of electrodynamics

The rest of this chapter is somewhat formal in nature. No new phenomena are involved and you cannot compute anything new using what follows. We will discuss certain questions of principle whose resolution reveals the stunning beauty of the formalism. You will also end up learning that electrodynamics is a *gauge theory,* like its cousins, the weak and strong interactions. Sheldon Glashow, Abdus Salam, and Steven Weinberg described electromagnetic and weak interactions as gauge theories. David Gross, Frank Wilczek, and David Politzer showed that only a gauge theory (quantum chromdynamics, or QCD) could describe the strong interactions which are known to become weak at short distances and strong at long distances.

You will find the answer to two common questions. If the magnetic force depends on the velocity of the charges, it is the velocity according to whom? When Maxwell's equations give the velocity of light as c, it is as measured by whom? The answer: both are velocities as measured by *any* inertial observer. Einstein assures you that you can apply the same laws of physics as if you were not moving, even when another inertial observer says you are.

15.3   Review of Lorentz transformations

The most familiar example of the Lorentz transformation is of the space-time coordinates:

$$x' = \frac{x - ut}{\sqrt{1 - u^2/c^2}} \tag{15.13}$$

$$y' = y \tag{15.14}$$

$$z' = z \tag{15.15}$$

$$t' = \frac{t - ux/c^2}{\sqrt{1 - u^2/c^2}} \tag{15.16}$$

where u denotes the velocity of the primed frame relative to the unprimed frame along the x-axis. (I use v to denote the velocity of a particle.)

We like to work with four-vectors whose components have the same dimension. Thus we introduce the position four-vector X:

$$X = (ct, \mathbf{r}) = (X_0, X_1, X_2, X_3) \equiv (x_0, x_1, x_2, x_3). \tag{15.17}$$

The Lorentz transformation is now more symmetric:

$$X_0' = \frac{X_0 - \beta X_1}{\sqrt{1 - \beta^2}} \tag{15.18}$$

$$X_1' = \frac{X_1 - \beta X_0}{\sqrt{1 - \beta^2}} \tag{15.19}$$

$$\beta = \frac{u}{c}. \tag{15.20}$$

The two components unaffected by the motion are suppressed. I will do this often and yet refer to the truncated vector as a four-vector. Four-vectors will not be in boldface. The boldface is reserved for the spatial parts of four-vectors, as in

$$X = (X_0, X_1, X_2, X_3) = (x_0, x, y, z) = (ct, \mathbf{r}) \tag{15.21}$$

$$P = (P_0, P_1, P_2, P_2) = \left(\frac{E}{c}, \mathbf{p} \right) \tag{15.22}$$

where P is the energy-momentum four-vector.

In general a four-vector V has components (V_0, V_1, V_2, V_3) that transform into linear combinations of each other exactly as (X_0, X_1, X_2, X_3) do. For motion in the 1-direction,

$$V \to V' \quad \text{where} \tag{15.23}$$

$$V_0' = \frac{V_0 - \beta V_1}{\sqrt{1 - \beta^2}} \tag{15.24}$$

$$V_1' = \frac{V_1 - \beta V_0}{\sqrt{1 - \beta^2}}. \tag{15.25}$$

Given this transformation law, it follows that the "dot product" of two four-vectors V and W is Lorentz invariant. That is, if

$$V = (V_0, V_1, V_2, V_3) \equiv (V_0, \mathbf{V}) \quad \text{and} \tag{15.26}$$

$$W = (W_0, W_1, W_2, W_3) \equiv (W_0, \mathbf{W}), \quad \text{then} \tag{15.27}$$

$$\begin{aligned} V \cdot W &= V_0 W_0 - V_1 W_1 - V_2 W_2 - V_3 W_3 \\ &= V_0 W_0 - \mathbf{V} \cdot \mathbf{W} \end{aligned} \tag{15.28}$$

$$\begin{aligned} &= V_0' W_0' - V_1' W_1' - V_2' W_2' - V_3' W_3' \\ &= V_0' W_0' - \mathbf{V}' \cdot \mathbf{W}' = V' \cdot W'. \end{aligned} \tag{15.29}$$

The minus signs in the dot products are part of life in space-time.

15.3.1 Implications for Newtonian mechanics

While the laws of nature must be the same for all inertial observers, *these need not be the laws in use before Einstein*. In particular, Newton's laws are not invariant in form under Lorentz transformations. The relativistic equations of dynamics take their place *because only they assume the same form in all inertial frames*.

This is why

$$\mathbf{F} = m \frac{d^2 \mathbf{r}}{dt^2} = \frac{d\mathbf{p}}{dt} \tag{15.30}$$

was replaced by

$$F = m \frac{d^2 X}{d\tau^2} = \frac{dP}{d\tau}. \tag{15.31}$$

Here τ is proper time, P is the energy-momentum four-vector

$$P = (E/c, \mathbf{p}), \tag{15.32}$$

and F is the four-force with components

$$F = (F_0, F_1, F_2, F_3) = \frac{dP}{d\tau} \tag{15.33}$$

$$= \frac{dP}{dt}\frac{dt}{d\tau} = \frac{1}{\sqrt{1-v^2/c^2}}\frac{dP}{dt} \qquad (15.34)$$

$$= \frac{1}{\sqrt{1-v^2/c^2}}\left[\frac{1}{c}\frac{dE}{dt}, \frac{d\mathbf{p}}{dt}\right] \qquad (15.35)$$

$$= \frac{1}{\sqrt{1-v^2/c^2}}\left[\frac{Power}{c}, \mathbf{F}\right]. \qquad (15.36)$$

Equation 15.31 satisfies Einstein's requirement that it assume the same form after a Lorentz transformation to another frame. This is true because τ is an invariant and F, P, and X all transform the same way under a Lorentz transformation.

Let us make sure we understand this statement. Suppose we know in one frame that the four-force and four-momentum are related as follows:

$$F_0 = \frac{dP_0}{d\tau} \qquad (15.37)$$

$$F_1 = \frac{dP_1}{d\tau}. \qquad (15.38)$$

Multiply the second equation by $-\beta = -u/c$ and add it to the first and divide by $\sqrt{1-\beta^2}$ to obtain

$$\frac{F_0 - \beta F_1}{\sqrt{1-\beta^2}} = \frac{\frac{dP_0}{d\tau} - \beta\frac{dP_1}{d\tau}}{\sqrt{1-\beta^2}} = \frac{d}{d\tau}\left[\frac{P_0 - \beta P_1}{\sqrt{1-\beta^2}}\right],$$

which means   (15.39)

$$F'_0 = \frac{dP'_0}{d\tau} \qquad (15.40)$$

where F'_0 and P'_0 are the components of the four-force F' four-momentum P' in the primed frame moving at velocity u. Doing the same thing for F'_1 we conclude that in the primed frame

$$F' = \frac{dP'}{d\tau}. \qquad (15.41)$$

Essential to this proof is the fact that τ is the same for both observers, the way t used to be in the Galiliean transformation, and that F and P transform the same way, as four-vectors.

Likewise,

$$F = m\frac{d^2 X}{d\tau^2} \tag{15.42}$$

will transform into

$$F' = m\frac{d^2 X'}{d\tau^2} \tag{15.43}$$

because m and τ are invariant.

Having seen the fate of mechanics after Einstein's revolution, it is time to ask what happens to the laws of electrodynamics. Of course they too must be the same in all inertial frames, but should they be the laws discovered well before Einstein, the laws I have described in the preceding chapters? Should they be modified as Newton's laws of mechanics were? Amazingly, no modifications are needed. The laws of electrodynamics before Einstein are fully compatible with relativity, though this was not known while they were being discovered. It turned out that one could simply rewrite the Maxwell equations 14.128 through 14.131 and the Lorentz force in terms of two new four-vectors J and A that emerge. The resulting equations have the same form in all Lorentz frames. The *same* velocity of light appears in all the frames.

I cannot show this in all its generality without going too far astray. I will occasionally have to state a few results without proof. I just want to give you a feeling for what happens and to prepare and encourage you to pursue the details on your own.

15.4 Scalar and vector fields

We have seen that vectors and scalars are *defined with respect to some specific transformations like rotations or Lorentz transformations*. For example, a *vector* $\mathbf{V}$ (which could be a particle's velocity) with two components (V_x, V_y) in one frame of reference will have two different components V'_x, V'_y in a rotated coordinate system. A scalar like $\mathbf{V} \cdot \mathbf{V}$ will have the same value in both frames.

But consider now not just one scalar but a *scalar field S*, which is a scalar function of the coordinates. At each point in space there is a value for S. Thus a field is a system with an infinite number of degrees of freedom. At any given point, the scalar field has the same value in the two coordinate systems, used by two different observers.

Here we consider the simplest case, a scalar with respect to rotations in $d = 2$. A scalar S is a number at each point, $S(x, y)$. A good example is the temperature at the point (x, y). If you change to a new set of coordinates related by a rotation,

$$x' = x\cos\theta + y\sin\theta \tag{15.44}$$

$$y' = -x\sin\theta + y\cos\theta, \tag{15.45}$$

as shown in Figure 15.2, the *same* temperature distribution will be described by a *different* function $S'(x', y')$ such that

$$S(x, y) = S'(x', y'). \tag{15.46}$$

In other words, (x, y) and (x', y') are two ways to refer to the same point and the temperature there is the same for both observers. The point may have a different name for different people but the temperature there has an objective, coordinate-independent meaning.

For example,

$$S(x, y) = e^{-(x-a)^2 - y^2} \tag{15.47}$$

describes a distribution peaked at $(x = a, y = 0)$. In the system rotated by $\theta = \frac{\pi}{2}$ (imagine the figure with $\theta = \frac{\pi}{2}$), the distribution will be peaked at

$$x' = a\cos\frac{\pi}{2} + 0\sin\frac{\pi}{2} = 0 \tag{15.48}$$

$$y' = -a\sin\frac{\pi}{2} + 0\cos\frac{\pi}{2} = -a. \tag{15.49}$$

The function peaked at $(x' = 0, y' = -a)$ is given by

$$S'(x', y') = e^{-x'^2 - (y'+a)^2}. \tag{15.50}$$

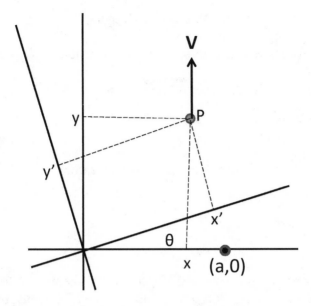

Figure 15.2   The same point P in two dimensions has coordinates (x,y) and (x',y') in two systems of coordinates related by a rotation θ. A scalar like temperature has the same numerical value at P for both observers. A vector $\mathbf{V}$ actually looks different to the two: it has only a y component in the unprimed frame and a bit of x' and y' components in the primed frame. When $\theta = \frac{\pi}{2}$, only the x' component of the vector $\mathbf{V}$ is non-zero. The coordinates of $(x = a, y = 0)$ become $(x' = 0, y' = -a)$.

The function $S'(x',y')$ is obtained from $S(x,y)$ by expressing (x,y) in terms of (x',y'). Both describe the same physical situation.

Imagine now a vector field $\mathbf{V}(x,y)$, like the wind velocity at (x,y). Whereas a scalar like temperature has the same value at a given point (labeled differently of course in the two frames) a vector (like velocity) will appear rotated to a rotated observer. In the rotated frame, *the vector at (x',y') will be the rotated version of the vector at the corresponding point* (x,y). In other words,

$$V'_x(x',y') = V_x(x,y)\cos\theta + V_y(x,y)\sin\theta \qquad (15.51)$$

$$V'_y(x',y') = -V_x(x,y)\sin\theta + V_y(x,y)\cos\theta. \qquad (15.52)$$

Figure 15.2 shows a vector **V**. It is entirely along the y axis. In the frame rotated by $\theta = \frac{\pi}{2}$, it is entirely along x'.

The inverse transformation is obtained by sending $\theta \to -\theta$:

$$V_x(x, y) = V'_x(x', y') \cos\theta - V'_y(x', y') \sin\theta \tag{15.53}$$

$$V_y(x, y) = V'_x(x', y') \sin\theta + V'_y(x', y') \cos\theta. \tag{15.54}$$

15.5 The derivative operator

Consider next the equation

$$\frac{\partial E_x}{\partial x} + \frac{\partial E_y}{\partial y} = \mathbf{V} \cdot \mathbf{E} = \frac{\rho}{\varepsilon_0} \tag{15.55}$$

in the unprimed frame. What will it look like in the primed frame? The right-hand side will be $\rho'(x', y')/\varepsilon_0$ since ρ is a scalar. Remember that in the equation

$$\rho(x, y) = \rho'(x', y') \tag{15.56}$$

(x, y) and (x', y') refer to the *same* point in space.

The left-hand side looks like a dot product of **V** and **E**, and we expect it to be invariant. That is, we expect the equation in the primed coordinates to read

$$\mathbf{V}' \cdot \mathbf{E}' = \frac{\partial E'_x}{\partial x'} + \frac{\partial E'_y}{\partial y'} = \frac{\rho'(x', y')}{\varepsilon_0}. \tag{15.57}$$

But this expectation needs to be verified because **V** is not an ordinary vector, with numerical components. Here is the verification.

Consider the left-hand side of Eqn. 15.55. We want to express all unprimed quantities in terms of primed ones and see if we get an identical expression. Let us start by replacing $\mathbf{E}(x, y)$ by the rotated $\mathbf{E}'(x', y')$ as per Eqns. 15.53 and 15.54:

$$\frac{\partial E_x}{\partial x} + \frac{\partial E_y}{\partial y} = \frac{\partial (E'_x \cos\theta - E'_y \sin\theta)}{\partial x}$$

$$+ \frac{\partial (E'_x \sin\theta + E'_y \cos\theta)}{\partial y} \tag{15.58}$$

where it is understood that the primed fields are functions of the primed coordinates. This is the easy part. Next we need to trade the unprimed derivatives for primed ones.

In general if $F'(x', y')$ is a function of x' and y', we can only define its derivatives with respect to x' and y' but not another *unrelated* pair (x, y). If, however, the pairs are related by some transformation, like the one we are considering,

$$x' = x\cos\theta + y\sin\theta \qquad\qquad (15.59)$$

$$y' = -x\sin\theta + y\cos\theta, \qquad\qquad (15.60)$$

then we can view F' as a function of (x, y)

$$F'(x', y') = F'(x'(x, y), y'(x, y)) \qquad\qquad (15.61)$$

and take its derivatives with respect to (x, y) using the chain rule:

$$\frac{dF(z(w))}{dw} = \frac{dF}{dz} \cdot \frac{dz}{dw}. \qquad\qquad (15.62)$$

From the transformation laws Eqns. 15.59 and 15.60 we have

$$\frac{\partial x'}{\partial x} = \cos\theta \qquad \frac{\partial x'}{\partial y} = \sin\theta \qquad\qquad (15.63)$$

$$\frac{\partial y'}{\partial x} = -\sin\theta \qquad \frac{\partial y'}{\partial y} = \cos\theta. \qquad\qquad (15.64)$$

For any function $F'(x', y')$ we have then

$$\frac{\partial F'(x', y')}{\partial x} = \frac{\partial F'}{\partial x'}\frac{\partial x'}{\partial x} + \frac{\partial F'}{\partial y'}\frac{\partial y'}{\partial x}$$

$$= \frac{\partial F'}{\partial x'}\cos\theta + \frac{\partial F'}{\partial y'}(-\sin\theta) \qquad\qquad (15.65)$$

$$\frac{\partial F'}{\partial y} = \frac{\partial F'}{\partial x'}\frac{\partial x'}{\partial y} + \frac{\partial F'}{\partial y'}\frac{\partial y'}{\partial y}$$

$$= \frac{\partial F'}{\partial x'}\sin\theta + \frac{\partial F'}{\partial y'}\cos\theta. \qquad\qquad (15.66)$$

Since this is true for *any F'*, one writes

$$\frac{\partial}{\partial x} = \frac{\partial}{\partial x'}\cos\theta - \frac{\partial}{\partial y'}\sin\theta \tag{15.67}$$

$$\frac{\partial}{\partial y} = \frac{\partial}{\partial x'}\sin\theta + \frac{\partial}{\partial y'}\cos\theta \tag{15.68}$$

where the equality means *the left- and right-hand sides will give the same result when acting on any F'*. If you can overlook the fact that the two sides contain not numbers, but derivatives, you will notice that

$$\left(\frac{\partial}{\partial x'}, \frac{\partial}{\partial y'}\right) \leftrightarrow (x, y) \tag{15.69}$$

where $\leftrightarrow$ means the two transform the same way into primed objects under rotations. This is what ensures that

$$\mathbf{\nabla} \cdot \mathbf{E} = \mathbf{\nabla}' \cdot \mathbf{E}' \tag{15.70}$$

is rotationally invariant, like any dot product. Let us verify this explicitly by going back to Eqn. 15.58 and making the substitution:

$$\frac{\partial E_x}{\partial x} + \frac{\partial E_y}{\partial y} = \frac{\partial(E'_x\cos\theta - E'_y\sin\theta)}{\partial x}$$
$$+ \frac{\partial(E'_x\sin\theta + E'_y\cos\theta)}{\partial y} \tag{15.71}$$

$$= \left(\frac{\partial}{\partial x'}\cos\theta - \frac{\partial}{\partial y'}\sin\theta\right)(E'_x\cos\theta - E'_y\sin\theta)$$

$$+ \left(\frac{\partial}{\partial x'}\sin\theta + \frac{\partial}{\partial y'}\cos\theta\right)(E'_x\sin\theta + E'_y\cos\theta) \tag{15.72}$$

$$= \frac{\partial E'_x}{\partial x'} + \frac{\partial E'_y}{\partial y'}. \tag{15.73}$$

We have therefore established that $\mathbf{\nabla} \cdot \mathbf{E}$ not only looks like a scalar but also transforms like one. (This was also implied by Eqn. 15.55, which equated it to a scalar ρ/ε_0.) Similarly, $\nabla \times \mathbf{E}$ looks like and transforms like a vector under rotations.

15.6 Lorentz scalars and vectors

An analogous procedure exists for defining and manipulating scalars and vectors when we pass from rotations to Lorentz transformations.

A *Lorentz scalar* is invariant under the Lorentz transformations. Examples are $X \cdot X$ or $X \cdot P$.

A *Lorentz vector*, or simply, four-vector $V = (V_0, V_1)$ transforms as

$$V_0' = \frac{V_0 - \beta V_1}{\sqrt{1 - \beta^2}} \tag{15.74}$$

$$V_1' = \frac{V_1 - \beta V_0}{\sqrt{1 - \beta^2}}. \tag{15.75}$$

The inverse transformation is obtained by reversing β:

$$V_0 = \frac{V_0' + \beta V_1'}{\sqrt{1 - \beta^2}} \tag{15.76}$$

$$V_1 = \frac{V_1' + \beta V_0'}{\sqrt{1 - \beta^2}}. \tag{15.77}$$

Applying the chain rule as before, now to functions of $(x_0, x_1) \equiv (ct, x)$, yields

$$\frac{\partial}{\partial x_0} = \frac{\partial}{\partial x_0'} \frac{1}{\sqrt{1 - \beta^2}} - \frac{\partial}{\partial x_1'} \frac{\beta}{\sqrt{1 - \beta^2}} = \frac{\frac{\partial}{\partial x_0'} - \beta \frac{\partial}{\partial x_1'}}{\sqrt{1 - \beta^2}} \tag{15.78}$$

$$\frac{\partial}{\partial x_1} = \frac{\partial}{\partial x_1'} \frac{1}{\sqrt{1 - \beta^2}} - \frac{\partial}{\partial x_0'} \frac{\beta}{\sqrt{1 - \beta^2}} = \frac{\frac{\partial}{\partial x_1'} - \beta \frac{\partial}{\partial x_0'}}{\sqrt{1 - \beta^2}}. \tag{15.79}$$

By comparing this result the resuls to Eqns. 15.76 and 15.77 we see that $(\partial/\partial x_0, \partial/\partial x_1)$ does not transform like (V_0, V_1); there are some problematic minus signs. But if we define

$$\nabla = (\nabla_0, \nabla_1) = \left(\frac{\partial}{\partial x_0}, -\frac{\partial}{\partial x_1} \right) \tag{15.80}$$

then the components of ∇ transform like a four-vector:

$$\nabla_0 = \frac{\nabla_0' + \beta \nabla_1'}{\sqrt{1 - \beta^2}} \tag{15.81}$$

$$\nabla_1 = \frac{\nabla_1' + \beta \nabla_0'}{\sqrt{1 - \beta^2}}. \tag{15.82}$$

The following operator is therefore Lorentz invariant:

$$\nabla \cdot \nabla = \nabla' \cdot \nabla' \quad \text{which means} \tag{15.83}$$

$$\nabla_0^2 - \nabla_1^2 = (\nabla_0')^2 - (\nabla_1')^2, \tag{15.84}$$

as may be verified from Eqns. 15.81 and 15.82. What this means is that for any $F'(x_0', x_1')$

$$\frac{\partial^2 F'(x_0', x_1')}{\partial x_0^2} - \frac{\partial^2 F'(x_0', x_1')}{\partial x_1^2}$$
$$= \frac{\partial^2 F'(x_0', x_1')}{\partial (x_0')^2} - \frac{\partial^2 F'(x_0', x_1')}{\partial (x_1')^2}. \tag{15.85}$$

This is generally true no matter what F' is. If, however, it is a scalar field obeying

$$F'(x_0', x_1') = F(x_0, x_1) \tag{15.86}$$

we may rewrite Eqn. 15.85 as the statement of the Lorentz invariance of $\nabla \cdot \nabla F$:

$$\frac{\partial^2 F(x_0, x_1)}{\partial x_0^2} - \frac{\partial^2 F(x_0, x_1)}{\partial x_1^2} = \frac{\partial^2 F'(x_0', x_1')}{\partial (x_0')^2} - \frac{\partial^2 F'(x_0', x_1')}{\partial (x_1')^2}$$
$$\tag{15.87}$$

In more familiar terms $\nabla \cdot \nabla = \nabla' \cdot \nabla'$ stands for

$$\frac{1}{c^2}\frac{\partial^2}{\partial t^2} - \frac{\partial^2}{\partial x^2} = \frac{1}{c^2}\frac{\partial^2}{\partial t'^2} - \frac{\partial^2}{\partial x'^2}. \tag{15.88}$$

If we bring in all four coordinates, the entity that transforms as a four-vector is

$$\nabla = \left(\frac{\partial}{\partial x_0}, -\frac{\partial}{\partial x}, -\frac{\partial}{\partial y}, -\frac{\partial}{\partial z}\right) \tag{15.89}$$

and

$$\nabla \cdot \nabla = \nabla' \cdot \nabla' \qquad \text{stands for} \qquad (15.90)$$

$$\frac{1}{c^2}\frac{\partial^2}{\partial t^2} - \frac{\partial^2}{\partial x^2} - \frac{\partial^2}{\partial y^2} - \frac{\partial^2}{\partial z^2} = \frac{1}{c^2}\frac{\partial^2}{\partial t'^2} - \frac{\partial^2}{\partial x'^2} - \frac{\partial^2}{\partial y'^2} - \frac{\partial^2}{\partial z'^2}.$$
$$\tag{15.91}$$

15.7   The four-current J

I have mentioned that electrodynamics can be written in four-vector notation in terms of two four-vectors J and A. Here is the first.

Consider the charge density ρ and current density $\mathbf{j}$. I will drop the vector symbol for the current when discussing motion along just x, just as I replaced $(x, y, z, t) = (\mathbf{r}, t)$ by (x, t) to introduce and study the Lorentz transformations in Volume I. The main point is that the mixture of space and time is evident with just x and t around, and usually y and z just come along for the ride.

Consider a tiny lump of charge at rest with density ρ_0. The current associated with it is zero. If we see the lump from a frame in which it moves at velocity v, we will find an increased density due to length contraction *in the direction of motion*, and a non-zero current because the charge is now moving:

$$\rho = \frac{\rho_0}{\sqrt{1 - v^2/c^2}} \tag{15.92}$$

$$j = \rho v = \frac{\rho_0 v}{\sqrt{1 - v^2/c^2}}. \tag{15.93}$$

This is identical in form to

$$\frac{E}{c^2} = \frac{m_0}{\sqrt{1 - v^2/c^2}} \tag{15.94}$$

$$p = \frac{m_0 v}{\sqrt{1 - v^2/c^2}} = \frac{E}{c^2} v \tag{15.95}$$

where, just to highlight the analogy, I have referred to the invariant mass as m_0 instead of just m.

Thus we have the correspondence between two four-vectors

$$\left[\frac{E}{c^2}, p\right] \leftrightarrow [\rho, j] \tag{15.96}$$

where the symbol $\leftrightarrow$ means that the two objects will transform the same way under Lorentz transformations.

As usual let us introduce factors of c so that all the components of the four-vectors have the same dimensionality. This leads to the correspondence of the transformation laws of the energy momentum four-vector P, the four-current J, and the position vector in space-time X:

$$P = \left[\frac{E}{c}, p\right] \leftrightarrow J = [\rho c, j] \leftrightarrow X = (ct, x). \tag{15.97}$$

The components of J are labeled in many equivalent ways:

$$J = (J_0, J_1, J_2, J_3) \equiv (\rho c, j_x, j_y, j_z) \equiv (\rho c, \mathbf{j}). \tag{15.98}$$

15.7.1 Charge conservation and the four-current J

The fact that electric charge is conserved can be expressed in the form of the following equation. Imagine a closed surface S bounding a volume V. The surface integral of the current density $\mathbf{j}$, which is the charge flowing per unit area per second, must equal the rate of decrease of charge in V if charge is neither created nor destroyed. So we may write

$$\oint_{S=\partial V} \mathbf{j} \cdot d\mathbf{S} = -\frac{d}{dt} \int_V \rho(\mathbf{r}, t) d^3 r = \int_V \left[-\frac{\partial \rho}{\partial t}\right] d^3 r. \tag{15.99}$$

Recall that the integral Maxwell equation

$$\oint_{S=\partial V} \mathbf{E} \cdot d\mathbf{S} = \int_V \frac{\rho}{\varepsilon_0} d^3\mathbf{r} \tag{15.100}$$

implied the following differential Maxwell equation

$$\mathbf{\nabla} \cdot \mathbf{E} = \frac{\rho}{\varepsilon_0}. \tag{15.101}$$

So the corresponding *continuity equation*, relating $\mathbf{j}$ and ρ, is

$$\mathbf{\nabla} \cdot \mathbf{j} + \frac{\partial \rho}{\partial t} = 0. \tag{15.102}$$

Amazingly this is a four-dimensional equation, invariant under Lorentz transformations because it may be written as a four-dimensional dot product:

$$\nabla \cdot J = 0 \quad \text{where} \tag{15.103}$$

$$\nabla = (\nabla_0, \nabla_1, \nabla_2, \nabla_3) = \left(\frac{\partial}{\partial ct}, -\frac{\partial}{\partial x}, -\frac{\partial}{\partial y}, -\frac{\partial}{\partial z} \right) \tag{15.104}$$

$$J = (\rho c, j_x, j_y, j_z). \tag{15.105}$$

Whereas the components of other four-vectors like J or P or X are the scalar and vector parts with the same sign, e.g., $X = (X_0, X_1, X_2, X_3) = (ct, x, y, z)$, the components of ∇ have negative signs in front of the spatial derivatives as part of its definition. As a result, its four-dimensional dot product with J does not have the usual minus signs found in Lorentz invariant dot products of typical vectors like X or P.

15.8 The four-potential A

Potentials are introduced into electrodynamics for many reasons, one of which is to facilitate the solution of Maxwell's equations.

Let us begin with the Maxwell equations (which cannot be written too often), in the *differential* version:

$$\nabla \cdot \mathbf{E} = \frac{\rho}{\varepsilon_0} \qquad \text{I;} \qquad (15.106)$$

$$\text{I;} \nabla \cdot \mathbf{B} = 0 \qquad \text{II;} \qquad (15.107)$$

$$\nabla \times \mathbf{E} + \frac{\partial \mathbf{B}}{\partial t} = 0 \qquad \text{III;} \qquad (15.108)$$

$$\nabla \times \mathbf{B} = \mu_0 \mathbf{j} + \mu_0 \varepsilon_0 \frac{\partial \mathbf{E}}{\partial t} \qquad \text{IV;} \qquad (15.109)$$

$$\mathbf{F} = q(\mathbf{E} + \mathbf{v} \times \mathbf{B}) \quad \text{Lorentz force.} \qquad (15.110)$$

Our goal is to solve for $\mathbf{E}$ and $\mathbf{B}$ given the charges and currents and then find the force on charges using the Lorentz force law.

Notice that the middle two equations do not depend on ρ and $\mathbf{j}$. There is a way to parameterize $\mathbf{E}$ and $\mathbf{B}$ so that these two equations are identically satisfied. Here is a simpler example of this strategy before I get to the real problem. Suppose $A(t)$ and $B(t)$ are two variables that are required to obey

$$A^2(t) + B^2(t) = 5 \qquad (15.111)$$

$$A^2(t) - B^2(t) = 5 \cos 12t \qquad (15.112)$$

for all times. If we parameterize them by an angle $\theta(t)$ as

$$A(t) = \sqrt{5} \sin\theta(t) \quad B(t) = \sqrt{5} \cos\theta(t), \qquad (15.113)$$

Eqn. 15.111 will be identically satisfied *no matter what* $\theta(t)$ *is*. If all we had was Eqn 15.111, we are done. But we still have Eqn. 15.112, which can restrict $\theta(t)$. We stick our parameterization Eqn. 15.113 into Eqn. 15.112 and we find the constraint

$$5(\cos^2\theta - \sin^2\theta) = 5\cos 2\theta = 5\cos 12t \qquad (15.114)$$

with a solution $\theta(t) = 6t \pm m\pi$, where m is any integer.

Now for the actual problem. Here is the parameterization for $\mathbf{E}$ and $\mathbf{B}$ that makes the middle two Maxwell equations into identities:

$$\mathbf{B} = \nabla \times \mathbf{A} \qquad (15.115)$$

$$\mathbf{E} = -\nabla V - \frac{\partial \mathbf{A}}{\partial t}. \qquad (15.116)$$

The *vector potential* **A** generates **B** as its curl,

$$\mathbf{B} = \nabla \times \mathbf{A}, \tag{15.117}$$

very much like the way the scalar potential V gave us the electrostatic field **E** as its gradient: $-\nabla V = \mathbf{E}$. In the time-dependent case we know **E** is not conservative and its circulation is controlled by the changing magnetic field. This is incorporated by the inclusion of $-\frac{\partial \mathbf{A}}{\partial t}$ in Eqn. 15.116.

Let us see how the parameterization in terms of V and **A** renders the middle Maxwell equations into identities.

First we see that

$$\nabla \cdot \mathbf{B} = \frac{\partial B_x}{\partial x} + \frac{\partial B_y}{\partial y} + \frac{\partial B_z}{\partial z} \tag{15.118}$$

$$= \frac{\partial}{\partial x}\left(\frac{\partial A_z}{\partial y} - \frac{\partial A_y}{\partial z}\right) + \frac{\partial}{\partial y}\left(\frac{\partial A_x}{\partial z} - \frac{\partial A_z}{\partial x}\right)$$

$$+ \frac{\partial}{\partial z}\left(\frac{\partial A_y}{\partial x} - \frac{\partial A_x}{\partial y}\right) \tag{15.119}$$

$$\equiv 0 \tag{15.120}$$

due to the cancellation of mixed derivatives.

Next plug Eqn. 15.116 and 15.115 into the left-hand side of Maxwell equation *III*:

$$\nabla \times \mathbf{E} + \frac{\partial \mathbf{B}}{\partial t} = \nabla \times \left(-\nabla V - \frac{\partial \mathbf{A}}{\partial t}\right) + \frac{\partial(\nabla \times \mathbf{A})}{\partial t} \tag{15.121}$$

$$= -\nabla \times \nabla V + \nabla \times \left(-\frac{\partial \mathbf{A}}{\partial t}\right) + \frac{\partial(\nabla \times \mathbf{A})}{\partial t} \tag{15.122}$$

$$= -\mathbf{i}\left(\frac{\partial^2 V}{\partial y \partial z} - \frac{\partial^2 V}{\partial z \partial y}\right) - \mathbf{j}\left(\frac{\partial^2 V}{\partial z \partial x} - \frac{\partial^2 V}{\partial x \partial z}\right)$$

$$-\mathbf{k}\left(\frac{\partial^2 V}{\partial x \partial y} - \frac{\partial^2 V}{\partial y \partial x}\right)$$

$$\equiv 0. \tag{15.123}$$

Along the way I have used

$$\mathbf{\nabla} \times \left(\frac{\partial \mathbf{A}}{\partial t} \right) = \frac{\partial (\mathbf{\nabla} \times \mathbf{A})}{\partial t}. \tag{15.124}$$

Whether or not you followed all this, remember the bottom line: *the two Maxwell equations that do not involve charges and currents are identically satisfied if we write* **E** *and* **B** *in terms of the scalar and vector potentials as follows:*

$$\mathbf{B} = \mathbf{\nabla} \times \mathbf{A} \tag{15.125}$$

$$\mathbf{E} = -\mathbf{\nabla} V - \frac{\partial \mathbf{A}}{\partial t}. \tag{15.126}$$

This is the analog of setting $A(t) = \sqrt{5}\sin\theta(t)$ and $B(t) = \sqrt{5}\cos\theta(t)$. There we saw that $\theta(t)$ could be anything if all we cared about was $A^2 + B^2 = 5$. Likewise V and $\mathbf{A}$ are arbitrary as long as we only care about the middle two Maxwell equations, unrelated to charges and currents. We now turn to the other two Maxwell equations (the analog of $A^2 - B^2 = 5\cos 12t$ that determined $\theta(t)$) to find the equations obeyed by V and $\mathbf{A}$.

Before we substitute Eqns. 15.126 and 15.125 in the Maxwell equations with charge and current densities in the right-hand side, there is one issue we have to confront.

15.8.1 Gauge invariance

Whereas I can walk into a room with some test charges and measure **E** and **B** at every point, this is not so for V and $\mathbf{A}$. The reason is that *they are not unique*: if a certain V and $\mathbf{A}$ lead to an **E** and **B**, another pair $(\tilde{V}, \tilde{\mathbf{A}})$, depending on an arbitrary function χ,

$$\tilde{\mathbf{A}} = \mathbf{A} + \mathbf{\nabla}\chi \tag{15.127}$$

$$\tilde{V} = V - \frac{\partial \chi}{\partial t}, \tag{15.128}$$

will lead to the same **E** *and* **B** . You should verify this. The change of V and $\mathbf{A}$ by derivatives of χ is called a *gauge transformation*. The pairs

$(V, \mathbf{A})$ and $(\tilde{V}, \tilde{\mathbf{A}})$ are said to be *gauge equivalent* or the *gauge transforms* of each other. Long ago we learned that the potential V is defined only up to an additive constant in electrostatics and gravity. Gauge invariance reflects the even greater latitude of the potentials in the general case of time-dependent electromagnetism.

Remember how we used the freedom to add a constant to V to suit our purpose? For celestial problems we chose the constant such that $V(r \to \infty) = 0$. For problems near the earth we chose V to vanish at its surface $r = R_E$: $V(r = R_E) = 0$.

Likewise we use gauge freedom to simplify some calculations by choosing from the family of physically equivalent V and $\mathbf{A}$ one representative by imposing an extra condition, called *the gauge condition*, on them. For example, we can demand that we trade the original $\mathbf{A}$ for a gauge transform $\tilde{\mathbf{A}}$ obeying

$$\frac{\partial \tilde{A}_x}{\partial x} + \frac{\partial \tilde{A}_y}{\partial y} + \frac{\partial \tilde{A}_z}{\partial z} \equiv \mathbf{\nabla} \cdot \tilde{\mathbf{A}} = 0. \tag{15.129}$$

This is called the *Coulomb gauge*. It is true, though I will not prove it, that any $\mathbf{A}$ can be gauge transformed (by a judicious choice of χ) to $\tilde{\mathbf{A}}$ obeying the Coulomb gauge condition.

The gauge we want to use in this discussion of relativistic invariance is the *Lorentz gauge*

$$\frac{\partial \tilde{A}_x}{\partial x} + \frac{\partial \tilde{A}_y}{\partial y} + \frac{\partial \tilde{A}_z}{\partial z} + \frac{1}{c^2} \frac{\partial \tilde{V}}{\partial t} = 0, \tag{15.130}$$

which for use in the near future I will rewrite as follows (dropping the tilde since this will be the only gauge for A from now on):

$$\mathbf{\nabla} \cdot \mathbf{A} = -\frac{1}{c^2} \frac{\partial V}{\partial t}. \tag{15.131}$$

With this condition, the Maxwell equations with charges and currents become wave equations that determine $(V, \mathbf{A})$ in terms of them. I will derive one of them and leave the rest to you.

15.9 Wave equation for the four-vector A

Start with

$$\nabla \cdot \mathbf{E} = \frac{\rho}{\varepsilon_0} \qquad (15.132)$$

and introduce the definition of $\mathbf{E}$ in terms of V and $\mathbf{A}$

$$\mathbf{E} = -\nabla V - \frac{\partial \mathbf{A}}{\partial t} \qquad (15.133)$$

to obtain

$$-\nabla \cdot \nabla V - \frac{\partial \nabla \cdot \mathbf{A}}{\partial t} = \frac{\rho}{\varepsilon_0} \qquad (15.134)$$

upon using

$$\nabla \cdot \frac{\partial \mathbf{A}}{\partial t} = \frac{\partial \nabla \cdot \mathbf{A}}{\partial t}. \qquad (15.135)$$

At this stage, V and $\mathbf{A}$ are entangled by Eqn. 15.134. Now remember that

$$\nabla \cdot \nabla = \left(\mathbf{i}\frac{\partial}{\partial x} + \mathbf{j}\frac{\partial}{\partial y} + \mathbf{k}\frac{\partial}{\partial z} \right) \cdot \left(\mathbf{i}\frac{\partial}{\partial x} + \mathbf{j}\frac{\partial}{\partial y} + \mathbf{k}\frac{\partial}{\partial z} \right) \quad (15.136)$$

$$= \left(\frac{\partial^2}{\partial x^2} + \frac{\partial^2}{\partial y^2} + \frac{\partial^2}{\partial z^2} \right) \qquad (15.137)$$

and

$$\frac{\partial}{\partial t}\nabla \cdot \mathbf{A} = -\frac{1}{c^2}\frac{\partial^2 V}{\partial t^2}$$

(time derivative of Lorentz gauge, Eqn. 15.131). $\quad (15.138)$

Put all this into Eqn. 15.134 and obtain *an equation that involves just* V:

$$\frac{\partial^2 V}{\partial x^2} + \frac{\partial^2 V}{\partial y^2} + \frac{\partial^2 V}{\partial z^2} - \frac{1}{c^2}\frac{\partial^2 V}{\partial t^2} = -\frac{\rho}{\varepsilon_0}. \qquad (15.139)$$

Similar manipulations with the equation for $\nabla \times \mathbf{B}$ will yield an equation involving just $\mathbf{A}$ upon imposing the Lorentz gauge condition.

Here is the final set of equations coming from rewriting the equations for $\mathbf{E}$ and $\mathbf{B}$ that involve ρ and $\mathbf{j}$ in terms of V and $\mathbf{A}$:

$$\frac{\partial^2 V}{\partial x^2} + \frac{\partial^2 V}{\partial y^2} + \frac{\partial^2 V}{\partial z^2} - \frac{1}{c^2}\frac{\partial^2 V}{\partial t^2} = -\frac{\rho}{\varepsilon_0} \tag{15.140}$$

$$\frac{\partial^2 \mathbf{A}}{\partial x^2} + \frac{\partial^2 \mathbf{A}}{\partial y^2} + \frac{\partial^2 \mathbf{A}}{\partial z^2} - \frac{1}{c^2}\frac{\partial^2 \mathbf{A}}{\partial t^2} = -\mu_0 \mathbf{j}. \tag{15.141}$$

These are called the *inhomogeneous wave equations* or the *wave equations with sources*. Their solutions will exhibit the retardation demanded by relativity: $A(t,\mathbf{r})$ will receive contributions from $J(t',\mathbf{r}')$ where $t' = t - |\mathbf{r} - \mathbf{r}'|/c$.

These equations were written down well before Einstein. What was realized after him was that V and $\mathbf{A}$ combine to form the *four-potential*

$$A = \left(\frac{V}{c}, \mathbf{A}\right) \equiv (A_0, A_1, A_2, A_3) \tag{15.142}$$

and that *Eqns. 15.140 and 15.141 could be combined into a single wave equation relating the four-vector A to the four-vector J.*

To verify this

(i) divide the first equation for V by c;

(ii) remember ρc is the 0-th component of J, and V/c is the 0-th component of A; and

(iii) finally, invoke $1/(\varepsilon_0 c^2) = \mu_0$.

This will lead to

$$\frac{\partial^2 A}{\partial x^2} + \frac{\partial^2 A}{\partial y^2} + \frac{\partial^2 A}{\partial z^2} - \frac{1}{c^2}\frac{\partial^2 A}{\partial t^2} = -\mu_0 J, \tag{15.143}$$

with A defined as in Eqn. 15.142. We may rewrite this equation as

$$\nabla \cdot \nabla A = \mu_0 J. \tag{15.144}$$

This equation implies that $A = (V, \mathbf{A})$ is a four-vector. The reason is that the right-hand side is the four-vector J and the combination of derivatives on the left-hand side (given by the dot product of ∇ with ∇) is invariant under Lorentz transformations. So A must transform like J, which is a four-vector.

The key to Lorentz invariance is the choice of the Lorentz gauge, introduced earlier in Eqn. 15.130,

$$\frac{\partial A_x}{\partial x} + \frac{\partial A_y}{\partial y} + \frac{\partial A_z}{\partial z} + \frac{1}{c^2}\frac{\partial V}{\partial t} = 0, \tag{15.145}$$

because it too may be rewritten in four-dimensional notation as

$$\nabla \cdot A = 0 \tag{15.146}$$

because

$$\frac{1}{c^2}\frac{\partial V}{\partial t} = \frac{\partial(V/c)}{\partial(ct)} = \frac{\partial V_0}{\partial x_0}. \tag{15.147}$$

As with $\nabla \cdot J$, there are no minus signs in the dot product since ∇ contains them in its definition.

Thus all the key equations can be written in terms of four-vectors and all equations have the same form in all Lorentz frames with the same velocity c appearing. The key point is that *this did not call for changing any of pre-Einstein electrodynamics, only re-expressing it in terms of four-vectors.*

When Maxwell came up with the wave equation a question that arose was this: "For whom is the wave velocity c?" Generally the velocity of a wave is with respect to the medium that supports it. Assuming that light was supported by a medium called *ether,* it was assumed that the value c would be measured only by an observer at rest with respect to the ether. It then seemed obvious, to people thinking in terms of the Galilean transformation

$$x = x' + ut' \tag{15.148}$$

$$t = t', \tag{15.149}$$

which implied the velocity transformation law

$$\frac{dx}{dt} = \frac{dx'}{dt} + u, \tag{15.150}$$

that the velocity of light would be different from c for any observer moving relative to the ether. One could hope to find one's velocity relative to this ether by measuring the velocity of light and subtracting c.

Of course, the same light velocity c was obtained no matter when or where or by whom it was measured. This led to some real confusion till Einstein finally arrived on the scene and banished ether as an unnecessary concept (which never had to be invoked in deriving the wave equation). If one used the Lorentz transformation to change space-time coordinates, the wave equation would remain invariant and the same value of c would appear for all inertial observers.

Before Einstein, Hendrik A. Lorentz (1853–1928), Joseph Larmor (1857–1942), and others had suggested that motion against the ether causes clocks to slow and rods to shrink in exactly the manner that Einstein later deduced. Henri Poincare (1854–1912) even wrote down the Lorentz transformations in the modern form and showed that it preserved the form of the wave equation for light. However, in the view of Lorentz and others, length contraction and time dilatation were real effects caused by absolute motion with respect to the all-pervasive medium, the ether. It was Einstein who explained these effects were relative and required by relativistic invariance.

15.9.1 Why work with V and A ?

Why bother with $A = (V/c, \mathbf{A})$, given that they are not unique and need to be constrained by an arbitrary gauge condition? For one thing, A is a four-vector and we could cast the Maxwell equations in terms of it to demonstrate their Lorentz invariance. Why not start with the Maxwell equations and the Lorentz force law in terms of $\mathbf{E}$ and $\mathbf{B}$ and show that they are Lorentz invariant? The reason is that $\mathbf{E}$ and $\mathbf{B}$ do not become parts of four-vectors, but parts of a *tensor*, an idea that may not be familiar to you. If you do not want to prove Lorentz invariance you are indeed free to avoid V and $\mathbf{A}$ and work with $\mathbf{E}$ and $\mathbf{B}$.

When we come to quantum theory the situation changes. We find we have no choice but to work with V and **A**. There is no known formalism that directly works with **E** and **B** fields.

An experiment suggested by the work of Yakir Aharanov (1932–) and David Bohm (1917–1992) provides a dramatic illustration of why we need to work with **A**. Imagine particles moving in a plane, say the plane of the page, pierced perpendicularly by an infinitely long impenetrable solenoid carrying some magnetic flux. Outside the solenoid **B** = 0 but **A** $\neq$ 0. (In other words, **A** is non-zero inside and outside the solenoid, but it has a curl only inside the solenoid.) As the particles are forbidden from going into the solenoid, they should not be sensitive to the flux inside. Yet they are! Without ever entering the solenoid, moving only in an area where **B** = 0, they are able to sense the flux inside the solenoid. Understanding this experiment requires a quantum mechanical treatment that unavoidably invokes **A**.

Finally, as I mentioned earlier, the theories of electromagnetic, weak and strong interactions are all gauge theories. This is one reason I dragged you through this.

15.10   The electromagnetic tensor $\mathcal{F}$

Let us return to classical electrodynamics. Suppose you do not want to work with the four-potential A and prefer **E** and **B** fields. How is relativistic invariance of electrodynamics demonstrated in terms of **E** and **B** if they do not team up with some other scalars to form four-vectors and instead their six-components combine to form a *tensor*? To answer this we have to bite the bullet and get acquainted with tensors.

15.10.1   Tensors

Recall that a scalar in three dimensions has just one (3^0) component. A vector **V** has $3^1 = 3$ components, denoted as V_x, V_y, V_z or V_1, V_2, V_3. A *second rank tensor* T, which is the only tensor I will discuss, has $3^2 = 9$ components.

What are the components of T and how do they transform under rotations?

As you can guess, the components of T are labeled either $T_{11}, T_{12}, \ldots T_{33}$ or $T_{xx}, T_{xy} \ldots T_{zz}$. Under a rotation of axes the 9 components of T will transform into linear combinations of each other, analogous to the way the 3 components of a vector do. What are the transformation rules? One way to find them is to make up a tensor by gluing together two vectors $\mathbf{V} = (V_x, V_y, V_z)$ and $\mathbf{W} = (W_x, W_y, W_z)$ as follows:

$$T_{xx} = V_x W_x \tag{15.151}$$

$$T_{xy} = V_x W_y \tag{15.152}$$

$$\cdots \quad \cdots \tag{15.153}$$

$$T_{zz} = V_z W_z. \tag{15.154}$$

The components of the rotated tensor are then found from the components of the rotated vectors. For example, under a rotation by θ around the z-axis we know

$$V'_x = V_x \cos\theta + V_y \sin\theta \tag{15.155}$$

$$V'_y = -V_x \sin\theta + V_y \cos\theta \tag{15.156}$$

$$V'_z = V_z \tag{15.157}$$

and similarly for $\mathbf{W}$. Thus we know that

$$T'_{xx} = V'_x W'_x \tag{15.158}$$

$$= (V_x \cos\theta + V_y \sin\theta)(W_x \cos\theta + W_y \sin\theta) \tag{15.159}$$

$$= T_{xx} \cos^2\theta + T_{xy} \cos\theta \sin\theta + T_{yx} \sin\theta \cos\theta$$

$$+ T_{yy} \sin^2\theta \tag{15.160}$$

$$T'_{yy} = V'_y W'_y \tag{15.161}$$

$$= (-V_x \sin\theta + V_y \cos\theta)(-W_x \sin\theta + W_y \cos\theta) \tag{15.162}$$

$$= T_{xx} \sin^2\theta - T_{xy} \sin\theta \cos\theta - T_{yx} \cos\theta \sin\theta$$

$$+ T_{yy} \cos^2\theta \tag{15.163}$$

$$T'_{zz} = T_{zz}, \tag{15.164}$$

and so on. *We now demand that these transformation rules are true for all second rank tensors, even if they were not obtained by fusing two vectors.* For example,

$$T'_{yy} = T_{xx} \sin^2\theta - T_{xy} \sin\theta\cos\theta$$
$$\quad - T_{yx}\cos\theta\sin\theta + T_{yy}\cos^2\theta \tag{15.165}$$

is true for all tensors for rotations around the z-axis.

Although the nine components rotate into linear combinations of each other as indicated above, some linear combinations of them may form smaller sets that rotate into each other. Here is an example you will recognize. Consider the combination

$$S = T_{xx} + T_{yy} + T_{zz} \tag{15.166}$$

when T is made out of **V** and **W**. You know that this will go into itself under rotations, i.e.,

$$S = T_{xx} + T_{yy} + T_{zz} = T'_{xx} + T'_{yy} + T'_{zz} = S' \tag{15.167}$$

because the sums above are just the dot products obeying $\mathbf{V} \cdot \mathbf{W} = \mathbf{V'} \cdot \mathbf{W'}$. But the result holds even if T is not composed of two vectors, because the answer only depends on the transformation rules for T, which apply to all tensors. I urge you to verify this from Eqns. 15.158 to 15.164 for the special case of z-rotations.

We learn another very deep result from the above: *if you set two tensor indices equal and sum over them, the tensor drops down in rank by two.*

Thus T_{ij} in general is a second rank tensor but

$$S = \sum_i T_{ii} = T_{xx} + T_{yy} + T_{zz} \tag{15.168}$$

has rank $2 - 2 = 0$, and is hence a scalar.

From the components of T we can also form the following linear combinations to produce an *antisymmetric tensor* $\mathcal{A}$, where components change sign under exchange of indices

$$\mathcal{A}_{ij} = -\mathcal{A}_{ji}. \tag{15.169}$$

They are generated from T_{ij} as follows:

$$\mathcal{A}_{ij} = T_{ij} - T_{ji} = -\mathcal{A}_{ji} \quad \text{in general, and as per} \tag{15.170}$$

$$A_{xy} = V_x W_y - V_y W_x = -A_{yx} \tag{15.171}$$

$$A_{yz} = V_y W_z - V_z W_y = -A_{zy} \tag{15.172}$$

$$A_{zx} = V_z W_x - V_x W_z = -A_{xz} \tag{15.173}$$

when A is composed of $\mathbf{V}$ and $\mathbf{W}$.

Components like A_{xx} vanish because the candidate $A_{xx} = V_x W_x - W_x V_x \equiv 0$. Finally components A_{yx}, A_{zy}, and A_{xz} are simply negatives of A_{xy}, A_{yz}, and A_{zx} and hence not independent.

So an antisymmetric tensor in three dimensions has only three independent components. But so does a vector in three dimensions! Indeed we recognize the three combinations Eqns. 15.171 to 15.173 as the components of the cross product $\mathbf{V} \times \mathbf{W}$:

$$A_{xy} = V_x W_y - V_y W_x = (\mathbf{V} \times \mathbf{W})_z \tag{15.174}$$

$$A_{yz} = V_y W_z - V_z W_y = (\mathbf{V} \times \mathbf{W})_x \tag{15.175}$$

$$A_{zx} = V_z W_x - V_x W_z = (\mathbf{V} \times \mathbf{W})_y. \tag{15.176}$$

Not surprisingly the three components of A transform into linear combinations of each other under rotations. (After all they are the components of the vector $\mathbf{V} \times \mathbf{W}$.)

It is only in three dimensions that we have this luxury of two equivalent descriptions: use two (necessarily) unequal indices to label the components an antisymmetric tensor (e.g., A_{xy}) or use the unique third index (z) to label a vector component $(\mathbf{V} \times \mathbf{W})_z$.

The torque $\boldsymbol{\tau} = \mathbf{r} \times \mathbf{F}$ is an example of an antisymmetric second rank tensor, which has just the right number of components to be a vector in $d = 3$. The same goes for the angular momentum $\mathbf{L} = \mathbf{r} \times \mathbf{p}$.

The curl of the vector potential is also an antisymmetric tensor of rank 2 but with a twist: the first factor in the cross product is not an ordinary vector but a set of derivatives:

$$\nabla \leftrightarrow \left(\frac{\partial}{\partial x}, \frac{\partial}{\partial y}, \frac{\partial}{\partial z} \right) \tag{15.177}$$

$$\mathbf{A} \leftrightarrow (A_x, A_y, A_z) \tag{15.178}$$

$$\nabla \times \mathbf{A} \leftrightarrow \left(\frac{\partial A_z}{\partial y} - \frac{\partial A_y}{\partial z}, \frac{\partial A_x}{\partial z} - \frac{\partial A_z}{\partial x}, \frac{\partial A_y}{\partial x} - \frac{\partial A_x}{\partial y} \right) \tag{15.179}$$

$$= (B_x, B_y, B_z). \tag{15.180}$$

Once again, not every antisymmetric tensor has to be formed from two vectors **V** and **W**. It is defined simply by its antisymmetry and transformation properties.

In three dimensions we can think of the curl either as a vector or an antisymmetric tensor. But if we are to generalize it to the four dimensions of space-time, the curl should be viewed as an antisymmetric tensor, as we shall see presently.

15.10.2 The electromagnetic field tensor $\mathcal{F}$

A general four-tensor $T_{\mu\nu}$ will have 16 components. Its response to a Lorentz transformation follows from the way four-vectors transform. Consider first the special case where T is composed of two four-vectors V and W:

$$T_{\mu\nu} = V_\mu W_\nu. \tag{15.181}$$

In this case

$$T'_{\mu\nu} = V'_\mu W'_\nu \tag{15.182}$$

and we know how V' and W' are related to V and W. For motion along the 1-direction,

$$T'_{01} = V'_0 W'_1 \tag{15.183}$$

$$= \frac{(V_0 - \beta V_1)}{\sqrt{1-\beta^2}} \frac{(W_1 - \beta W_0)}{\sqrt{1-\beta^2}} \tag{15.184}$$

$$= \frac{T_{01} - \beta T_{00} - \beta T_{11} + \beta^2 T_{10}}{1-\beta^2}. \tag{15.185}$$

As with rotations, we demand that all tensors transform this way, whether or not they were constructed from two vectors.

The antisymmetric tensor

$$A_{\mu\nu} = T_{\mu\nu} - T_{\nu\mu} \quad \mu = 0\ldots 3, \ \nu = 0,\ldots 3. \tag{15.186}$$

will have 6 independent components, which transform into combinations of each other. (This is another result I state but do not prove.)

The antisymmetric tensor of interest to us is $\mathcal{F}$, the *electromagnetic field tensor*. In analogy with $\mathbf{B} = \nabla \times \mathbf{A}$ (Eqn. 15.179), $\mathcal{F}$ is defined as the *four-dimensional curl of the four-vector potential A*:

$$\nabla \leftrightarrow \left(\frac{\partial}{\partial ct}, -\frac{\partial}{\partial x}, -\frac{\partial}{\partial y}, -\frac{\partial}{\partial z} \right) \equiv (\nabla_0, \nabla_x, \nabla_y, \nabla_z) \quad (15.187)$$

$$A \leftrightarrow (V/c, A_x, A_y, A_z) \equiv (A_0, A_x, A_y, A_z) \quad (15.188)$$

$$\mathcal{F}_{\mu\nu} = \nabla_\mu A_\nu - \nabla_\nu A_\mu \quad \text{(or, more explicitly)} \quad (15.189)$$

$$\mathcal{F}_{0x} = \nabla_0 A_x - \nabla_x A_0 = \frac{\partial A_x}{c \partial t} + \frac{\partial V/c}{\partial x} = -E_x/c \quad (15.190)$$

$$\mathcal{F}_{0y} = \nabla_0 A_y - \nabla_y A_0 = -E_y/c \quad (15.191)$$

$$\mathcal{F}_{0z} = \nabla_0 A_z - \nabla_z A_0 = -E_z/c \quad (15.192)$$

$$\mathcal{F}_{xy} = \nabla_x A_y - \nabla_y A_x = -B_z \quad (15.193)$$

$$\mathcal{F}_{yz} = \nabla_y A_z - \nabla_z A_y = -B_x \quad (15.194)$$

$$\mathcal{F}_{zx} = \nabla_z A_x - \nabla_x A_z = -B_y. \quad (15.195)$$

Unlike other three-vectors like $\mathbf{j}$ or $\mathbf{p}$ that combine with scalars like $c\rho$ or E/c to form four-vectors, $\mathbf{E}$ and $\mathbf{B}$ combine with each other to form the six independent components of the antisymmetric Lorentz tensor $\mathcal{F}$. They transform into combinations of each other under Lorentz transformations.

Here is a trivial example. In one-space and one-time dimension, the only non-zero component is $\mathcal{F}_{01} = -\mathcal{F}_{10} = -E_x/c$. (There can be no magnetic field in one-space dimension.) Being the sole component, it has to go into itself under a Lorentz transformation. Let us verify this using Eqn. 15.185:

$$\mathcal{F}'_{01} = \frac{\mathcal{F}_{01} + \beta^2 \mathcal{F}_{10}}{1 - \beta^2} \quad (15.196)$$

$$= \frac{\mathcal{F}_{01}(1 - \beta^2)}{1 - \beta^2} = \mathcal{F}_{01} \quad (15.197)$$

using $\mathcal{F}_{01} = -\mathcal{F}_{10}$.

Equations 15.116 and 15.115,

$$\mathbf{E} = -\nabla V - \frac{\partial \mathbf{A}}{\partial t} \tag{15.198}$$

$$\mathbf{B} = \nabla \times \mathbf{A}, \tag{15.199}$$

which expressed $\mathbf{E}$ and $\mathbf{B}$ in terms of V and $\mathbf{A}$ and ensured that the Maxwell equations not involving charges and currents are identically satisfied, are now replaced by a single tensor equation 15.189:

$$\mathcal{F}_{\mu\nu} = \nabla_\mu A_\nu - \nabla_\nu A_\mu. \tag{15.200}$$

When the other Maxwell equations involving charges and currents are written in terms of A, they relate it to J in a manner we have already encountered:

$$\nabla \cdot \nabla A = \mu_0 J. \tag{15.201}$$

The four-force F on a charge q may be written in terms of $\mathcal{F}$ and the four-velocity

$$V = (V_0, V_x, V_y, V_z) = \left[\frac{c}{\sqrt{1 - v^2/c^2}}, \frac{\mathbf{v}}{\sqrt{1 - v^2/c^2}} \right] \tag{15.202}$$

as follows:

$$F_\mu = q \left[\mathcal{F}_{\mu 0} V_0 - \mathcal{F}_{\mu x} V_x - \mathcal{F}_{\mu y} V_y - \mathcal{F}_{\mu z} V_z \right]$$
$$\mu = 0, x, y, z. \tag{15.203}$$

This gives, in one stroke, the Lorentz force and the power for $\mu = x, y, z$ and $\mu = 0$ respectively. I leave the verification to you.

On the left side of Eqn. 15.203 we have a one-index object, a vector, the four-force F. On the right we have a potentially three-index object: two indices from $\mathcal{F}_{\mu\nu}$ and one from V_ν. However, the index ν drops out from both because it is repeated and summed over (with the usual sign difference between the 0-0 and space-space terms required in any four-dimensional dot product). Thus the right-hand side also transforms like a vector. A relation equating two vectors will of course have the same form after a Lorentz transformation since both sides respond the same way.

Earlier I asked according to which observer is the velocity $\mathbf{v}$ in $\mathbf{v} \times \mathbf{B}$ to be measured. The answer, we see above, is "according to any inertial observer." We can rewrite Eqn. 15.203 as

$$F = q \, \mathcal{F} \cdot V, \qquad\qquad (15.204)$$

where the dot product is again a sum over a repeated index with the usual minus signs. If $\mathcal{F}$ had been a vector, the dot product with V would have yielded a scalar. However, $\mathcal{F}$ is a tensor with two indices and only one of them is neutralized in the dot product with V, while the other survives and matches the index of F.

We begin to understand why the Lorentz force does not contain higher powers of velocity. The only velocity we can use is the four-velocity and its square is c^2. So functions of $V \cdot V$ that could modify the answer are trivial. Potential corrections linear in $\mathcal{F}$ and cubic in V, like $(V \cdot \mathcal{F} \cdot V)V$ (where the two dot products on either side of $\mathcal{F}$ kill both its indices), vanish identically because $\mathcal{F}$ is antisymmetric under the exchange $\mu \leftrightarrow \nu$ while $V_\mu V_\nu$ is symmetric.

Here is all of electrodynamics in a nutshell (in the Lorentz gauge):

$$\nabla \cdot \nabla A = \mu_0 J \qquad\qquad (15.205)$$

$$\nabla \cdot A = 0 \quad \text{Lorentz gauge} \qquad\qquad (15.206)$$

$$\mathcal{F}_{\mu\nu} = \nabla_\mu A_\nu - \nabla_\nu A_\mu \qquad\qquad (15.207)$$

$$F = q \, \mathcal{F} \cdot V \quad \text{Lorentz 4-force} \qquad\qquad (15.208)$$

Writing $\mathcal{F}$ as the four-dimensional curl of A reduces half the Maxwell equations to identities. The other two determine A in terms of J via Eqn. 15.205 in the Lorentz gauge.

Remember the procedure for using these equations:

- Given J, solve for A (in the Lorentz gauge) from Eqn. 15.205. (I have not told you how to handle this purely mathematical problem.)
- Work out $\mathcal{F}$ as the four-dimensional curl of the four-potential A as per Eqn. 15.207.
- Use $\mathcal{F}$ in the Lorentz force law Eqn. 15.208 to find the fate of any charge q.

CHAPTER 16

Optics I: Geometric Optics Revisited

We just finished with Maxwell's theory of light. We took Ampère's law, Faraday's law, displacement current, and so on and produced the dramatic result: electromagnetic waves can exist on their own, travel away from charges and currents, and actually describe light. Light is an oscillatory phenomenon, but what is oscillating is not a medium like a string or water on a lake, but electric and magnetic fields. The field is a condition at a certain point that you can determine with test charges. You sit at that point and measure it, and you find that sometimes the field points up, sometimes the field points down, sometimes it is strong, and sometimes it is weak. It's that condition in space that travels in an electromagnetic wave.

16.1 Geometric or ray optics

The preceding point of view came near the second half of the nineteenth century, after many centuries of studying light. What I'm going to do next is present a simpler version of optics, discovered long before Maxwell. It is relevant when the wavelength of light is much smaller than the scale of observation. In daily life, for example, we are thinking in terms of centimeters and meters, whereas the wavelength of light is of the order $5 \cdot 10^{-7}$ m. In this situation you may forget about Maxwell's waves and use this simplified theory called *geometric optics*, very much the way you can forget relativistic mechanics and use its Newtonian version for small velocities, $v \ll c$. In geometric optics light goes in a straight line from start to finish, say from the source to your eye, unless it hits something. This is why it is also called *ray optics.* If you apply Maxwell's theory to a situation where the wavelength is very small you arrive at this ray approximation. (We will not derive this approximation.) When I say "very small," you always have to ask, "Very small compared to what?" Do you understand that? Just saying that the wavelength is small has no meaning. I can pick units in which the same wavelength is 1 million or 1 over a million. Small and large can be changed by change of units.

What you need is another relevant length in play, to serve as the reference for λ. This is illustrated in Figure 16.1. I take an opaque partition with a hole. On one side I place a source of light and on the other I place a screen at a distance L. The light from the source, assumed to be at a distance $\gg d$, goes through the hole and forms an image on the screen. Now I can tell you what I mean by saying the wavelength λ is small or large: geometric optics works only if $\frac{\lambda}{d} \cdot \frac{L}{d} \ll 1$. *Let us keep L/d fixed* so that the condition becomes $\frac{\lambda}{d} \ll 1$. If $\lambda \ll d$ we may use geometric or ray optics. In this limit the image on the screen is found by drawing straight lines, as shown in the upper half of Figure 16.1. The screen behind the hole is illuminated in a region that has the same shape and size as the hole.

If $\lambda \ll d$ is not satisfied, the light fans out of the hole and illuminates a region on the screen much bigger than the geometric shadow. The smaller

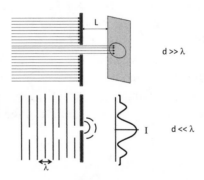

Figure 16.1   Top: geometric optics. The beam of parallel rays crosses a hole of dimension $d \gg \lambda$ and forms an image of the same shape and size as the hole. (We are keeping L/d fixed.) Bottom: When a plane wave (whose crests and troughs are shown by solid and broken lines) hits a hole with $d \ll \lambda$, it fans out to illuminate a region of the screen that is much bigger than the hole. The oscillatory curve depicts I, the intensity or brightness, as we move off center.

the hole, the more the light fans out. The degree of brightness does not fall monotonically as we move off center but oscillates. The oscillatory curve depicts the intensity I or brightness as we move off center. Though geometric optics cannot describe all this, it reigned for centuries because its limitations became apparent only for apertures small comparable to the wavelength of visible light $\lambda \simeq 5 \cdot 10^{-7}$ m.

Why did I not start with geometric optics and work my way up, culminating in the final description of light due to Maxwell, instead of going back in time to geometric optics? There are many reasons. The first is that this course is focused on electromagnetic theory and light came out as a surprise at the end. The next is that I do not plan to simply go over all the ideas of geometric optics. Instead I am going to show you a single overarching principle from which all those seemingly unrelated results can be derived. Finally, Maxwell's theory isn't the last word on light either. It fails when you consider light of very low intensity. If light becomes really dim you might think that all that will happen is that the magnitudes of $\mathbf{E}$ and $\mathbf{B}$ (whose squares measure intensity I) become smaller and smaller. But something else happens. You find that light energy is not coming in continuously like it should in a wave, but in discrete packets. These are called *photons*. You will not be aware of photons if the light is very intense because there will be so many of them coming at you, just as you are unaware that water is made of molecules when you take a shower or when you look at the ocean and study its waves. You don't see the molecules and you don't need them for describing ocean waves. Likewise, you do not need to deal with photons unless the light is very feeble. But such is the condition in the microscopic world, and we will talk about photons at length as part of quantum mechanics.

In short, having dealt with Maxwell, we will first go back in time to geometric optics and then forward to the quantum theory of photons.

16.2   Brief history of c

What did people of antiquity know about light? After some false starts, they figured out that anything bright or shiny emits something called light and we can see it. It seemed to travel in a straight line, and, for the longest time, people did not know how fast it traveled. It looked like it traveled instantaneously from source to sensor, because observers couldn't measure its travel time in daily life. This is unlike sound. You know sound travels at a finite speed, because if you yell at a mountain, it yells back with a delay you can time even with your pulse. From the delay you can find the velocity of sound.

Galileo tried to find the velocity of light in a similar manner. He asked one of his buddies to stand on top of one mountain, while he stood on top of another mountain a mile away. Each had a lantern with a shutter. First Galileo would open his shutter, and the instant his friend saw the flash of light, he would open *his* shutter to signal back to Galileo, who was timing the round trip. He soon realized that he was not measuring the time for the round trip but the sum of their reaction times because he observed the same delay when his friend was very close. It was clear that to measure the speed of light, if it were finite, would require a very long distance of travel. Given the accuracy of time measurement in those days, we can see that even the distance equal to the circumference of the earth would not have been enough because light would take only about $\frac{1}{7}$-th of a second to traverse that.

The first successful scheme for measuring the velocity of light came in 1676 from Olaf Römer (1644–1710). His brilliant strategy is depicted in Figure 16.2. You see the earth and Jupiter, initially located at E_1 and J_1 in their journey around the sun (S). Jupiter has a moon called Io, which Newtonian physics assures us will orbit with a definite time period T. Every time he saw Io go over a certain position (involving an eclipse) relative to Jupiter, Römer noted the time in his lab book. Let me call this notation a pulse, as though Io were visible only when it was at this specific location relative to Jupiter. The first pulse (shown as a solid line in the inset) reaches the earth when it is in position E_1. Let us call that time $t = 0$. If nothing but Io moved, the subsequent pulses should arrive at $t = T$, $t = 2T$, and so forth, as shown by more solid lines. But Römer observed that as the earth made its journey around the sun, the actual pulses were delayed relative to the expectations. The spacing between the expected and actual pulses, shown by dotted lines, grew. (The growth is exaggerated in the figure.) Römer found that when the earth had reached the diametrically opposite point E_2 six months later, the delay was about 22 minutes. Let us assume Jupiter hardly moves during this period, though Römer could easily account for that motion. He attributed the 22-minute delay to the extra time light takes to cross the diameter of the earth's orbit. Using the best estimate for the diameter (close to 200 million miles) he obtained a velocity of roughly $200,000 \ km/s$, which is 2/3 that of the correct answer of $300,000 \ km/s$. (Had he used the correct delay of about 16.7 minutes, he would have come a lot closer.) His theory was

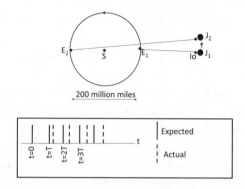

Figure 16.2   Römer's experiment. The first pulse of light from Jupiter's moon Io (when it assumes a specific position in its orbit) arrives at earth when it is at E_1 and Jupiter is at J_1. This pulse is shown by the very first solid line at $t = 0$ in the inset at the bottom. The next pulse should arrive after one time period T, but it is delayed and shown as the dotted line. The delay between the expected and actual signals, exaggerated for clarity, keeps increasing and reaches a maximum of about 22 minutes after six months, when the earth is at the diametrically opposite point E_2. (The motion of Jupiter to J_2 will be ignored in our discussion.) The delay was correctly attributed by Römer to the extra time taken to traverse the diameter of the earth's orbit.

initially greeted with disbelief but his stunned colleagues saw he was right when Io moved according to his predictions. Nonetheless, it took a while for his result to be generally accepted. Though Römer was off by some 30 percent in the value of c, his was a spectacular achievement given that before he came along people had no clue about the speed of light, not even whether it was finite. After him, people started doing laboratory experiments to measure the velocity of light, knowing its approximate value.

16.3   Some highlights of geometric optics

As mentioned earlier, my intent is not to discuss in depth all the results of geometric optics, but to introduce you to a principle from which all the results of geometric or ray optics follow. Though I will give only a few illustrative examples, you may rest assured that every result of geometric

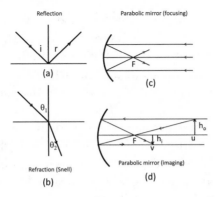

Figure 16.3   Geometric optics description of (a) reflection, (b) refraction (Snell's law), (c) focusing, and (d) image formation by a mirror.

optics involving mirrors and lenses may be deduced from this single principle.

Here are the results that I will derive to illustrate the point. Look at the four parts of Figure 16.3.

- (a) When light bounces off a plane mirror, $i = r$, where i and r are the angles of incidence and reflection measured from the normal. I will refer to this as the "$i = r$" law, though the actual angles may carry other names like α or β.
- (b) When light goes from a medium with *refractive index* n_1 and velocity c/n_1 to a medium with refractive index n_2 and velocity c/n_2, the angles of incidence and refraction obey Snell's law:

$$n_1 \sin\theta_1 = n_2 \sin\theta_2. \tag{16.1}$$

Thus when light goes from a rare medium (small n) to a dense medium (large n), it will bend closer to the normal to the interface. If you run the ray backward, from the dense medium to the rare medium, it will bend further away from the normal. It is understood that in addition to the refracted ray, there is in general a reflected ray obeying $i = r$.
- (c) If a parallel beam of light coming from infinity or from far (what does *far* mean in this context?) is incident on a parabolic mirror along its symmetry axis, the rays converge at the *focal point* F at a distance f measured from the middle of the mirror. Every ray parallel to the axis goes through the focal point. Your TV dish receives the parallel

beam from the satellite and gathers it all at the focal point, where it is picked up by the receiver. If you reverse the rays, you have a car headlight with the bulb at F emitting rays that hit the mirror at several angles and emerge as a parallel beam along the symmetry axis.

- (d) If the source of light, the *object*, were not the point at infinity, but an upright arrow of height h_0 a *finite* distance u away, the location v and height h_i of the image are found as follows, in the simplest case. From the tip of the arrow you draw a line parallel to the axis toward the mirror. It gets reflected and goes through the focal point. Then you draw a line from the tip through the focal point, and that hits the mirror and comes out parallel to the axis. The crossing point of these two reflected rays is where the image of the tip of the arrow lies.

  The relation of the various distances is

$$\frac{1}{u} + \frac{1}{v} = \frac{1}{f} \tag{16.2}$$

$$\frac{h_i}{h_o} = \frac{v}{u}. \tag{16.3}$$

What I call v and u, others may call i and o, for image and object.

- (e) A similar formula holds for lenses. Consider for example a *convex lens* or focusing lens, which is a piece of glass with the property that when you shine rays of light parallel to the axis from one side, they all meet at the focal point a distance f away on the other side (not shown in the figure because we will revisit it later). If you have an object on one side a distance u from the lens, the image will be on the other side at a distance v assuming $u > f$. The image will be upside down and the various distances will obey the same Eqn. 16.2.

  Of course, in more complicated cases some of these lengths, say f or v, could be negative, the image upright and virtual, etc.

Now for the single unifying principle from which these assorted facts can all be derived. It is called *Fermat's principle of least time* due to none other than Pierre Fermat (1601–1665), who made the famous conjecture (only recently proven by Andrew Wiles) about the non-existence of integer-valued solutions to $x^n + y^n = z^n$ for $n > 2$. His principle says:

Light will go from start to finish on a path that takes the least amount of time.

I hope you will share my delight in deriving so many diverse results from this single principle, in not having to carry all that miscellaneous baggage in your head. Let us start applying the principle.

16.4 The law of reflection from Fermat's principle

Let us say I am at B and you are at A in Figure 16.4. You send me a light signal. What path will it take? What is the path of least time? Everybody knows it is a straight line. No point going any other way. So that tells you that light travels in straight lines when there's no other obstacle because the straight line is the shortest path between A and B.

Next I want the light to hit the mirror and then come to me. It is like a race where the racers have to leave A, touch the wall (mirror), and reach B. Whoever gets to B first wins. Now there are different tactics open to the racers. Some may meander like crazy as they head for the wall. These are sure losers. We ignore them for it is obviously best to go on a straight line to the wall. Even then there remains a question: where to touch the wall? One person may say, "Look, I was told to touch the wall, so I'm going to get that out of the way first. I'm going to run straight to the wall, touch the point right in front of A, and then run straight to B." Fine, that's a possibility. Another person can say, "Let me touch the wall right in front of B, then run straight to B." There are infinitely many options open. We must find from all these possible paths, made of two straight segments, the one of least time.

So the only question the ray of light (or racer) has to ask is the following: "Where should I hit that mirror (or wall)?" Let's call a generic

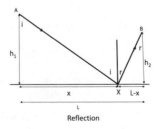

Reflection

Figure 16.4   The path of least time for the ray to leave A and reach B after reflecting off the mirror at X is found by minimizing $T(x)$, the sum of the lengths AX and XB as a function of x. The minimum is attained only at the point where $i = r$.

reflection point X and let x be its horizontal distance from A (measured parallel to the mirror). Let L be the horizontal distance between A and B and let h_1 and h_2 be their perpendicular distances from the mirror.

I will simply calculate and add the lengths of AX and XB, divide by c to get the time $T(x)$ as a function of x, and then minimize $T(x)$. By the Pythagoras theorem

$$T(x) = \frac{\sqrt{h_1^2 + x^2}}{c} + \frac{\sqrt{h_2^2 + (L - x)^2}}{c}. \tag{16.4}$$

We equate the x derivative to 0 to find the least time:

$$0 = \frac{dT}{dx} = \frac{x}{c\sqrt{h_1^2 + x^2}} - \frac{(L - x)}{c\sqrt{h_2^2 + (L - x)^2}}. \tag{16.5}$$

Thus the optimal x satisfies

$$\frac{x}{\sqrt{h_1^2 + x^2}} = \frac{(L - x)}{\sqrt{h_2^2 + (L - x)^2}} \tag{16.6}$$

$$\sin i = \sin r \tag{16.7}$$

$$i = r, \tag{16.8}$$

which is the desired result. This is the first victory for the principle of least time.

16.5 Snell's law from Fermat's principle

Next I'm going to reproduce Snell's law, which applies when light changes mediums. Look at Figure 16.5. The ray has to go from A, which is in the medium with index n_1 at a distance h_1 from the interface, to B, which is in medium 2 of index n_2 at distance h_2 from the interface. The separation between A and B (measured parallel to the interface) is L.

Here is the racer analogy. Imagine you are A, a lifeguard on the beach, and B is a person screaming for help in the ocean. How do you get there in the least amount of time? One point of view is to say, "Let me go in a

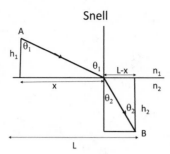

Figure 16.5   Using the principle of least time to find Snell's law for refraction. The problem is similar to reflection except for the two different velocities in the two segments. The figure assumes $n_2 > n_1$, i.e., that light travels slower in medium 2. The reversed ray going from B to A will bend *away* from the normal ($\theta_1 > \theta_2$). The condition $\theta_1 \le \frac{\pi}{2}$ places an upper limit on θ_2 for refraction. Beyond this limit there will be total internal reflection back to medium 2.

straight line all the way from A to B because I heard somewhere that the straight route is a winning strategy." But that may not be so good when you change mediums, because maybe you want to spend less time in the water, where you are slower. Another point of view is to say, "Let me go as far as I can on land, till I am in front of the victim, and then swim perpendicular to the shore to B." There are infinitely many options. To find the winning strategy we will again minimize the travel time, remembering that this is no longer synonymous with least distance due to the difference in velocities.

We must now divide the distance in each medium by the velocity in that medium (c/n) to find the time spent there, add the two times, and minimize the total. The subsequent steps are very similar to reflection. First,

$$T(x) = \frac{n_1 \sqrt{h_1^2 + x^2}}{c} + \frac{n_2 \sqrt{h_2^2 + (L-x)^2}}{c}. \qquad (16.9)$$

We equate the x derivative to 0 to find the least time:

$$0 = \frac{dT}{dx} = n_1 \frac{x}{c\sqrt{h_1^2 + x^2}} - n_2 \frac{(L-x)}{c\sqrt{h_2^2 + (L-x)^2}}. \qquad (16.10)$$

The x for least time satisfies

$$n_1 \frac{x}{\sqrt{h_1^2 + x^2}} = n_2 \frac{(L - x)}{\sqrt{h_2^2 + (L - x)^2}} \qquad (16.11)$$

$$n_1 \sin\theta_1 = n_2 \sin\theta_2, \qquad (16.12)$$

which is Snell's law.

Here is some practical advice based on the material discussed above. If you are a lifeguard, you should keep in readiness the ratio of your running and swimming speeds, i.e., n_1/n_2, so that you know where to hit the water when a victim calls.

Next, if you are at the bottom of a lake and you shine a flashlight to get help, remember the emergent light rays will bend away from the normal. This will be the case if you trace the ray *backward* from B to A in Figure 16.5. Some light will get reflected (with $i = r$) and the rest transmitted as per Snell's law. If, however, the angle θ_2 from your side exceeds a certain value, there will be no acceptable angle for the light to emerge because Snell's law will make the impossible requirement $\sin\theta_1 > 1$. In this case the beam will suffer *total internal reflection* and no light will make it out of the water.

16.6 Reflection off a curved surface by Fermat

We have seen that when a ray reflects off a planar surface, Fermat's principle leads to $i = r$. Imagine now a generalization in which it bounces off a *curved surface*. It is intuitively clear that it will still obey $i = r$, provided the angles are measured from the *local normal*. In other words, the tangent approximates the curved surface by a plane near the point of incidence, and the local normal lies perpendicular to it.

Reading the proof of this claim is optional but not remembering the result, which will be invoked here and there.

Consider the situation depicted in Figure 16.6. Whereas the plane mirror was represented by a straight line in Figure 16.4, the non-planar surface is now portrayed by a curve $\mathbf{r}(t)$, where t is a parameter that labels points on it. You can pretend $\mathbf{r}(t)$ is the path traced out by a particle as a function of time t. Consider now light that leaves the point $\mathbf{r}_1$, hits the mirror at $\mathbf{r}(t)$, and goes to $\mathbf{r}_2$. We want to vary t or $\mathbf{r}(t)$ and look for a path

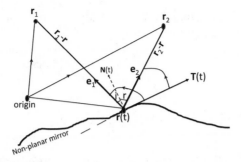

Figure 16.6   Ray going from $\mathbf{r}_1$ to $\mathbf{r}_2$ after bouncing off a non-planar mirror defined by the curve $\mathbf{r}(t)$. The unit vectors $\mathbf{e}_1$ and $\mathbf{e}_2$ are shown to have opposite projections along the tangent vector $\mathbf{T}(t) = \frac{d\mathbf{r}}{dt}$ *on the path of least time.*
(The path shown is not one.) The normal $\mathbf{N}(t)$ is shown by a dotted line, as is the continuation of the tangent vector $\mathbf{T}(t)$ in the opposite direction.

of least time. The *parameter t* has nothing to do with the time taken by light to travel from $\mathbf{r}_1$ to $\mathbf{r}_2$ after reflection.

Let us define the unit vectors

$$\mathbf{e}_1 = \frac{\mathbf{r}_1 - \mathbf{r}}{|\mathbf{r}_1 - \mathbf{r}|} \tag{16.13}$$

$$\mathbf{e}_2 = \frac{\mathbf{r}_2 - \mathbf{r}}{|\mathbf{r}_1 - \mathbf{r}|} \tag{16.14}$$

and $\mathbf{T}(t)$, the local tangent to the mirror at $\mathbf{r}(t)$, which is just the velocity vector of the fictitious particle

$$\mathbf{T}(t) = \frac{d\mathbf{r}}{dt}. \tag{16.15}$$

The local normal $\mathbf{N}(t)$ is shown by dotted lines.

We want to vary t (and through it the point of reflection $\mathbf{r}(t)$) and show that the path of least time obeys $i = r$ where the angles i and r lie between $\mathbf{e}_1$ and $\mathbf{N}(t)$ and $\mathbf{e}_2$ and $\mathbf{N}(t)$ respectively.

Instead of measuring the angles i and r from the normal, let us measure the angles

$$\theta_1 = \frac{\pi}{2} + i \qquad (16.16)$$

$$\theta_2 = \frac{\pi}{2} - r \qquad (16.17)$$

from the tangent in the counterclockwise sense, as indicated. What we need to show is then

$$\theta_2 + \theta_1 = \frac{\pi}{2} + i + \frac{\pi}{2} - r = \pi \qquad (16.18)$$

since $i = r$.

The total distance $D(t)$ traveled by the ray is a function of the parameter t, which determines the point of reflection:

$$D(t) = |\mathbf{r}_1 - \mathbf{r}(t)| + |\mathbf{r}_2 - \mathbf{r}(t)|. \qquad (16.19)$$

Remember that only $\mathbf{r}(t)$ depends on t: $\mathbf{r}_1$ and $\mathbf{r}_2$ are fixed. The travel time is $D(t)/c$.

Now let us rewrite $D(t)$ in terms of dot products and proceed as follows:

$$D(t) = \sqrt{(\mathbf{r}_1 - \mathbf{r}(t)) \cdot (\mathbf{r}_1 - \mathbf{r}(t))}$$
$$+ \sqrt{(\mathbf{r}_2 - \mathbf{r}(t)) \cdot (\mathbf{r}_2 - \mathbf{r}(t))} \qquad (16.20)$$

$$\frac{dD(t)}{dt} = \frac{2(\mathbf{r}_1 - \mathbf{r}(t)) \cdot \left(\frac{d(\mathbf{r}_1 - \mathbf{r}(t))}{dt}\right)}{2|\mathbf{r}_1 - \mathbf{r}(t)|}$$

$$+ \frac{2(\mathbf{r}_2 - \mathbf{r}(t)) \cdot \left(\frac{d(\mathbf{r}_2 - \mathbf{r}(t))}{dt}\right)}{2|\mathbf{r}_2 - \mathbf{r}(t)|} \qquad (16.21)$$

$$= -\frac{(\mathbf{r}_1 - \mathbf{r}(t))}{|\mathbf{r}_1 - \mathbf{r}(t)|} \cdot \frac{d\mathbf{r}(t)}{dt} - \frac{(\mathbf{r}_2 - \mathbf{r}(t))}{|\mathbf{r}_2 - \mathbf{r}(t)|} \cdot \frac{d\mathbf{r}(t)}{dt} \qquad (16.22)$$

$$= -(\mathbf{e}_1(t) \cdot \mathbf{T} + \mathbf{e}_2(t) \cdot \mathbf{T}) \qquad (16.23)$$

where $e_1(t)$ and $e_2(t)$ are the unit vectors from $r(t)$ to the starting and ending points r_1 and r_2. The least-time condition (the minimum of $D(t)/c$) is

$$\frac{dD(t)}{dt} = 0, \quad \text{which means} \tag{16.24}$$

$$(e_1(t) \cdot T + e_2(t) \cdot T) = 0. \tag{16.25}$$

Thus, $e_1(t)$ and $e_2(t)$ have equal and opposite projections along the tangent T,

$$\theta_1 = \pi - \theta_2, \tag{16.26}$$

which proves Eqn. 16.18.

16.7 Elliptical mirrors and Fermat's principle

Light is supposed to take the path of least time. This is exactly what happened in the case of reflection and refraction (Snell): there was a unique path of least time and light took it. But what if, in addition to one path of obviously least time (obeying $i = r$), there are many more paths that take the same (least) time between the same two end points? That's what we're going to talk about now.

Take an elliptical room, shown in Figure 16.7 with reflecting walls. You stand at one of the focal points F_1. Your task is to send a laser beam that hits the wall and goes to a person at the other focal point F_2. You know what you have to do. The tangent to the wall at X is like a horizontal mirror, as was proved in the last section. It is clear that if you send the ray there, it will end up at F_2 because it will satisfy $i = r$. (The light ray doesn't care if the mirror bends away from the point X. As far as the ray is concerned, it could be reflecting off the infinite tangent plane.)

While this is a correct answer, it is not the only correct answer. *It turns out that no matter in what direction you send your ray from F_1, no matter what the reflection point P is, it will arrive at F_2.* In other words, I am asserting that no matter where P is, the angle of incidence α will equal the angle of reflection β, both measured from the normal at P.

If you wanted only to use ray optics, one way to verify this claim is to establish analytically $\alpha = \beta$ at every point P on the ellipse. You could start with the equation for the ellipse in terms of its semimajor and semiminor axes a and b,

$$\frac{x^2}{a^2} + \frac{y^2}{b^2} = 1, \tag{16.27}$$

compute the normal to the ellipse at a generic point $P = (x, y)$, and verify that it bisects $\angle F_1 P F_2$.

I will now show that Fermat's principle allows us to finesse this tedious calculation.

My argument rests on the fact that *every path from F_1 to F_2 via any point P on the ellipse has the same length as $F_1 X F_2$, itself a path of least time*. Then according to Fermat's principle, all such paths obey the laws of geometric optics, in particular $\alpha = \beta$.

The fact that all the paths have the same length and hence take the same time follows from the definition of the ellipse as the *locus of points the sum of whose distances to the two focal points is constant*. Remember this is how an ellipse is drawn. If you drive two thumbtacks into the paper to anchor the ends of a string, stretch it taut with a pencil, and move the tip around, it will trace out an ellipse. In the notation of Figure 16.7 this means

$$r_1 + r_2 = \text{constant.} \tag{16.28}$$

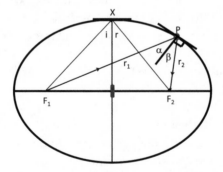

Figure 16.7   With elliptical walls, every path $F_1 P F_2$ is a path of least time. It is obvious that the path via X takes the least time and also obeys $i = r$. Not so obvious but true is that these two features hold for all the paths going via any point P on the ellipse. This is why all rays leaving F_1 focus at F_2. (The tiny vertical barrier at the center excludes the direct path between the focal points.)

Since the time taken by light to go from F_1 to F_2 is $(r_1 + r_2)/c$, every path takes the same time, no matter what P is. This common time is also the least time since one of the paths is the symmetric path that goes via X, known to be a path of least time from the plane mirror example.

Recall that the path of least time is characterized by the fact that if you change the point of reflection slightly there will be no change in the travel time (or distance) to first order. Now take any path F_1PF_2. If you modify it slightly by changing P a bit, the time taken *will not change at all* (not just to first order) because on either side of this path are paths of *exactly* the same time.

Now for another practical tip. Imagine that this elliptical wall that reflects light is replaced by a steel wall that reflects bullets. You have a gun with just one bullet left. You are at F_1 and your mortal enemy is at F_2, similarly armed. In what direction will you fire? One student said, "At your enemy," and I had to concede he was right. I was so in love with the complicated solution I had in mind that I had overlooked the obvious. So I said, "Imagine there is a small steel partition between you two. Now what will you do?" Now everyone agreed that they would aim at X. That will certainly work, but you know now that *you can fire in any direction and still hit the enemy.* The strategy works because bullets are like light. They obey $i = r$. The bullet bouncing off X obviously obeys $i = r$. I have just shown this is true for any point of reflection P. You will thank me if you ever have to use this rule. While your opponent, who took only Physics 101, is standing at F_2 wasting valuable time aiming for the midpoint X, you, who took Physics 201, will fire immediately from F_1 in any old direction and score a hit. (Obviously you will not aim exactly *away* from F_2 because on the rebound the bullet will first hit you and then the steel partition.)

The strategy also works for sound of small wavelength and high frequency (geometric acoustics): if you are at F_1, you may summon your dog at F_2 by blowing the whistle in any direction (again assuming a partition at the center that obstructs direct sound propagation).

The take-away message is this: *if there are many paths that take the same time between two given end points, rays leaving the first in many different directions will converge at the second upon reflection.*

Remember that not only do the rays leaving F_1 *converge* at F_2, they *take the same time* to do this via every P. (I want you to contrast this with a case where the rays converge at F_2 but after taking different periods of time.) The equal travel time means that if there were a candle at F_1 forming an image at F_2 and you suddenly extinguished it, the image would disappear abruptly, after the delay $(r_1 + r_2)/c$, and not gradually as it would if the different rays took different times. If you emitted a flash, all the radiated energy would converge at F_2 at the same time. But for this, the satellite dish would not focus the incident energy and your dog at F_2 would not hear you.

16.8 Parabolic mirrors

Having seen that focusing becomes possible when there is more than one path of least time, let us try to understand a focusing mirror shown in Figure 16.8 in these terms. Light comes in parallel rays from some source at infinity. You want to put a mirror of profile $y(x)$ in the way of the parallel beam so that every one of these parallel rays will hit the mirror and come to the focal point F after traveling the same distance.

(The actual mirror will be its surface of revolution of $y(x)$ and reside in three dimensions. For example, if this curve were a parabola, the actual mirror would be a parabolic dish, the type used for satellite TV.)

Let us now design the mirror.

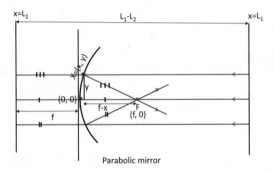

Parabolic mirror

Figure 16.8   The parabolic mirror has its focal point at F, a distance f to the right of the center of the mirror, with coordinates $(0,0)$. The race begins at line L_1 perpendicular to the symmetry axis. The line L_2 is also perpendicular to the axis and is at a distance f to the left of the center, $(0,0)$. Each ray hits the mirror at some (x, y) and goes to F. Since the distance from (x, y) to F is the same as the distance to the line L_2 (the defining property of the parabola), every ray travels the same distance $L_1 - L_2$. Three such equal distances are shown with one, two, and three vertical bars.

Since the rays begin at the point at infinity, they will all take infinite time to get to F no matter how they get there. There is no way to pick one or more of these paths as the ones taking the least time. So let us measure all distances traveled by the parallel rays from a fixed line L_1 normal to the axis, rather than from the point at infinity, to make a meaningful comparison of *finite* distances. Let us first take the ray that goes along the symmetry axis. It passes through F, goes a distance f, hits the mirror at $(0,0)$, and comes back a distance f to the point F. This is obviously a path of least time and obeys $i = r = 0$. Now draw a line L_2 perpendicular to the axis, but behind the mirror a distance f away. *The total distance traveled by this ray is the distance between the two parallel lines, L_1 and L_2.* This is true because once the ray hits the mirror, the distance it travels upon reflection back to F is the same as the distance to the line L_2 behind the mirror. Therefore the time to go from L_1 to F after reflection is the time to go from L_1 to L_2. Now consider a second parallel ray above the axis that hits the mirror at some $X = (x, y)$. We want it to travel to F taking the same time as the first ray. This will happen *if the distance from X to F is the same as the distance from X to the line L_2.* Make sure you understand this. If there were no mirror, all rays parallel to the axis would take the same time to go from L_1 to L_2. Instead, they go the same distance by first hitting the mirror and then, rather than going straight on to L_2, go to F, *which is equidistant.* Three such segments of equal lengths are shown by one, two, and three crossbars.

We now have a condition on the shape of the mirror: it is a curve $y(x)$ with the property that every point on it is equidistant from a point F and a line L_2. (The distance to the line is the perpendicular or shortest distance.) If you can find such a curve (or its surface of revolution in three dimensions) that's the curve or surface you want to take to your mirror maker.

We know such a curve: it is a *parabola*, which is the locus of points equidistant from a given point (F in our problem) and a given line (L_2 in our problem). Let us find the equation for the mirror surface using this property.

With the origin of coordinates $(0,0)$ at the center of the mirror, let us label a generic point on the mirror as (x, y). What we want is the expression for the function $y(x)$. We will derive it simply by imposing the definition of the parabola.

Equating the distances from the point (x, y) to the focal point F and to the line L_2 we find for the upper ray:

$$\sqrt{(f - x)^2 + y^2} = x + f. \tag{16.29}$$

Squaring both sides and simplifying,

$$x^2 + f^2 - 2fx + y^2 = x^2 + f^2 + 2fx \tag{16.30}$$

$$y^2 = 4fx, \tag{16.31}$$

which defines a parabola. (You may be more used to a parabola of the form $y = ax^2$, which rises above the x-axis quadratically, symmetric about the y-axis. What I have above is a rotated version with x and y interchanged.)

CHAPTER 17

Optics II: More Mirrors and Lenses

It is easy to summarize the last chapter in a sentence: Light obeys Fermat's principle of least time. From this principle we derived the laws of reflection, refraction (Snell's law), the unusual reflecting properties of the elliptical mirror, and the equation for the parabolic mirror. We understood how focusing occurs when there are multiple paths of least time connecting the object and image.

17.1 Spherical approximations to parabolic mirrors

If you want a mirror that will focus a beam parallel to the axis, no matter how wide, you need a parabolic mirror. A parabolic mirror is what you would order for the Hubble telescope. But if you cannot afford a parabola, there is a cheap alternative: a sphere. Now, a sphere is not quite a parabola, but I am sure you will appreciate that a slice from the sphere can mimic the parabola up to some distance, as shown in Figure 17.1. Beyond that of course it will deviate. But if you consider only beams very near the axis, then the two are equivalent, except for the cost.

We will often refer to spherical mirrors in this chapter.

Let us begin by asking ourselves the following question: if we take a hollow sphere of radius R, slice a part of it, and paint the convex side with silver, what will be the focal length of the concave mirror?

Look at Figure 17.1. We choose as the origin of coordinates the leftmost point of the circle, because this point will serve as the origin of

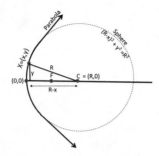

Figure 17.1   Part of the sphere of radius R (dotted line) approximates a parabola (solid line) if we do not go too far off the symmetry axis. The spherical mirror has $f = \frac{R}{2}$ in this approximation.

the approximate parabola. In these coordinates the equation for the circle of radius R centered at $(R,0)$ is

$$(R - x)^2 + y^2 = R^2, \tag{17.1}$$

which simplifies to

$$y^2 = 2xR - x^2. \tag{17.2}$$

First let us ignore the x^2 on the right and compare the equation to that of the parabola, $y^2 = 4xf$. We find the focal length of the spherical mirror is

$$f = \frac{R}{2}. \tag{17.3}$$

But we're not done yet, because we just threw away the x^2 term. We need a reason for that. It involves the following notion of big and small lengths. Whenever we deal with a mirror or a lens, lengths like u, v, f are all going to be treated as big numbers. Lengths like y that take you off the axis are going to be considered relatively small numbers. The reason is that if this is not so, an object of height y will not lead to a well-defined image. (Even a parabolic mirror, which exactly focuses a parallel beam of any width, will not produce a sharp image of an object at a finite distance if its height violates the smallness condition.) Lengths like x that are proportional to y^2 are even smaller. So the hierarchy is this: u, v, f are big, y is small, x is small squared. Now look at the two terms on the right in Eqn. 17.2. One is x times R, the other is x times x. So x times R beats x times x by a factor x/R. So we're going to drop the x^2 term and get the parabola as an approximation. In this approximation we have

$$y^2 = 2xR. \tag{17.4}$$

This equation is consistent with the notion that x is quadratic in the already small quantity y or, equivalently, that it is smaller by an additional factor of y/R:

$$x = y \cdot \frac{y}{2R}. \tag{17.5}$$

We know a sphere can mimic a parabola only for small deviations from the axis. This can be quantified now: if the rays come so far off the axis that x^2 is not negligible comparable to xR, the spherical mirror will neither look like a parabola nor focus like one.

Whereas the parabolic mirror has only one privileged point, namely F, the spherical mirror has a second one: the center of the sphere C, at a distance $R = 2f$.

17.2   Image formation: geometric optics

We have confirmed that a parabolic mirror and its spherical approximation can focus parallel light rays coming from infinity starting with just the principle of least time. Good! But mirrors are expected to do more than just focus parallel beams emanating from an object at infinity. They are also supposed to form sharp images of objects of finite height placed at a finite distance.

In this section I will analyze this problem using ray tracing (for objects that are not too tall) just to illustrate how it is done. I will derive the relation between the object location u and size h_0 (which we may take as input) and the image location v and height h_i, which we may take as output. In the next section I will re-derive the same final formulas using Fermat's principle.

Consider an arrow of height h_0 at a distance u from the mirror as shown in Figure 17.2. We want to know where the image of its tip will be formed. We find it by drawing two rays whose fate we know from Fermat. The first travels to the mirror horizontally and upon reflection

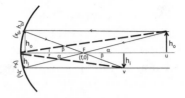

Figure 17.2   A spherical mirror. The solid lines are two rays used to find the relation between u, v, and f for an object placed a finite distance away. The rays leave (u, h_0), hit the mirror at (x_0, h_0) or $(x_i, -h_i)$ and meet at the image $(v, -h_i)$. The dotted line is a third ray that hits the mirror at $(0, 0)$ and meets the other two at the image point.

goes through the focal point. The second one goes to the mirror through the focal point and emerges horizontal. We know this because if we run this ray backward, we see the horizontal ray hit the mirror and go through the focal point. But if while going backward it is a path of least time connecting the end points, it's also a path of least time going forward.

The two reflected rays meet at a point that defines the tip of the inverted image of the arrow. Let this occur at a distance v at a height h_i. (By convention if h_i is positive the image is at $(v, -h_i)$.) We want to determine h_i and v in terms of h_0 and u that we may choose at will.

Equating the tangents of the two opposite angles α we find

$$\frac{h_0}{u - f} = \frac{h_i}{f - x_i} \simeq \frac{h_i}{f}, \tag{17.6}$$

where x_i has been ignored compared to f. Similarly from the two equal β's we have

$$\frac{h_i}{v - f} = \frac{h_0}{f - x_0} \simeq \frac{h_0}{f}, \tag{17.7}$$

where x_0 has been ignored compared to f. Equating the products of the left-hand sides to the product of the right-hand sides and canceling $h_i h_0$, we find

$$(u - f)(v - f) = f^2, \tag{17.8}$$

which can be rearranged to the familiar form

$$\frac{1}{u} + \frac{1}{v} = \frac{1}{f}.$$

(17.9)

The ratio of the image size to the object size h_i/h_0 follows from Eqn. 17.7:

$$\frac{h_i}{h_0} = \frac{v-f}{f} = \frac{v}{f} - 1 = v\left[\frac{1}{u} + \frac{1}{v}\right] - 1 = \frac{v}{u}.$$

(17.10)

The ratio of the object size to the image size is just the ratio of the object distance to the image distance.

Equations 17.9 and 17.10 determine h_i and v in terms of the parameters h_0 and u that we may choose at will. Of course, we may use these equations to find any two parameters given the other two. For example, in some problem we may want the image to be at a given v, in which case the same formula can be used to find the requisite u.

The *magnification M* is defined to be

$$M = -\frac{v}{u}.$$

(17.11)

In this case of positive u and v, the minus sign signifies that the image will be inverted. In some other mirrors, you will find v is negative because the image is virtual. Then M will be positive, meaning the image is upright. When you look into the bathroom mirror, your image has the same orientation as your face, not upside down.

Although we only considered the tip of the object and its image, these arguments hold for any other point on the object because v, the image location, is independent of h_0, the height of the tip. In other words, the image of the upright arrow will be an inverted arrow.

17.2.1 A midlife crisis

Some years after learning geometric optics, long after all the exams were over, I began to ask myself the following. We draw just two rays and claim their intersection point is the location of the image (of the tip of the arrow). But any two rays will always meet somewhere unless they are parallel. (This will happen even if you distort the mirror from its spherical or parabolic shape.) Why should that point be the image point? What if I draw another ray? How do I know it too will come to the same spot? In other words, two rays crossing somewhere is inevitable, but three or more crossing at one (image) point would be more compelling evidence of image formation by focusing of rays.

So I considered one more ray (dotted line in Figure 17.2) reflecting off the center, where the tangent is vertical. We know it must obey $i = r$, which imposes an additional constraint on h_i, h_0, u, and v:

$$\frac{h_0}{u} = \frac{h_i}{v}. \tag{17.12}$$

If this constraint is not satisfied it will mean that the third ray cannot obey the law of reflection and also pass through the intersection point of the first two rays. Luckily this condition is satisfied; see Eqn. 17.10.

17.3 Image formation by Fermat's principle

So we have three reflected rays meeting at one point. That is reassuring but not enough. Maybe it worked out because the third ray hit the mirror at a special (symmetric) point. If I draw yet another one, hitting the mirror at a more generic point, how will I know it too will also come to the same image point?

Using ray optics we can show once and for all that all reflected rays converge at the image point, regardless of the height y at which they hit the mirror (provided it is small compared to f or, in the spherical case, R).

But I want to derive the same result using the principle of least time. In this approach the rays leave the object in many directions, hit the mirror at various heights y, and meet at the image, after having traveled the same distance or having taken the same time, which should also be the least time. I will show this in two stages.

1. I will show that the ray that hits the mirror at $y = 0$, namely at the origin $(0,0)$, is a path of least time (and obeys $i = r$).
2. I will show that the same time is taken by other rays that hit the mirror at neighboring values of y. That is to say, the travel time is y-independent for small y.

Before I embark on this I must warn you that the y-independence of travel time is not exact. It will be valid only when powers of y higher than y^2 are neglected. This is not because the sphere is an approximation to a parabola; it is true even for the parabola. The latter may perfectly focus a parallel beam of arbitrary width, but it will not form perfect images of objects at a finite distance unless y is small.

Here are two results we will need. The first is the equation of the mirror surface:

$$y^2 = 4xf. \tag{17.13}$$

The next is the binomial approximation to $(A^2 + a)^{1/2}$ for $\frac{a}{A^2} \ll 1$:

$$(A^2 + a)^{1/2} = A\left(1 + \frac{a}{A^2}\right)^{\frac{1}{2}} = A + \frac{a}{2A} + \dots \tag{17.14}$$

Now look at Figure 17.3, which shows an object of height h_0 at u and an inverted image of height h_i at v.

We are free to choose h_0 and u at will, and we need a way to determine h_i and v, the image height and location.

Our strategy will be to compute $D(y)$, the path length as a function of y, and see if it can be made y-independent by judicious choice of h_i and v.

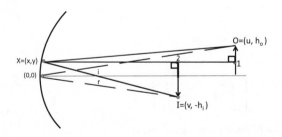

Figure 17.3   The path length for the ray that goes from object O to image I via the point $X = (x, y)$ on the mirror is the sum of the hypotenuses of the two right triangles $OX1$ and $IX2$. It is independent of y to order y^2. All paths therefore take the same time, which is also the least time.

The distance traveled by the ray that hits the mirror at height y is the sum of the hypotenuses of the two right triangles $OX1$ and $IX2$:

$$D(y) = \left[(u-x)^2 + (h_0 - y)^2\right]^{1/2}$$

$$+ \left[(v-x)^2 + (h_i + y)^2\right]^{1/2}. \qquad (17.15)$$

We are going to keep terms of quadratic order in y and h and linear order in x since $x = y^2/4f$ on the mirror surface. In this approximation

$$D(y) = \left[u^2 - 2ux + (h_0 - y)^2\right]^{1/2}$$

$$+ \left[(v^2 - 2vx + (h_i + y)^2\right]^{1/2} \qquad (17.16)$$

$$\simeq u + \frac{-2ux + (h_0 - y)^2}{2u} + v$$

$$+ \frac{-2vx + (h_i + y)^2}{2v} \quad \text{using Eqn.17.14} \qquad (17.17)$$

$$= u - \frac{y^2}{4f} + \frac{(h_0 - y)^2}{2u} + v$$

$$- \frac{y^2}{4f} + \frac{(h_i + y)^2}{2v} \quad \text{upon using } x = y^2/4f. \qquad (17.18)$$

Let us write this as

$$D(y) = u + v + \frac{h_0^2}{2u} + \frac{h_i^2}{2v}$$

$$+ y\left[-\frac{h_0}{u} + \frac{h_i}{v}\right]$$

$$+ y^2\left[\frac{1}{2u} + \frac{1}{2v} - \frac{1}{2f}\right]$$

$$+ \text{ higher powers of } x \text{ and } y \qquad (17.19)$$

$$\equiv D(0) + y\left.\frac{dD(y)}{dy}\right|_{y=0} + \frac{y^2}{2}\left.\frac{d^2 D(y)}{dy^2}\right|_{y=0} + \dots \qquad (17.20)$$

We want the result not to depend on y.

The first line is $D(0)$, the path length for reflection off the vertical tangent at $y = 0$. It serves as the reference. The coefficient of the linear term can be made to vanish if

$$\frac{dD(y)}{dy}\bigg|_{y=0} = 0. \tag{17.21}$$

That is,

$$\frac{h_0}{u} = \frac{h_i}{v}. \tag{17.22}$$

If this condition of vanishing first derivative of $D(y)$ at $y = 0$ is satisfied, it follows that the path through $y = 0$ is a path of least time. Not surprisingly, from the figure we can see that this just says that $i = r$ for reflection off the vertical tangent. This only determines the fate of a single ray, one that hits $y = 0$.

This condition alone does not fix the image location because we can slide the image along the reflected ray, keeping h_i/v or the angle subtended fixed. This is to be expected because it is not enough that the paths infinitesimally close to $y = 0$ all have the same path length. This will be true even if the mirror were flat. What we are looking for is a *mirror that is curved in just the right way that the path length is constant over a wider region,* to achieve the *focusing* of rays reflected off a continuum of points near $y = 0$.

To this end we turn to the quadratic terms. Demanding that they vanish requires the vanishing of the second derivative:

$$\frac{d^2 D(y)}{dy^2}\bigg|_{y=0} = 0. \tag{17.23}$$

This means

$$\frac{1}{u} + \frac{1}{v} = \frac{1}{f}, \tag{17.24}$$

which is the second of the equations obtained by ray optics.

Any powers of y that survive in a more accurate calculation must be of order y^3 or higher. We cannot do anything about them since we have used up the two degrees of freedom at our disposal, namely h_i and v to kill the first two powers. Unless these higher order terms are anomalously large, the graph of $D(y)$ versus y will be extremely flat near $y = 0$. The (approximate) constancy of path length means the constancy of travel time. Since $y = 0$ corresponds to path of least time (reflecting off a vertical tangent at the origin) all are paths of least time.

If h or y gets large enough for the neglected terms to become significant, the image will be blurred.

To summarize, we have imposed the condition of least time for *paths with a continuous range of y* to ensure that rays hitting the mirror not only at $y = 0$ but nearby all meet to produce a focused image. This in turn gave us two equations (the vanishing of the first and second derivatives of $D(y)$ at $y = 0$) that determined h_i and v as a function of h_0 and u. These equations agreed with those of ray optics.

17.4 Tricky cases

There are countless applications of the equation

$$\frac{1}{u} + \frac{1}{v} = \frac{1}{f} \tag{17.25}$$

for mirrors and lenses. As mentioned earlier, some of these variables can be negative in complicated cases. For example, in the case of a lens where the (virtual) image is on the same side as the object, v is negative. A convex mirror, which spreads out a parallel beam instead of focusing it, has a negative f. Only u and h_0 can always be chosen to be positive by convention. While the equation above will work in all situations if you pay attention to the signs, it is interesting to look at one or two cases where the standard recipe does not work, either in ray optics or least time. Both examples involve virtual images.

The first involves a convex mirror with negative f, that is to say, the image of a point at infinity is a virtual focal point.

The second involves a virtual image of an object in front of a concave mirror with $u < f$.

We will see how the standard recipe fails in both cases and how it is modified to give results that agree with Eqn. 17.25.

17.4.1 Fermat's principle for virtual focal points

Consider a convex mirror shown in Figure 17.4. When a parallel beam (solid lines) is incident on it from the right, the reflected rays diverge rather than converge.

It is known in ray optics that the reflected rays will seem to be coming from a virtual focal point behind the mirror at a distance $|f|$, that is, they will meet at F *if continued backward, into the mirror.*

We want to prove this result using Fermat's principle. It looks like we cannot even get started because the rays that come to the mirror from infinity diverge upon reflection and never really meet again. How are we to compare their travel times and pick the one of least time if there is no common end point? This seems to be a result in geometric optics we cannot derive from the principle of least time. Actually we can, but in two stages.

First assume the mirror is reflective on both sides. Look at the dotted lines on the concave side of the mirror. We see rays parallel to the axis come from the left. By Fermat's principle, applicable on the concave side, these rays will meet at F and also obey $\alpha = \beta$. Now continue the dotted lines through the mirror to the convex side as solid lines and reverse the arrows

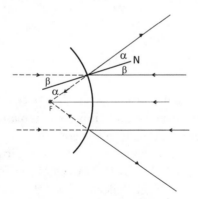

Figure 17.4 On the concave side we see a parallel beam (dotted rays) incident on a concave mirror. We know from Fermat (applicable here) that these will focus at F and obey $\alpha = \beta$. If we continue the dotted lines to the convex side as solid lines and reverse the arrows, we see rays come in parallel to the axis from the right and bounce off the mirror, obeying $\alpha = \beta$, *as a physical ray would*. It follows that all outgoing physical rays, if continued back into the mirror, would pass through F.

on them. At this stage, these are just some lines and may not correspond to physical rays. But if we extend the normal through to the convex side, we see rays coming from the right parallel to the axis and bouncing off the mirror obeying $\alpha = \beta$. *This is exactly what a real physical ray bouncing off a curved surface would do.* It then follows that if the outgoing physical ray on the convex side is continued to the concave side, it will pass through F. Thus F will be the virtual focal point for the convex mirror.

17.4.2   Ray optics for virtual images

Now we consider a problem for which the standard recipe from ray optics fails. Consider a concave mirror and an object that is at a distance $u > 2f$. We are supposed to find the image by drawing two rays, one that goes in parallel and comes out of F and the other that goes in via F and emerges parallel. Their intersection will be at a distance $v < 2f$, as shown in Figure 17.2. As you reduce u, you increase v till you hit $u = 2f$. Then $v = 2f$ as well, and the object and its inverted image are equidistant from the mirror. Suppose I move the object closer. The image will move out with $v > 2f$. When the object is at F, then $1/v = 0$, which means $v = \infty$. This is just the reverse of an incoming parallel beam converging at F. Now let us push our luck and let $u < f$. Where is the image? From

$$\frac{1}{v} = \frac{1}{f} - \frac{1}{u} \qquad\qquad (17.26)$$

we find that $v < 0$. But a negative v means the image is behind the mirror! How can there be an image behind an impenetrable mirror? The answer of course is that there isn't an image there but *it will look that way.* To understand this we can try some ray tracing, as indicated in Figure 17.5. First we draw a ray from the tip (T) of the object that hits the mirror parallel to the axis and then gets reflected through F. This ray, call it ray-1, is supposed to intersect the ray that first crossed F, hit the mirror, and emerged parallel. But if you draw a ray from the tip T to F, it points *away from the mirror!* It is never going to reflect off the mirror at all. So what we are told to do is this: draw the second ray, ray-2, *from F to T.* If you continue it till it hits the mirror, it will emerge parallel upon reflection. (This is true because if you reverse the entire ray, you have an incoming parallel ray hitting the mirror and ending up at F.) So we have successfully drawn two rays obeying the laws of reflection, but they are moving away

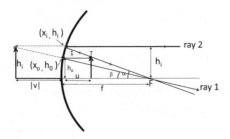

Figure 17.5 Virtual image formation for concave mirror when the naive construction fails. First, the ray from the tip T and passing through F does not hit the mirror. So it is continued the other way till it does and emerges parallel (ray 2). But it does not intersect ray 1 which hits the mirror parallel to the axis and gets reflected through F. However, if continued to the dark side, rays 1 and 2 appear to come from the virtual image on the left. The standard equation holds with $v = -|v|$ negative.

from each other and will not intersect in the right side of the mirror. On the other hand, if continued backward through the mirror (dotted lines in the figure), the rays will meet. The meeting point is the virtual image of the tip because the rays of light will seem to be coming from there.

A question still remains: will the virtual image location $|v|$ to the left of the mirror obey Eqn. 17.25? It will, despite the fact that the image is virtual. Equate the two expressions for $\tan\beta$, one from the small right triangle with vertical side h_0 and the other from the larger one with vertical side h_i:

$$\tan\beta = \frac{h_i}{f+|v|} = \frac{h_0}{f-x_0} \simeq \frac{h_0}{f}. \tag{17.27}$$

Now do the same for angle α:

$$\tan\alpha = \frac{h_i}{f-x_i} \simeq \frac{h_i}{f} = \frac{h_0}{f-u}. \tag{17.28}$$

We have seen this pair before and know that they lead to Eqn. 17.25 and the magnification formula. For example, equating the quotients of the two sides we find

$$\frac{f}{f+|v|} = \frac{f-u}{f} \quad\text{or,} \tag{17.29}$$

$$(f - u)(f - v) = f^2 \qquad\qquad\qquad (17.30)$$

because $|v| = -v$. Equation 17.30 is the same as Eqn. 17.8 for the real image.

The least time approach can also reproduce these results, but only after we invoke some clever tricks.

17.5 Lenses à la Fermat

Look at the lens in Figure 17.6. I have an object O, which is infinitesimally tall at the distance u to the left of the center of the lens. I want the lens to form its image I on the other side, at a distance v, through the convergence of many rays, all taking the least time. The shortest route goes straight from object to image a distance $u + v$, while others that cross the lens at higher altitudes are longer. They still have a chance to be competitive because least distance no longer means least time: light travels in the lens at a reduced velocity of c/n. This means that as far as travel time goes, 1 cm of lens is equal to n cm of air.

The figure shows one ray going in a straight line through the thickest part of the lens. Another skips the lens entirely and grazes over the topmost point P. We will simply equate the times for two extremes. That the times are equal for intermediate heights (in the usual approximation of dropping higher powers of small quantities) is harder to show and will not be attempted here.

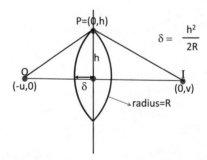

Figure 17.6   The focusing lens whose faces are part of spheres of radius R. The time taken to go from the object (O) to the image (I) is the same along the two indicated paths. The shorter path straight through the lens takes the same time as the one that grazes the top (P) because light travels at speed c/n in the lens.

Assuming that each lens surface is part of a sphere of radius R, the old formula

$$x = \frac{y^2}{2R} \tag{17.31}$$

from our study of spherical mirrors tells us that the distance δ in the figure is

$$\delta = \frac{h^2}{2R} \tag{17.32}$$

where h is the height of the lens.

The ray that goes straight travels a distance $u - \delta$ to the lens, then a distance 2δ inside the lens, and finally a distance $v - \delta$ on the other side to the image point. If we divide the distances in air by c and the distances in the glass by c/n we find the total time is

$$T(\text{straight}) = \frac{(u - \delta)}{c} + \frac{(v - \delta)}{c} + \frac{2\delta}{c/n}. \tag{17.33}$$

Multiplying both sides by c we find the part in the lens counts for a distance $2n\delta$ as advertised:

$$c \cdot T(\text{straight}) = u + v + 2\delta(n - 1)$$

$$= u + v + \frac{2(n - 1)h^2}{2R} \quad \text{using Eqn. 17.32 for } \delta. \tag{17.34}$$

For the path that grazes the topmost point at P,

$$c \cdot T(\text{via } P) = \sqrt{u^2 + h^2} + \sqrt{v^2 + h^2} \tag{17.35}$$

$$= u + \frac{h^2}{2u} + v + \frac{h^2}{2v} \quad \text{using Eqn. 17.14.} \tag{17.36}$$

For the paths to take the same time, we need

$$\frac{2(n - 1)h^2}{2R} = \frac{h^2}{2u} + \frac{h^2}{2v} \quad \text{which may be rewritten as} \tag{17.37}$$

$$\frac{1}{f} = \frac{1}{u} + \frac{1}{v} \quad \text{where I } define \frac{1}{f} \text{ by} \tag{17.38}$$

$$\frac{1}{f} = \frac{2(n-1)}{R}.$$
                                                                                    (17.39)

We are free to call the combination $2(n-1)/R$ what we want, but in calling it $\frac{1}{f}$ we are implying it is the inverse focal length. And it is. If we set $u = \infty$, that is, place the object at infinity, the image location v must equal f. This is true in Eqn. 17.38.

Notice that there are two ways to find the focal length. One is to do an experiment with parallel rays and see where they meet. But what we have here is a *calculation* of the focal length in terms of R and n. The resulting formula will tell you what to do if your lens has a focal length that is too big or too small. You can vary it by varying either n or R. You can take different materials or you can take different radii of curvature.

17.6 Principle of least action

Fermat's principle, which describes light, can be generalized to describe particles. Consider a Newtonian particle that goes from (x_1, t_1) to (x_2, t_2). Generically, there is only one Newtonian trajectory passing through both. (Two points on a trajectory replace the initial position and velocity at one point in fixing the trajectory.) It follows a trajectory $x_N(t)$ shown in Figure 17.7, determined by solving $F = ma$. How is that path $x_N(t)$ different from all possible paths $x(t)$ one could draw between those two end points? It is true that at each point it obeys $md^2x/dt^2 = -dV/dx$. But that is a very local statement. Is there anything *global we can say about the trajectory as a whole?*

The answer is that we can: it is the path of least *action,* where the action S is defined as

$$S = \int_{t_1}^{t_2} \left[\frac{1}{2} m \left(\frac{dx(t)}{dt} \right)^2 - V(x(t)) \right] dt.$$
                                                                                    (17.40)

If I give you a path $x(t)$, you can find its action $S[x(t)]$ as a function of the path by integrating the difference of the kinetic and potential energies on this path. Thus S is a function of the entire path $x(t)$ under consideration. A function of a function is called a *functional* and this one is written $S[x(t)]$.

The claim is that the Newtonian path is the path of least action.

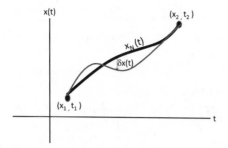

Figure 17.7   The thick line is the actual path taken by the particle on the path of least action. The thinner line shows a neighboring path that deviates by $\delta x(t)$ at time t. The total change in action δS due to this change is required to vanish to first order in $\delta x(t)$. This yields the Euler-Lagrange equation, which reduces to $F = ma$.

In the case of reflection, in looking for the path of least time, we considered very simple paths. They were made of two straight segments and specified by just one variable: the value of x where the ray hit the mirror. If x_{min} led to the path of least time, it had to be the solution to

$$\left.\frac{dT}{dx}\right|_{x_{min}} = 0. \tag{17.41}$$

Equivalently we could say that since

$$dT = \frac{dT}{dx} \cdot dx + \mathcal{O}(dx^2), \tag{17.42}$$

dT, the change in T to first order in dx, vanishes at x_{min}. So it should really be called the path of *stationary* time and not least time. (Whether or not it is a minimum is determined by the second derivative, which we did not examine.) But the name has stuck in optics and mechanics.

The problem of minimizing S in mechanics is more difficult than minimizing $T(x)$ for reflection. Whereas $T(x)$ is a function of the only variable at hand, the location x of the point of reflection, the action S depends on an *entire function* $x(t)$. The path is not assumed to be made of straight lines. Given the two end points (x_1, t_1) and (x_2, t_2) we need to consider every conceivable path joining them. We have to select the path of least action from among these. We may still use the fact that if the path of least action is altered infinitesimally by an amount $\delta x(t)$ at time t, as shown in Figure 17.7, the action should not change to first order in $\delta x(t)$. This is a problem in the *calculus of variations*. The analog of $dT/dx = 0$ are the *Euler-Lagrange equations. These turn out to be just Newton's laws.* In other words, to follow the path of least action, the particle need not compute S in advance for each possible path and then pick the winner; it has only to obey Newton's law at each instant.

Why do we bother with the principle of least action if the equations we get in the end are equivalent to Newtonian mechanics? There are many advantages that were appreciated even in the years following this development. You can find them in any book on advanced mechanics or online. More recently, Richard Feynman (1918–1988) discovered a simple way to state the laws of quantum mechanics in terms of the action of classical paths. There is no such route starting with $F = ma$. All of modern quantum field theory is stated in this language of action. For example, if you want to formulate the theory of how quarks and gluons interact, you do not seek the analog of $F = ma$, you seek the correct action written in terms of quark and gluon variables. (The action is not the time-integral of the difference of kinetic and potential energies except in the simplest problems of mechanics. For a particle in a magnetic field, the action is cooked up so that the Euler-Lagrange equations reduce to $\mathbf{F} = q(\mathbf{E} + \mathbf{v} \times \mathbf{B})$. Incidentally, the action can be written only in terms of $\mathbf{A}$ and V and not $\mathbf{E}$ and $\mathbf{B}$.)

17.7 The eye

The human eye and visual system are very very impressive. You place an object in front of it and the lens in the eye forms an inverted image on the retina. The inversion of the image is a well-defined, objective, and verifiable claim. If I look into your retina as you look at an upright candle, the image of the candle will be inverted. But this inversion doesn't seem to bother us. After all, at what stage do we really see something? It's not very clear. Do we see it in the retina or do we see it in the brain? There is simply a 1:1 correlation between what's registered in my retina and what I run into in my real life. I'm walking around, and I bump into the upside-down table. The fact that it's upside-down in the retina is not relevant. I know how to place an upside-down cup of coffee on the upside-down table. The brain has learned how to translate the image on the retina into what we will encounter. In fact, the brain's ability to manipulate images has been demonstrated in a bizarre experiment, where participants wore glasses that inverted retinal images one more time. After a few days, those guys were just fine. So the visual system has a lot of software behind it. I learned that the hard way when I had an eye operation and the doctor pulled off the bandage with a flourish. I was in a panic: I could not focus and was seeing double. My doctor didn't seem particularly worried. He said, "You'll be fine in a few days." Now I've heard that line before and was not comforted. But slowly, my vision got better and better, not due to any more surgery. Slowly my brain began to reprogram itself with respect to the new parameters, to once again correlate the images in the eye with what was actually out there.

So much for the impressive software. But let us look at the hardware responsible for the imaging on the retina. We can see a potential problem there. The retina is at a fixed distance from the lens, equal to the diameter of the eyeball. That means v is fixed. Even as we vary u, the location of the object, we want a sharp image on the retina at fixed v. But we know from the lens equation

$$\frac{1}{v} = \frac{1}{f} - \frac{1}{u} \tag{17.43}$$

that v is determined by u and should vary with it. How does the eye manage? The answer is that something we have always held constant so far, the focal length f, changes. That is the amazing thing about the human lens. It's made out of some jelly-like stuff, and there are some muscles pulling it. If the muscles pull it, it will become longer and thinner, and it will have one focal length. If the muscles relax, it will have another focal length. When I look at an object far away, the muscles are in the relaxed state. As the object comes closer, it takes a certain effort to focus on it.

Let us move on to the image so produced on the retina. I will use another principle from ray optics best suited for this discussion: the ray through the center of the lens goes straight through. (At the center the opposite faces of the lens are parallel, if we do not go too far off-axis, and light refracts coming in and going out as if through a glass slab with parallel faces. In this case we know it emerges in the same direction and with a lateral displacement that is negligible in this context.) From the upper half of Figure 17.8 we see that the apparent size of anything, the size of its image on the retina, is decided by the angle subtended by the object,

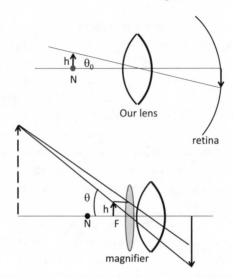

Figure 17.8   Top: Without the magnifying lens the biggest subtended angle θ_0 is obtained by placing the object at the near point N. Bottom: With the magnifying lens, the subtended angle is larger because the object is now closer than N, but the virtual image (with the same opening angle θ) is far away and easier on the eye.

$\theta \simeq \tan\theta = h/u$. If the object is small, we just bring it closer till h/u is big enough. If we could do this indefinitely, we would not need a microscope. If we want to see little bacteria, we just pull the little guys really close to our eye. But this does not work beyond some point. The eye cannot focus anything that is closer than the *near point*, which is roughly $N = 25$ *cm* away. The muscles pulling on the lens cannot deform it any more. For an object of fixed height h, the best you can do is therefore

$$\theta_0 = \frac{h}{N}. \tag{17.44}$$

Suppose you bring it closer anyway. There is good news and bad news. The good news is that the image on the retina is bigger. The bad news is that the big image is blurry. Your lens cannot focus the image on the retina because it is too close. The solution is to view it through a magnifying lens. As shown in the figure, it produces a virtual image *with the same subtended angle, but at a convenient distance.* Usually the preferred

v is large. In the $v = \infty$ limit we have

$$\frac{1}{u} = \frac{1}{f}, \tag{17.45}$$

i.e., you place the object at the focal point of the lens. The angular size of the object (now at $u = f$) is

$$\theta = \frac{h}{f}. \tag{17.46}$$

The magnification is defined as the ratio of subtended angles

$$M = \frac{\theta}{\theta_0} = \frac{h/f}{h/N} = \frac{N}{f}. \tag{17.47}$$

Thus a lens with $f = 2.5$ cm will cause a magnification of $25/2.5 = 10$.

It is possible to improve on the magnification beyond N/f a bit if we are willing to suffer a little. We begin with the formula for the general placement of the object (not necessarily at F):

$$\theta = \frac{h}{u} = h\left[\frac{1}{f} - \frac{1}{v}\right] = h\left[\frac{1}{f} + \frac{1}{|v|}\right] \tag{17.48}$$

$$\theta_0 = \frac{h}{N} \tag{17.49}$$

$$M = \frac{\theta}{\theta_0} = N\left[\frac{1}{f} + \frac{1}{|v|}\right]. \tag{17.50}$$

Clearly M increases as $|v|$ decreases, but we cannot make it too small: we require $|v| \geq N$ for us to see the virtual image clearly. Thus the best magnification possible, with the virtual image at the near point, is

$$M_{\text{best}} = N\left[\frac{1}{f} + \frac{1}{N}\right] = \frac{N}{f} + 1. \tag{17.51}$$

The price you pay for the extra 1 in magnification is that you have to look at an image at the limit of your capabilities, which can be very tiring after a while. You might gladly give up this gain of 1 in magnification for the comfort of an image at infinity if you were a jeweler or watchmaker.

Can you really see something at infinity? You can, if it is infinitely large. The idea is that if you make the object bigger and bigger as it recedes, keeping the angle subtended constant, you will see it no matter how far away it is. In practice, infinity really means far enough that the rays that come from it are very nearly parallel.

CHAPTER 18

Wave Theory of Light

We now go forward in time from geometric optics. As always, it was overthrown by the one authority we all must bow to: experiment.

If your theory doesn't agree with the experiment it's over, even if your first name is Isaac or Albert. And conversely, if you're an unknown newcomer who makes predictions that agree with experiment you become a rock star. Everything is based on experiment. That's the only way we change our minds. Now you might say, "Why do you keep doing this to us? We believe everything you tell us. We write everything down. We do the problem sets and then you say, 'Oh, the theory I taught you the other day is inadequate. Here's a better theory.' What's going on? Are physicists really wrong so often?" I have to be very careful when I say we are wrong because the news leaks to the press and the media who will say, "Physicists think they're always wrong." In fact, I've gone on record saying, "We're always wrong." What I mean is that no matter how many laws we find, one day we will find some new experiments that are not explained by these laws. That's not really bad news. That's what keeps us in business. We *want* to find something that doesn't fit anything we know. For example, Newtonian mechanics is wrong in the sense that it doesn't work when velocities approach the speed of light, but it's not wrong in the sense that the predictions it made in its proper domain do not work anymore. It was supposed to work in a limited range of experimental observations. If you cross the limit, if you build accelerators that send particles at very high speeds, you may find the particles don't obey Newtonian mechanics. Then you need Einstein's special relativity.

Now if you have a theory like Einstein's, a new theory that overthrows the old theory and explains new phenomena, there is still one extra requirement. Can you guess what that might be? It is that the old experiments, which were explained by the old theory, must be explained by the new theory as well. In fact the new theory, when it's a good one, will also explain why people fell for the old theory for so many years. Relativity does just that. It works for all velocities up to the speed of light, but if you let $\frac{v}{c} \to 0$ you will get back Newtonian mechanics. Similarly, quantum mechanics is essential for very, very tiny objects at the atomic scale, but if you apply it to big things you'll find that the world begins to look Newtonian. The quantum equations of motion reduce to Newton's laws for macroscopic objects.

We now turn to the experiments that signaled the limits of geometrical optics. I will tell you in due course why we didn't realize there was something wrong with it for so long.

Consider the experiment I already mentioned at the beginning of geometric optics. In it there is an opaque partition with light coming from one side, going through a hole and illuminating a screen on the other side. The illuminated region is of the same shape and size as the hole. As you shrink the hole, the light begins to spread out to bigger and bigger areas. It no longer forms the geometric image of the aperture. In addition, the intensity of light oscillates and dies off as we move off the center. These features are not going to come from geometric optics. Something has to take its place. A major clue came from the experiment performed in 1801 by Thomas Young (1773–1821). That dramatic experiment really demolished the ray theory of light. It involves a phenomenon called *interference.*

Young's experiment is depicted in Figure 18.1. We are looking down at a rectangular experimental region. Light is emitted by a source E on the left wall. At the right wall is a screen to receive the light. (The screen and the rest of the apparatus have a dimension perpendicular to the page.) Between the two walls is an opaque barrier with two slits S_1 and S_2, which can be open or closed. The top half of the figure shows the intensity $I_1(y)$ as a function of the coordinate y measured along the back wall, with just S_1 open. The pattern for $I_1(y)$ is almost flat with a slight peak in front of slit 1. Not shown is a similar pattern for $I_2(y)$ with only slit S_2 open.

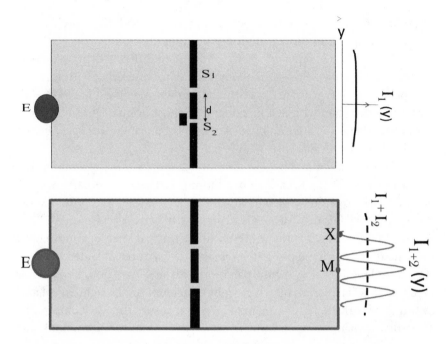

Figure 18.1   Top: Light from the emitter E at the left passes via one open slit S_1 in the opaque partition and illuminates the screen behind with an intensity pattern $I_1(y)$ that is almost flat and slightly peaked in front of slit 1. A similar pattern $I_2(y)$ obtains when only slit S_2 is open (not shown). Bottom: When both slits are open the result is the oscillatory pattern I_{1+2} instead of $I_1 + I_2$, the sum of the two intensities (shown by a dotted curve). At the point X, there is less light with both slits open than with one.

What intensity I_{1+2} do you expect when both slits are open? If this was a case of sunlight streaming through two windows, your expectation would be that you will simply get the sum of the intensities. If you sat in a region illuminated by both windows, you expect to have more light and warmth than with just one window open. This expectation is shown by a dotted line labeled $I_1 + I_2$ in the lower half of the figure. This expectation would indeed be realized if you were talking about sunlight streaming through two windows.

However, this is not what happened in Young's experiment. He found an intensity pattern labeled as I_{1+2}, where

$$I_{1+2} \neq I_1 + I_2. \tag{18.1}$$

In the screen (which extends perpendicular to the page), these oscillations in I_{1+2} correspond to vertical stripes of bright and dark.

This is called *interference.* Here are the main differences between I_{1+2} and $I_1 + I_2$. At a place like M the intensity is *four* times that with just one slit open and not twice. Then there are places where you get *less* light with both slits open than with just one open, the most dramatic of these being *places like X that got some light when one or the other slit was open, but no light at all with both slits open.*

The oscillatory graph I_{1+2} does not make any sense in ray optics: it explains neither the broad regions illuminated when one narrow slit is open nor why opening a second slit can make it darker at a point like X. But I_{1+2} is very familiar from experiments with waves, say in water. (Take a peek at Figure 18.4 if you want. We will return to it later in some detail.) Imagine that the source of light is replaced by a source of water waves of some wavelength λ. The opaque partition with slits is replaced by a barrier with two slits. The screen on which light is incident on the right is replaced by a line of corks bobbing up and down to measure the wave amplitude. Then the intensities, which are squares of these amplitudes, will look just like I_1, I_2, and I_{1+2}. More importantly, they can be readily explained by wave theory, as we shall see.

This is why Young's interference experiment convinced everyone that light was a wave. He could even find the wavelength λ, though he had no idea what was waving. That had to await Maxwell.

Why do we not see interference of light when we open a second window? Does not every place get brighter? If you were warm at X due to the light coming from window S_1 and you told somebody, "Hey, this is good, open the other window," you will end up getting more light and warmth. You will not get less, because the interference pattern as described by the oscillatory graph I_{1+2} will be realized only under the following conditions.

1. The light must have definite wavelength. The reason is that the locations of the maxima and minima vary with λ and these features will get washed out in a mixture. Sunlight, which is a mixture of many λ's, does not qualify.
2. The wavelength λ must be comparable to or larger than the size of the slits or the spacing between them. This is also not true for sunlight entering through two windows.

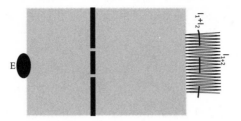

Figure 18.2   If the oscillations are too rapid, any probe (such as your eyes) would sense the average of I_{1+2}, the dotted curve $I_1 + I_2$.

If monochromatic light (light of definite λ) were coming in from two windows to a certain place, there could, in principle, be oscillations due to interference, but the spacing between the crests and troughs would be so small none of your senses would discern them. Your eyes would spatially average the pattern I_{1+2} over many cycles and only see the dotted line $I_1 + I_2$ shown in Figure 18.2.

18.1   Interference of waves

Let us look at interference in general, for any kind of wave. Let us denote by ψ whatever is oscillating. It can be the deviation from equilibrium of the position of a string, air pressure, or the height of water. The *inhomogeneous* linear wave equation in one-space dimension is

$$\frac{1}{v^2}\frac{\partial^2 \psi}{\partial t^2} - \frac{\partial^2 \psi}{\partial x^2} = S(x,t) \quad \text{where} \tag{18.2}$$

$S(x,t)$ is called the *source term* or *driving term*. This could describe a string that is responding not only to its internal forces, but also an external force, say a violin bow.

In higher dimensions, it is

$$\frac{1}{v^2}\frac{\partial^2 \psi}{\partial t^2} - \frac{\partial^2 \psi}{\partial x^2} - \frac{\partial^2 \psi}{\partial y^2} - \frac{\partial^2 \psi}{\partial z^2} = S(\mathbf{r},t)$$

which may be written compactly as $\qquad\qquad\qquad\qquad$ (18.3)

$$\nabla \cdot \nabla \psi = S(\mathbf{r},t). \tag{18.4}$$

The wave equation is linear. This implies that if

$$\nabla \cdot \nabla \psi_1 = S_1(x, t) \quad \text{and} \tag{18.5}$$

$$\nabla \cdot \nabla \psi_2 = S_2(x, t) \quad \text{then} \tag{18.6}$$

$$\nabla \cdot \nabla (A\psi_1 + B\psi_2) = AS_1 + BS_2$$

$$\text{where } A \text{ and } B \text{ are constants} \tag{18.7}$$

because we can take the derivatives in $\nabla \cdot \nabla$ through A and B. This means that you can take two solutions, ψ_1 and ψ_2, which are the responses to S_1 and S_2, multiply by constants A or B, and add them to get the response to $AS_1 + BS_2$. That's the principle of superposition. Clearly it holds for the *homogeneous* case with no sources $S_1 = 0$ and $S_2 = 0$.

What if someone tried to convince you that the ψ in a linear homogeneous problem is always positive? How will you refute that argument? The answer is that if ψ were such a positive solution, then $(-1)\psi$, which is always negative, would also have to be a solution.

If ψ can be positive or negative, it cannot stand for things that are always positive such as the brightness of light or the energy in a wave. The worst thing you can have is no light. You cannot have negative brightness. That is why in the electromagnetic theory of light, brightness is not measured by the electric (or magnetic) field, which can have either sign, but by something quadratic in the field, which we have seen is the intensity I. The intensity in general (say of sound) is proportional to the square of whatever is oscillating. It is the measure of the energy contained in the wave.

Therefore when you have one source producing some light and a second source producing some light, together they'll produce a *field* that is the sum of the two fields. What you can add or superpose is ψ, which in this case happen to be **E** and **B**. You cannot add the intensities. But there *is* a definite rule for total intensity, which follows from its definition. If

$$I_1 = \psi_1^2 \tag{18.8}$$

$$I_2 = \psi_2^2 \quad \text{then} \tag{18.9}$$

$$I_{1+2} = (\psi_1 + \psi_2)^2 = \psi_1^2 + \psi_2^2 + 2\psi_1\psi_2 \tag{18.10}$$

$$= I_1 + I_2 + 2\psi_1\psi_2. \tag{18.11}$$

In the last equation, the first two terms are positive, but the last can have either sign. It is the one that causes the oscillations. However, the oscillations cannot be more negative than the sum of the first two terms since the total, being a square, must be positive.

So there are two levels at which things happen with waves. There is the thing that actually oscillates and obeys the wave equation, and then there is its square, which represents energy or brightness. The superposition principle applies to the thing that oscillates and not to its square. That's why in the experiment when you open a second slit, some places like X can become darker than they were with just one slit open.

We will shortly begin our study of quantum mechanics. There too we will encounter a ψ called the *wave function* . Now that's a very bizarre object. Let me just say for now that it is *intrinsically complex*. Remember the harmonic oscillator where we viewed the physical variable $x = A\cos\omega t$ as the real part of $x = Ae^{i\omega t}$? We brought in the complex exponential because it made it easier to solve certain equations. In the end we took just the real part. But in quantum mechanics all of ψ, its real part and imaginary parts, is needed. In fact there is an i right in the Schrödinger equation, the analog of Newton's law. There is no escaping complex numbers in quantum mechanics.

Clearly in the quantum case the analog of intensity cannot be ψ^2 because it is not always positive or even real. For example, if $\psi = 3 + 4i$, then $\psi^2 = 9 - 16 + 24i$. You must know enough about complex numbers to guess the corresponding intensity:

$$I = |\psi|^2 = \psi^*\psi = (Re\,[\psi])^2 + (Im\,[\psi])^2. \qquad (18.12)$$

In general you may take $I = |\psi|^2$, for if ψ happened to be real, you will simply find $|\psi|^2 = \psi^2$.

18.2   Adding waves using real numbers

Let us start our study of interference of waves in the following simple context. Imagine that you just sit at one point and let two waves come to you. We are looking not at the wave as a function of x and t (in one dimension) but as a function of just t at your location. For example, you could be on a lake and someone could be rocking a boat somewhere and sending out ripples with a frequency ω. At your location $\psi_1(t)$, the water level relative to the tranquil lake is

$$\psi_1(t) = A\cos\omega t \quad \text{where } \omega = 2\pi f = \frac{2\pi}{T}. \tag{18.13}$$

Now someone else starts sending a second wave of the same amplitude, at the same frequency, but out of step, with a phase difference ϕ:

$$\psi_2(t) = A\cos(\omega t + \phi). \tag{18.14}$$

Remember the phase *difference* between two waves is physically significant and cannot be eliminated by resetting your clock. We assume that

$$\pi \leq \phi < \pi. \tag{18.15}$$

With both sources sending waves, the height of the water (relative to the tranquil lake) will be simply the sum of the two heights:

$$\psi_{1+2}(t) = \psi_1(t) + \psi_2(t) = A[\cos(\omega t) + \cos(\omega t + \phi)]. \tag{18.16}$$

This is not obvious but true. It follows from the linearity of the wave equation. (Water waves sometimes obey a non-linear equation in which case the heights will not be additive when both sources are on.) To proceed further we need a trig identity:

$$\cos\alpha + \cos\beta = 2\cos\left(\frac{\beta - \alpha}{2}\right)\cos\left(\frac{\alpha + \beta}{2}\right). \tag{18.17}$$

If you have some doubts about the formula you can test special cases. For example, if $\alpha = \beta$, we have $2\cos\alpha$ on the left and $2\cos 0 \times \cos\alpha = 2\cos\alpha$ on the right. If $\beta = 0$, we get $\cos\alpha + 1$ on the left and $2\cos^2\frac{\alpha}{2}$ on the right, which also matches. Finally both sides are invariant under $\alpha \leftrightarrow \beta$. While these successful tests do not mean the formula is right, a failure would immediately condemn it.

Anyway, this is the right formula, and in our case it gives

$$\psi_{1+2}(t) = 2A\cos\frac{\phi}{2}\cos\left(\omega t + \frac{\phi}{2}\right) \tag{18.18}$$

$$= \tilde{A}\cos\left(\omega t + \frac{\phi}{2}\right) \quad \text{where} \tag{18.19}$$

$$\tilde{A} = 2A\cos\frac{\phi}{2} \quad \text{is the amplitude.} \tag{18.20}$$

The sum therefore is a signal with amplitude $\tilde{A} = 2A\cos\frac{\phi}{2}$ and phase $\frac{\phi}{2}$.

Let us check the formula in a couple of special cases. If $\phi = 0$, the two signals are identical. We don't need any fancy stuff to know the answer is $2A\cos\omega t$ and indeed this is true for our answer. Another case you can do very easily in your head is $\phi = \pi$. Since $\cos(\theta + \pi) = -\cos\theta$, the second signal is exactly the opposite of the first and must cancel it at all times. This agrees with our result since $\tilde{A} = 2A\cos\frac{\pi}{2} = 0$.

This was a special case where the two waves had the same amplitude. If they had different amplitudes the formula would be messier but the main features would persist except for some inevitable differences, such as the impossibility of perfect cancellations.

18.3 Adding waves with complex numbers

I am now going to re-derive Eqn. 18.18 using complex numbers. One of the reasons is that I want you to get used to complex numbers in readiness for quantum mechanics.

Recall that a complex number z may be written in two ways as indicated in Figure 18.3(a). These are the Cartesian and polar forms:

$$z = x + iy \quad \text{(Cartesian form)} \tag{18.21}$$

$$= |z|(\cos\theta + i\sin\theta) = |z|e^{i\theta} \quad \text{(polar form) using} \tag{18.22}$$

$$e^{i\theta} = \cos\theta + i\sin\theta. \tag{18.23}$$

We should know how to express the polar coordinates in terms of the Cartesian:

$$|z| = \sqrt{x^2 + y^2} \tag{18.24}$$

$$\theta = \tan^{-1}\frac{y}{x} \tag{18.25}$$

and vice versa

$$x = |z|\cos\theta \equiv Re\,[z] \tag{18.26}$$

$$y = |z|\sin\theta \equiv Im\,[z]. \tag{18.27}$$

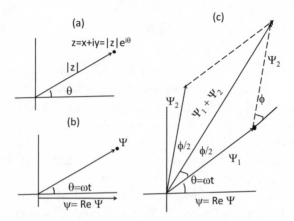

Figure 18.3   (a) Polar and Cartesian forms of z. (b) The complex Ψ and its real part, the physical ψ. (c) Adding two complex Ψ's.

A real function $\psi = A\cos\omega t$ may therefore be written as

$$\psi = Re\left[Ae^{i\omega t}\right]. \tag{18.28}$$

Let us define a complex Ψ

$$\Psi = Ae^{i\omega t} \tag{18.29}$$

whose real part is our ψ:

$$\psi = Re[\Psi] \tag{18.30}$$

as shown in Figure 18.3(b). One can visualize Ψ as a rotating complex number of length A and phase angle $\theta = \omega t$, and ψ as its instantaneous real part. As time goes by, the vector Ψ rotates and its projection on the real axis describes the physical and real variable ψ. If there were a light source shining down the y-axis, the shadow of Ψ on the real axis would be our ψ.

Suppose we want to add two such real waves

$$\psi_1 = A\cos\omega t \tag{18.31}$$

$$\psi_2 = A\cos(\omega t + \phi). \tag{18.32}$$

Since the sum of the real parts is the real parts of the sum, we may *first* add the complex Ψ's and *then* take the real part:

$$\psi_1 + \psi_2 = Re\left[Ae^{i\omega t}\right] + Re\left[Ae^{i\omega t + \phi}\right] \qquad (18.33)$$

$$= Re\left[\Psi_1\right] + Re\left[\Psi_2\right] = Re\left[\Psi_1 + \Psi_2\right]. \qquad (18.34)$$

Adding the complex Ψ's is easy. From Figure 18.3 we can see that the sum $\Psi_1 + \Psi_2$ is the sum of two planar vectors. That the sum vector bisects the angle between Ψ_1 and Ψ_2 follows from the congruence of the two triangles that make up the parallelogram. This means the phase of the sum is $\omega t + \frac{\phi}{2}$. The length of $\Psi_1 + \Psi_2$ follows from vector analysis

$$|\mathbf{A} + \mathbf{B}|^2 = (\mathbf{A} + \mathbf{B}) \cdot (\mathbf{A} + \mathbf{B}) = |\mathbf{A}|^2 + |\mathbf{B}|^2 + 2\mathbf{A} \cdot \mathbf{B}. \qquad (18.35)$$

In our problem this means

$$|\Psi_1 + \Psi_2|^2 = A^2 + A^2 + 2A^2 \cos\phi = 2A^2(1 + \cos\phi)$$

$$= 4A^2 \cos^2 \frac{\phi}{2}. \qquad (18.36)$$

Consequently, the sum in polar form is

$$\Psi_1 + \Psi_2 = |2A\cos\frac{\phi}{2}| e^{i(\omega t + \phi/2)} \qquad (18.37)$$

and

$$\psi_1 + \psi_2 = Re\left[\Psi_1 + \Psi_2\right] = |2A\cos\frac{\phi}{2}| \cos\left[\omega t + \frac{\phi}{2}\right] \qquad (18.38)$$

in agreement with Eqn. 18.18. As $-\pi \le \phi \le \pi$, $\cos\frac{\phi}{2}$ is never negative and the modulus sign in Eqns. 18.37 and 18.38 may be dropped.

Finally I am going to do the addition $\Psi_1 + \Psi_2$ *algebraically*, without the aid of any pictures:

$$\Psi_1 + \Psi_2 = Ae^{i\omega t} + Ae^{i(\omega t + \phi)} \qquad (18.39)$$

$$= Ae^{i(\omega t + \frac{\phi}{2})}\left[e^{-i\frac{\phi}{2}} + e^{i\frac{\phi}{2}}\right] \qquad (18.40)$$

$$= Ae^{i(\omega t + \frac{\phi}{2})} \cdot 2\cos\frac{\phi}{2} \qquad\qquad (18.41)$$

using

$$\frac{e^{i\theta} + e^{-i\theta}}{2} = \cos\theta. \qquad\qquad (18.42)$$

Taking the real part of Eqn. 18.41 we obtain Eqn. 18.18 once again.

18.4 Analysis of interference

Let us now analyze the double-slit experiment armed with all these results. A plane wave comes from the left and hits a partition with two slits and some of it escapes to the other side where we have placed a screen, as shown in Figure 18.4. We want to understand the variation of intensity $I(y)$ as a function of the variable y measured along the screen. If you look at the partition from where the screen is, you will see two glowing slits. It is intuitively clear that each acts as a source of light. (This idea was formalized by Christian Huygens [1629–1695], who used it to propagate the wave in time by treating each point on the instantaneous wave front

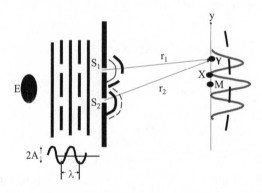

Figure 18.4   A distant emitter of light E produces the incoming plane wave, which in turn sets off synchronized radial waves from the two slits. The radial waves then reach different points on the screen labeled by y, with different phase differences. Their sum squared will determine the intensity there. It is a maximum (constructive interference) at points like M and a minimum (destructive interference) at points like X.

as a point source of light.) The waves from each slit will radiate outward from that slit. They will be emitted in step because the crests and troughs of the incoming wave that generate them reach the slits at the same time. The functions describing the radial waves are

$$\psi_1 = A\cos(kr_1 - \omega t) \tag{18.43}$$

$$\psi_2 = A\cos(kr_2 - \omega t) \tag{18.44}$$

$$k = \frac{2\pi}{\lambda}. \tag{18.45}$$

Unlike a plane wave traveling along y, which behaves as $\cos(ky - \omega t)$, these waves go out radially from the slits with a phase that changes with r as kr, where r is measured from the slits. Whereas the plane wave had the same phase at a given y (at some fixed time, say $t = 0$), these radial waves have the same phase at a given r. I show a couple of crests and troughs of these radial waves.

The radial waves then reach different points on the screen labeled by y, with different phase differences. The square of their sum will determine the intensity there. We want to compute $I_{1+2}(x)$.

Let us think in terms of water waves rather than electromagnetic waves, since we can visualize them more easily and the ideas are the same. Looking down at the shallow tank of water shown in Figure 18.4 we see crests and troughs of the incoming plane wave, shown by solid and dotted lines. I am sitting at the far right, at a point Y, which is at a distance r_1 from slit S_1 and a distance r_2 from S_2. The water at Y will be bobbing up and down by an amount equal to what the signal from S_1 tells it to do plus what the signal from S_2 tells it to do.

Let us add the two contributions at my location

$$\psi_1 = A\cos(kr_1 - \omega t) \tag{18.46}$$

$$\psi_2 = A\cos(kr_2 - \omega t) \tag{18.47}$$

$$\psi_{1+2} = A\cos(kr_1 - \omega t) + A\cos(kr_2 - \omega t) \tag{18.48}$$

$$= 2A\cos\left[\frac{k(r_2 - r_1)}{2}\right]\cos\left[\frac{k(r_2 + r_1)}{2} - \omega t\right], \tag{18.49}$$

using Eqn. 18.17 for the addition of cosines. The final answer is unaffected by the change $r_1 \leftrightarrow r_2$, because the cosine is an even function. This means that whatever happens at a certain point above the symmetric point M will have to happen at the same distance below M.

Equation 18.49 tells us that the signal at point Y has amplitude $2A\cos\frac{k(r_2-r_1)}{2}$, frequency ω, and (an inconsequential) phase $k(\frac{r_2+r_1}{2})$.

What we really care about is the intensity at that point, given by the square of the amplitude:

$$I_{1+2}(r_1, r_2) = 4A^2 \cos^2\left[\frac{k(r_2 - r_1)}{2}\right]. \tag{18.50}$$

From now on I will drop the subscript on I_{1+2} since we will only consider the case with both slits open.

Let us analyze how $I(r_1, r_2)$ varies as we move up and down the screen. Let δ denote the difference in path lengths from the two slits to a generic point on the screen:

$$\delta = r_2 - r_1. \tag{18.51}$$

Then what we want to study is the behavior of

$$I = 4A^2 \cos^2\frac{k\delta}{2}. \tag{18.52}$$

First consider a point M symmetrically located, equidistant from the slits. For this point $\delta = 0$ and

$$I = 4A^2. \tag{18.53}$$

At this point the total Ψ_{1+2} has double the amplitude due to each slit and an intensity I that is four times as big. This is a point of *constructive interference*.

We can understand this in real time as follows. The two slits produce waves that are synchronized because the crests and troughs of the plane wave that generate them hit the slits simultaneously. These radial crests and troughs then travel the same distance to reach M and arrive in sync. Thus the total signal at every instant is double that due to one slit, the amplitude is double, and the intensity is quadruple.

As we move off center, δ, the difference in path length, will grow and the cosine in Eqn. 18.52 will fall. It will hit zero when

$$\frac{k\delta}{2} = \frac{\pi}{2} \tag{18.54}$$

$$\delta = \frac{\pi}{k} = \frac{\pi}{(2\pi/\lambda)} = \frac{\lambda}{2}. \tag{18.55}$$

At this location the signals from the two slits exactly cancel each other. The signal from S_2 is delayed by half a time period (because it has to travel half a wavelength more). You can see in the cosine that if you move by half a period or wavelength, you reverse its sign. Thus when ψ_1 tells the water to go up, ψ_2 tells it to go down by the same amount. At every instant the signals are negatives of each other and the net result is identically zero.

This is a point of *destructive interference.*

As we move further away from M, the pattern repeats itself. If we go up in y till $r_2 - r_1$ is a full wavelength, it is as if the difference were zero and we hit another maximum. Above that is the second minimum, where the path difference is $3\lambda/2$, and so on. What happens at a certain distance above M also happens at the same distance below M.

The following formula says it all:

$$\delta = r_2 - r_1 = 0, \pm\lambda, \pm2\lambda, \ldots \text{constructive interference} \tag{18.56}$$

$$= \pm\frac{\lambda}{2}, \pm\frac{3\lambda}{2}, \ldots \text{destructive interference.} \tag{18.57}$$

Consider the fact that opening a second slit can make a spot like X dark, whereas it used to be bright with just one slit open. This can happen only with waves. In ray theory, the rays that come with two slits open will be the sum of the rays that come from each one. Likewise in Newton's corpuscular theory of light, the number of light corpuscles coming to a point on the screen with two slits open will be the sum of the numbers coming from each. The contribution from one slit cannot cancel the contribution from another. It is like mosquitoes: if you have two holes in the mosquito net, you get twice as many mosquitoes. Had negative mosquitoes been possible, two holes could have led to fewer mosquitoes than one hole. But it is not and they do not.

But this can happen with waves. The thing with waves is that what is additive is ψ, which is not always positive. This leaves room for cancellations and interference. Young's experiment with light exhibited interference. He did not know what light was. He did not know about electromagnetic waves, but he didn't need to. If you shine light through two slits and you get the dark and bright and dark and bright fringes, you (and everyone else) are convinced you are dealing with a wave.

His experiment also determined the wavelength of light as follows.

Let us begin with the conditions for constructive and destructive interference, Eqns. 18.56–18.57, expressed in terms of $\delta = r_2 - r_1$. Let us trade them for expressions in terms of the parameters more readily measured, and defined in Figure 18.5A: d, the spacing between slits, and (L, y), the coordinates of the point in question relative to the origin $(0, 0)$. We will write down an exact result for the distances first and then approximate it using the smallness of d/L, *keeping only the first power of d* (as we did with the dipole field). The distance y could be of the same order as L and is not considered small.

$$r_1 = \sqrt{L^2 + \left(y - \frac{d}{2}\right)^2} \tag{18.58}$$

$$\simeq \sqrt{L^2 + y^2 - yd} \equiv \sqrt{r^2 - yd}$$

$$\text{where } r^2 = L^2 + y^2 \tag{18.59}$$

$$= r\left(1 - \frac{yd}{r^2}\right)^{\frac{1}{2}} = r - \frac{yd}{2r} \quad \text{(to this order in } d\text{)} \tag{18.60}$$

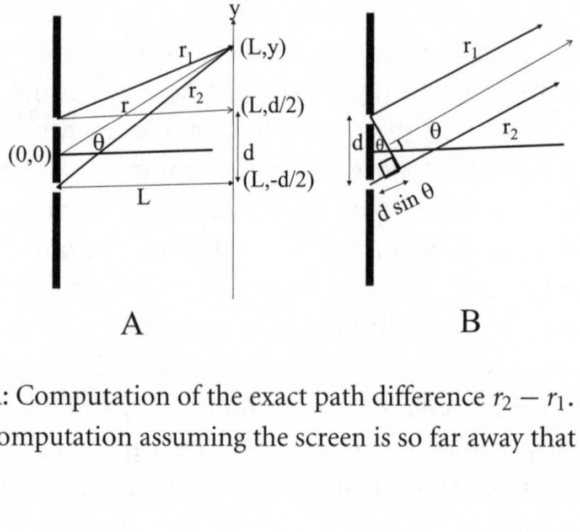

Figure 18.5   A: Computation of the exact path difference $r_2 - r_1$. B: Approximate computation assuming the screen is so far away that the rays are parallel.

$$r_2 = r + \frac{yd}{2r} \tag{18.61}$$

$$r_2 - r_1 = \frac{yd}{r} = d\sin\theta \tag{18.62}$$

where θ is the angle of the vector $\mathbf{r}$ joining the origin $(0,0)$, midway between the slits, to the point (L, y).

The nature of the approximation can be made more transparent by deriving the same result more quickly in another way. Look at Part B of the figure. It shows $\mathbf{r}_1$ and $\mathbf{r}_2$ as parallel lines going to a very distant screen. The extra distance associated with $\mathbf{r}_2$ is $d\sin\theta$.

In terms of d and θ we may now write Eqns. 18.56 and 18.57 as

$$d\sin\theta = 0, \pm\lambda, \pm 2\lambda \ldots \equiv m\lambda$$

$$\text{(constructive interference)} \tag{18.63}$$

$$= \pm\frac{\lambda}{2}, \pm\frac{3\lambda}{2}, \ldots \equiv \left(m + \frac{1}{2}\right)\lambda$$

$$\text{(destructive interference)} \tag{18.64}$$

where m is an integer. Beyond some m we cannot satisfy the equations. For example, there will be no maxima with $m\lambda > d$, for this would require $\sin\theta = m\lambda/d > 1$. As you can see from the figure, the longest path difference is d (for both $\mathbf{r}_1$ and $\mathbf{r}_2$ pointing straight up or straight down with $\theta = \pm\frac{\pi}{2}$).

These equations can be used to locate the (angles of the) various maxima (besides the central maximum) and minima. Conversely, from the measured angles one can infer λ. One can do this for light, without knowing it is the $\mathbf{E}$ and $\mathbf{B}$ that are oscillating.

Now you can see why people in the old days, working with light on a macroscopic scale, got fooled by geometric optics. If, say, $d = 1$ mm and $\lambda \simeq 5 \cdot 10^{-7} m$, the angular difference between the central maximum and first minimum is of order $5 \cdot 10^{-4}$ radians. If you see this from a distance of 1 cm, the spacing between maxima will be around $5 \cdot 10^{-4}$ cm. If the interference pattern is this dense, the eye only sees an average of I_{1+2} over many cycles, in which case it reduces to $I_1 + I_2$, the result expected in geometric optics.

Here is another practical tip from Physics 201. Imagine you have some oceanfront property and are relaxing on your yacht. There is an oil rig out there sending waves that rock your boat when you are trying to relax, so you build a wall to keep the ocean waves out. Then one day there is a breach in the wall (S_1) and the waves start coming in. You have two options. One is to try to plug the hole, but let's say you've got no bricks, no mortar, no time, no patience. But you have a sledgehammer. With that you can make another hole. If these water waves are of a definite wavelength, you can locate the second hole (S_2) so that the sum of the waves cancel at your yacht location. (They are now twice as big on your neighbor's yacht, but that is not your problem: you did not rise to the top of your profession by worrying too much about your rivals.)

18.5 Diffraction grating

A diffraction grating is a generalization of the double slit: instead of just two slits, imagine a very large number of equally placed slits, assumed infinite in the following analysis. Figure 18.6 shows a finite segment of this grating. One way to makes gratings used to involve taking a piece of glass covered with soot and then drawing evenly spaced lines on it that allow light to pass when illuminated from one side. Nowadays there are far superior techiniques.

With two slits we saw that their contributions sometimes add and sometimes cancel depending on the angle. With an infinite number of such slits, the only way to get all the slits to contribute constructively is for all the path differences to be multiples of λ. One such case is simply the forward direction, when all the emitted waves go an equal distance to the screen (assumed to be far away) and therefore add in step. (Sometimes they are focused by a lens to form an image on a nearby screen.) The next maximum occurs when each path length differs from its neighbor by λ. From Figure 18.6 you can see that this means $d\sin\theta = \lambda$. More generally the maxima will occur at angles where the path difference obeys

$$d\sin\theta = m\lambda \quad m = \pm 1, \pm 2 \ldots \tag{18.65}$$

At any other angle, the path difference between neighboring paths will be some δ. In terms of the complex Ψ (whose real part we will take at the end) we are trying to add

$$\Psi = \Psi_1 + \Psi_2 + \ldots$$
$$= A(e^{i0} + e^{ik\delta} + e^{-ik\delta} + e^{2ik\delta} + e^{-2ik\delta} + \ldots) \qquad (18.66)$$

where the first term is some reference contribution chosen by convention to have zero phase. Let us rearrange the sum as follows:

$$\Psi = A(1 + e^{ik\delta} + e^{2ik\delta} + e^{3ik\delta} + \ldots)$$
$$+ A(1 + e^{-ik\delta} + e^{-2ik\delta} + e^{-3ik\delta} + \ldots) - A. \qquad (18.67)$$

Each term in the first bracket is a number of modulus A and phase $mk\delta$. Imagine adding a string of vectors of length A, which slowly twist in orientation. Their sum will go round and round and give something of order A. The same goes for the second bracket and the constant $-A$. If, however,

$$k\delta = 0, \pm 2\pi, \pm 4\pi \ldots \qquad (18.68)$$

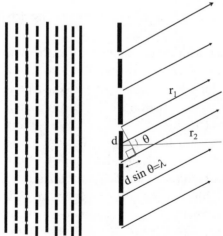

Figure 18.6 Part of a diffraction grating. If $d\sin\theta$, the path difference between adjacent slits is a multiple of λ, and the signals from all the slits add constructively.

the arrows will be parallel and add up to a length NA if there are N slits. Of course the condition

$$k\delta = 0, \pm 2\pi + \dots \quad \text{is just the condition} \tag{18.69}$$

$$\delta = 0, \pm\lambda, \pm 2\lambda, \dots \tag{18.70}$$

encountered in Eqn. 18.65. Unlike the maximum, the condition for the minima (which lie between the maxima) is harder to find for N slits.

Suppose you send in white light to a grating. The central maximum, which corresponds to $\delta = 0$, is a maximum for all colors or all λ's since all rays travel the same distance in the forward direction to the screen. The central maximum ($m = 0$) will therefore also be white. However, the secondary maxima, with $m \neq 0$, *will be at angles determined by* λ. What is a path difference of one wavelength for red will not be one wavelength for blue. So the colors in the incoming white light will split into different directions, with the maxima determined by the following condition on the path difference δ:

$$\delta = d\sin\theta = m\lambda_c \tag{18.71}$$

where λ_c is the wavelength of color c. The grating acts like a prism that splits the colors.

If you look at white light coming from the sun you'll find some colors missing, some lines of darkness in a broad band of sunlight split into its various colors. For example, these could be at the wavelengths emitted by hydrogen. Now, hydrogen, like all atoms, not only likes to emit light of certain wavelengths, it also likes to absorb only those wavelengths. Consequently when white light makes its way out of the sun's interior, the hydrogen atoms on the way absorb these colors, leading to the observed dark lines. These lines in the *absorption spectrum* are as good a fingerprint of hydrogen as the lines in the emission spectrum. They tell us the sun contains hydrogen. That is how people know what elements are present on different planets or stars. It was not at all clear in the ancient days that stars and planets were made of the same stuff we see on earth. Now we know that the same elements as here are out there, because we can identify the atoms by the missing lines, the colors they gobble up as white light makes its way out of the interior, or by the colors they emit.

18.6 Single-slit diffraction

We have already seen in the double-slit experiment that when light emerges from a slit, it fans out radially, with the slit as a point source. So what is there to study with just one slit? The answer is that the point source description holds only if the wavelength is much bigger than the slit width. In the double-slit experiment, I only specified the slit *separation d* and not *width*, which was assumed to be zero.

However, every real slit will have some width, which I will call D to distinguish it from the slit separation d in the double-slit experiment. The slit will behave like a point only if its width D is much smaller than λ. Now consider a case when D is comparable to or bigger than λ, and a fresh analysis is called for.

When seen from the dark side the slit will be glowing. Let us *mentally* divide the single slit into many adjacent mini-slits, each small enough to be approximated by a point. In the forward direction, all mini-slits will make contributions in phase. As we move away the contributions will begin to go out of step and the sum will diminish. We can find the angle θ at which they will add up to zero. It is given by

$$D\sin\theta = \lambda. \tag{18.72}$$

This is not a typo. Look at Figure 18.7A. Suppose, for simplicity, that there are N mini-slits. Number 1 and N are in step if $D\sin\theta = \lambda$. This may seem wrong for a minimum. But these two are not the only mini-slits we have to worry about. We have to account for all N. So let me pair them as follows. The first and the $\frac{N}{2} + 1$-th mini-slit are out of step by $\lambda/2$ and neutralize each other. The same goes for the second and the $\frac{N}{2} + 2$ mini-slit, and so on. So when the end-to-end path difference is λ, I can organize the mini-slits into canceling pairs, with a path difference of $\lambda/2$ within each duo. After that first zero, there are further oscillations, but usually it is pretty much all over as shown in part B of the figure. The half-width of the central maximum is

$$\theta = \sin^{-1}\frac{\lambda}{D}. \tag{18.73}$$

If $\lambda/D \ll 1$, then $\theta \simeq \lambda/D$, the angular width of the emergent beam, is negligible and we are in the realm of geometrical optics. As D becomes

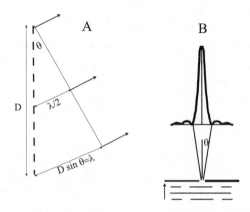

Figure 18.7   A: The condition for $N \to \infty$ mini-slits (square dots) making up a single slit of width D to cancel is that the path difference between the first and last is λ. This means each mini-slit can be paired with one with path difference $\lambda/2$, and the duo cancel each other. If the mini-slits are labeled 1 to N, we pair 1 with $\frac{N}{2} + 1$, 2 with $\frac{N}{2} + 2$, etc. B: The resulting intensity on a screen. Only a few oscillations are shown.

smaller and comparable to λ, the emergent beam fans out more and more. For example, when $D = 2\lambda, \theta = \frac{\pi}{6}$. Now you definitely need wave optics.

You can now understand the common statement that in order to see an object of size D clearly you need light of wavelength $\lambda \ll d$. Let the object be a hole of size D (not necessarily circular, but with some sharp features) in an opaque screen. We "see" the hole by shining light from one side and looking at the illuminated part of the screen on the other side. If $\lambda \ll D$, the diffraction peak is very narrow, geometric optics applies, and the bright image of the hole is directly in front of it and shows its fine features. As we lower λ the diffraction peak broadens out and the image starts getting fuzzy. When $\lambda \simeq D$, the beam spreads out by a half-angle of $\pi/2$, the light fans out completely, and we have lost any semblance of a sharp image.

18.7   Understanding reflection and crystal diffraction

Look at a line of atoms shown in Figure 18.8 that form a regular lattice, as in a crystal. Light is incident on them at an angle θ_1 relative to the normal to the line of atoms. You know the surface will reflect it such that $i = r$. But why? If you thought of light as made of particles, and the surface crystal as continuous, $i = r$ corresponds to just an elastic collision in which momentum perpendicular to the surface is reversed. But we are committed

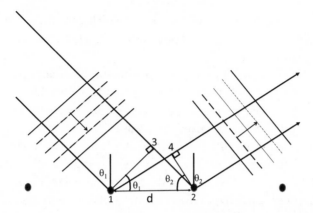

Figure 18.8 Part of an array of atoms reflecting an incident beam. Atom 1 gets the incident light before 2, but its emitted light has to travel farther in the reflected direction. (The common delay between absorption and emission cancels out.)

to waves after Young's experiment. And we also know the surface is made up of an array of atoms and is not a continuum. We have to explain $i = r$ in these terms.

To this end we need to know how atoms "reflect" light. It turns out they first absorb it and then re-emit it. An atom that re-emits light does so isotropically, with no memory of the incoming direction. So how can all this *isotropic* emission end up producing a strong signal along just one direction obeying $i = r$? It must be that this is the direction in which the emitted waves add in phase. To see that this is so, we need consider just two adjacent atoms numbered *1* and *2* in Figure 18.8. You can see that atom 1 gets hit first by the incoming wave fronts and then atom 2 a little later, because the light has to travel an extra distance, the side d_{32} in the right triangle 132. Assume the atoms re-emit instantaneously (the common delay drops out). In the outgoing direction, the emitted light from 2 has a head start, a distance d_{41} in the right triangle 241. So atom 1 gets the incident light sooner than 2, but its reflected light has to travel a longer distance than the light emitted by 2. For the final waves to be in step, we need the two distances to be equal:

$$d_{32} = d_{41} \tag{18.74}$$

$$d \sin\theta_1 = d \sin\theta_2 \tag{18.75}$$

$$\theta_1 = \theta_2, \tag{18.76}$$

and so $i = r$. (I am assuming $\lambda > 2d$, in which case there are no other solutions. Otherwise, there could be solutions in which the path difference is a multiple of λ and $\theta_2 \neq \theta_1$.)

Consider now X-rays incident on a crystal-like diamond, or matter waves (which you will learn about later) incident on nickel. The crystal has many layers of atoms, periodically stacked one below the other. All the layers can receive and re-transmit the waves. Then we need to ask how the reflected wave from *different* layers will interfere. This is considered in Figure 18.9. The upper layer reflects waves obeying $i = r$ for reasons just discussed. The reflected wave from the second layer lags because it has to travel an extra distance $d_{A2} + d_{2B} = d\sin\theta + d\sin\theta$. For this not to make any difference we require the *Bragg condition*:

$$2d\sin\theta = m\lambda \qquad m = 1,2,\ldots \tag{18.77}$$

where θ is the angle between the incident beam and *the line of atoms, not its normal.* Once this condition for two successive layers to be in phase is satisfied, all the layers will also scatter in phase because the relevant path differences will also be multiples of λ.

Whereas reflection by one layer (obeying $i = r$) will take place for any angle of incidence, coherent diffraction by all the layers will occur only for certain values of incident angle θ obeying the Bragg condition. These special angles can be achieved either by changing the direction of the beam

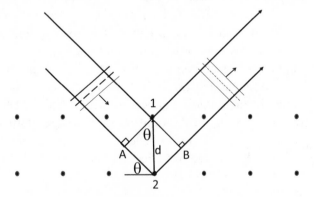

Figure 18.9   The condition for all layers to reflect waves in phase is that the path difference between successive layers is a multiple of λ. Note θ is the angle between the incident beam and the line of atoms and not the normal to the line.

incident on a fixed crystal or by rotating the crystal illuminated by a fixed beam.

18.8   Light incident on an oil slick

If there is an oil slick on a wet street, you see many colors in the reflected light. Here is what is happening. We have three regions: air at the top, next oil of thickness δ, and water below that. The incident light can reflect off the two interfaces and the sum of the two reflected waves, seen by someone looking down on the slick, can interfere constructively or destructively.

18.8.1   Normal incidence

First consider the simpler case where white light strikes the air-water interface along the normal, as shown in Figure 18.10. Some of it gets reflected and some transmitted. The reflected signal is ψ_1 in Figure 18.10. The transmitted signal hits the oil-water interface and some of it gets reflected. This reflected light then crosses back to the air as ψ_2 and interferes with ψ_1. Their sum is what you see looking down. (I have displaced ψ_1 and ψ_2 from the normal for clarity.)

The sum depends on the phase difference between ψ_1 and ψ_2. Say we want ψ_1 and ψ_2 to interfere destructively. The requisite condition depends on the wavelength. Suppose the color blue is suppressed this way. This means that looking down on the slick we will see white minus the blue. If the thickness of oil varies, the color we see will also vary. If ψ_1 and ψ_2 interfered constructively, the color in question (say blue) would be

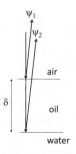

Figure 18.10   Normal incidence on a thin layer of oil on top of a layer of water. (Rays reflected from the first and second interface are displaced slightly from the normal and each other for clarity.)

enhanced relative to the others in what we see. In any event, the initial ratio of colors in the incident white light gets altered, with some colors getting suppressed and some enhanced, leading to multiple colors in what we see.

Let us find the condition for constructive and destructive interference for incident light of wavelength λ. There are two ingredients to consider.

The first, which I do not expect you to know, is that when light reaches the interface to a denser medium (i.e., greater refractive index n) it will suffer an extra phase shift of π upon reflection back to the rarer medium. In our example this phase shift will occur when oil ($n_0 = 1.5$) reflects light back to air ($n = 1$) but not at the next interface when the water ($n_w = 1.33$) reflects it back to oil.

The second ingredient has an obvious part, that ψ_2 has to travel an extra distance of 2δ compared to ψ_1, and a non-obvious part, that the corresponding phase shift is 2π for every wavelength λ_0 *in oil.*

The wavelength λ_0 in oil will not be the wavelength λ in air. We may understand this in terms of two defining relations:

$$c = f\lambda \quad \text{in air;} \tag{18.78}$$

$$\frac{c}{n_0} = f\lambda_0 \quad \text{in oil.} \tag{18.79}$$

Notice I use the same f in oil and in water but not the same λ. The reason is that light is generated by some source at some frequency f and this can lead only to waves of that f even if it crosses from one medium to another. In Huygens's approach, the light in the first medium acts as the source for the light in the second medium and so it will drive it at the same frequency. The lower velocity will be due to the shorter wavelength.

Or think in terms of water waves. Suppose some mechanical vibrator is producing waves on water with some f. Let us say the wave velocity depends on the depth of water and this depth suddenly changes when the waves enter a second region. The waves in the second region will still rise and fall at the same frequency as the driving vibrator even if they propagate more slowly. The reduced velocity will be due to the reduction in λ.

Equations 18.78 and 18.79 tell us that

$$\lambda_0 = \frac{\lambda}{n_0}. \tag{18.80}$$

So an extra distance of 2δ is worth $\frac{2\delta}{\lambda_0} = 2n_0\delta/\lambda$ wavelengths and a phase delay in wave ψ_2 of

$$\Delta\phi_2 = 2\pi \frac{2n_0\delta}{\lambda}. \tag{18.81}$$

For ψ_1 we have

$$\Delta\phi_1 = \pi. \tag{18.82}$$

Thus the total phase difference between the two waves reaching the observer (including the extra π from the air-oil interface) is

$$\Delta\phi_2 - \Delta\phi_1 = \frac{4\pi n_0\delta}{\lambda} - \pi. \tag{18.83}$$

We want this to be a multiple of 2π for constructive interference

$$\frac{4\pi n_0\delta}{\lambda} - \pi = 2\pi m$$

$$\frac{2n_0\delta}{\lambda} = \left(m + \frac{1}{2}\right) \quad m = 0, 1, 2, \dots \text{ constructive} \tag{18.84}$$

and an odd multiple of π for destructive interference:

$$\frac{4\pi n_0\delta}{\lambda} - \pi = \pi(2m - 1) \quad m = 0, 1, 2, \dots$$

$$\frac{2n_0\delta}{\lambda} = m \quad m = 0, 1, 2, \dots \text{ destructive.} \tag{18.85}$$

If the formulas appear strange, it is due to the extra π coming from the first reflection.

Here is an example with some numbers. Suppose a film of oil produces constructive interference for $\lambda = 400nm$ and destructive interference for $\lambda = 500nm$. What is δ? The data given may be written as follows. For constructive interference of light with $\lambda = 400$ nm we need

$$2n_0\delta = 400nm \cdot \left(\frac{1}{2}, \frac{3}{2}, \frac{5}{2}, \dots\right)$$

$$= 200nm, 600nm, 1000nm, \dots \tag{18.86}$$

For destructive interference of light with $\lambda = 500\ nm$ we need

$$2n_0\delta = 500nm \cdot (1,2,3,\ldots)$$

$$= 500nm, 1000nm, 1500nm, 2000nm. \tag{18.87}$$

We see that the first point of agreement, the smallest value of δ for which both conditions are satisfied, is

$$2n_0\delta = 1000nm, \tag{18.88}$$

which corresponds to

$$\delta = \frac{1000}{2 \cdot 1.5} = 333.33nm. \tag{18.89}$$

If we go further down the two sequences, we will find a second common value: $2n_0\delta = 3000nm$, which translates into $\delta = 1000nm$. But $\delta = 333.33nm$ is the smallest.

I leave it to you to construct variations on this theme: change the media so that a phase change of π occurs at both interfaces or neither.

18.8.2   Oblique incidence

A non-trivial variation occurs when the incoming light strikes the first interface at an angle θ_1 relative to the normal as shown in Figure 18.11. It then enters the second medium at an angle θ_2 determined by Snell's law, reflects off the second interface obeying $i = r$, and finally re-enters the first

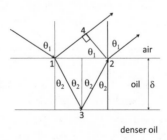

Figure 18.11   Oblique incidence on a thin layer of oil on top of a layer of denser oil. (There is phase shift of π at *each* interface, so we may ignore their combined effect.)

medium at an angle θ_1 to the normal. The phase shifts of π are the same as before but the path differences are more complicated.

Consider the simple case where there is no uncanceled π. This is true in Figure 18.11, when the refractive index gets bigger at both interfaces and there is a shift of π at each interface for a total of 2π, which may be ignored. Let the refractive index of the middle medium be n. The conditions for constructive and destructive interference will turn out to be

$$2n\delta \cos\theta_2 = m\lambda \quad \text{(constructive)} \tag{18.90}$$

$$= \left(m + \frac{1}{2}\right)\lambda \quad \text{(destructive).} \tag{18.91}$$

Here are the details. The wave ψ_1 has to travel an extra distance

$$d_{14} = d_{12}\sin\theta_1 \tag{18.92}$$

$$= 2\delta\tan\theta_2\sin\theta_1 \tag{18.93}$$

$$= 2n\delta\tan\theta_2\sin\theta_2 \quad \text{(Snell's law: } 1\cdot\sin\theta_1 = n\sin\theta_2\text{).} \tag{18.94}$$

The wave ψ_2 travels an extra *optical* distance (equivalent distance in air)

$$2nd_{13} = \frac{2n\delta}{\cos\theta_2}. \tag{18.95}$$

The net path difference is

$$\text{Path difference} = \frac{2n\delta}{\cos\theta_2} - 2\delta n\tan\theta_2\sin\theta_2$$

$$= 2n\delta\cos\theta_2, \tag{18.96}$$

which leads to Eqns. 18.90 and 18.91.

CHAPTER 19

Quantum Mechanics: The Main Experiment

We are going to focus on quantum mechanics from now on till the end. I've got bad news and good news. The bad news is that it is going to be hard for you to follow the physics intuitively, and the good news is that nobody can. Richard Feynman, one of the leading physicists of our time, used to say that no one understands quantum mechanics. Here then is my modest goal. Right now, I'm the only one who doesn't understand quantum mechanics. After these lectures, every one of you will be unable to understand it.

I want you to think about this as a real adventure. Try to think beyond the exams and grades. It's one of the greatest and deepest discoveries in physics and in all of science. It is remarkable how people figured out the underlying laws from the experiments.

I will not follow the historical route. It is pedagogically not the best way. You go through all the wrong tracks and false starts. When the dust settles down, a certain picture emerges and that's the picture I want to give you from the beginning. I will describe experiments that were perhaps not done in the manner (or sequence) in which I describe them, but rest assured that if they were performed, the results would be as described. The central experiment is the *double-slit experiment*, which Feynman has identified as exhibiting the heart of quantum mechanics. It not only shows in the clearest possible way the failure of Newtonian mechanics and Maxwell's wave theory of light, but it also gives us clues on how to go forward. How can this experiment, which proved wave theory unambiguously and paved the way for Maxwell's triumph, also lead to his theory's downfall? The answer is the usual one: because we pushed the experiment to a new range of parameters.

19.1 Double-slit experiment with light

Recall the highlights of the standard double-slit experiment. There is light of some wavelength λ coming from the left and incident on an opaque partition with two slits and emerging on the other side where it is detected. Assume a photographic plate is used for detection. It is made of tiny little pixels that change color when light hits them and forms a picture. It is a detector particularly suited for the variant that follows.

You measure the intensity I_1 with slit S_1 open, the intensity I_2 with just slit S_2 open, and then, with both slits open, the intensity I_{1+2}, which exhibits interference. A dramatic aspect of interference is that there are points that are bright when one or the other slit is open but dark when both are. The reason is that the ψ in this problem, the one which can be superposed, is the electric or magnetic field. When two slits are open, you add the fields, not the intensities that are proportional to the square of the total field. The two fields that add can have any relative sign or phase, and they can even cancel each other out.

We do not see this kind of interference with sunlight streaming into a room through two windows because that light is a mixture of many λ's and any surviving interference pattern I_{1+2} oscillates so rapidly that our senses can only detect its spatial average, which is just $I_1 + I_2$. Interference of waves was a familiar phenomenon before Young came along: just dropping two rocks into a tranquil lake allows one to see the interference of the two concentric waves produced. So when Young demonstrated the interference of light, it was clear to one and all that light was a wave. And then Maxwell derived his wave equations and that seemed to be the last step in a complete theory of optics.

19.2 Trouble with Maxwell

The interference pattern looks good for Maxwell's wave theory till you implement the following change: *You make the source of light dimmer and dimmer.* If you are unable to turn down the brightness enough, you can always move the source far away.

Imagine you insert a new photographic film, turn on a bright source, and call it a day. The next morning you find a pattern of light and dark

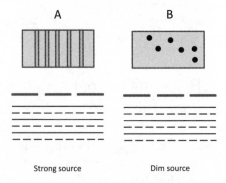

Figure 19.1   The figure shows the top view of the wave and the slits and a frontal view of the film. A: Pattern on film with a strong source. The light is coming from below toward the double-slit. B: Pattern with dim source, showing six exposed pixels.

stripes as shown in Figure 19.1A. When you repeat the experiment with a dimmer source, you get fainter stripes (not shown). Then you make a *drastic* reduction in the source brightness and wait overnight. You find no pattern, just six pixels that have been exposed, at seemingly random locations, as shown in Figure 19.1B. If you make the source weak enough, you can have a situation in which you just get one hit during the whole night. Now all this is very strange. If light were a wave, no matter how weak, it should illuminate the entire screen. It cannot hit just certain pixels. So something else is hitting that screen and it's not a wave.

You continue the experiment at this low intensity of one hit per night to probe this further. You make further observations and measure the momentum imparted to the film on each exposed pixel.

You find that each exposed pixel has received exactly the same amount of momentum p. By varying λ you establish that the value of this momentum is related to it as follows:

$$p = \frac{2\pi \hbar}{\lambda} \tag{19.1}$$

where

$$\hbar = 1.05 \cdot 10^{-34} \ J \cdot s \tag{19.2}$$

is *called* Planck's constant. (In the old days the name was reserved for $h = 2\pi\,\hbar$.) In terms of the wave number (phase change per unit length)

$$k = \frac{2\pi}{\lambda},$$
(19.3)

Eqn. 19.1 becomes

$$p = \hbar k.$$
(19.4)

Next you find that *each exposed pixel receives a fixed amount of energy related to the frequency of the incident light as follows:*

$$E = \hbar\omega.$$
(19.5)

The most natural interpretation of these results is that light of frequency ω or, equivalently, wave number k, is made up of particles, the *photons*, with the following energy and momentum:

$$E = \hbar\omega$$
(19.6)

$$p = \hbar k.$$
(19.7)

Since $\omega = kc$ it follows that the energy and momentum of the photons are related as follows:

$$E = pc$$
(19.8)

or $E^2 = c^2 p^2$, which, when compared to

$$E^2 = c^2 p^2 + m^2 c^4,$$
(19.9)

tells us photons are massless particles. The only way they manage to have momentum without mass, given the formula

$$p = \frac{mv}{\sqrt{1 - v^2/c^2}},$$
(19.10)

is by moving at the speed of light.

If you keep the extremely dim source on for many many days, you find that the spots, which initially appeared to be random, gradually fill out to form the stripes of the earlier experiments.

Amazingly then, what the incident beam of low intensity reveals is that light, which you thought was a continuous wave, is actually made up of discrete particles. If you turn on a bright light source you miss this aspect because millions of these photons rush in and form the interference pattern instantaneously. You see the dark and bright fringes right away, and you think it's due to a wave that hits the entire film instantaneously. But if you look under the hood, you find every pattern is formed by tiny little dots that appear individually.

Now if all somebody told you was that light was made of particles, that it was not continuous, that in itself would not be so disturbing. You are used to that notion. For example, you know that water, which you perceive as continuous, is actually made of water molecules. Many things that you think of as continuous are made up of little molecules. That's not the surprise. The surprise is that *these photons are not and cannot be your standard classical particles of the type that appear in Newtonian or Einsteinian mechanics, following continuous trajectories decided by the applied forces.* The reason behind this conclusion is the interference pattern I_{1+2}. Let us understand why.

Suppose the photon were a *classical particle,* by which I mean governed by the laws of Newton or Einstein. What do we expect it to do in the double-slit experiment? Look at Figure 19.2.

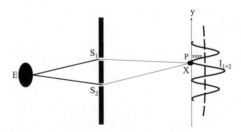

Figure 19.2   Two paths a photon can take, via slits 1 or 2. I try to show that 4 photons come to X, the zero of the interference pattern I_{1+2}, via *one or the other* slit, by drawing 4 x's at a nearby point P. When both are open I get zero at X instead of 8.

Say only slit S_2 is open. The photons will take some path going through the slit S_2 on their way to the pixel on the film. A similar result applies when only S_1 is open. What should happen when both are open? The answer is that the number arriving at any point has to be the sum of the numbers that came through each slit. Classical particles on the trajectory headed for one slit have no idea whether the second slit is open or closed or that it even exists. So the number arriving at some point with two slits open must be the sum of the numbers with either one open. In other words $I_{1+2} = I_1 + I_2$ is a *logical necessity* for classical particles.

Consider in particular the point X, which is a zero of I_{1+2}. Say on a given day 4 photons come to X with just S_1 open and 4 photons come with just S_2 open. (I have shown these 4 photons by x's at a nearby point P.) We expect 8 to arrive with both open but we know that no photons will arrive at X. How can you cancel a positive number of particles coming through one slit with more positive number of particles coming from a second slit? It is impossible to understand this in terms of classical particles. They cannot know how many slits are open and they cannot produce a pattern *that depends on the separation between the two slits.* The fact that this happens is proof that photons are not classical particles.

By contrast, a wave has no trouble knowing how many slits are open and how far apart they are, because it is not localized. The wave comes and hits both the slits simultaneously and knows their spacing. There is room for cancellations when two slits are open due to destructive interference. So maybe we should go back to the view that light is just a wave, as Young convinced us? But that too is no longer an option in view of what we just learned: a wave cannot deposit energy and momentum on just one pixel.

So the photon has particle-like and wave-like attributes. We may summarize the data as follows:

- Light of wave number k and frequency $\omega = kc$ is made up of particles (photons) each of which carries the same energy $E = \hbar\omega$ and the same momentum $p = \hbar k$. The energy and momentum are localized in these particles.
- The distribution of a large number of photons in the double-slit experiment is given by the interference pattern produced by a wave of that k or λ.

There is no point in asking if light is a particle or a wave. These words are inadequate to describe light. It is what it is as described above.

If we send in a million photons, one at a time, each will land at a definite pixel of the film and they will collectively produce the interference pattern.

Suppose a million photons have formed an interference pattern of bright and dark lines on the film, whose form we can predict from a simple interference calculation with waves of this λ. Now we send in the $1,000,001$-th photon. Where will it go?

We do not know for sure. We only know that if we repeat the experiment a million times, we get this pattern. We cannot anticipate the outcome of a single trial with just one photon. We just know that the odds are high where the function I_{1+2} is large, and the odds are low where the function is small, and the odds are zero where the function is zero. *So the role of the wave is to determine, via its intensity, the probability $P(\mathbf{r})$ that the photon, a particle with localized energy and momentum, will be absorbed by a pixel at* $\mathbf{r}$. The probability is computed by adding the waves from the two slits and then squaring.

19.3   Digression on photons

I want to digress briefly to clarify a historical fact: photons were not really found by looking at the pixels of a photographic plate. They were first predicted by Einstein based on fairly complicated thermodynamic arguments. He showed that radiation of frequency ω behaved *as if* it were made of particles, each of energy $E = \hbar\omega$. Einstein dropped the characterization "as if" and argued for the actual existence of these particles. He showed in 1905 that he could explain the *photoelectric effect* very easily in terms of these photons. We will see how in just a moment. Later, in 1927, very direct evidence of photons was provided by Arthur Compton (1892–1962), who showed that the scattering of light of wave number k and frequency ω by an electron could be described simply as a relativistic elastic collision between the electron and a massless particle, the photon, with energy $E = \hbar\omega$ and momentum $p = \hbar k$. Einstein got the Nobel Prize for his work on the photon, rather than for either theory of relativity.

19.3.1   Photoelectric effect

Now for the first experiment that is explained by photons, the *photoelectric effect*. Recall that in a metal some electrons are communal. Say each atom donates one electron to the whole metal. They can run all over the metal.

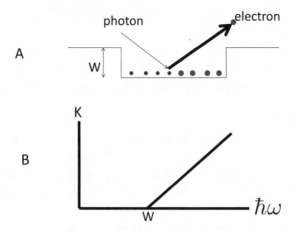

Figure 19.3   A: Electrons in a metal are in a well of depth W. A photon is able to liberate them if it has enough energy. B: The plot of the kinetic energy of the ejected electron versus photon energy $\hbar\omega$ (for $\hbar\omega \geq W$).

They don't have to be near their parent nuclei. But they cannot leave the metal. They are trapped in an electrostatic well, as shown in Figure 19.3A. It costs a minimum energy W, called the *work function*, to get them out with zero kinetic energy. (Imagine a well of depths h, at the bottom of which are objects of mass m. To pull them out [at rest] you need to supply a minimum energy $W = mgh$. If you give more than the minimum, they will come out with some kinetic energy.)

There is a natural way to furnish this energy. Since the electron has an electric charge you can apply an electric field to act on it and rip it out after doing the requisite work. Since light is nothing but electric and magnetic fields, you can try shining light at the metal. The electric field should grab the electron and shake it loose. And once it escapes, it can take off.

You do this and find nothing comes out. Since the force eE on the electron grows with intensity $I \propto |E|^2$, you crank up the intensity of light and still nothing happens. Then you discover that if you increase the *frequency* of light, suddenly electrons start coming out. They come out even if the light at this increased frequency is very feeble. A feeble source of light leads to fewer electrons coming out, but they *do* come out now. You measure K, the kinetic energy of the emergent electrons, and plot K versus

ω and find the graph in Figure 19.3B. The graph is simplicity itself:

for $\hbar\omega < W$ no emitted electrons

for $\hbar\omega > W$ electrons emitted with $K = \hbar\omega - W$.   (19.11)

This graph makes no sense within Maxwell theory. How can feeble light (with a tiny field $\mathbf{E}$) of high frequency liberate electrons while strong light (with a large $\mathbf{E}$) at low frequencies cannot? But it makes perfect sense in terms of photons. The low-frequency beam consists of a large number of photons, each of which carries less energy than it takes to liberate the electrons. It is like sending a large number of toddlers (working independently) to lift a suitcase. They just cannot do it. On the other hand, even a single adult can. This is analogous to what happens when a feeble high-frequency light composed of high-energy photons is used.

The graph is readily understood as follows. If $\hbar\omega < W$, no electrons come out. If $\hbar\omega > W$, out of the photon's energy $\hbar\omega$, a share W goes to pull the electron out of the well of depth W, and the rest, $\hbar\omega - W = K$, goes to the kinetic energy of the liberated electron. By 1905 it was known that the energy of photoelectrons increases with increasing frequency of incident light and is independent of the intensity of the light. However, the precise manner of the increase was not experimentally determined until 1914, when Robert Millikan (1868–1953) showed that Einstein's prediction was correct.

19.3.2   Compton effect

Now for Compton's 1927 experiment, which provided very direct evidence of photons. Imagine shining X-rays, i.e., light of some λ (or wave number $k = 2\pi/\lambda$) along the x-axis on a free and static electron, as shown in the left half of Figure 19.4.

(In reality the electron is bound to an atom. However, the incident X-ray photons have so much energy that the initial electron may be treated as free and at rest.) The electron scatters the light into some direction and recoils in some other direction, as shown in the right half of the figure. Forget the electron and just observe the scattered light. The light scattered in a direction θ relative to the x-axis is found to have a wavelength λ'

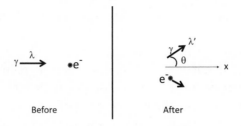

Before | After

Figure 19.4   Left: Light (γ) of wavelength λ or wave number $k = 2\pi/\lambda$ (photons of momentum $\hbar k$ and energy $\hbar\omega$) is incident in the x-direction on an electron e^- at rest. Right: The light (photon) comes out at a direction θ with a shifted wavelength λ', and the electron recoils conserving energy and momentum.

obeying

$$\lambda' - \lambda = \left(\frac{2\pi\hbar}{mc}\right)(1 - \cos\theta) \equiv \lambda_C(1 - \cos\theta) \qquad (19.12)$$

where

$$\lambda_C = \frac{2\pi\hbar}{mc} \qquad (19.13)$$

is called the *Compton wavelength* of the particle of mass m. This result can be derived very simply if you do the following:

1. Treat the incoming light as made of photons of energy $E = \hbar\omega$ and momentum $\hbar k$. The four-momentum of the photon is $K = (\hbar\omega, \hbar k)$. The initial electron has a four-momentum $P_e = (mc, 0)$.
2. Assume energy and momentum are conserved in the collision and solve for the final photon four-momentum $K' = (\hbar\omega', \hbar k')$. This was done in Volume I and I will not repeat it here. If you translate the final k' to λ' you get Eqn. 19.12.

Notice how we go back and forth between waves and particles. Light is characterized by a wavelength and by the corresponding photon momentum and by the frequency and the corresponding photon energy. When you think about the particles, you think of the energy and momentum. When you think about the waves, you think of frequency and wave number. After Compton's experiment one could not doubt the reality of the photons.

You may have heard that Einstein was very unhappy with quantum mechanics and did not join the chorus. There is even an impression that he had become just another conservative in his old age. This is utterly false. If you look at the history, you will find he made enormous contributions to quantum mechanics from the very outset. Even Planck was equivocal about the reality of the photons that were implied by his own formula. Einstein took their existence seriously and applied it to the photoelectric effect. He computed the specific heat of solids using oscillators of quantized energy to represent lattice vibrations. Schrödinger acknowledges his debt to Einstein for his wave equation. So when you hear that Einstein didn't like quantum mechanics, do not think he couldn't do the problem sets. It's that he had problems with the problem sets. He did not like the probabilistic nature of quantum mechanics, but he had no trouble divining what was going on. Indeed he himself ushered in probabilities in his treatment of induced radiation. If someone says "I don't like that joke" there can be two reasons: he or she they didn't get the joke or got it but didn't think it was funny. It was the latter for Einstein and quantum mechanics. He certainly understood all the complexities of quantum mechanics. He has said he had spent far more time wrestling with quantum mechanics than either the special or the general theory of relativity. It is true that till the end he didn't find a formalism that satisfied him. The formalism I'm giving you certainly works in the sense that its every prediction has been correct. Until something better comes to replace it, we will keep using it.

19.4   Matter waves

Now came the French physicist, Louis de Broglie (1892–1987), who argued as follows in his PhD thesis. If light, which we thought was a wave, is actually made up of particles, perhaps things that we always thought of as particles, like electrons, must have a wave associated with them, with a wavelength related to their momentum as follows

$$\lambda = \frac{2\pi\hbar}{p}. \tag{19.14}$$

If this is right, we should see interference in the double-slit experiment with electrons.

Equation 19.14 is of course the same as

$$p = \frac{2\pi \hbar}{\lambda} \qquad (19.15)$$

for photons with one conceptual difference. For light, λ is a natural quantity and the photon and its momentum p are the surprises, while for electrons the momentum p is a natural quantity and the associated wavelength λ is the surprise.

In the case of the electron or other massive particles like protons or neutrons, $\lambda = \frac{2\pi \hbar}{p}$ is called the *de Broglie wavelength*. The double-slit experiment for electrons aimed at testing de Broglie's hypothesis is designed in pretty much the same way as for photons, but with some obvious and inevitable differences. First, the source of electrons is different—it could be some electrode that boils off electrons with negligible kinetic energy K. These are then accelerated to some fixed momentum p by allowing them to fall through a potential V such that

$$K = \frac{p^2}{2m} = eV. \qquad (19.16)$$

Notice that I use the non-relativistic expression for the kinetic energy of the particle. This will be the case except for photons, which always travel at c and obey $E = pc$.

A velocity filter may be used to ensure that all electrons reaching the slits have the same p and hence the same de Broglie wavelength. All this was simply accomplished in the case of light by a monochromatic source.

To detect electrons, you replace the photographic film with a row of electron detectors or a single detector that can slide along the right edge as in in Figure 19.5. These detectors can amplify a single electron that hits them into an avalanche that leads to a macroscopic current. You generate a histogram of events triggered by the detected electron as shown by x's in Figure 19.5. After several hits, the histogram of the number of electrons arriving in some fixed time will form the pattern I_1 with just S_1 open, I_2 with just S_2 open, and I_{1+2} with both slits open. The period of the

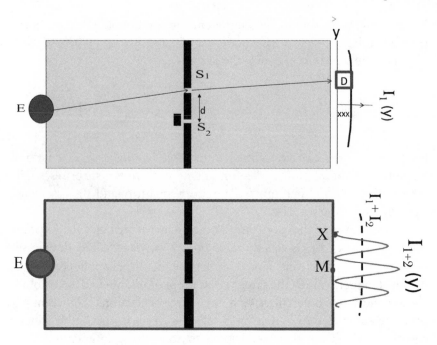

Figure 19.5   Top: Electrons go from emitter E to a sliding detector D with just slit S_1 open. The figure shows a possible classical trajectory connecting the two end points. The histogram $I_1(y)$ is generated by recording the arrivals (shown by x's) over a fixed time period. Not shown is a similar pattern with S_2 open. Bottom: As with photons, an interference pattern $I_{1+2} \neq I_1 + I_2$ is seen with both slits open. In particular, no electrons come to a point like X with both slits open, though they did come with one or the other open.

oscillations will be determined by the de Broglie wavelength $\lambda = 2\pi \hbar/p$ of the electrons, confirming its wave-like nature.

Now the surprise is not that the electron hits only one detector, depositing all its charge, energy, and momentum there. It is supposed to do that; it is after all a particle with localized attributes. What is surprising is that when two slits are open, you get the interference pattern. At a place like M you get four times as many electrons as with one slit open, and not double. Even more dramatic are locations like X where you don't get any electrons with both slits open, whereas you used to get some with just one slit open.

This is the end of Newtonian mechanics for particles like electrons. If an electron were a Newtonian particle it would go from the emitter E to the detector D via one or the other slit. Opening a second slit would have no effect on the number going through the first. A particle is aware only of the space right next to it and cannot sense or respond in any way to another slit far from the one it is headed for. *The number coming with two slits open had to be the sum of the numbers coming in with either one open.* You cannot explain points like X where no electrons come with both slits open, while some did with just one.

Now some people may say, "Well, if you have a lot of electrons coming in, maybe these guys coming out of S_1 bumped into those guys coming out of S_2 so that the final intensity was not $I_1 + I_2$. Nobody came to X because these collisions diverted the electrons headed for X to some other direction." This is wrong on many counts.

First, it is very unlikely that random collisions of this type can produce a nice and repeatable interference pattern correlated with the incoming electron momentum p and the slit separation d.

Second, you can lay the notion of a classical electron to rest by making the electron source so feeble that, at a given time, there's only one electron in the experimental region. We know when it left the emitter E and when it came to the detector D. It cannot collide with itself. *And yet it knows two slits are open* because after many runs, the interference pattern emerges. A Newtonian particle cannot know that two slits are open. So the electron must have an associated wave that knows how many slits are open, knows what their spacing is, and can interfere with itself.

For completeness let me mention that the de Broglie hypothesis was originally confirmed not with the double-slit but by diffraction off a nickel crystal in 1927 by Clinton Davisson (1881–1958) and Lester Germer (1896–1971). If you shine a beam of mono-energetic electrons (which have been accelerated to a fixed momentum and are hence associated with a definite de Broglie wavelength λ) at a crystal, you find that the electrons scatter only for incident angles θ relative to the plane of atoms that satisfy the Bragg condition $2d\sin\theta = n\lambda$ where d is the spacing between layers of atoms and n is an integer. This experiment had been presaged by Walter Elasser in the early 1920s and finally succeeded after some serendipitous incidents and accidents.

To summarize, light, which we thought was a wave, is made of particles, and electrons, which we thought were particles, are guided by waves. Everything exhibits *wave-particle duality* in the microscopic world.

19.5   Photons versus electrons

In the double-slit experiment described so far, photons have behaved very much like the electron (which is a stand-in for all other particles like protons, neutrons, pions, etc.). Let me remind you of the similarities.

1. Both exhibit wave-like interference: the function $I_{1+2}(y) \equiv I(y)$ oscillates on the line of detection parameterized by a coordinate y.
2. The λ of the underlying wave may be deduced from the spacing between maxima and minima, the slit separation d, and distance L to detectors or film *without knowing what the wave actually describes*. For photons λ would be just the wavelength of the incident electromagnetic wave, while for electrons it would be the the de Broglie wavelength.
3. In both cases $I(y)$ gives the likelihood of a photon or electron triggering a pixel or detector at the point y.

But as we go forward and develop the quantum theory, the photon ends up being treated very differently from the electron. This is the case because the photon *is* different.

First of all, the photon can never be at rest. Being massless, it has to travel at c. By contrast an electron can be brought to rest and there is a regime where non-relativistic kinematics applies. Next, the number of electrons is conserved (in the non-relativistic regime): an electron never appears out of nowhere, nor does it disappear into nothing. By contrast, the number of photons can change, and even does so during this experiment, increasing by one during emission by the source and decreasing by one during absorption by the pixel.

This leads to a different interpretation of the intensity $I(y)$ and of the underlying wave that produces it.

1. The wave underlying the photon is just the electromagnetic wave, described by $\mathbf{E}$ and $\mathbf{B}$. The intensity is $I(y) \propto |\mathbf{E}(y)|^2 + |\mathbf{B}(y)|^2$, dropping constants like ε_0, μ_0, and c. (Go back and consult Eqn. 14.87 and the ones leading to it.) In the case of electrons, the underlying wave, called the *wave function* $\psi(y)$, does not correspond to any classical field. It is an entity we are forced to introduce to explain the double-slit experiment. All we know is that in the experiment with mono-energetic electrons, it is attributed a wavelength $\lambda = 2\pi\hbar/p$. The intensity is taken to be $I(y) = |\psi(y)|^2$, and not $I(y) = \psi^2(y)$, just in case ψ is complex.

2. In the case of the electrons $I(y) = |\psi(y)|^2$ encodes the *probability $P(y)$ of finding the electron at y*. Hence we write

$$P(y) = |\psi(y)|^2. \tag{19.17}$$

This relation between $|\psi(y)|^2$ and the probability of finding an electron at y was proposed by Max Born (1882–1970); it is one of the pillars of quantum mechanics. If the experiment is repeated many times, $P(y) = |\psi(y)|^2$ will be proportional to the density of electrons found at y.

In the case of photons we *do not* identify $I(y) \propto |\mathbf{E}(y)|^2 + |\mathbf{B}(y)|^2$ as the probability of *finding* a photon at y. Instead we identify it with the *probability of its being absorbed by an atom or pixel at y*. What is the big difference between the photon being absorbed at y and the photon being found at y? The answer is that the absorption of the photon has a very precise location, namely of the pixel that changed color or the atom that absorbed it. This is not so for the location of the photon, because there is no trace of the photon after detection. If $I(y)$ is the probability of *finding* the photon at y, then where is it? It is gone after detection.

By contrast, the detected electron is actually there, rattling around inside the detector as a distinct entity carrying charge $-e$ and mass m. So we *can* meaningfully say $I(y)$ is proportional to the probability of an electron being *found* at y, of it actually being at y upon detection.

There is a fundamental problem with assigning any probability function $P(y)$ for a photon being at y. Consider a macroscopic electromagnetic field. Its energy density is proportional to the product of $P(y)$ (the probability the photon is at y) and the energy $\hbar\omega$ of *each* photon:

$$|\mathbf{E}(y)|^2 + |\mathbf{B}(y)|^2 \propto \hbar\omega(y)P(y), \text{which means} \tag{19.18}$$

$$P(y) \propto \frac{|\mathbf{E}(y)|^2 + |\mathbf{B}(y)|^2}{\omega(y)} \tag{19.19}$$

$$\propto (|\mathbf{E}(y)|^2 + |\mathbf{B}(y)|^2) \cdot \lambda(y), \tag{19.20}$$

dropping all constants. But there is no meaning to $\lambda(y)$, the "wavelength at y." It appears that $P(y)$, the proposed probability of finding the photon at y, depends not just on the values of $\mathbf{E}(y)$ and $\mathbf{B}(y)$ *at y*, but also on the non-local quantity, the wavelength. This can only be inferred from the values of fields over a distance comparable to the wavelength, which need not be small. So the only reasonable candidate for $P(y)$, the probability of finding a photon at y, is a non-starter.

The bottom line is that unlike electrons, photons do not have an associated wave function $\psi(y)$ from which we can obtain $P(y) = |\psi(y)|^2$ following Born.

The rest of this book will deal only with the quantum mechanics of massive particles like electrons, for which we can define a wave function $\psi(y)$ and for which $P(y) = |\psi(y)|^2$ is the probability of finding them at y.

I will also limit myself to non-relativistic quantum mechanics, which means the (kinetic) energy and momentum of the particles are related by the approximate formula

$$E = mc^2 + \frac{p^2}{2m} \quad \text{or} \tag{19.21}$$

$$K = E - mc^2 = \frac{p^2}{2m} \tag{19.22}$$

and not the exact one $E = \sqrt{m^2 c^4 + c^2 p^2}$. The photon will enter here and there and affect the dynamics of the electron, as in Compton scattering or the emission and absorption of light by atoms. It can have an energy and momentum but not its own wave function $\psi(y)$.

19.6 The Heisenberg uncertainty principle

The fact that particles are described by waves that control the probability of their being somewhere and that a particle in a state of momentum p has an associated de Broglie wave of wavelength

$$\lambda = \frac{2\pi\hbar}{p} \tag{19.23}$$

implies the celebrated Heisenberg uncertainty principle.

19.6.1   There are no states of well-defined position and momentum

There are many ways to state the principle and let us begin with one:

It is impossible to prepare a particle in a state in which its momentum and position (along one axis) are exactly known. The product of the uncertainties Δx and Δp is required to obey

$$\Delta x \Delta p \geq \frac{\hbar}{2}. \tag{19.24}$$

This formula, as written, is applicable only if Δx and Δp conform to the precise definition of uncertainties in quantum theory. Postponing this definition for later, we will instead identify in each context a reasonable measure of what we could call the uncertainties in position and momentum. The products of these heuristic uncertainties naturally need not be bounded below by $\hbar/2$. However, they will always be of the same order:

$$\Delta x \Delta p \gtrsim \hbar \tag{19.25}$$

where factors like 2 or π are not guaranteed to match on both sides, and it is understood that the Δp that multiplies Δx is Δp_x. The main point is that $\hbar \simeq 10^{-34} \, J \cdot s$ sets the overall scale for these quantum effects and a factor of π here and there does not change this. (The only 2π I will rigidly retain is in de Broglie's formula $\lambda = 2\pi \hbar/p$, where all the quantities are precisely defined.)

Let us now try (in vain) to produce a state of well-defined position and momentum. There is no problem doing this in classical mechanics: we let a particle roll down a slope till its momentum reaches a value $\mathbf{p}_0$ at some point $\mathbf{r}_0$ and label the state by the pair $(\mathbf{r}_0, \mathbf{p}_0)$.

Let us try something similar in the quantum case for motion along the y-axis. We first accelerate the electron by letting it gain kinetic energy

$$\frac{p_0^2}{2m} = eV \tag{19.26}$$

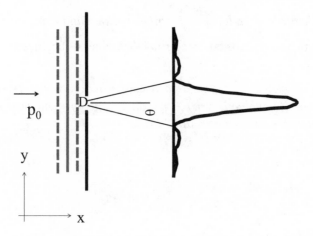

Figure 19.6   In an attempt to localize the electron's position and momentum in the y-direction, we send it along the x-axis through a slit of width D in the y-direction. The emergent electron has positional uncertainty $\Delta y \simeq D$ and a y-momentum that has angular spread at least as great as that of the first diffraction peak $2\theta \simeq 2\lambda/D$.

using an accelerator of voltage V. We fire it in the x-direction toward a slit of width D in the y-direction as shown in Figure 19.6. An electron just emerging on the other side has a position right in front of the slit. We may reasonably take as the uncertainty in its y-position

$$\Delta y \simeq D. \tag{19.27}$$

A different definition may change this by a factor of order unity, which is why we use the $\simeq$ symbol. We can make Δy as small as we want by reducing D.

What is its y-momentum? It came in with momentum p_0 in the x-direction and nothing in the y-direction. Classically it would have the same y-momentum (zero) just after crossing the slit. Since this momentum is known exactly, it looks like $\Delta p_y = 0$ and $\Delta y \Delta p_y = 0$. However, this is not so in the quantum theory. The incoming electron has an associated wave with

$$\lambda = \frac{2\pi\hbar}{p_0}, \tag{19.28}$$

and when such a wave hits a slit, it diverges on the other side due to diffraction. We have seen that the wave has a significant amplitude not just in the forward direction, but up to the first zero, which occurs at an angle

$$D\sin\theta \simeq D\cdot\theta = \lambda \quad \text{or } \theta \simeq \frac{\lambda}{D}. \tag{19.29}$$

The opening angle of this diffraction cone is

$$2\theta \simeq 2\frac{\lambda}{D} = \frac{4\pi\hbar}{p_0\,D}. \tag{19.30}$$

A particle capable of landing anywhere in the central maximum must be endowed with the requisite y-momentum that will take it there from the slit. Although the wave is significant only within this central maximum, it is not strictly zero outside. So the y-momentum also has a probability distribution of angular width no smaller than 2θ, which translates into

$$\Delta p_y \gtrsim p_0(2\theta) = \frac{4\pi\hbar}{D}. \tag{19.31}$$

Cross multiplying by D, which is just Δy, we arrive at

$$\Delta y\cdot\Delta p_y \gtrsim 4\pi\hbar. \tag{19.32}$$

The $\hbar$ on the right-hand side is solid, but the 4π is not, since it can be easily changed by a slight and reasonable redefinition of Δp_y and Δy. (For example, we could say Δp_y is larger because the diffraction pattern is not strictly zero outside the central peak.) This is why we drop the numerical factors of order unity and write

$$\Delta y\cdot\Delta p_y \gtrsim \hbar. \tag{19.33}$$

I emphasize: it is not that *we do not know* the p_y of the emergent electron; *it does not have a definite p_y* when it emerges from the slit because there is a non-zero probability of getting any answer in the diffraction peak. A particle whose momentum measurement has a probability of giving a range of answers cannot be said to have a definite momentum.

Do not be fooled by the fact that at various times we may know various things that seem to contradict the uncertainty principle. At the outset we knew the momentum exactly: it was p_0 in the x-direction and 0 in the y-direction. We had no idea where it was in the y-direction. Just after it crossed the slit, we knew its y-location to within an uncertainty $\Delta y \simeq D$. But in this state it had an indefinite y-momentum, with a non-zero probability for pointing anywhere in the central diffraction peak. Later, when that electron hit a particular detector, we could work backward to find out *what momentum it must have had* to arrive at this location starting from the slit. This retroactive knowledge that we obtain only *after* it hits the detector does not describe a property of the electron when it emerged from the slit.

It is an inescapable property of waves that you cannot confine them spatially with a slit without forcing them to fan out. The result $\theta \simeq \frac{\lambda}{D}$ was known well before quantum mechanics in the study of single-slit diffraction. The new input from quantum theory is that λ now describes a particle of momentum $p = \frac{2\pi \hbar}{\lambda}$ and the fanning out translates into an uncertainty in y-momentum.

The role of probability here is very different from classical mechanics. Suppose I sprayed a stream of classical particles at an opening and they came out in a range of angles on the other side and hit a screen. Here too I can give the odds that particles will arrive at some point on the screen. But this use of probability is a practical strategy and not mandated by fundamental principles of classical mechanics, which in fact allow us to predict where each and every particle would land. On a given trial each particle that was fired had to go to one particular spot on the screen. In the quantum case we are talking about *just one electron*, not a beam. That *single* electron is capable of arriving at a range of points on the screen, each with some probability. This is the sense in which it does not have a definite momentum when it leaves the slit. In the case of the classical particle, its measured momentum might have been given by a probability distribution, but it *had* a definite momentum, We just did not know it. In the quantum case the electron coming out of the slit *did not have* a definite momentum. Assuming it must have had a definite momentum is like assuming it must have gone through one particular slit in the double-slit experiment. This idea will be discussed more as we go along, so do not worry if you cannot digest it right now.

19.6.2 Heisenberg microscope

We have seen above that given the underlying wave, a state of well-defined position and momentum simply does not exist. In trying to prepare an electron with a narrow range of positions, we ended up giving it a spread in its momentum. This was understood using wave theory, in which diffraction of a wave is very natural.

We would like to understand this in the particle picture. Let us say an electron is in a state of definite momentum p and we want to locate it. If we could do this without altering its momentum in any way, we would have a state of definite position and momentum. This is forbidden by the following version of the uncertainty principle:

The act of locating the position of an electron to within Δx will introduce an uncertainty in its momentum Δp satisfying

$$\Delta x \Delta p \gtrsim \hbar. \tag{19.34}$$

Here we turn to a simple experiment whose sole aim is to find the position of the electron. The electron lives on the x-axis as shown in Figure 19.7. We hope to locate it by shining light along the x-axis and observing it from above with a microscope of aperture D, capable of sliding along x.

To proceed, we first need to derive an expression for the *resolving power of a microscope*, which is its ability to distinguish nearby objects. Look at Figure 19.7A. Consider two point-like objects on the x axis a distance Δx apart and at a distance f in front of the aperture (where f is typically but not necessarily the focal length of the lens used).

In geometric optics the rays through the center of a lens go straight and form two sharp images separated by an angle 2α where, for small α,

$$\tan \alpha \simeq \alpha \simeq \frac{\Delta x}{2f}. \tag{19.35}$$

In wave optics, the images inside the microscope are not point-like but spread over the angle $\pm\theta$ where

$$D \sin \theta \simeq \lambda \tag{19.36}$$

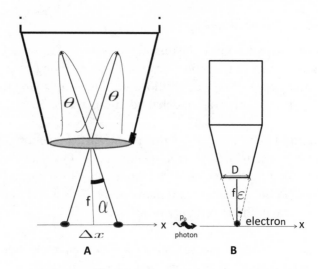

Figure 19.7   The Heisenberg microscope. Left: Two points a distance Δx apart form two images of width θ (due to diffraction) inside the microscope. For them to be distinguished we require $\alpha > \theta$. Right: Light from the left illuminates the electron and enters the microscope within a cone of angle $\varepsilon \simeq D/2f$, where D is the aperture. This makes the final photon momentum and hence the final electron momentum in the x-direction uncertain.

due to diffraction through the aperture. For θ small, this becomes

$$\theta \simeq \frac{\lambda}{D}. \tag{19.37}$$

For the objects to be clearly resolved into two distinct entities we need the peak separation to exceed the peak width:

$$\alpha > \theta \tag{19.38}$$

or

$$\frac{\Delta x}{2f} > \frac{\lambda}{D} \quad \text{or} \tag{19.39}$$

$$\Delta x > \frac{2f\lambda}{D}. \tag{19.40}$$

Now, according to the right half of Figure 19.7, the scattered photon can enter the microscope in a cone of opening angle 2ε given by

$$\tan \varepsilon \simeq \varepsilon = \frac{D}{2f}. \tag{19.41}$$

Thus we arrive at the resolving power of the microscope in terms of λ and ε:

$$\Delta x \gtrsim \frac{\lambda}{\varepsilon}, \tag{19.42}$$

a well-known result in classical optics. (A more accurate one is $\Delta x \gtrsim \frac{\lambda}{\sin \varepsilon}$.) There is no lower limit on just Δx: at fixed ε we can reduce it arbitrarily by reducing λ. Since two point-particles cannot be distinguished if they come closer than Δx, we may rightly call Δx the uncertainty in their location.

How about the electron's momentum? Look at Figure 19.7B. Assume the electron had a well-defined momentum before the position measurement. (The uncertainty principle does not forbid *one* variable, in this case p, from having a well-defined value.) The photon comes in with momentum

$$p_0 = \frac{2\pi \hbar}{\lambda} \tag{19.43}$$

in the x-direction, scatters off the electron, and enters the microscope (assumed to be with the same *magnitude* of momentum). It can enter it anywhere in the cone of half-angle ε. So its final x-momentum has an uncertainty of order

$$\Delta p_x = p_0 (2\varepsilon) \quad \text{for small } \varepsilon. \tag{19.44}$$

This uncertainty in the photon momentum translates into the same uncertainty in the electron's final momentum by the conservation of momentum.

In summary, we have used the microscope to produce an electron in a state with uncertainties Δx and Δp_x obeying

$$\Delta x \cdot \Delta p_x \gtrsim \frac{\lambda}{\varepsilon} \cdot p_0 2\varepsilon = 2\lambda p_0 = 4\pi \hbar. \tag{19.45}$$

With the approximate Δp_x and Δx, and with factors of order unity ignored, we write

$$\Delta p_x \Delta x \gtrsim \hbar. \tag{19.46}$$

Since it takes at least one photon to detect the electron, the uncertainty product can only get bigger if more photons are involved.

Here is a point worth repeating: it's not the fact that the photon came in with a large momentum or that it transferred a large momentum to the electron that causes the uncertainty in the final electron momentum; it is the fact that the photon went into the microscope with an *uncertainty* in its angle. This uncertainty in the angle turns into the uncertainty in the x-component of the photon's final momentum and hence the electron's final momentum. (Remember, when we say the photon or electron has an uncertainty in its momentum, we are not speaking of our ignorance; we are saying *it does not have* a definite momentum.) Once again the argument requires us to pass deftly between the wave and particle pictures. This sleight of hand can be avoided once the full theory is mastered. Then it will be possible to define Δx and Δp precisely and derive a precise lower bound for the uncertainty product

$$\Delta x \Delta p \geq \frac{\hbar}{2}. \tag{19.47}$$

19.7 Let there be light

Let us take stock of the double-slit experiment. Consider just one electron. We know it was emitted when the emitter recoiled. We also know it was subsequently detected by a detector. These are undeniable facts. Even quantum mechanics cannot change them. But what happened between these two observations? We cannot say, because we did not see the electron in between. *It seems reasonable to assume that it followed a specific trajectory that went from the emitter E to the detector D via one or the other slit.* We may not know which path it took, but surely it *must* have taken one of the two paths. This reasonable assumption is flatly contradicted by the interference pattern. If in each case the electrons followed a definite trajectory passing through one or the other slit, they cannot be aware of the other slit, and $I_{1+2} = I_1 + I_2$ would be an inevitable consequence.

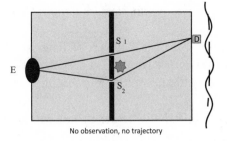

No observation, no trajectory

Figure 19.8   When a lightbulb is placed near the slits to see which slit the electron took, we find that the ones that were observed showed no interference while the ones that escaped detection produce interference oscillations on top of the featureless $I_1 + I_2$ (dotted line).

Suppose we do not buy this notion that an electron does not go through a particular slit. We place a glowing lightbulb right after the two slits as shown in Figure 19.8. Whenever an electron makes it past the slits, we will *see* for ourselves which slit it went through. Then there can be no talk about not going through a definite slit or not having a definite trajectory. Every electron that registered a click at a detector is then classified as having passed through S_1 or S_2, as having followed a definite trajectory. By sheer logic we have to add the numbers through each slit to get the total number: we must have $I_{1+2} = I_1 + I_2$.

Indeed this is what will happen if every electron that was picked up by the detector was also seen on its way to the detector. But once in a while some electrons may make it to the detector without being observed near the slits. So in addition to electrons labeled as coming via S_1 or S_2, there is a third species of electrons: those which were not observed, which slipped by. *The reasonable assumption that they too would behave like the others we saw is wrong.* They profoundly alter the distribution. Let us say that of the electrons that triggered the detectors, 10% escaped undetected by the lightbulb. They are the ones to which we cannot ascribe a particular slit, a particular trajectory. We now find that the distribution I_{1+2} looks like $I_1 + I_2$ plus a $\simeq 10\%$ wiggle. In other words, the numbers of electrons that we caught and identified as going through slit 1 or slit 2 add up the way they do in Newtonian mechanics, but the electrons that slipped by without detection, which we cannot associate with a specific slit or a specific trajectory, show the interference pattern. *They must know about both slits to produce an interference pattern that depends on the slit separation d.*

Think about this: the ones that were seen near either slit act as if they followed a definite trajectory (through a specific slit) while the ones that slipped by act as if they did not follow any specific trajectory, because they knew about both slits.

It is very surprising that whether or not we see the electron makes such a difference. When we study an object in Newtonian mechanics, we don't care if the object is observed or not at every stage. We shoot two billiard balls at each other and predict the outcome given the initial data. As they collide, we may be watching them or may not be watching them. The outcome is independent of our watching. We believe in an objective reality described by natural laws; our observing it at intermediate stages is incidental and does not influence the outcome. *Newtonian mechanics allows for an ideal observer who can observe without affecting the outcome.*

Why then does observation make such a difference to the electrons? To answer this let us ask how we observe the electron to see which slit it passed through. Look at Figure 19.9. The slits are a distance d apart in the y-direction. We need to obtain an image that can resolve distances of order d. We have seen (Eqn. 19.45) that to locate a particle to a precision Δx, we need to employ photons that will necessarily transfer an *indefinite* amount

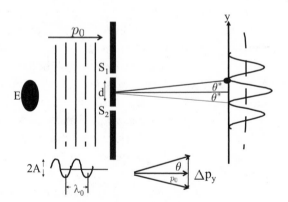

Figure 19.9   The photon used to determine which slit the electron took produces an uncertainty in its y-momentum of order $\Delta p_y \gtrsim \frac{4\pi\hbar}{d}$ and in its direction of order $\theta = \Delta p_y/p_0 = \frac{2\lambda_0}{d}$, which is of the same order as the angular separation $2\theta^*$ between two successive minima.

of momentum Δp_x given by

$$\Delta p_x \gtrsim \frac{4\pi\hbar}{\Delta x}. \tag{19.48}$$

Although I will keep the factor of 4π in the following discussion, only the $\hbar$ matters.

Let us apply that formula here with some obvious modifications. We want to know which slit the electron took. So we want to be able to tell an electron at slit 1 from an electron at slit 2. The slits are a distance d apart *in the y-direction*. So we swap x for y in Eqn. 19.48 and let d, the distance between the slits, play the role of Δx. We deduce that determining which slit the electron took will introduce the following uncertainty in its y-momentum:

$$\Delta p_y \gtrsim \frac{4\pi\hbar}{d}. \tag{19.49}$$

Now, the incoming electron had a momentum p_0 in the horizontal (x) direction and a wavelength $\lambda_0 = 2\pi\hbar/p_0$. The uncertainty Δp_y introduced by the position measurement will cause an angular uncertainty in p_y of size

$$2\theta = \frac{\Delta p_y}{p_0} \tag{19.50}$$

$$= \left(\frac{4\pi\hbar}{d}\right)/p_0 \quad \text{using Eqn. 19.49} \tag{19.51}$$

$$= \frac{4\pi\hbar}{d \cdot p_0} \tag{19.52}$$

$$= \frac{2\lambda_0}{d}. \tag{19.53}$$

On the other hand the angular spacing $2\theta^*$ between successive maxima and minima (Figure 19.9) is deduced from

$$d\sin\theta^* \simeq d \cdot \theta^* = \frac{\lambda_0}{2} \tag{19.54}$$

to be

$$2\theta^* = \frac{\lambda_0}{d}. \tag{19.55}$$

The factor of 2 between θ^* and θ (Eqn. 19.55 for the separation between minima and Eqn. 19.53 for the angular uncertainty caused by the position measurement) is not important. What matters is that they are *of the same order*, which is enough to wash out the interference pattern.

The act of observation by photons is dramatic and traumatic for the electron but not for you and me. Right now, I'm getting slammed by millions of photons, but I'm taking it like a man. But for the electron, it is a different story. A single collision with a photon can be like getting hit by a truck. The key is not just the huge momentum of the photon but the fact that the momentum transfer is *unknown* by amount of order $\Delta p_y \gtrsim \frac{4\pi\hbar}{d}$. Dimming the light source will not help; it will just reduce the number of photons and the likelihood of detection but not the punch delivered by the photons that do collide with the electron.

Why do undetected electrons exhibit interference but not macroscopic objects like bullets? Suppose the electron gun is replaced by a machine gun and the opaque barrier by a concrete wall with a hole in it. "They" have tied you to a post on the other side and are firing bullets at the hole from the left. In other words, you are the "detector." You are naturally anxious as you dodge the bullets coming through the hole and now a "friend" offers to help you by making a second hole at a location that ensures destructive interference. You refuse, because in the double-slit experiment with bullets the second hole will not help. Why does something that works at the atomic level fail at the macroscopic level? There are two reasons.

The first has to do with the wavelength $\lambda = 2\pi\hbar/p$. If in the equation $p = mv$ you set $m = 1$ g and $v = 10^3$ m/s, you get a λ of order 10^{-34} m. That means these oscillations at your location, a few meters down the road, will be also of this order, give or take a few powers of 10. For reference, the size of a single proton is about 10^{-15} m and there will be around 10^{19} oscillations over the size of a proton. No macroscopic sensor (like you, tied to the back wall) can detect that. Only the spatial average, which looks like $I_1 + I_2$, will be perceptible. The probabilities for getting shot will be additive over the two slits, and life with two slits open will be roughly twice as much at risk as with just one open.

The second reason interference is hard to see in the macroscopic scale is that macroscopic systems are being constantly (and often unintentionally) observed: by ambient light, by air molecules that bump into them, by cosmic rays, and possibly by dark matter. If you could isolate your system from all these and could detect oscillations of an absurdly small spatial period, you would see interference effects even in macroscopic systems. Starting from the atomic scale, experimentalists have been systematically trying to get bigger and bigger systems to display such interference, suspended in limbo between more than one classical state at a given time.

19.8 The wave function ψ

Let us compare the kinematics of quantum mechanics and classical mechanics. In classical mechanics the state of a particle is specified by its position $\mathbf{r}$ and momentum $\mathbf{p}$.

In quantum theory the particle is described at any one time by the *wave function* $\psi(\mathbf{r})$. Remember $\psi(\mathbf{r})$ describes one particle, not a swarm of particles. Thus we have gone from just two variables $(\mathbf{r},\mathbf{p})$ to a whole function $\psi(\mathbf{r})$. What does the function tell us about the particle? From the probabilistic picture that emerged from the double-slit experiment we have learned that $|\psi(\mathbf{r})|^2$ gives the odds of finding the particle at $\mathbf{r}$. We will assume this interpretation of ψ holds in all situations. An example is depicted in Figure 19.10. It describes a particle in one spatial dimension (described by the coordinate x) unlike the particle in the double-slit experiment, which was moving in two dimensions, with coordinate $\mathbf{r} = (x, y)$.

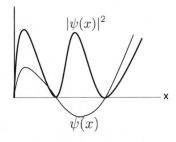

Figure 19.10   An example of a generic wave function $\psi(x)$ of a particle confined to the x-axis and the corresponding probability distribution $|\psi(x)|^2$ over a part of the x-axis.

Here is one more thing we know about wave functions from the interference pattern in the double-slit experiment: *The ψ describing the incoming particle of momentum p is associated with a definite wavelength $\lambda = \frac{2\pi\hbar}{p}$.*

We do not know any more about the actual functional form of ψ. (Remember that in Young's experiment it was possible to extract the wavelength of light from the interference pattern without knowing anything more about the wave, in particular that it represented oscillating **E** and **B** fields.) For example, it could be that $\psi(x) = A\cos\frac{2\pi x}{\lambda} = A\cos\frac{px}{\hbar}$. There is no way to deduce the functional form of ψ from just the double-slit experiment.

Instead it is given by a postulate:

A particle of momentum p in the x-direction is described by a wave function

$$\psi_p(x) = A\exp\left[\frac{ipx}{\hbar}\right]. \tag{19.56}$$

The subscript p reminds us it is a state of momentum p and the constant A in front will remain undetermined for now.

Observe that this is a complex wave function. It obeys the uncertainty principle

$$\Delta x\Delta p \geq \frac{\hbar}{2} \tag{19.57}$$

in the following manner. By definition $\Delta p = 0$ because this is a state of definite momentum p. So we need $\Delta x = \infty$. This is indeed true for this $|\psi_p(x)|^2$, *which is flat (independent of x) and gives no indication of where the particle is*:

$$|\psi_p|^2 = |A|^2\exp\left[\frac{ipx}{\hbar}\right]\cdot\exp\left[-\frac{ipx}{\hbar}\right] = |A|^2. \tag{19.58}$$

By being complex ψ_p has managed to meet two seemingly incompatible demands: *it has a wavelength λ associated with it (to encode the* *particle momentum) and yet its absolute value (squared) is constant, giving no information on the position.*

The oscillations in $\psi(x)$ ended up being erased when we multiplied it by $\psi^*(x)$ to compute $|\psi(x)|^2$. Does this mean that any interference pattern $I_{1+2}(y)$ produced by this complex wave will also be flat and non-oscillatory?

No! When this complex plane wave hits the two slits, the slits will give rise to *two* complex interfering radial waves on the other side, *of the same wavelength.* (That is, diffraction through the slits preserves p, the magnitude of $\mathbf{p}$, and all uncertainties refer to its direction.)

$$\psi(r_1, r_2) = A' \left(\exp\left[\frac{ipr_1}{\hbar} \right] + \exp\left[\frac{ipr_2}{\hbar} \right] \right) \tag{19.59}$$

where r_1 and r_2 are the distances from the two slits as displayed in Figure 19.11 and serve as coordinates. This function $\psi(r_1, r_2)$ will in fact show oscillations in $|\psi(\mathbf{r})|^2$ where the particle is detected:

$$|\psi(r_1, r_2)|^2 = |A'|^2 \left(\exp\left[-\frac{ipr_1}{\hbar} \right] + \exp\left[-\frac{ipr_2}{\hbar} \right] \right)$$
$$\times \left(\exp\left[\frac{ipr_1}{\hbar} \right] + \exp\left[\frac{ipr_2}{\hbar} \right] \right) \tag{19.60}$$

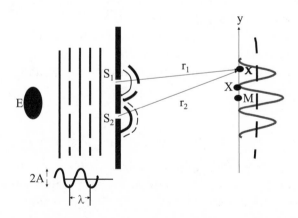

Figure 19.11 A single complex plane wave producing two radial waves upon hitting the two slits. The crests and troughs of wavelength λ correspond to the real part.

$$= |A'|^2 \left[1 + 1 + \exp\left[\frac{ip(r_1 - r_2)}{\hbar} \right] \right.$$

$$\left. + \exp\left[\frac{-ip(r_1 - r_2)}{\hbar} \right] \right] \qquad (19.61)$$

$$= 2|A'|^2 \left(1 + \cos\left[\frac{p(r_1 - r_2)}{\hbar} \right] \right) \qquad (19.62)$$

$$= 4|A'|^2 \cos^2\left[\frac{p(r_1 - r_2)}{2\hbar} \right]$$

$$= 4|A'|^2 \cos^2\left[\frac{k(r_1 - r_2)}{2} \right] \qquad (19.63)$$

exactly as in the Young experiment, Eqn. 18.50.

It's fair to say that if you did not know complex exponentials, you wouldn't have gone beyond this point in the development of quantum mechanics. The wave function of an electron of definite momentum is a complex exponential. Complex functions enter quantum mechanics in an essential way. It's not that the function $\psi_p(x)$ is really $A\cos(px/\hbar)$ and we are trying to write it as a real part of a complex exponential to simplify some calculation. We need this complex beast if we want to describe a particle of definite momentum p and totally unknown location.

I emphasize that I did not derive the result that $\psi_p \simeq e^{ipx/\hbar}$ describes a particle with momentum p in the x-direction. It is a postulate. Arguments based on the double-slit experiment were merely espoused to make the final answer seem reasonable. You cannot derive the postulates of quantum mechanics by pure logic or mathematics. You have to guess postulates of the underlying theoretical structure from the data and see how well they work.

19.9   Collapse of the wave function

Consider a particle with a wave function $\psi(x)$ and the associated probability $P(x) = |\psi(x)|^2$. Suppose now we catch it at some point, say $x = 5$. *If there is any reality to this detection, the particle must be found at $x = 5$ if its position is measured immediately afterward.* This means that right after the measurement, both $\psi(x)$ and $P(x)$ must collapse to narrow spikes at $x = 5$. This *collapse of the wave function* is postulated to happen and is found to happen. In the double-slit experiment, the oscillatory

$I_{1+2}(y)$ at the detectors describes the situation *before* the electron is detected. Once it triggers a detector, both ψ and P collapse to the detector.

Of course, as time passes the collapsed wave function may evolve and broaden out. The collapsed function applies only immediately after the measurement.

19.10 Summary

Here is a summary of what we have discussed so far.

1. At the microscopic level all entities exhibit wave-particle duality.
2. Light, which was believed to be a wave, is actually made of individual particles called photons, which represent bundles of energy and momentum. Monochromatic light consists of photons all of which have exactly the same momentum p given by de Broglie's formula

$$p = \frac{2\pi\hbar}{\lambda} \quad \text{where } \hbar = 1.05 \cdot 10^{-34} J \cdot s \tag{19.64}$$

and the same energy determined by the frequency of light

$$E = \hbar\omega. \tag{19.65}$$

3. Electrons, protons, and the like, which were known to be particles with localized energy, momentum, and charge, exhibit wave-like qualities in a double-slit experiment. The de Broglie relation between momentum and wavelength is the same as for photons:

$$\lambda = \frac{2\pi\hbar}{p}, \tag{19.66}$$

though I write it with λ on the left-hand side to indicate that λ and not p is the unexpected feature for a particle.
4. Each (massive) particle is associated with a wave function $\psi(\mathbf{r})$ whose absolute value squared $|\psi(\mathbf{r})|^2$ gives the likelihood of finding the particle at $\mathbf{r}$. If the particle is detected at $\mathbf{r}$, $\psi(\mathbf{r})$ collapses to a spike at $\mathbf{r}$ just after measurement. An immediate position remeasurement will give the same answer. It may of course change as time goes by.
5. The possibility of interference implies that the particles referred to above are not classical: they do not follow a definite trajectory (for

example, through a particular slit in the double-slit experiment) between observations. Assuming they do implies $I_{1+2} = I_1 + I_2$, which contradicts experiment.

6. The interference pattern is destroyed if the slit the particle took is determined, say by shining light. This occurs because the photons employed introduce a minimum uncertainty in momentum, which is enough to wash out the pattern.

7. Macroscopic bodies do not show interference because they are constantly being bombarded by the environment and because any pattern that survives would exhibit absurdly rapid spatial oscillations.

8. The wave function associated with a particle of definite momentum p in the x-direction, or simply, *a state of definite momentum p*, is

$$\psi_p(x) = A \exp\left[\frac{ipx}{\hbar}\right]. \tag{19.67}$$

9. In every quantum state, the roughly estimated uncertainties Δx and Δp have to obey the Heisenberg uncertainty principle:

$$\Delta x \Delta p \gtrsim \hbar \tag{19.68}$$

(up to factors like 2π etc., and likewise in the y-direction).

    If Δx and Δp are the precisely defined uncertainties, and not the heuristic estimates, we may write

$$\Delta x \Delta p \geq \frac{\hbar}{2}. \tag{19.69}$$

The uncertainty principle merely reflects the fact that trying to localize a wave in one direction (say by passing it through a narrow slit) makes it fan out.

    You have been exposed to so many new results in this chapter. There is not much I can do to relieve the information load or to make it appear more natural, because it is not natural. However, I want to extract from these results what I consider to be postulates, notions that cannot be deduced by logic or from other postulates. The list is not rigorous and I will enlarge and amend it as we go along.

Even though the particles in the double-slit experiment moved in two dimensions, I want, for pedagogical purposes, to extract postulates for a particle moving in just one dimension, described by a coordinate x.

Postulate 1. The state of a particle living on the *x*-axis is completely specified by a wave function $\psi(x)$ (generally complex) that contains all the information about it.

Postulate 2. The relative probability of finding the particle at *x* is given by

$$P(x) = |\psi(x)|^2.$$

If the particle is detected at *x*, $\psi(x)$ collapses to a spike at *x* just after measurement.

Postulate 3. A particle in a state of momentum *p* is described by

$$\psi_p(x) = A \exp\left[\frac{ipx}{\hbar}\right].$$

If we bring $\psi_p(x)$ to the standard form of a wave written in terms of λ

$$\exp\left[\frac{ipx}{\hbar}\right] \equiv \exp\left[\frac{2\pi\, ix}{\lambda}\right] \tag{19.70}$$

we see that not only does this postulate subsume the de Broglie formula

$$\lambda = \frac{2\pi\hbar}{p} \tag{19.71}$$

relating the wavelength to the momentum *p*, but it also goes beyond, by specifying the actual functional form.

What about the uncertainty principle? It is not a postulate; it follows from combining what is postulated (relating momentum to wavelength) with results from classical wave theory.

CHAPTER 20

The Wave Function and Its Interpretation

Even though the last chapter ended with a summary, I will go over the facts again since they are quite bizarre and talking about them often is one effective way to digest them.

Electrons, photons, protons, neutrons are all particles. I will simply refer to them collectively as electrons in this discussion. Let there be no doubt about what I mean by a particle here: if one of them hits your face, you will feel it in only one tiny region, in just one spot. The electron will dump all its charge, all its momentum, all its energy to one little part of your face. There's nothing extended about the impact, the kind you would expect from getting hit by a wave front. If it is a particle in all these ways, where is the problem? The problem appears when you do the double-slit experiment. That's what puts the nail in the coffin for Newtonian or classical physics. Recall the essentials. There is a source, like an electron gun, that emits electrons on the left, there is a partition with two slits in the middle, and an array of detectors (or a sliding detector) on the right. The electron gun has been engineered to send electrons of a definite momentum and energy by accelerating them down a definite potential. If this gun is far away to the left, then the only way electrons are going to hit the slits is if they are essentially moving in the horizontal direction. What do we really know when we do the experiment? Once in a while the gun will emit an electron and recoil like a rifle. That's when we know the electron has left. Then we don't know anything for a while, and then one of the counters goes "click." That means the electron has arrived there. This is all we really know. Everything else we say about the electron is conjecture at this point. We know it was here first, and we know it was there later. The question is, what was it doing in between? We might say, "We don't know the trajectory it followed because we did not track it, but it must have followed *some* trajectory, either through slit 1 or through slit 2." This reasonable assumption contradicts experiment: it predicts $I_{1+2} = I_1 + I_2$, which is false. A dramatic illustration is seen at the point that was labeled X, a zero of the interference pattern. We used to get N electrons per hour with one slit open, and N electrons per hour with the second slit open, and none with both open, instead of $2N$. This was not the result of electrons from one slit colliding with electrons from the other and deflecting them away from X, because the same result is obtained if the experiment is done with just one electron in the region at any given time.

That is the great mystery. That is the end of Newtonian physics.

It then gets even more mysterious if we try to see which slit the electron took by placing a glowing lightbulb near the slits. Now we find that the ones that were detected are additive over the slits, while the ones that slipped by produce an interference pattern. So the behavior of the electron is affected by whether we see it or not. This is true because the light used to see which slit the electron took necessarily transfers to the electron some momentum whose uncertainty was estimated to be of the order $\Delta p \gtrsim \frac{2\pi\hbar}{d}$, where d is the slit separation. This in turn translates into an uncertainty in the electron direction by an angle comparable to that which separates successive maxima and minima in the interference pattern. The pattern gets washed out upon detection.

We do not see such interference patterns on a macroscopic scale because macroscopic objects are constantly being bombarded, intentionally or otherwise, and any interference pattern that miraculously survives will be of an absurdly small wavelength and escape detection. Only the spatial average, which reduces to $I_1 + I_2$, will be detected.

What are we to make of the oscillatory pattern I_{1+2} in the double-slit experiment? A trained physicist like you will say, "Hey, this reminds me of interference, which I have encountered with water waves and sound. Obviously there is some underlying wave and some wavelength. The minute you give me the wavelength and a slit separation, I can calculate this pattern using $d\sin\theta = m\lambda$. Conversely, from the angles at which maxima occur I can infer λ." You go on and find that the wavelength is some number $2\pi\hbar = 2\pi \cdot 1.05 \cdot 10^{-34} J \cdot s$ divided by the momentum p of the incoming electrons:

$$\lambda = \frac{2\pi\hbar}{p}. \tag{20.1}$$

In other words, you find that if you sent in more energetic electrons, that is, accelerate them through bigger voltage to increase their p, the wavelength λ goes down inversely, with $2\pi\hbar$ as the constant of proportionality.

So you can successfully predict this pattern given the electron momentum and the wave of corresponding λ, but what does it tell you about what's going on? What good is that pattern? The pattern tells you that if you repeated the experiment with this electron gun a million or a billion times and plotted the histogram of electrons registering at different counter locations, the histogram would eventually fill out and take the shape I_{1+2} produced by wave interference. However, this wave is not associated with a stream of electrons. The *single* electron in the experimental region is controlled by this wave. It's not a wave of charge or of matter as in water or a string. It's a mathematical function, and you are driven to it as the only way you know to get this wiggly graph I_{1+2}: give the wave a definite wavelength and let it interfere. And what does it mean for the individual trial? It gives you the odds of where the electron will land on that screen.

So there seems to be a function whose square at a point $\mathbf{r}$ gives you the probability of finding the electron at $\mathbf{r}$. That function is called the *wave function* $\psi(\mathbf{r})$. Given the wave and the relation $p = 2\pi\hbar/\lambda$, the uncertainty principle $\Delta x \Delta p \gtrsim \hbar$ follows. One way to arrive at it is to try to engineer a situation in which the product of the uncertainties in the electron position and momentum is arbitrarily small. We will find it is possible in classical mechanics but not quantum mechanics.

First consider classical mechanics. We send a beam of classical particles of definite momentum p_0 in the horizontal or x-direction and let it strike a partition with a slit of width D in the transverse y-direction. Any particle emerging from the slit has $p_y = 0$ (since it is still moving horizontally), and a y coordinate with uncertainty $\Delta y = D$. Since $\Delta p_y = 0$, the uncertainty product vanishes. Besides, we can make Δy as small as we want by reducing D.

This is of course not true for a quantum particle like the electron. It is still true that the electron just emerging from the slit has $\Delta y \simeq D$.

However, its fate is governed by a wave with $\lambda = 2\pi\hbar/p_0$. The wave fans out by diffraction to an angle θ given by $D\sin\theta = \lambda$. The final electron has a non-zero probability of hitting points on the screen at any angle within the principal maximum of the diffraction pattern, i.e., within $\pm\theta$. To get there it needs to have a y component of momentum with an uncertainty $\Delta p_y \gtrsim 2p_0\sin\theta = 2p_0\lambda/D = 4\pi\hbar/D$, in accord with the uncertainty principle.

The uncertainty principle is valid in the macroscopic scale but irrelevant. Consider an object of mass 1 kilogram whose location is known to the accuracy of the size of 1 proton, which is $10^{-15}m$. So we have here an object made of $\simeq 10^{26}$ protons and we know its location to the width of 1 proton. That is good enough for most imaginable purposes. The corresponding $\Delta p = 10^{-19}kg \cdot m/s$ translates into uncertainty in velocity of $10^{-19}m/s$. Now how bad is that? Suppose I knew the velocity to this accuracy, and I let the body travel for one year. Since a year is roughly $10^7 s$, that becomes a position uncertainty of $10^{-12}m$, which is one-hundredth the size of an atom. So you see, these uncertainties are not important in daily life.

What is the incoming wave that produces this interference pattern in the double-slit experiment conducted with electrons of definite momentum p in the x-direction? We know the wavelength is $\lambda = 2\pi\hbar/p$, but many functions can oscillate with this λ. The correct answer is the complex exponential

$$\psi_p(x) = A\exp\left[\frac{ipx}{\hbar}\right] \tag{20.2}$$

where the constant A is unspecified at this stage. The label p reminds us that it is not just any old ψ; it is one that describes a particle of momentum p. Such nomenclature is common and will be employed often in what follows. This function manages to encode the oscillations of the correct wavelength in its phase and yet possess an absolute value squared $|\psi_p(x)|^2$ that is y-independent. So the particle location is completely unknown and the probability distribution absolutely flat, which means $\Delta x = \infty$, as required of a state of definite momentum.

I emphasize that the preceding ψ_p with its flat $P(\mathbf{r})$ describes the incoming wave *before* it hits the two slits. After passing through them, the wave emerges as two radial waves that interfere and produce an oscillating $P(y)$ along the line of detectors.

20.1 Probability in classical and quantum mechanics

Suppose you flip a coin and ask, "Which way will it land?" This calls for a very difficult calculation. But it can be done in principle, because once released from your hand, the coin can land in only one way. That is the determinism of Newtonian mechanics. If you knew the exact initial position, velocity, linear momentum and angular momentum, the viscosity of air, and so on, you could predict whether it was going to land heads or tails. There is no fundamental need to resort to probability. In practice, no one can do the calculation. What you do in practice is throw the same coin 5,000,000 times, you find out the odds for heads or tails, and you say, "I predict that when you throw it next time, it will be heads with probability 0.56." That is how you make statistical predictions. You did not *have* to use statistics, but you did so as a practical strategy when faced with an impractical calculation.

Next, suppose I toss a coin and when it lands on my palm, I close my palm without looking. When I uncover my palm and look at the coin, it may be heads or it may be tails. Suppose I got heads. It was heads even before I opened my hand, right? The measured outcome preexisted inside my hand. What I saw was what it was doing even before I looked. This is how probability works in classical mechanics.

I'll give another example, with a continuum of possible outcomes. Figure 20.1 shows the probability of locating me somewhere. It is peaked near my home in Cheshire, and near Yale, and has some sizable value on the infamous Route 10 connecting the two. Somebody has studied me for a long time and said, "If you look for this guy, here are the odds of finding him at various locations. Either he's working at home, working

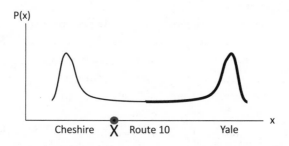

Figure 20.1   The probability of finding me somewhere during a typical day at home in Cheshire, along Route 10, or at Yale.

at Yale, or driving on Route 10." The first thing to understand is that the spread-out probability does not mean I am myself spread out, unless I got into a terrible accident on Route 10. I'm in only one place at any one time. Only the graph of the odds is extended. Well, suppose you catch me at the point X on one of your many trials. If you catch me only once, you don't know if the prediction for $P(x)$ is any good, so you repeat it. You locate me many times and plot the histogram and you get the graph that looks like this $P(x)$. The important thing is, every time you catch me somewhere, I was already there; you just happened to catch me there. My location was not known to you, but I had one. I had a definite location because in the macroscopic world I'm moving in, my location is being constantly measured. You didn't ask or you didn't find out, but I'm plowing through air molecules. I've slammed into them. They remember that. I ran over this ant. The last thing the ant did was measure my position, at considerable cost to itself.

Now let us look at Figure 20.2. It is no longer me that is being described by $P(x)$, but an electron that is shared by two nuclei, N_1 and N_2. Whereas in classical physics there is just a probability function $P(x)$, *in quantum theory there is an underlying wave function $\psi(x)$*, which in turn determines $P(x) = |\psi(x)|^2$. The electron is described by $\psi_1(x)$ when it is centered around nucleus N_1, with a corresponding probability $P_1(x) = |\psi_1(x)|^2$. It does not have a precise location since it can be found anywhere

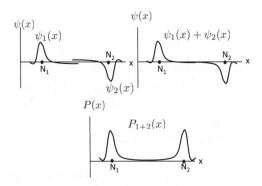

Figure 20.2   Top left: The wave functions ψ_1 for an electron centered near nucleus N_1 or ψ_2 centered around nucleus N_2. Top right: Their sum $\psi_1 + \psi_2$, which is also a possible wave function. Bottom: The corresponding probability $P_{1+2}(x)$.

the function is non-zero. Let us say we are only going to make crude position measurements that only tell us near which nucleus the electron is. If the wave function is ψ_1, we know we will find it near nucleus N_1, that it belongs to N_1. Likewise $\psi_2(x)$ is centered around nucleus N_2 with a corresponding probability $P_2(x) = |\psi_2(x)|^2$ and describes an electron known to be near N_2 and belonging to N_2.

Now quantum theory also allows for another state $\psi_{1+2} = \psi_1 + \psi_2$. It is peaked near both nuclei, as is $P_{1+2} = |\psi_{1+2}|^2$. The electron is now in a state that belongs to neither nucleus. Once again, if you catch the electron, you will catch all of it in one place, near one or the other nucleus. It is the odds that are spread out across both nuclei.

All this looks just like the probability of my being at my home, on Route 10, or at Yale. But there is a big difference: if you catch the electron in state ψ_{1+2} near nucleus N_2, it is wrong to think that it was there before you located it. So where was it? It was not near either nucleus. *It had no location till you found its location.* To assume it was definitely near one nucleus or the other *before* locating it is like assuming the electron went through one or the other slit when you did not detect it using light. Assuming so leads to consequences at odds with experiment.

I repeat: finding the electron near N_1 or N_2 is not quite like finding me in Cheshire or finding me at Yale, because in those cases, on a given day on a given measurement, you could have only gotten one answer, depending on where I *actually* was. Right now if you look for me, you can only find me here, slaving on this book. You cannot find me anywhere else. But in the case of the electron, *the one and the same electron, on a given trial, at a given instant, is fully capable of being here or there.*

There is one common feature between classical and quantum probability. Once a particle given by some $P(x)$ is actually caught at some point, say $x = 5$, we know for sure it will be there at least for an infinitesimal time after detection. The function $P(x)$ collapses to the point $x = 5$. In the quantum case, the underlying wave function also undergoes collapse.

In short, probability enters classical mechanics to make up for our imperfect knowledge of the state of a particle. In quantum mechanics, even given the maximal information allowed by the theory, i.e., the wave function $\psi(x)$, one still needs probabilities in an unavoidable way. The second crucial difference is that while in both cases there is a non-negative function $P(x)$, in the quantum case, there is a layer beneath $P(x)$, namely the wave function, which can be negative or even complex and can be superposed to produce interference.

Classically we think of measurement as revealing a preexisting property of the object, like its position. But in quantum theory, it's not that you don't know the particle location, but that it does not have a location. It is not anywhere. *It's the act of position measurement that confers a definite location on the electron.* Until you detect it, it could have been anywhere $P(x)$ did not vanish. That state of being, where something can be simultaneously here or there, in the sense that on that single occasion, it could be found in either location, has no analog in the classical world. If anybody tries to give you an example of this phenomenon from daily life, don't believe it, because there are no examples in the macroscopic world that look like this. No analogies should satisfy you, because this has no analog in the macroscopic world.

How small does an object have to be before it exhibits quantum mechanical behavior, before it can be doing two things at the same time? That is an experimental question being probed vigorously these days. We know that if it's really small like an electron, it's always quantum mechanical. If it is large like a bowling ball, it will seem to have a well-defined position and momentum at all times. People are trying to build bigger and bigger systems that can be in this state of limbo. Creating a situation when an object is capable of being found here and there, or doing this and doing that simultaneously, requires that you isolate the object from the outside world. This gets harder as the object gets bigger. Whereas an electron in an atom is usually in a vacuum, macroscopic objects are under constant bombardment by the environment. That's what ruins everything.

This is a major problem in building quantum computers. A quantum computer, you might know, has qubits. Unlike the classical bits in your laptop, which are in either one of the classical states traditionally called 0 or 1, a single qubit can be in a state where it can be found in either 0 or 1 *on a given trial.* It's like the electron going through both the slits.

So a quantum bit can explore both classical possibilities at the same time. If you build a computer with 10 qubits it can be exploring 2^{10} classical states at the same time, in the sense that a measurement can yield any of the 2^{10} classical answers. And if it has a million bits, it is exploring $2^{1,000,000}$ classical states at the same time. This allows it to solve certain problems exponentially faster than what is currently known to be possible on a classical computer. This means that if the classical computer needs 10^{17} seconds (rough age of the universe) to solve a problem, the quantum computer can do so in 17 seconds. However, it first needs to be built and needs to be programmed to do this. *As of now, not only do we not have a quantum computer with more than a handful of qubits, but very few problems are known that can be solved exponentially faster on a quantum computer and for which we have the requisite program.* There is one celebrated program due to Peter Shor that can factorize a huge number into its two prime factors in a few seconds, while a classical one could take the age of the universe to do it. This may surprise you. You can multiply a 100-digit prime number by a 100-digit prime number on your computer almost instantaneously. But if I gave you the 200-digit product and asked you to find the two prime factors with a hundred digits each, you won't find it in years. That's why one of the ways to securely send your credit card information on the internet is to use very large numbers obtained by multiplying two primes to encrypt them. Decryption requires its prime factors. These cannot be found even though the large number (the product) is broadcast openly. But if you have a quantum computer, made up of these qubits, and used Shor's algorithm, you can actually factor the number in a few seconds.

This gives you two options if you have secretly managed to build a quantum computer. Either you can become famous and win the Nobel Prize (or even get an NSF grant), or you can go on the biggest shopping spree of your life, because you can get anyone's credit card number. When you come to that fork, you can decide which way you want to go. Maybe you can take both choices, if you are small enough.

There are many quantum systems that can do one of two things, which can be in a state that is both this and that. These are all potential qubits. The problem is that they cannot be in contact with the outside world, because even a single contact with them can destroy the quantum state of limbo, the way the photon used to locate the electron in the double-slit experiment can destroy the interference pattern. So you have to keep your quantum computer fully isolated. But a computer that is not talking to the outside world, unfortunately, is also not talking to you. This means you cannot ask it any questions, and if it knows the answer, it cannot tell you. So sometimes you want it to talk. Sometimes you don't want it to talk. What should you do? You have to build a quantum system with which you sometimes make contact in a controlled way to give it the problem. Then you want to leave it alone while it quantum-computes. Finally you make a measurement to find out the answer.

20.2 Getting to know ψ

Let us continue our study of quantum mechanics. Recall that in classical mechanics the pair (x, p) is the full story. Given that, I know everything I need to know at any one instant. The kinetic energy is $K = p^2/2m$, the angular momentum is (in higher dimensions) $\mathbf{r} \times \mathbf{p}$, and so on. Everything is given in terms of the coordinates and momenta. In quantum theory, we don't even know where the particle is. We have a wave function $\psi(x)$ describing the state, and $|\psi(x)|^2$ gives the likelihood of finding the particle at x.

What are the conditions on the function ψ? The first is that it should be continuous and single-valued so that at each point it gives a unique $P(x)$. Another technical requirement is that it should be *square-integrable*: the integral of $|\psi(x)|^2$ over all of space must be finite. This is the principle that allows us to place restrictions on the allowed values of energies in bound-state problems. These are problems where E, the total energy of the particle, is less than $V(\pm\infty)$, the potential energy at infinity, and escape to infinity is classically forbidden. (If the particle escaped to infinity, it would be required to have negative kinetic energy $K(\pm\infty) = E - V < 0$.) In the quantum case, we will find that at generic energies the integral of $|\psi|^2$ blows up *exponentially* in L, the size of the universe. These states are simply dismissed and the allowed energies are identified as those at which the square integral is bounded.

There are two exceptions to the square integrability and unfortunately they are rather commonplace. The first are the states of definite momentum with constant $|\psi_p(x)|^2$. Their square integral grows *linearly* with the size L of the universe. This *linear* divergence is a borderline case we can handle using what are called *delta functions*, to be discussed near the end of the book. The other exceptions are states of exactly known position, which I have loosely referred to as "spikes" centered at some point. These are also described by delta functions. For now you must accept these functions despite their not having a finite square integral. You can also view them as limits of functions that are square-integrable.

Apart from these restrictions, ψ can be whatever you like. In particular if $\psi_1(x)$ and $\psi_2(x)$ are two allowed wave functions so is the linear combination $A\psi_1(x) + B\psi_2(x)$. The linear *superposition* describes a state in which it can be found doing what it does in ψ_1 (peaked near nucleus N_1) or what it does in ψ_2 (peaked near nucleus N_2), with relative probabilities $|A|^2$ and $|B|^2$.

I have been saying that $|\psi(x)|^2$ gives the probability of finding the particle at x. This statement needs to be refined. I'll tell you why. Consider a statistical event that has a countable number of outcomes. For example, when you throw a die, there are 6 possible outcomes. You can measure or assign $P(1)$, the probability for obtaining 1, $P(2)$ the probability for 2, and so on. These are the odds for getting any number from 1 to 6. Since *some* number has to come up, we require that the probability that *some* number will show up equals unity:

$$\sum_{i=1}^{6} P(i) = 1. \tag{20.3}$$

This is called the *normalization* condition. An example of a normalized probability distribution is depicted in Figure 20.3A:

$$P(1) = 0.2, P(2) = 0.2, P(3) = 0.05, P(4) = 0.25,$$

$$P(5) = 0.15, P(6) = 0.15. \tag{20.4}$$

The same information is contained in the unnormalized *relative* probabilities:

$$P'(1) = 20, P(2) = 20, P'(3) = 5, P'(4) = 25,$$

$$P'(5) = 15, P'(6) = 15. \tag{20.5}$$

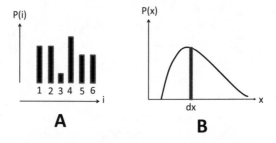

$P(i)$ $P(x)$

1 2 3 4 5 6 i dx x

A   **B**

Figure 20.3   A: The normalized probability distribution $P(i), i = 1 \ldots 6$ for 6 discrete outcomes. B: The probability density $P(x)$ for a continuous probability distribution. The shaded area $P(x)dx$ is the probability of finding the particle between x and $x + dx$.

From the unnormalized P''s I can get the normalized P's by rescaling:

$$P(i) = \frac{P'(i)}{\sum_{j=1}^{6} P'(j)}. \tag{20.6}$$

Now suppose the set of outcomes is not countable as with a die, but continuous, like the location of an electron. Then you cannot give a *finite* probability for any particular x. (In physics we often use "finite" to mean "not infinitesimal" rather than "not infinite.") If the probability for any one point was some finite number, the sum over an infinity of such points will be infinite and cannot be rescaled to unity, i.e., cannot be normalized. So we introduce the notion of a *probability density* $P(x)$ defined as follows:

$$|\psi(x)|^2 dx \equiv P(x)dx = \text{the probability the particle}$$
$$\text{is found between } x \text{ and } x + dx. \tag{20.7}$$

That means that if you draw a graph of $|\psi(x)|^2 = P(x)$ and take a sliver of width dx at x, the area of the rectangle $P(x)dx$ is the probability of finding the electron between x and $x + dx$, as indicated in Figure 20.3B. So you assign an infinitesimal probability to an infinitesimal region. The statement that the particle has to be somewhere, namely, that all the probabilities add up to 1, becomes the normalization condition

$$\int_{-\infty}^{\infty} |\psi(x)|^2 dx = \int_{-\infty}^{\infty} P(x)dx = 1. \tag{20.8}$$

Now a ψ that did not obey this condition also contains the same information: the odds are big where $|\psi|^2$ is big, small where it is small, zero where it is zero, and so on. So when you multiply $\psi(x)$ by any number, you don't change the predictions of the theory, namely the *relative* odds. It is just that if your original $\psi(x)$ had a square integral of unity, the new one will not.

Thus the wave function ψ of quantum mechanics is very different from other ψ's you may have encountered elsewhere. For example, if $\psi(x)$ stood for the displacement of a vibrating string, $2\psi(x)$ is a totally different configuration of the string. If you took the electric field and made it twice as big, that's a different situation, because the forces on the charges are doubled and the energy density quadrupled. *But in quantum mechanics, $\psi(x)$ and any multiple of it stand for the same physical state.* The only job of ψ is to give you the *relative* odds. If you wish, you can rescale it so it is normalized and gives the absolute probability density.

Here is an analogy. Suppose you are a cop asking a witness which way the burglar ran, and she wants to say at 45 degrees to the x-axis; she can say "Along $\mathbf{i} + \mathbf{j}$." She can also say "Along $96\mathbf{i} + 96\mathbf{j}$." What she is trying to convey is not a vector but a *ray* or a direction. For this reason one says that in quantum mechanics the wave function is a *ray*. Of all the rays obtained by rescaling a given $\psi(x)$, it is common to pick one that is normalized to unity.

Look at the simple example depicted in Figure 20.4. Let

$$\psi(x) = A \quad \text{for } |x| < a \tag{20.9}$$

$$= 0 \quad \text{for } |x| > a. \tag{20.10}$$

This ψ describes an electron that is going to be found with equal probability anywhere within $|x| < a$ and never outside. *This fact will not change if you replace A by 10A or $\psi(x)$ by $10\psi(x)$.*

(This function is not single-valued at $x = \pm a$ where it abruptly plunges from A to 0. I still use it because it is easy to work with in the following illustrative calculation. If you wish you can think of it as the limit of a single-valued function that drops very very rapidly to zero near $|x| = a$.)

Of this family of physically equivalent functions, we are now going to pick one that is normalized, i.e, obeys

$$\int_{-\infty}^{\infty} |\psi(x)|^2 dx = 1 \tag{20.11}$$

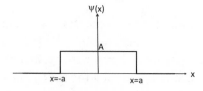

Figure 20.4   An unnormalized wave function that is non-zero and constant for $-a < x < a$. Its height A may be chosen to normalize it.

by a judicious choice of A. Since

$$\int_{-\infty}^{\infty} |\psi(x)|^2 dx = |A|^2 (2a) \tag{20.12}$$

we choose

$$A = \frac{1}{\sqrt{2a}} \quad \text{so that} \tag{20.13}$$

$$\psi(x) = \frac{1}{\sqrt{2a}} \quad \text{for } |x| < a \tag{20.14}$$

$$= 0 \quad \text{for } |x| > a \tag{20.15}$$

is normalized. (Actually we can still multiply A by a pure phase $e^{i\theta}$ without affecting $|A|^2$. It is common to choose the normalization factor A to be real whenever possible.)

Another popular example is the bell-shaped *Gaussian*:

$$\psi(x) = Ae^{-x^2/2\Delta^2}. \tag{20.16}$$

This function starts out with a value A at $x = 0$ and falls off to negligible values when $x \gg \Delta$, where Δ is called the *width of the Gaussian*. Because this ψ is real, normalization requires the integral of $\psi^2(x)$ to equal unity:

$$|A|^2 \int_{-\infty}^{\infty} e^{-x^2/\Delta^2} dx = 1. \tag{20.17}$$

Setting $\alpha = 1/\Delta^2$ in the tabulated integral

$$\int_{-\infty}^{\infty} e^{-\alpha x^2} dx = \sqrt{\frac{\pi}{\alpha}}, \tag{20.18}$$

we arrive at the following normalized wave function:

$$\psi(x) = \frac{1}{(\pi \Delta^2)^{1/4}} e^{-x^2/2\Delta^2}. \tag{20.19}$$

20.3 Statistical concepts: mean and uncertainty

Here are some basic ideas from statistics that will be needed in our study of quantum mechanics.

Suppose there is a variable v that can take on many values v_i, $[i = 1 \ldots N]$ that are supposed to occur with normalized probability $P(i)$ according to some theory or hypothesis. For example, we may be talking about a die, with the following normalized probabilities $P(i)$ for obtaining one of the six numbers

$$P(1) = 0.2, P(2) = 0.2, P(3) = 0.05, P(4) = 0.25,$$
$$P(5) = 0.15, P(6) = 0.15. \tag{20.20}$$

To verify this statistical description of the die we must toss it N times or toss N identical dice once and see if $N(i)$, the number of times a given value i occurs, obeys

$$\lim_{N \to \infty} \frac{N(i)}{N} = P(i). \tag{20.21}$$

A collection of such identical dice is called an *ensemble.*

The most complete statistical description of the dice is given by the full list of $P(i)$. This can be quite tedious if the die has, say, 30,000 faces. Then one provides, as a first attempt at description, just one number called the *mean,* which is the weighted average of the possible values:

$$\langle v \rangle = \sum_i P(i) v_i. \tag{20.22}$$

(If $P'(i)$ is an unnormalized distribution, we must divide the weighted sum by $\sum_j P'(j)$.)

For the die the mean is

$$\langle v \rangle = \sum_{i=1}^{6} P(i) v_i \tag{20.23}$$

$$= 1 \cdot 0.2 + 2 \cdot 0.2 + 3 \cdot 0.05 + 4 \cdot 0.25$$
$$+ 5 \cdot 0.15 + 6 \cdot 0.15 \tag{20.24}$$

$$= 3.4. \tag{20.25}$$

Now, you can get the same mean with two different $P(i)$'s, one with a very narrow spread in the possible values and one with a broad spread. To tell them apart, we provide a measure of how wide the distribution is using the *standard deviation*

$$\Delta v = \sqrt{\sum_i P(i)(v_i - \langle v \rangle)^2}. \tag{20.26}$$

So first we take the weighted average of the *squares* of the deviations. Then we take the square root to obtain a quantity with the same dimension as v. (Without the square, the average of the deviations $v_i - \langle v \rangle$ will be zero. I invite you to show this.)

For the die

$$\Delta v = \left[(1 - 3.4)^2 \cdot 0.2 + (2 - 3.4)^2 \cdot 0.2 + (3 - 3.4)^2 \cdot 0.05 \right.$$
$$+ (4 - 3.4)^2 \cdot 0.25 + (5 - 3.4)^2 \cdot 0.15$$
$$\left. + (6 - 3.4)^2 \cdot 0.15 \right]^{1/2} = 1.74. \tag{20.27}$$

For a continuous variable like x with a normalized probability density $P(x)$ we make the expected modifications:

$$\langle x \rangle = \int P(x)x\,dx \tag{20.28}$$

$$\Delta x = \sqrt{\int P(x)(x - \langle x \rangle)^2\,dx}. \tag{20.29}$$

In the context of quantum theory $P(x) = |\psi(x)|^2$, $\langle x \rangle$ is called the *expectation value* and Δx the *uncertainty*. I have mentioned that the precise uncertainty principle

$$\Delta x \Delta p \geq \frac{\hbar}{2} \tag{20.30}$$

holds only if Δx is the precisely defined uncertainty. Eqn. 20.29 provides that definition. A similar definition holds for Δp in terms of the probabilities for getting different values for momentum. These probabilities will be discussed in the next chapter.

Consider as an example the normalized wave function shown in Figure 20.4:

$$\psi(x) = \frac{1}{\sqrt{2a}} \quad \text{for } |x| < a \tag{20.31}$$

$$= 0 \quad \text{for } |x| > a \tag{20.32}$$

$$P(x) = \frac{1}{2a} \quad \text{for } |x| < a \tag{20.33}$$

$$= 0 \quad \text{for } |x| > a. \tag{20.34}$$

It is visually obvious that the expectation value vanishes by symmetry. This can be easily confirmed:

$$\langle x \rangle = \int P(x)x\,dx \tag{20.35}$$

$$= \frac{1}{2a} \int_{-a}^{a} x\,dx = 0. \tag{20.36}$$

The uncertainty can be crudely estimated to be $\Delta x \simeq 2a$. More precisely, the uncertainty squared is

$$(\Delta x)^2 = \frac{1}{2a} \int_{-a}^{a} (x-0)^2\,dx \tag{20.37}$$

$$= \frac{a^2}{3} \tag{20.38}$$

and the uncertainty is

$$\Delta x = \frac{a}{\sqrt{3}}. \tag{20.39}$$

To test these predictions we again need an ensemble of a large number of particles all prepared in the same quantum state $\psi(x)$ with the same probability density in position $P(x)$. There is one difference between the classical and quantum ensembles. In the classical ensemble with N identical particles, roughly $N \cdot P(x)\,dx$ particles would be between x and

$x + dx$ just before and just after the measurement. In the quantum case each particle would have been in a state of limbo prior to measurement, in which it *could* be caught at any x where $P(x)$ did not vanish. The particles in the ensemble would acquire a definite position only after the position measurement.

CHAPTER 21

Quantization and Measurement

As usual, let us begin with a quick review of recent material. We have been studying a particle living in one spatial dimension, described by the coordinate x. Everything we need to know about that particle at one instant is contained in the wave function $\psi(x)$, which could be complex. This is quantum kinematics, the analog of the statement in classical mechanics that (x, p) describe the state of the particle. Given this pair, all other dynamical variables like kinetic energy and angular momentum (in higher dimensions) have fixed values. For example, $K = \frac{1}{2}mv^2 = p^2/2m$. (Dynamics is the question of how this state changes with time and is given by Newton's laws. Later we will study the equation that governs the time evolution of ψ.)

Whereas in classical physics two numbers tell you the whole story, quantum theory requires a whole function $\psi(x)$. We know a function is really an infinite amount of information because at every point x the function has a value $\psi(x)$ and we have to specify all those values.

What sort of information does this $\psi(x)$, which is supposed to tell us everything, contain? How is that information to be extracted? We have seen that

$$|\psi(x)|^2 = P(x) \tag{21.1}$$

is the probability density for finding the particle at the point x. By that I mean $P(x)\,dx$ is the probability that the particle will be detected between

x and $x + dx$. We like to impose the requirement that the total probability to find the particle anywhere add up to 1. That is a convention and not a fundamental requirement. It's up to you to define probability. You may tell your friend that the odds that you will get through this course is 50:50. These numbers add up to 100 and not 1, but they convey correctly that your chances of passing and failing are equal. The normalized probabilities are $P(\text{pass}) = 50/(50 + 50) = .5$ and $P(\text{fail}) = 50/(50 + 50) = .5$.

Likewise in quantum theory the wave function you are given need not be normalized and obey instead

$$\int_{-\infty}^{\infty} |\psi(x)|^2 dx = N. \tag{21.2}$$

If you choose, you can rescale the given ψ by a factor $1/\sqrt{N}$ and obtain a normalized ψ. We will generally do this. Of course it is understood that N should be finite so that this rescaling is possible. So we want the wave function ψ to be *normalizable* or square-integrable, and not necessarily normalized. We also require that $\psi(x)$ be single-valued: it must have only one value at each point x. So jumps are not allowed. (The illustrative examples at the end of the last chapter that violated this requirement could be viewed as a limit of an allowed function.) Apart from that you can write your own ticket.

The ψ of quantum mechanics is not like the ψ's you have seen before, say in the vibrating string or water waves. If the displacement $\psi(x)$ of a string or a body of water is multiplied by 6, it describes a different state. On the other hand ψ and a multiple of it describe the same physical state and give the same *relative* probabilities.

We have seen one particular ψ that describes a particle of definite momentum p:

$$\psi_p(x) = A \exp\left[\frac{ipx}{\hbar}\right]. \tag{21.3}$$

Though I gave some arguments in favor of it, this is really a postulate. The subscript on $\psi_p(x)$ tells us that this is not any old ψ, but one that describes a particle with a special attribute: it has definite momentum p. We use such labels all the time. We don't go to a party and say, "Hi, I am human." We say something like "I am Alexey" or "I am Barry," because that says a little more about us than just which species we belong to.

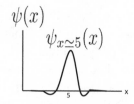

Figure 21.1   Wave function of a particle localized near $x = 5$.

The probability density in this state of precisely known momentum, $(\Delta p = 0)$, is independent of x and equals the constant $|A|^2$. We have no idea where it is and $\Delta x = \infty$. This is in accord with the uncertainty principle.

Here is another example of a wave function with some characteristic. Suppose a particle is known to be close to $x = 5$ as a result of a crude position measurement (using photons of small momentum) that did not determine x to arbitrary precision. What function will describe that particle? You cannot of course come up with the precise form with just the information I gave, because many functions can be peaked around $x = 5$. However, you should not be surprised if the answer is something like the one in Figure 21.1. Conversely if you were given this wave function, you should be able to see right away that it describes a particle very likely to be found near $x = 5$.

21.1   More on momentum states

Suppose I give you a state

$$\psi(x) = Ae^{96ix}.$$
(21.4)

What can you say about the particle? By comparing it with the prototype

$$\psi_p(x) = A\exp\left[\frac{ipx}{\hbar}\right]$$
(21.5)

you can deduce it has a definite momentum

$$p = 96\hbar.$$
(21.6)

(Don't worry about units; they are contained in the "96," which is really $96m^{-1}$.)

Similarly if an electron has been accelerated from rest by a voltage V_0, its momentum is fixed by

$$\frac{p^2}{2m} = eV_0 \tag{21.7}$$

and the wave function of the electron coming out of the accelerator (along the positive x-axis) is

$$\psi(x) = A \exp\left[\frac{i\sqrt{2meV_0}x}{\hbar}\right]. \tag{21.8}$$

Now we must deal with the pre-factor A, which has remained arbitrary because its value has no physical significance. It is conventional to choose A to normalize the wave function, by demanding

$$1 = \int_{-\infty}^{\infty} |\psi(x)|^2 dx = \int_{-\infty}^{\infty} |A|^2 dx = |A|^2 \cdot \infty. \tag{21.9}$$

There is no choice of A that will work because $|A|^2$ is multiplied by the size of our universe, which extends from $-\infty$ to $+\infty$. A common way out of this predicament is to pretend our universe is large but finite and has no boundaries. (This may even be the case in reality.) In the one-dimensional case we may take it to be a circle of radius R and circumference

$$L = 2\pi R. \tag{21.10}$$

You can form such a circle by taking a line of length L and gluing its two ends together. If the line is parameterized by a coordinate $0 \leq x \leq L$ the circle is obtained by joining $x = 0$ and $x = L$, as shown in Figure 21.2. In this closed universe if you throw a rock it will come back and hit you from behind. You can even see this happen if you wait for the light to come all the way back to your eyes. But such peculiarities in the cosmic scale will not matter to the quantum mechanics of a tiny atom or electron. These little guys don't care if the universe does not exist beyond this room, any more than you care in your daily life that the earth is not flat.

While we introduced the circumference of the universe as an artifact to normalize ψ_p, there are many present-day experiments in which the

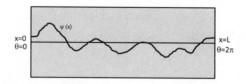

Figure 21.2   A typical periodic function on the circle, which has been opened out to a line of length L with the understanding that $x = 0$ ($\theta = 0$) and $x = L$ ($\theta = 2\pi$) describe the same point. Note how ψ joins with itself smoothly when the ends are glued.

electron actually lives on a ring, not just of finite size, but of radius R on the scale of a micron (10^{-6} m).

In any event, we can finally write down the normalized ψ_p on this circle:

$$\psi_p = \frac{1}{\sqrt{L}} \exp\left[\frac{ipx}{\hbar}\right].$$

(21.11)

21.2   Single-valuedness and quantization of momentum

Since the universe is a circle, it could also be parameterized by an angle θ restricted to $0 \leq \theta \leq 2\pi$. It is obvious that a point labeled by an angle θ is identical to the one labeled by $\theta + 2\pi$. The single-valued condition for ψ assumes the form

$$\psi(\theta) = \psi(\theta + 2\pi).$$

(21.12)

This just means that ψ is a periodic function of θ with period 2π. We are also free to use a linear coordinate x that runs around the circle

$$x = R\theta \qquad [x : 0, 2\pi R = L].$$

(21.13)

In terms of x, the single-valuedness condition becomes

$$\psi(x) = \psi(x + L),$$

(21.14)

as illustrated in Figure 21.2.

Let us consider in this universe a normalized state of definite momentum

$$\psi_p(x) = \frac{1}{\sqrt{L}} \exp\left[\frac{ipx}{\hbar}\right]. \tag{21.15}$$

The probability density $P(x)$ is x-independent:

$$P(x) = |\psi|^2 = \psi^*(x)\psi(x)$$

$$= \frac{1}{\sqrt{L}} \exp\left[-\frac{ipx}{\hbar}\right] \frac{1}{\sqrt{L}} \exp\left[\frac{ipx}{\hbar}\right] = \frac{1}{L} \tag{21.16}$$

and is normalized as promised:

$$\int_0^L P(x)dx = \int_0^L \frac{1}{L}dx = 1. \tag{21.17}$$

So far there has been no restriction on p: it could be any real number. This changes when the single-valued requirement

$$\psi(x+L) = \psi(x), \tag{21.18}$$

imposed on all wave functions on the circle, is imposed on ψ_p:

$$\frac{1}{\sqrt{L}} \exp\left[\frac{ipx}{\hbar}\right] = \frac{1}{\sqrt{L}} \exp\left[\frac{ip(x+L)}{\hbar}\right] \tag{21.19}$$

$$= \frac{1}{\sqrt{L}} \exp\left[\frac{ipx}{\hbar}\right] \exp\left[\frac{ipL}{\hbar}\right]. \tag{21.20}$$

This means

$$\exp\left[\frac{ipL}{\hbar}\right] = 1. \tag{21.21}$$

As $e^{i\theta}$ lies on the unit circle in the complex plane and has period 2π,

$$\exp\left[\frac{ipL}{\hbar}\right] = 1 \quad \text{means} \tag{21.22}$$

$$\frac{pL}{\hbar} = 2\pi m \quad \text{where } m = \ldots -2, -1, 0, 1, 2, \ldots \tag{21.23}$$

In the unlikely event you are shaky about this, here is a second chance. Starting with

$$\exp\left[\frac{ipL}{\hbar}\right] = \cos\left[\frac{pL}{\hbar}\right] + i\sin\left[\frac{pL}{\hbar}\right] = 1, \qquad (21.24)$$

and equating the real and imaginary parts of the two sides, we find two conditions:

$$\cos\left[\frac{pL}{\hbar}\right] = 1 \qquad (21.25)$$

$$\sin\left[\frac{pL}{\hbar}\right] = 0. \qquad (21.26)$$

There are infinitely many solutions to this pair:

$$\frac{pL}{\hbar} = 2\pi m \quad \text{where } m = \ldots -2, -1, 0, 1, 2, \ldots \qquad (21.27)$$

The allowed values of momentum are

$$p = \frac{2\pi \hbar m}{L} \quad \text{where } m = \ldots -2, -1, 0, 1, 2, \ldots \qquad (21.28)$$

I will often use the symbol p_m to denote the momentum associated with the integer m:

$$p_m \equiv \frac{2\pi \hbar m}{L} \quad \text{where } m = \ldots -2, -1, 0, 1, 2, \ldots \qquad (21.29)$$

and use m instead of p_m as the label for the state. The label m has the nice feature that it runs over the integers.

Pictorially, these allowed values $p = p_m$ ensure that the cosine and sine, the real and imaginary parts of the wave function, complete an integer number (m) of full cycles as we go around the circle and join smoothly on to themselves. For example, the real part of the wave function varies as

$$\cos\left[\frac{px}{\hbar}\right] = \cos\left[\frac{2\pi m\hbar x}{\hbar L}\right] = \cos\left[\frac{2\pi mx}{L}\right] \qquad (21.30)$$

and as x grows from 0 to L, the argument of the cosine changes by $2\pi m$ and it completes m full cycles. The same goes for the imaginary part, the sine. The case of $m = 0$ is special, because $e^{i0} = 1$. This constant wave function is also periodic but completes zero cycles. (When $m = 0$, the real part [cosine] equals 1 and the imaginary part [sine] vanishes.)

21.2.1 Quantization

Now this is a very big moment in your life. Why? Because you have just encountered the *quantization* of (the allowed values of) a dynamical variable, which happens to be the momentum p in this instance. This is the *quantum* of quantum mechanics. Classically, a particle living in a circle can travel with any momentum, but quantum mechanically, only the values given by p_m are allowed. The quantization came from demanding that the wave function be single-valued. The origin of quantization is often a mathematical requirement: single-valuedness in this case and normalizability in some others.

In the limit in which L is very, very large, on the macroscopic scale, the spacing between the allowed values of $p = 2\pi m\hbar/L$ becomes very, very small, and you may not even realize that p is taking only discrete values. The difference between two adjacent allowed values of p is

$$dp = \frac{2\pi (m+1)\hbar}{L} - \frac{2\pi m\hbar}{L} = \frac{2\pi \hbar}{L}. \qquad (21.31)$$

When m changes to $m+1$, p changes by a number of order $10^{-34} kg \cdot m/s$ assuming L is of the order of a meter. At this scale, the quantum world will appear classical. By contrast, in a quantum ring of radius, say $1\mu m$, the quantization of p will be a very real effect that needs to be reckoned with.

For a particle moving on a circle, it is natural to rewrite the quantization of p

$$p = \frac{2\pi \hbar m}{L} = \frac{2\pi \hbar m}{2\pi R} \qquad (21.32)$$

in a more appealing way:

$$pR = m\hbar, \qquad (21.33)$$

as the quantization of *angular momentum pR in multiples of* $\hbar$, a condition you may have encountered earlier without proof. Now you see it is a consequence of demanding single-valuedness.

Let us rewrite the function ψ_p in terms of m and θ:

$$\psi_p(x) = \frac{1}{\sqrt{L}} \exp\left[\frac{ipx}{\hbar}\right] \tag{21.34}$$

$$= \frac{1}{\sqrt{L}} \exp\left[\frac{i\frac{2\pi \hbar m}{L}x}{\hbar}\right] \tag{21.35}$$

$$= \frac{1}{\sqrt{L}} \exp\left[i\frac{2\pi m}{2\pi R}x\right], \quad \text{or since } x = R\theta \tag{21.36}$$

$$= \frac{1}{\sqrt{L}} \exp\left[im\theta\right] \equiv \psi_m(\theta). \tag{21.37}$$

The state is the same, whether we refer to it by its momentum and write it as a function of x, or by its angular momentum, and write it as a function of θ. *I will go back and forth between these two equivalent labels p and m for the state.* Later on m will stand for the mass of the particle, but in this chapter it will label the allowed values of momentum and angular momentum.

21.2.2 The integral of $\psi_p(x)$

Here is one important result that you should commit to memory: *The integral of every $\psi_p(x)$ vanishes except for $p = 0$.*

Here is the proof:

$$\int_0^L \psi_p(x)\,dx = \int_0^L \frac{1}{\sqrt{L}} \exp\left[\frac{ipx}{\hbar}\right] dx \tag{21.38}$$

$$= \int_0^L \frac{1}{\sqrt{L}} \left[\cos\frac{2\pi mx}{L} + i\sin\frac{2\pi mx}{L}\right] dx$$

$$= 0 \tag{21.39}$$

because the sine and cosine complete m full cycles. The case $p = m = 0$ is special:

$$\int_0^L \frac{e^{i0}}{\sqrt{L}} dx = \sqrt{L}. \tag{21.40}$$

Let us re-derive this directly with the complex exponential so you get used to it. We need the result

$$\int_a^b e^{\alpha x} dx = \frac{1}{\alpha} \left[e^{\alpha b} - e^{\alpha a} \right], \tag{21.41}$$

which is valid even if α is complex, and in particular purely imaginary. (In the latter case you may prove it yourself by using Euler's formula to convert the exponential to the sines and cosines, integrating them and rewriting the answer in terms of complex exponentials.) Proceeding, we find

$$\frac{1}{\sqrt{L}} \int_0^L \exp\left[\frac{ipx}{\hbar} \right] dx = \frac{1}{\sqrt{L}} \int_0^L \exp\left[\frac{2\pi\, imx}{L} \right] dx \tag{21.42}$$

$$= \frac{\sqrt{L}}{2\pi\, im} \left[e^{2\pi\, im} - e^{i0} \right] \tag{21.43}$$

$$= 0, \tag{21.44}$$

assuming $m \neq 0$. If $m = 0$, the preceding formula yields the indeterminate form $0/0$. It is then best to go back to the integral and find readily that

$$\frac{1}{\sqrt{L}} \int_0^L e^{i0} dx = \sqrt{L}. \tag{21.45}$$

21.3 Measurement postulate: momentum

Let us now consider a particle on a ring described by some generic wave function $\psi(x)$ depicted earlier in Figure 21.2. Of course, $\psi(x)$ meets itself smoothly when you go around the circle: it is a single-valued function of period L. But it is not a state of definite momentum or angular momentum because it is not an oscillating exponential of definite λ.

What can we say about the particle in such a generic state?

The first is old stuff: $|\psi(x)|^2 = P(x)$ gives the probability density as a function of x. That means that if you take a million particles on a million rings each in exactly this quantum state and make the position measurements (using a Heisenberg microscope to locate x to arbitrarily high precision, with no concern for p) the resulting histogram will look like $P(x)$.

But there is more to life than just knowing the answer to "Where is the particle?" In classical mechanics you can also ask, "What is its momentum?" The only time we seem to know the answer for sure in quantum mechanics is if $\psi(x) = \psi_p(x)$, the complex exponential with a definite period $\lambda = 2\pi\hbar/p$. What about a general single-valued wave function not of this form? What will we find if we measure momentum? Will the theory again give the probabilities for the different outcomes? Will we need to introduce *another* wave function $A(p)$ that varies with p and gives the probability for obtaining a value p by the relation $P(p) = |A(p)|^2$?

(Bear in mind that $\psi_p(x)$ is a function of x *labeled* by the momentum p you are guaranteed to get upon measurement, while $A(p)$ is a function of p, whose mod-square gives the odds for measuring various values of p.)

I do not expect you to answer these questions because they are not decided by logic or mathematics. We need a postulate like the one that said $|\psi(x)|^2$ is $P(x)$. The postulate would tell us how to get $P(p)$, the odds for getting the value p in a momentum measurement.

Let us work toward the general case by first considering a simple example

$$\psi(x) = A(p_1)\frac{1}{\sqrt{L}}\exp\left[\frac{ip_1 x}{\hbar}\right] + A(p_2)\frac{1}{\sqrt{L}}\exp\left[\frac{ip_2 x}{\hbar}\right]$$

$$\equiv A(p_1)\psi_{p_1}(x) + A(p_2)\psi_{p_2}(x) \qquad (21.46)$$

where $A(p_1)$ and $A(p_2)$ are constants independent of x, while $p_1 = \frac{2\pi m_1 \hbar}{L}$ and $p_2 = \frac{2\pi m_2 \hbar}{L}$ are two allowed momenta, and $\psi_{p_1}(x)$ and $\psi_{p_2}(x)$ the corresponding *normalized* wave functions.

This is a *superposition* of two *normalized* wave functions $\psi_{p_1}(x)$ and $\psi_{p_2}(x)$ each describing a state of definite momentum (p_1 or p_2). We only know that if $A(p_2)$ were zero, we would surely get p_1 and if $A(p_1)$ were zero we would surely get p_2. But suppose neither is zero. Will we get a momentum that is some weighted average of p_1 and p_2? What if this average is not one of the allowed values of p on the circle? What will be the state right after the measurement? Will it be single-valued on the circle?

The answer is given by the two-part measurement postulate:

- **Part 1.** The result of a momentum measurement will yield p_1 with *relative* probability $P'(p_1) = |A(p_1)|^2$ and p_2 with relative probability $P'(p_2) = |A(p_2)|^2$. (We use probability and not probability density because the allowed values of p are discrete and labeled by the integer m.)
- **Part 2.** The state right after the measurement will be a state of the momentum that was obtained in the measurement.

There are many points to note in this postulate.

1. The only values a momentum measurement will yield correspond to the two associated with the wave functions in the superposition, namely p_1 or p_2, *not some kind of average of the two*. The possible momenta correspond to single-valued wave functions.
2. From the *relative* probabilities $P'(p_1) = |A(p_1)|^2$ and $P'(p_2) = |A(p_2)|^2$, we may extract the *absolute* probabilities in an obvious way:

$$P(p_1) = \frac{|A(p_1)|^2}{|A(p_1)|^2 + |A(p_2)|^2} \tag{21.47}$$

$$P(p_2) = \frac{|A(p_2)|^2}{|A(p_1)|^2 + |A(p_2)|^2}. \tag{21.48}$$

3. It is crucial that the functions $\psi_p(x)$ in Eqn. 21.46 be normalized for Part 1 to be valid. Why do we suddenly care how $\psi_p(x)$ is normalized after saying repeatedly that the number A in $Ae^{ipx/\hbar}$ does not matter? The answer is that the *overall* scale of any given wave function is unphysical, but not the *relative* scales of two wave functions. Thus we may rescale the $\psi(x)$ on the left-hand side of Eqn. 21.46, say by a factor of 10, and simultaneously *both* $\psi_{p_1}(x)$ and $\psi_{p_2}(x)$ on the right-hand side also by 10. (Thus, instead of using normalized $\psi_p(x)$, you may rescale all of them by some common amount. For example, you can drop the $1/\sqrt{L}$ in front of both of them.)

Here is an analogy. In a world of rays, where only directions matter, we can use **i** or 10**i** to indicate east and **j** or 13**j** to indicate north. But to indicate northeast, we may use **i** + **j** or 10**i** + 10**j** or 13**i** + 13**j** but not 10**i** + 13**j**.

Hereafter every state of definite momentum $\psi_p(x)$ will be assumed to be normalized to unity.

4. If the measurement yields the value p_1, the state ψ, which used to be a sum over $\psi_{p_1}(x)$ and $\psi_{p_2}(x)$, collapses to just one term, namely $\psi_{p_1}(x)$. A similar result holds if p_2 is obtained.

5. If the measurement yields the value p_1, an *immediate* remeasurement will again yield p_1. This has to be true if there is any sense to saying that the particle was found in a state of momentum p_1 when momentum was measured. As time goes by the state may change, but the value p_1 should persist at least for an infinitesimal time. The same thing happens in a position measurement: the wave function collapses to the point where the particle was found right after measurement.

6. These results generalize in an obvious manner when the superposition describing the state includes an arbitrary number of terms:

$$\psi(x) = A(p_1)\frac{1}{\sqrt{L}}\exp\left[\frac{ip_1x}{\hbar}\right] + A(p_2)\frac{1}{\sqrt{L}}\exp\left[\frac{ip_2x}{\hbar}\right]$$

$$+ A(p_3)\frac{1}{\sqrt{L}}\exp\left[\frac{ip_3x}{\hbar}\right]\ldots$$

$$= \sum_j A(p_j)\frac{1}{\sqrt{L}}\exp\left[\frac{ip_jx}{\hbar}\right], \tag{21.49}$$

where j is a label that runs over all allowed values of momenta.

A natural label for p_j is the integer m (the angular momentum in units of $\hbar$) that enters the quantization condition:

$$p_m = \frac{2\pi m\hbar}{L} \quad m = 0, \pm1, \pm2\ldots \tag{21.50}$$

In terms of m, the most general such superposition assumes the form

$$\psi(x) = \sum_{m=-\infty}^{\infty} A(p_m)\frac{1}{\sqrt{L}}\exp\left[\frac{ip_mx}{\hbar}\right]$$

$$= \sum_{m=-\infty}^{\infty} A(p_m)\frac{1}{\sqrt{L}}\exp\left[\frac{2\pi imx}{L}\right]. \tag{21.51}$$

In this notation $|A(p_m)|^2$ gives the relative probability for obtaining p_m and the state collapses to the one particular m that was measured. To find the absolute probability we use

$$P(p_m) = \frac{|A(p_m)|^2}{\sum_{m'} |A(p_{m'})|^2}. \qquad (21.52)$$

(The label m' being summed over in the denominator is just like m and runs over the same values.)

Equation 21.51 does not mean there is always an infinite number of terms in the superposition. We can always restrict the sum. For example, we obtain the simple example we began with by choosing just $A(p_1)$ and $A(p_2) \neq 0$. Even simpler is the case with just one of them, say $A(p_{43}) \neq 0$. This would correspond in our notation to $\psi_{p_{43}}(x)$ where

$$p_{43} = \frac{2\pi \cdot 43 \cdot \hbar}{L}. \qquad (21.53)$$

7. (This is an optional topic. Come back to it later if you are feeling overwhelmed.) There is a different way to get the normalized probabilities. *Instead of rescaling the P's as in Eqn. 21.52 we could normalize the given $\psi(x)$ to reach the same goal.* In other words, if we *first* normalized the given $\psi(x)$ and *then* computed the coefficients $A(p)$ for the normalized $\psi(x)$, these coefficients would automatically give the normalized probabilities: $P(p) = |A(p)|^2$. I state this without proof and invite you to check this for the simple case with just two non-zero $A(p)$'s. If you start with the $\psi(x)$ in Eqn. 21.46, compute the integral of $|\psi(x)|^2$ and do the appropriate rescaling to normalize ψ, you will find that the rescaled $\tilde{A}(p)$'s are given in terms of the original $A(p)$ by

$$\tilde{A}(p_1) = \frac{A(p_1)}{\sqrt{|A(p_1)|^2 + |A(p_2)|^2}} \qquad (21.54)$$

and likewise for $\tilde{A}(p_2)$. This ensures that

$$\sum_{m=1,2} |\tilde{A}(p_m)|^2 = 1. \qquad (21.55)$$

So that is the complete answer to the question of what we will get when we measure momentum *for wave functions of the form in Eqn. 21.51.* We have graduated from wave functions of the form $\psi_p(x)$, which were guaranteed to yield the value p, to functions that are superpositions of such functions with coefficients $A(p)$. In this case the measurement postulate tells us we could get any p that was present in the sum with relative probability $|A(p)|^2$.

What about functions not of this form? This is a reasonable question. While it is obvious that every superposition of $\psi_p(x)$ with coefficients $A(p)$ is periodic in L (because each term is) and therefore represents an allowed wave function on the circle, the converse is not obvious. Is every allowed wave function ψ that obeys $\psi(x) = \psi(x+L)$ such a superposition? If not, what is the corresponding measurement postulate?

Here is the great news: *There are no other allowed wave functions besides such superpositions!* This is a purely mathematical result due to Joseph Fourier (1768–1830). (In learning quantum mechanics it is important to distinguish between postulates deduced from experiment and theorems deduced by mathematical reasoning.)

Fourier's theorem I. *Every* allowed wave function $\psi(x)$ obeying $\psi(x) = \psi(x + L)$ may be written as a superposition of $\psi_p(x)$'s with suitable coefficients $A(p)$:

$$\psi(x) = \sum_{m=-\infty}^{\infty} A(p_m)\frac{1}{\sqrt{L}}\exp\left[\frac{ip_m x}{\hbar}\right], \text{ where } p_m = \frac{2\pi m\hbar}{L}. \quad (21.56)$$

Fourier's theorem II. *The coefficients $A(p)$ corresponding to a given $\psi(x)$ are given by the following integrals:*

$$A(p) = \int_0^L \psi_p^*(x)\psi(x)\,dx. \quad (21.57)$$

Consider the first theorem. On the left is a generic wave function $\psi(x)$ obeying $\psi(x) = \psi(x + L)$. It has a period L. On the right are the functions describing particles with a definite momentum, $\psi_p(x)$, where $p = 2\pi m\hbar/L$. These too are periodic in L, but *in addition* they also complete m full trigonometric cycles *within* the length L. Fourier's theorem assures us that any periodic (single-valued) function on the circle may be written as a linear superposition of such oscillatory $\psi_p(x)$ with some coefficients $A(p_m)$.

This result may be more familiar to some of you if rewritten in terms of states of definite angular momentum $pR = m\hbar$ defined earlier and $\theta = x/R$:

$$\psi_p(x) = \frac{1}{\sqrt{L}} \exp\left[i\frac{2\pi m}{2\pi R}x\right] = \frac{1}{\sqrt{L}} \exp[im\theta] \equiv \psi_m(\theta). \quad (21.58)$$

The above postulate now becomes

$$\psi(\theta) = \sum_{m=-\infty}^{\infty} A(m)\frac{1}{\sqrt{L}} \exp[im\theta]. \quad (21.59)$$

(You may have encountered Fourier series written in terms of the real and imaginary parts of $e^{im\theta}$.)

While the first theorem assures us that every legitimate function $\psi(x)$ on the ring *can* be written as a sum over $\psi_p(x)$ with coefficients $A(p)$,

$$\psi(x) = \sum_p A(p)\psi_p(x) = \sum_p A(p)\frac{1}{\sqrt{L}} \exp\left[\frac{ipx}{\hbar}\right], \quad (21.60)$$

the second tells us *how* to determine the expansion coefficients $A(p)$ for a given $\psi(x)$:

$$A(p) = \int_0^L \psi_p^*(x)\psi(x)\,dx. \quad (21.61)$$

Without explicit knowledge of these coefficients, we cannot give the probabilities for obtaining the different p's.

For now I ask you to accept Fourier's theorems. I will say a few things in a later section that may help you understand them in terms of more familiar ideas from elementary vector analysis.

21.3.1 An example solvable by inspection

I begin with an example where finding the $A(p)$ ends up being very easy. The state is

$$\psi(x) = A\cos\left[\frac{6\pi x}{L}\right] \tag{21.62}$$

where A is some real constant. This is a legitimate wave function because it obeys $\psi(x) = \psi(x + L)$. We know from Fourier's theorem that this function *may* be written as a series of the form Eqn. 21.60.

We can always find the $A(p)$ using

$$A(p) = \int \psi_p^*(x)\psi(x)\,dx, \tag{21.63}$$

but it turns out that in this case we can read off the $A(p)$ by inspection if we first cast $\psi(x)$ in a suggestive form using Euler's identity. Here are the details.

$$\psi(x) = A\cos\left[\frac{6\pi x}{L}\right] \tag{21.64}$$

$$= \frac{A}{2}\left[\exp\left[\frac{6\pi ix}{L}\right] + \exp\left[\frac{-6\pi ix}{L}\right]\right] \tag{21.65}$$

$$= \frac{A\sqrt{L}}{2}\left[\frac{1}{\sqrt{L}}\exp\left[\frac{6\pi ix}{L}\right] + \frac{1}{\sqrt{L}}\exp\left[\frac{-6\pi ix}{L}\right]\right] \tag{21.66}$$

$$= \frac{A\sqrt{L}}{2}\left[\psi_{p=6\pi\hbar/L}(x) + \psi_{p=-6\pi\hbar/L}(x)\right] \tag{21.67}$$

$$\equiv \frac{A\sqrt{L}}{2}\left[\psi_{m=3}(x) + \psi_{m=-3}(x)\right] \tag{21.68}$$

where in the last equation I employ the integer m as a label instead of the corresponding momentum $p = 2\pi m\hbar/L$.

We have managed to write the given $\psi(x)$ in the form of a Fourier series

$$\psi(x) = \sum_p A(p)\psi_p(x) = \sum_p A(p)\frac{1}{\sqrt{L}}\exp\left[\frac{ipx}{\hbar}\right]. \tag{21.69}$$

By comparing Eqns. 21.68 and 21.69 we can see that the only possible momenta are

$$p = \pm \frac{3 \cdot 2\pi \hbar}{L} \tag{21.70}$$

and the coefficients are

$$A(p_3) = \frac{A\sqrt{L}}{2} \tag{21.71}$$

$$A(p_{-3}) = \frac{A\sqrt{L}}{2} \tag{21.72}$$

$$A(p_m) = 0 \quad |m| \neq 3. \tag{21.73}$$

It should be obvious that since the two non-zero $A(p_m)$'s are equal, the normalized probabilities are

$$P(p_3) = P(p_{-3}) = \frac{1}{2}. \tag{21.74}$$

If measurement yields a value $m = 3$, the state will reduce to $\psi_{m=3}(x)$.

Let us now re-derive the same $A(p_m)$ using Fourier's theorem of Eqn. 21.61:

$$A(p) = \int_0^L \psi_p^*(x)\psi(x)\,dx$$

$$= \int_0^L \frac{1}{\sqrt{L}} \exp\left[\frac{-ipx}{\hbar}\right] \psi(x)\,dx \tag{21.75}$$

$$= A \int_0^L \frac{1}{\sqrt{L}} \exp\left[\frac{-ipx}{\hbar}\right] \cos\left[\frac{6\pi x}{L}\right] dx \tag{21.76}$$

$$= \frac{A}{2\sqrt{L}} \int_0^L \exp\left[\frac{-ipx}{\hbar}\right]$$

$$\times \left[\exp\left[\frac{6\pi i x}{L}\right] + \exp\left[\frac{-6\pi i x}{L}\right]\right] dx. \tag{21.77}$$

Let us now write $p = \frac{2\pi\, m\hbar}{L}$ and continue

$$A(p) = \frac{A}{2\sqrt{L}} \int_0^L \exp\left[-i\frac{2\pi\, mx}{L}\right]$$

$$\times \left[\exp\left[\frac{6\pi\, ix}{L}\right] + \exp\left[\frac{-6\pi\, ix}{L}\right]\right] dx \qquad (21.78)$$

$$= \frac{A}{2\sqrt{L}} \int_0^L \left[\exp\left[\frac{2\pi\,(3-m)ix}{L}\right]\right.$$

$$\left. + \exp\left[\frac{2\pi\, i(-3-m)x}{L}\right]\right] dx. \qquad (21.79)$$

Both exponentials describe states of definite momentum. I have already shown that their integral is zero unless the momentum vanishes. This happens in the first term when $m = 3$ and the exponential becomes $e^0 = 1$ and integrates to L, giving

$$A(p_3) = \frac{A\sqrt{L}}{2}. \qquad (21.80)$$

Likewise the second exponential survives integration when $m = -3$ and leads to

$$A(p_{-3}) = \frac{A\sqrt{L}}{2}. \qquad (21.81)$$

If $m \neq \pm 3$, both exponentials complete an integral number of oscillations and integrate to zero. So $A(m \neq \pm 3) = 0$.

These are just the values we obtained earlier by simply writing the given ψ (a cosine) in terms of states of momentum with $m = \pm 3$ and reading off the coefficients by inspection.

21.3.2 Using a normalized ψ

I mentioned earlier that if the original ψ is normalized, $|A(p)|^2$ will be absolute probabilities. Let us verify this for the case where we are given an unnormalized ψ

$$\psi(x) = A\cos\frac{6\pi\, x}{L}. \qquad (21.82)$$

We must choose A so that

$$A^2 \int_0^L \cos^2\left[\frac{6\pi x}{L}\right] dx = 1. \tag{21.83}$$

As x goes from 0 to L, the angle within the cosine changes by 6π. It completes three full cycles. We have seen many times that the average of $\cos^2\theta$ over any number of full cycles is $\frac{1}{2}$. So we find

$$A^2 \frac{L}{2} = 1 \tag{21.84}$$

$$A = \sqrt{\frac{2}{L}} \quad \text{so that the normalized } \psi \text{ is} \tag{21.85}$$

$$\psi(x) = \sqrt{\frac{2}{L}} \cos\frac{6\pi x}{L}. \tag{21.86}$$

To find $A(p)$ we simply rewrite the cosine in terms of exponentials:

$$\psi(x) = \sqrt{\frac{2}{L}}\frac{1}{2}\left[\exp\left[\frac{6\pi ix}{L}\right] + \exp\left[\frac{-6\pi ix}{L}\right]\right] \tag{21.87}$$

$$= \frac{1}{\sqrt{2}}\left[\frac{1}{\sqrt{L}}\exp\left[\frac{6\pi ix}{L}\right] + \frac{1}{\sqrt{L}}\exp\left[\frac{-6\pi ix}{L}\right]\right] \tag{21.88}$$

$$= \frac{1}{\sqrt{2}}\left[\psi_{p=6\pi\hbar/L}(x) + \psi_{p=-6\pi\hbar/L}(x)\right] \tag{21.89}$$

$$\equiv \frac{1}{\sqrt{2}}\left[\psi_{m=3}(x) + \psi_{m=-3}(x)\right]. \tag{21.90}$$

Comparison to Eqn. 21.60 tells us they are

$$A(p_3) = \frac{1}{\sqrt{2}} \tag{21.91}$$

$$A(p_{-3}) = \frac{1}{\sqrt{2}} \tag{21.92}$$

$$A(p_m, |m| \neq 3) = 0. \tag{21.93}$$

The non-vanishing absolute probabilities are given by the squares of these numbers

$$P(p_{\pm3}) = \frac{1}{2} \qquad (21.94)$$

and add up to 1 as promised.

The expectation value of p is evidently zero, but here are the steps:

$$\langle p \rangle = \sum_i P(p_i)p_i = \frac{1}{2}p_3 + \frac{1}{2}p_{-3} = 0 \qquad (21.95)$$

because $p_3 = -p_{-3} = \frac{6\pi\hbar}{L}$.

The uncertainty squared is

$$(\Delta p)^2 = \sum_i P(p_i)(p_i - \langle p \rangle)^2 \qquad (21.96)$$

$$= \sum_i P(p_i)(p_i - 0)^2 \qquad (21.97)$$

$$= \frac{1}{2}p_3^2 + \frac{1}{2}p_{-3}^2 = p_3^2 \quad \text{and the uncertainty is} \qquad (21.98)$$

$$\Delta p = p_3 = \frac{6\pi\hbar}{L}. \qquad (21.99)$$

21.4   Finding $A(p)$ by computation

Now we turn to the example where you actually have to do an integral to find $A(p)$ (and then square it to get $P(p)$).

We take the interval of length L to be in the range

$$\frac{-L}{2} \leq x \leq \frac{L}{2}. \qquad (21.100)$$

The ends $x = \pm L/2$ are to be glued to form the circle. The un-normalized wave function of interest

$$\psi(x) = Ae^{-\alpha|x|} \qquad (21.101)$$

is depicted in Figure 21.3 for the case $A = 1$ as a function of αx.

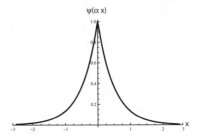

Figure 21.3   An exponential wave function that dies very rapidly as we approach the end points $\pm L/2$, which are glued to form the circle.

It is highest at the origin and falls exponentially at the same rate for positive and negative x due to the $|x|$ dependence of ψ. How far can we go from the origin before ψ becomes negligible? That happens when $\alpha|x|$ is large or when $|x| \gg 1/\alpha$. So this is a particle whose position has an uncertainty of order $1/\alpha$:

$$\Delta x \simeq \frac{1}{\alpha}. \tag{21.102}$$

We can vary the width Δx by varying α but it is understood that even if ψ is broad near the origin it is negligible at the points $\pm L/2$ that are glued to form the circle. In other words we assume

$$\alpha L \gg 1 \quad \text{and} \quad e^{-\alpha|\pm L/2|} \ll 1. \tag{21.103}$$

Let us explore the content of this wave function. The first question I can ask is, "If I look for its position what will I find?" The probability density is given by

$$P(x) = |\psi(x)|^2 = \psi^2(x) = A^2 e^{-2\alpha|x|}, \tag{21.104}$$

which is also an exponential but with double the slope as ψ.

Let us now normalize it by demanding

$$A^2 \int_{-L/2}^{L/2} e^{-2\alpha|x|} \, dx = 1. \tag{21.105}$$

To simplify life, I am going to extend the limits to $\mp\infty$. This is an insignificant modification because, by assumption, αL is large and $\psi^2(x)$ is dead long before we get to the ends at $x = \pm L/2$ (which are glued). So the normalization condition is

$$1 = A^2 \int_{-\infty}^{\infty} e^{-2\alpha|x|} dx \tag{21.106}$$

$$= 2A^2 \int_{0}^{\infty} e^{-2\alpha|x|} dx \quad \text{because } |x| \text{ is even} \tag{21.107}$$

$$= 2A^2 \frac{1}{2\alpha} \quad \text{which means} \tag{21.108}$$

$$A = \sqrt{\alpha} \quad \text{and the normalized } \psi \text{ is} \tag{21.109}$$

$$\psi(x) = \sqrt{\alpha}\, e^{-\alpha|x|}. \tag{21.110}$$

Before computing $A(p)$, let us pause to find the statistical properties of this state. By symmetry, $\langle x \rangle = 0$. The uncertainty squared and uncertainty are (setting $L = \infty$)

$$(\Delta x)^2 = \int_{-\infty}^{\infty} \alpha e^{-2\alpha|x|} x^2 dx \tag{21.111}$$

$$= 2\alpha \int_{0}^{\infty} e^{-2\alpha x} x^2 dx$$

$$= \frac{1}{2\alpha^2} \quad \text{on integrating by parts twice} \tag{21.112}$$

$$\Delta x = \frac{1}{\sqrt{2}\alpha}, \tag{21.113}$$

not far from our crude estimate $\Delta x \simeq 1/\alpha$ of Eqn. 21.102.

Now we compute $A(p)$. We cannot get them by inspection, because the ψ above is not a sum of complex exponentials corresponding to states of definite momentum. We have to deal with the general recipe

$$A(p) = \int_{-\infty}^{\infty} \psi_p^*(x)\psi(x)dx$$

$$= \int_{-\infty}^{\infty} \frac{1}{\sqrt{L}} e^{-ipx/\hbar} \sqrt{\alpha}\, e^{-\alpha|x|} dx. \tag{21.114}$$

The integral is a little tricky because $|x|$ equals x when $x > 0$ and $-x$ when $x < 0$. So let us break up the integral into two parts, one for $x > 0$ and one for $x < 0$:

$$A(p) = \int_{-\infty}^{\infty} \sqrt{\frac{\alpha}{L}} e^{-ipx/\hbar} e^{-\alpha|x|} dx \tag{21.115}$$

$$= \int_{0}^{\infty} \sqrt{\frac{\alpha}{L}} e^{-ipx/\hbar} e^{-\alpha x} dx$$

$$+ \int_{-\infty}^{0} \sqrt{\frac{\alpha}{L}} e^{-ipx/\hbar} e^{\alpha x} dx \tag{21.116}$$

$$\equiv I_{+} + I_{-}. \tag{21.117}$$

The evaluation of I_{+} is simple and only the lower limit contributes

$$I_{+} = \int_{0}^{\infty} \sqrt{\frac{\alpha}{L}} e^{-ipx/\hbar} e^{-\alpha x} dx \tag{21.118}$$

$$= \sqrt{\frac{\alpha}{L}} \frac{e^{(-\alpha - ip/\hbar)x}}{-\alpha - ip/\hbar} \Big|_{0}^{\infty} \tag{21.119}$$

$$= \sqrt{\frac{\alpha}{L}} \frac{1}{\alpha + ip/\hbar}. \tag{21.120}$$

I leave it to you to show (by changing x to $-x$ in the integral) that

$$I_{-} = \sqrt{\frac{\alpha}{L}} \frac{1}{\alpha - ip/\hbar}. \tag{21.121}$$

Therefore

$$A(p) = I_{+} + I_{-} = \sqrt{\frac{\alpha}{L}} \left[\frac{1}{\alpha + ip/\hbar} + \frac{1}{\alpha - ip/\hbar} \right] \tag{21.122}$$

$$= \sqrt{\frac{\alpha}{L}} \frac{2\alpha}{\alpha^2 + (p/\hbar)^2} \quad \text{and} \tag{21.123}$$

$$P(p) = \frac{4\alpha^3}{L(\alpha^2 + (p/\hbar)^2)^2}. \tag{21.124}$$

Strictly speaking we are interested in this function only at the quantized values of p. Let us assume here that L is very large and p essentially continuous. Ignoring overall constants, the function has the form

$$P(p) = \frac{\text{constant}}{(p^2 + (\alpha\hbar)^2)^2}. \tag{21.125}$$

It is peaked at $p = 0$ and falls off smoothly on a scale set by $\alpha\hbar$. To characterize its width roughly, we identify the point where it falls to a fourth of its maximum value because it is easy to locate. This happens at

$$p = \pm\alpha\hbar. \tag{21.126}$$

Thus the uncertainty in p is of the order

$$\Delta p \simeq 2\alpha\hbar, \tag{21.127}$$

which, combined with

$$\Delta x \simeq \frac{1}{\alpha}, \tag{21.128}$$

gives us

$$\Delta x \cdot \Delta p \simeq 2\hbar. \tag{21.129}$$

Thus we find that the narrower the function is in x, the bigger the spread in the possible momenta you can get. So squeezing it in x broadens it out in p and the opposite is also true. And that is the origin of the uncertainty principle.

Long before quantum mechanics, it was known in Fourier analysis that a function that is narrow in x needs many wave numbers k or wavelengths λ in its expansion. This was stated in the form

$$\Delta x \cdot \Delta k \gtrsim 1 \tag{21.130}$$

with no reference to $\hbar$. Quantum mechanics enters when we associate a momentum $p = \hbar k$ with the wave number k. Multiplying both sides by $\hbar$ we arrive at the uncertainty principle.

To apply the precise form of the uncertainty principle,

$$\Delta x \Delta p \geq \frac{1}{2}\hbar, \tag{21.131}$$

Δx and Δp have to be the uncertainties defined in Eqns. 20.26 and 20.29. We already have found $\Delta x = (1/\sqrt{2}a)$. Computing the uncertainty Δp from $P(p)$ is complicated by the fact that p takes on discrete values $p = p_m$. There is a simple way to proceed in the limit $L \to \infty$, when the allowed p's become very close. In that limit we may replace the sum over p of any function $f(p)$ by an integral

$$\sum_p f(p) \to \frac{L}{2\pi\hbar} \int f(p)\,dp. \tag{21.132}$$

Here is the logic behind this trick, which is used a lot in many advanced courses. If you want to skip the details and just use the result above, jump to Eqn. 21.136.

Remember how the integral of a function $f(x)$ is found. We plot $f(x)$ vertically at a dense set of points x_i separated by dx, and do the sum

$$\int f(x)\,dx = \lim_{dx \to 0} \sum_i f(x_i)\,dx. \tag{21.133}$$

In our case, we have functions like $P(p_m)$ defined at points $p_m = \frac{2\pi m\hbar}{L}$ with a spacing

$$dp = p_{m+1} - p_m = \frac{2\pi(m+1)\hbar}{L} - \frac{2\pi m\hbar}{L} = \frac{2\pi\hbar}{L}, \tag{21.134}$$

which vanishes as $L \to \infty$.

Let us verify that our probabilities add up to 1. By definition of the integral

$$\lim_{dp \to 0}\left[\sum_{m=-\infty}^{\infty} P(p_m)(dp = \frac{2\pi\hbar}{L}) \right] = \int_{-\infty}^{\infty} P(p)\,dp. \tag{21.135}$$

Therefore the sum over $P(p_m)$ that we want, because it lacks the dp to make it into an integral, is related to the integral as follows (in the limit $dp \to 0$ or $L \to \infty$):

$$\sum_m P(p_m) = \lim_{L \to \infty} \left(\frac{L}{2\pi\hbar} \int_{-\infty}^{\infty} P(p)\,dp \right) \tag{21.136}$$

$$= \lim_{L \to \infty} \left(\frac{L}{2\pi\hbar} \frac{4\alpha^3}{L} \int_{-\infty}^{\infty} \frac{1}{(\alpha^2 + (p/\hbar)^2)^2}\,dp \right) \tag{21.137}$$

$$= 1. \tag{21.138}$$

I invite you to verify the last step by doing the integral. Notice that L drops out and the sum of all the $P(p_m)$ is 1. This is to be expected because $\psi(x)$ was normalized to 1.

Continuing,

$$(\Delta p)^2 = \sum_m P(p_m)(p_m - 0)^2 \quad \text{(because } \langle p \rangle = 0) \tag{21.139}$$

$$= \lim_{L \to \infty} \left(\frac{L}{2\pi\hbar} \int_{-\infty}^{\infty} P(p)p^2\,dp \right) \tag{21.140}$$

$$= \lim_{L \to \infty} \left(\frac{L}{2\pi\hbar} \frac{4\alpha^3}{L} \int_{-\infty}^{\infty} \frac{p^2}{(\alpha^2 + (p/\hbar)^2)^2}\,dp \right) \tag{21.141}$$

$$= \hbar^2\alpha^2. \tag{21.142}$$

So that finally

$$\Delta p = \hbar\alpha, \tag{21.143}$$

not far from our rough estimate of $2\alpha\hbar$ of Eqn. 21.127. Continuing,

$$\Delta x \Delta p = \frac{1}{\sqrt{2}\alpha}\hbar\alpha = \frac{\hbar}{\sqrt{2}} \tag{21.144}$$

in accordance with the precise uncertainty principle $\Delta x \Delta p \geq \frac{1}{2}\hbar$.

21.5 More on Fourier's theorems

This is a mathematical digression for those who are not familiar with Fourier series. Consider an arbitrary vector $\mathbf{V}$ in three dimensions. We may write it in terms of the unit vectors $\mathbf{i}$, $\mathbf{j}$, and $\mathbf{k}$ as

$$\mathbf{V} = V_x\mathbf{i} + V_y\mathbf{j} + V_z\mathbf{k}. \tag{21.145}$$

Starting from the origin we can get to the tip of *any* vector $\mathbf{V}$ by moving along x by V_x, along y by V_y, and along z by V_z. No vector can evade this construction. One refers to the triad $\mathbf{i}$, $\mathbf{j}$, and $\mathbf{k}$ as a *basis* because we can synthesize any $\mathbf{V}$ in terms of them. For our purposes it is better to rename the three components as V_1, V_2, and V_3 and the three basis vectors as follows:

$$\mathbf{i} = \mathbf{e}_1 \tag{21.146}$$

$$\mathbf{j} = \mathbf{e}_2 \tag{21.147}$$

$$\mathbf{k} = \mathbf{e}_3 \tag{21.148}$$

and rewrite the expansion in Eqn. 21.145 as

$$\mathbf{V} = \sum_{i=1}^{3} V_i\mathbf{e}_i. \tag{21.149}$$

The reason we use numerical subscripts is that it is easy to sum over them (rather than over x, y, and z) and they do not fail us if we want to sum over more than 26 values.

Equation 21.149 is the vector analog of

$$\psi(x) = \sum_{m=-\infty}^{\infty} A(p_m) \frac{1}{\sqrt{L}} \exp\left[\frac{ip_m x}{\hbar}\right]$$

$$\text{where } p_m = \tfrac{2\pi m\hbar}{L}, \tag{21.150}$$

the Fourier series in Eqn. 21.56 or Eqn. 21.59 for periodic functions. In one case we express a generic vector $\mathbf{V}$ in terms of basis vectors $\mathbf{e}_i$ and in the other, a generic function $\psi(x)$ in terms of basis functions $\psi_{p_m}(x)$. The only difference is that in the latter case we sum over an infinite number of basis functions labeled by p or m.

Each basis vector $\mathbf{e}_i$ has unit length and is orthogonal to the other two. This *orthonormality* is written as follows:

$$\mathbf{e}_i \cdot \mathbf{e}_j = \delta_{ij} \tag{21.151}$$

where

$$\delta_{ij} = 1 \quad \text{if } i = j$$
$$= 0 \quad \text{if } i \neq j \tag{21.152}$$

is called the *Kronecker delta*. Instead of saying all the time "1 if i and j are equal, 0 if they're different," we use the symbol δ_{ij}. It is a very concise way to say that each basis vector $\mathbf{e}_i$ is of unit length perpendicular to the others.

Let us pursue this analogy to find a way to extract the coefficients $A(p)$ of the Fourier expansion. Suppose you have in mind a specific vector $\mathbf{V}$ (an arrow of definite length and orientation) and want to write it in terms of the basis vectors. For this you need the coefficients V_i. Suppose you want V_2. Then you take the dot product of both sides of Eqn. 21.149 with $\mathbf{e}_2$

$$\mathbf{V} \cdot \mathbf{e}_2 = \left(\sum_{i=1}^{3} V_i \mathbf{e}_i \right) \cdot \mathbf{e}_2 \tag{21.153}$$

$$= V_1 \mathbf{e}_1 \cdot \mathbf{e}_2 + V_2 \mathbf{e}_2 \cdot \mathbf{e}_2 + V_3 \mathbf{e}_3 \cdot \mathbf{e}_2 \tag{21.154}$$

$$= V_1 \cdot 0 + V_2 \cdot 1 + V_3 \cdot 0 \tag{21.155}$$

$$= V_2. \tag{21.156}$$

(The left-hand side is the length of $\mathbf{V}$ times the cosine of the angle it makes with the unit vector $\mathbf{e}_2 = \mathbf{j}$.) Only the second term in the sum survives because

$$\mathbf{e}_i \cdot \mathbf{e}_2 = 1 \text{ if } i = 2 \text{ and 0 if not.} \tag{21.157}$$

More generally if we have in n-dimensions n vectors that are mutually orthogonal and of unit length, their orthonormality may be written concisely in terms of the Kronecker delta as

$$\mathbf{e}_i \cdot \mathbf{e}_j = \delta_{ij}. \tag{21.158}$$

Every vector in this n-dimensional space may be written as

$$\mathbf{V} = \sum_{i=1}^{n} V_i \mathbf{e}_i. \tag{21.159}$$

To find the coefficient V_j, we take the dot product of both sides with $\mathbf{e}_j$ and find

$$\mathbf{V} \cdot \mathbf{e}_j = \left(\sum_{i=1}^{N} V_i \mathbf{e}_i \right) \cdot \mathbf{e}_j \tag{21.160}$$

$$= \left(\sum_{i=1}^{N} V_i \mathbf{e}_i \cdot \mathbf{e}_j \right) \tag{21.161}$$

$$= \sum_{i=1}^{N} V_i \delta_{ij} = V_j. \tag{21.162}$$

Try to remember this result for a while:

$$V_j = \mathbf{V} \cdot \mathbf{e}_j = \mathbf{e}_j \cdot \mathbf{V}. \tag{21.163}$$

I use the freedom to rewrite the dot product with the order of the vectors reversed so it will closely resemble an expression in Fourier theory that follows shortly.

Now turn to the Fourier expansion

$$\psi(x) = \sum_{p} A(p) \psi_p(x) = \sum_{p} A(p) \frac{1}{\sqrt{L}} \exp\left[\frac{ipx}{\hbar} \right]. \tag{21.164}$$

To find the coefficient $A(p')$ the way we found V_j we would like the analog of the orthonormality relation of the basis vectors

$$\mathbf{e}_i \cdot \mathbf{e}_j = \delta_{ij} \tag{21.165}$$

for the basis function $\psi_p(x)$.

Here it is. Given two basis functions $\psi_p(x)$ and $\psi_{p'}(x)$, we define their dot product to be a certain integral *because it equals* $\delta_{pp'}$:

$$\int_0^L \psi_p^*(x)\psi_{p'}(x)\,dx = \delta_{pp'}. \tag{21.166}$$

This will allow us to find the coefficients $A(p)$.

Proof of orthonormality of $\psi_p(x)$:

$$\int_0^L \psi_p^*(x)\psi_{p'}(x)\,dx = \frac{1}{L}\int_0^L \exp\left[\frac{i(-p+p')x}{\hbar}\right]dx$$

$$\left(\text{with } p = \tfrac{2\pi\,m\hbar}{L}, p' = \tfrac{2\pi\,m'\hbar}{L}\right)$$

$$= \frac{1}{L}\int_0^L \exp\left[\frac{2\pi\,i(-m+m')x}{L}\right]dx \tag{21.167}$$

$$= \delta_{mm'} = \delta_{pp'} \tag{21.168}$$

where in the last step I have invoked a result I asked you to memorize: the integral of every $\psi_p(x)$ over the circle is zero unless $p = 0$, in which case the integrand is a constant. This result applies here since the integrand in Eqn. 21.167 corresponds to $p = 2\pi(m'-m)\hbar/L$.

Now I claim the analog of $V_j = \mathbf{e}_j \cdot \mathbf{V}$:

$$A(p') = \int_0^L \psi_{p'}^*(x)\psi(x)\,dx. \tag{21.169}$$

Here are the steps in the proof, analogous to Eqns. 21.153 through 21.156.

$$\int_0^L \psi_{p'}^*(x)\psi(x)\,dx = \int_0^L \psi_{p'}^*(x)\left[\sum_P A(p)\psi_p(x)\right]dx \tag{21.170}$$

$$= \sum_P A(p)\int_0^L \psi_{p'}^*(x)\psi_p(x)\,dx \tag{21.171}$$

$$= \sum_P A(p)\delta_{p'p} = A(p'). \tag{21.172}$$

Table 21.1   Vector versus function expansions

Object	Vector	Function
Generic member	$\mathbf{V}$	$\psi(x)$
Basis label	i	p
Basis	$\mathbf{e}_i$	$\psi_p(x)$
Expansion	$\mathbf{V} = \sum_i V_i \mathbf{e}_i$	$\psi(x) = \sum_p A(p)\psi_p(x)$
Coefficient	V_i	$A(p)$
Orthonormality	$\mathbf{e}_i \cdot \mathbf{e}_j = \delta_{ij}$	$\int_0^L \psi_p^*(x)\psi_{p'}(x)dx = \delta_{pp'}$
Finding coefficient	$V_i = \mathbf{e}_i \cdot \mathbf{V}$	$A(p) = \int_0^L \psi_p^*(x)\psi(x)dx$

Table 21.1 gives the complete correspondence between vectors and functions.

Let us take stock of where we are. There is a particle on a ring described by some $\psi(x)$. We want to know what answers we will get if we measure its momentum. Single-valuedness dictated that the allowed momenta obey the quantization rule

$$p = \frac{2\pi m\hbar}{L} \quad m = \ldots -2, -1, 0, 1, 2, \ldots \tag{21.173}$$

The absolute probability $P(p)$ for finding any one of these allowed values is given by the following recipe:

$$P(p) = |A(p)|^2 = \left| \int_0^L \frac{1}{\sqrt{L}} \exp\left[\frac{-ipx}{\hbar}\right] \psi(x)dx \right|^2 \tag{21.174}$$

assuming $\psi(x)$ has been normalized. If not, $|A(p)|^2 = P'(p)$, the relative probability. If a value p_0 is obtained, $\psi(x)$ collapses to $\psi_{p_0}(x)$ immediately following the measurement.

21.6   Measurement postulate: general

After the concrete example of momentum, we are ready for a more general statement of the measurement postulate. Let α denote the set of allowed values of some dynamical variable $\mathcal{A}$ represented in classical mechanics by some function of x and p. For example, $\mathcal{A}$ can be the momentum itself and p one of its allowed values; $\mathcal{A}$ can be angular momentum and $m\hbar$

one of its allowed values; A could be the energy and E one of its allowed values (more on this variable in the following chapters). Let $\psi_\alpha(x)$ denote a *normalized* state in which A is guaranteed to yield a particular value α. (This is like $\psi_p(x)$ which is guaranteed to yield the value p for momentum. I will tell you later how to actually find the functions $\psi_\alpha(x)$ for each A. For now assume that for each variable A we know the corresponding $\psi_\alpha(x)$.)

First we have two *purely mathematical results* that generalize Fourier series.

1. We may expand *any* $\psi(x)$ as a linear combination

$$\psi(x) = \sum_\alpha A(\alpha)\psi_\alpha(x). \tag{21.175}$$

2. The coefficients of the expansion are given by

$$A(\alpha) = \int \psi_\alpha^*(x)\psi(x)\,dx. \tag{21.176}$$

The physics now enters in the form of the measurement postulate: *When the variable A is measured on a particle described by the $\psi(x)$ in Eqn. 21.175, the probability of obtaining a particular value of α, say α_0, is given by $P(\alpha_0) = |A(\alpha_0)|^2$. If measurement yields a value α_0, the state right after measurement collapses from the sum over α to just the one term $\psi_{\alpha_0}(x)$.*

This is very general: under measurement, the particle goes from being in a superposition of states with different possible values for some variable to the one state in the sum in which it was detected. It could go from being in many places to being in just the one place where it was detected, from being in many states of momentum to the one found in the momentum measurement, from being near either slit to being near the one where the photons from the lightbulb detected it. This collapse is due to the inevitable effect of measurement and it is one of the most dramatic postulates.

The measurement postulate gives the answer to a question that comes up often: How do we ever know what state a particle is in? Here is an answer that is often applicable: It is in a state corresponding to the value of some observable that was just measured. Thus, if we measured p and obtained $p = p_0$, the state right after measurement is $\psi_{p_0}(x)$. In addition, if we can compute the time-dependence of a known initial state using the laws of quantum dynamics (time-dependent Schrödinger equation) we will know the state at future times as well.

21.7 More than one variable

Let us briefly consider not one variable, but two, say x and p. Classically a particle can be in a state with definite values (x_0, p_0) for position and momentum. I can prepare such a state as follows: I push the particle till it picks up the desired momentum p_0 as it reaches some point x_0. At that instant, I assign to it the pair (x_0, p_0). If I remeasure position and momentum *immediately*, I will get the same pair (x_0, p_0). A series of rapid measurements of position and momentum will yield the string $(x_0, p_0, x_0, p_0, \ldots)$. In fact I can measure p first and then x or the other way around and it will not matter.

All this changes in the quantum case. Let us say the particle was in a state of definite momentum, $\psi_{p_0}(x)$. I measure momentum and I get p_0. I have no idea where the particle is. So I locate it using the Heisenberg microscope. Say I find it at $x = 5$. The wave function right after this position measurement becomes $\psi_{x=5}(x)$, which is peaked at $x = 5$. But I cannot say the particle is in a state $(x = 5, p = p_0)$ because if I measure p just to make sure, I will not necessarily get p_0. To see what I could get, I must first write $\psi_{x=5}(x)$ as a sum of states of definite momentum

$$\psi_{x=5}(x) = \sum_p A(p)\psi_p(x), \tag{21.177}$$

where

$$A(p) = \int \psi_p^*(x)\psi_{x=5}(x)\,dx. \tag{21.178}$$

I can get any p that is present in the sum, any p for which $A(p) \neq 0$. I could get $2p_0$. Or I could get $-p_0$ if such a term were present. Since an immediate remeasurement does not necessarily yield p_0, the particle can never be said to have been in a state $(x = 5, p = p_0)$. Suppose the second momentum measurement gave an answer $-p_0$. Is the particle in a state $(x = 5, p = -p_0)$? No, because in the subsequent position measurement (in this state of definite momentum $-p_0$) every point on the ring is equally probable, with no preference for, or memory of, $x = 5$. A string of rapid x and p measurements will therefore yield a string of generally unpredictable and non-repeating numbers as we alternately expand a function narrow in position in terms of functions of sharp momentum and vice versa. There is no sense in which the particle can be said to have a well-defined value of position and momentum.

The pair (x, p) happens to be maximally incompatible. There *are* other pairs of variables in quantum theory that *can* have simultaneously well-defined values for both, where the same pair of measured values will repeat upon successive remeasurement and where the order of measurement will not matter. We will encounter one example in the next chapter.

CHAPTER 22

States of Definite Energy

Let us continue our study of the measurement postulate in its general form. It has no analog in classical mechanics. There, if we know the state variables (x, p) (or its generalizations to higher dimensions) we need not measure any other dynamical variable. For example, the angular momentum (in three dimensions) is given by $\mathbf{L} = \mathbf{r} \times \mathbf{p}$. We could measure it directly, but if we knew $\mathbf{r}$ and $\mathbf{p}$, we could just compute the cross product.

We saw that in quantum theory $\psi(x)$ describes the state and plays the role of the pair (x, p). It contains all possible information on the particle at any given time. While the questions we ask are not too different from classical mechanics and take the form, "What will I get if I measure $\mathcal{A}$?," where $\mathcal{A}$ is some variable like position or momentum, the answer is generally probabilistic in nature. For example, given a generic $\psi(x)$ if we ask for the result of a position measurement we are told that the outcome x will occur with a probability density

$$P(x) = |\psi(x)|^2. \tag{22.1}$$

If the particle is found at $x = x_0$, the wave function $\psi(x)$ will collapse from whatever it was to $\psi_{x_0}(x)$, a spike at x_0.

The collapse of the probability occurs in classical mechanics as well. Recall the probability distribution $P(x)$ for finding me somewhere near my home or my office or en route. If you catch me somewhere, the classical distribution collapses to where I was caught. The difference is that I was where I was caught even before you caught me: I was being constantly observed by a stream of photons or air molecules, for example. At the quantum level, we have the underlying wave function in addition to $P(x)$. The spread-out ψ is not like the spread-out $P(x)$: it describes a particle that really is nowhere in particular. It has no position prior to measurement. This state of limbo has no classical analog.

Things get much more complicated if we ask the same question of a momentum measurement. The answer is longer and given in several stages.

1. By postulate, a state of momentum p is described by the wave function

$$\psi_p(x) = A\exp\left[\frac{ipx}{\hbar}\right].\tag{22.2}$$

In a finite universe of circumference L, the normalized state is

$$\psi_p(x) = \frac{1}{\sqrt{L}}\exp\left[\frac{ipx}{\hbar}\right]\tag{22.3}$$

and the condition of single-valuedness restricts the allowed values of momentum to

$$p = p_n = \frac{2\pi n\hbar}{L}\quad n = 0,\pm1,\pm2\ldots\tag{22.4}$$

From now on I will use n to denote the integer since m will be reserved for the mass. Despite this, I previously used m as the index for momentum and angular momentum because the angular momentum associated with rotations around an axis is traditionally written as $m\hbar$.

2. Given all this, the measurement postulate tells us that a momentum measurement will yield a result p_n with probability $P(p_n) = |A(p_n)|^2$ where $A(p_n)$ is the coefficient in the expansion

$$\psi(x) = \sum_n A(p_n)\frac{1}{\sqrt{L}}\exp\left[\frac{2\pi inx}{L}\right] \equiv \sum_p A(p)\psi_p(x)\tag{22.5}$$

where I use both p_n and p to denote one of the allowed momenta. So the sum over n and the sum over p stand for the same thing, the sum over all allowed momenta.

The expansion coefficient $A(p)$ may either be read off by inspection in some cases or computed in all cases by evaluating the integral

$$A(p) = \int_0^L \psi_p^*(x)\psi(x)\,dx. \qquad (22.6)$$

The coefficient $A(p)$ can be complex just like $\psi(x)$. But $P(p) = |A(p)|^2$ will always be real and non-negative.

3. If the measurement yields a value p_n, the state $\psi(x)$, which used to be a superposition, collapses to $\psi_{p_n}(x)$. An immediate remeasurement of p will yield p_n. The rest of ψ will get chopped out. It's like Polaroid glasses. The electric field $\mathbf{E}$ in the incoming light can be polarized in any direction perpendicular to the direction of propagation, but once it goes through the glasses it will be polarized along the axis of the lenses. The component of $\mathbf{E}$ in the perpendicular direction will be chopped off. So measurement is like a filtering process. Out of the sum over many terms, measurement filters the one term that corresponds to the one answer you got.

Let us be clear about the roles of $\psi(x)$ and $\psi_p(x)$ in Eqn. 22.5. The function $\psi(x)$ describes the state the particle is in. It is an arbitrary periodic function on the circle. We want to know the result of a momentum measurement on this state. The functions $\psi_p(x)$ are also functions on the ring, but they are postulated to have a definite momentum p associated with them. If a particle is in $\psi_p(x)$, a momentum measurement is guaranteed to give the value p. The outcome of a momentum measurement on $\psi(x)$ is more complicated. It is determined by writing the given $\psi(x)$ as a linear combination of $\psi_p(x)$ as in Eqn. 22.5. Unlike the case of $\psi_p(x)$, we can get any p that appears in the expansions and the probability for this is $P(p) = |A(p)|^2$. If the state had many p's in its expansion, it collapses to the one term that was found upon measurement. Of course one possible special case is that the sum has only one term, say only $A(p_3) \neq 0$. Then the outcome is certain to be p_3 and the state *is unaffected by the measurement* and remains $\psi_{p_3}(x)$.

The following analogy may help. We know that any three-dimensional vector $\mathbf{V}$ may be expressed as $\mathbf{V} = V_x\mathbf{i} + V_y\mathbf{j} + V_z\mathbf{k}$. A basis vector like $\mathbf{i}$ is every bit a vector like $\mathbf{V}$; it just happens to be aligned with one of the coordinate axes.

This recipe generalizes to all dynamical variables $\mathcal{A}$, by which I mean anything that is a function of x and p in classical mechanics. For example, the angular momentum $\mathbf{L}$ is a dynamical variable given by $\mathbf{L} = \mathbf{r} \times \mathbf{p}$. The energy of a particle of mass m attached to a spring of force constant k and undergoing simple harmonic motion is another:

$$E(x,p) = \frac{p^2}{2m} + \frac{1}{2}kx^2. \tag{22.7}$$

All this is for one variable at a time. Generally you may not have states in which two (or more) variables have well-defined or guaranteed values. If you try to prepare a particle with a definite value for one, it may be spread out in the other, in the sense that measurement of the second variable could give a range of answers. If you measure the second one and get some answer, there is no guarantee the first will give the old value with certainty. This was the case for position and momentum. However, in this chapter we will meet a pair of variables both of which can be specified simultaneously.

Back to $\mathcal{A}$. Let α denote the set of allowed values of some variable $\mathcal{A}$. Let $\psi_\alpha(x)$ denote a state in which the variable is guaranteed to yield a value α. For example, $\mathcal{A}$ can be the momentum, $p_n = 2\pi n\hbar/L$ one of its allowed values, and $\psi_{p_n}(x)$ the corresponding wave function. Mathematics tells us the following.

1. We may expand any $\psi(x)$ as a linear combination

$$\psi(x) = \sum_\alpha A(\alpha)\psi_\alpha(x). \tag{22.8}$$

2. The coefficients of the expansion are given by

$$A(\alpha) = \int \psi_\alpha^*(x)\psi(x)dx. \tag{22.9}$$

Physics then tells us the following: When the variable $\mathcal{A}$ is measured on a particle described by the $\psi(x)$ in Eqn. 22.8, the probability of obtaining a result α is given by $P(\alpha) = |A(\alpha)|^2$. If measurement yielded a particular value $\alpha = \alpha_0$, the state right after measurement collapses from the sum over α to just the one term $\psi_{\alpha_0}(x)$.

This recipe is not as complete as the one for momentum because *I have not given you the wave functions $\psi_\alpha(x)$ that are states of definite value for $\mathcal{A}$.* Till we have these, we cannot hope to express the given ψ in the form of Eqn. 22.8.

For this we need another postulate. Pick any observable $\mathcal{A}$ and the postulate will tell you how to find $\psi_\alpha(x)$. However, the equations determining ψ_α will depend on what $\mathcal{A}$ is. You change your mind on which $\mathcal{A}$ you want to measure, and you have a new equation to solve. Rather than deal with this procedure in all its generality, let us consider one case of the greatest importance, where $\mathcal{A}$ is the energy E and $\psi_E(x)$ the corresponding wave functions of definite E.

Energy plays a central role in the dynamics, when we ask how an arbitrary initial wave function $\psi(x, 0)$ evolves with time into $\psi(x, t)$. It will turn out that as long as the potential V is time-independent, the answer is most easily found by first expressing the initial $\psi(x, 0)$ as a linear combination of $\psi_E(x)$. A remarkable corollary will be that if a particle is found to have energy E in a measurement, not only does the state collapse to $\psi_E(x)$ and stay that way at least for an infinitesimal amount of time (as it would for any variable), it will stay that way forever! All this will be elaborated when we turn to dynamics in the next chapter. For now I just want you to take my word that of all possible variables, there are excellent reasons for focusing on the case $\mathcal{A} = E$ and the corresponding wave functions of definite energy $\psi_E(x)$.

Postulate for $\psi_E(x)$. The states of definite energy E are the normalizable single-valued solutions to the *time-independent Schrödinger equation*

$$-\frac{\hbar^2}{2m}\frac{d^2\psi_E(x)}{dx^2} + V(x)\psi_E(x) = E\psi_E(x). \tag{22.10}$$

This is the master formula. Do not worry, I will see you through this equation.

Now you might say, "Why don't you just give me $\psi_E(x)$ as you did $\psi_p(x)$ and be done with it?" The problem is that the time-independent Schrödinger equation depends on what $V(x)$ is. Every possible $V(x)$ has its own Eqn. 22.10 and its own family of solutions $\psi_E(x)$. There will be one family of functions $\psi_E(x)$ for $V(x) = kx^2$ and another for $V(x) = k'x^4$. (In three dimensions you can have an electron in the $1/r$ potential due to the proton and the corresponding wave function $\psi_E(\mathbf{r})$ of the hydrogen atom.)

We will often focus on *bound states*, states in which the energy E is less than $V(\pm\infty)$, the potential at $x = \pm\infty$. This is the case where classically the particle cannot escape to infinity, for if it did, its kinetic energy $K(\pm\infty) = E - V(\pm\infty)$ would have to be negative, which is impossible. In such cases we will find that solutions to the Schrödinger equationare possible only at some quantized values of energy E_n labeled by some integer n, with the corresponding wave functions $\psi_{E_n}(x) \equiv \psi_n(x)$. Solving Eqn. 22.10 will tell us both the allowed energies E_n and the $\psi_{E_n}(x)$. Then we can go on and find $A(E_n)$ and the probabilities. But first we have to solve the equation.

Why are there bound-state solutions only for some values $E = E_n$? We will see that at other energies the solutions blow up exponentially either as $x \to \infty$ or as $x \to -\infty$ or both, and hence they are not normalizable. When the particle is confined to a finite ring, we do not have the problem of blowup at spatial infinity, but that of single-valuedness. This requirement, which quantized the momenta, will be seen to quantize the energies on the ring.

22.1 Free particle on a ring

The first problem I want to solve involves a free particle, one for which $V(x) \equiv 0$. Let us imagine it lives on a circle of circumference $L = 2\pi R$. The Schrödinger equation(Eqn. 22.10) assumes the form

$$-\frac{\hbar^2}{2m}\frac{d^2\psi_E(x)}{dx^2} = E\psi_E(x). \tag{22.11}$$

Let us rearrange the equation to read:

$$\frac{d^2\psi_E(x)}{dx^2} + k^2\psi_E(x) = 0 \quad \text{where} \tag{22.12}$$

$$k^2 = \frac{2mE}{\hbar^2}. \tag{22.13}$$

Remember that the E we are looking for is now encoded in k via the preceding equation.

The solutions to this equation are of the form

$$\psi_E(x) = Ae^{ikx} + Be^{-ikx} \tag{22.14}$$

where A and B are some constants. Let us verify this. Every time we take a derivative of e^{ikx} we pull down an ik for a total of $(ik)^2 = -k^2$. The second exponential also yields the same factor because $(-ik)^2 = -k^2$ as well. Consequently $\psi_E(x)$ satisfies

$$\frac{d^2 \psi_E(x)}{dx^2} = -k^2 \psi_E(x). \tag{22.15}$$

Upon comparison to the prototype

$$\psi_p(x) = A \exp\left[\frac{ipx}{\hbar}\right] \tag{22.16}$$

the exponential functions in Eqn. 22.14 are seen to be states of definite momentum

$$p = \pm \hbar k. \tag{22.17}$$

If we now express k in terms of E using Eqn. 22.13 we find

$$p = \pm \hbar \sqrt{\frac{2mE}{\hbar^2}} = \pm \sqrt{2mE}. \tag{22.18}$$

Thus we may rewrite Eqn. 22.14 in terms of the energy label E as

$$\psi_E(x) = A e^{i\sqrt{2mE}x/\hbar} + B e^{-i\sqrt{2mE}x/\hbar}. \tag{22.19}$$

Look at Eqn. 22.18. It is exactly as in classical mechanics! In other words, if I told you a free particle of energy E was running around in a circle with kinetic energy E, you would say that its momentum is determined by

$$\frac{p^2}{2m} = E \quad \text{with solutions} \tag{22.20}$$

$$p = \pm \sqrt{2mE}. \tag{22.21}$$

So what is new in the quantum case?

There are two profound differences between classical and quantum mechanics.

1. The allowed values of p are restricted in the quantum theory to the values

$$p = p_n = \frac{2\pi n\hbar}{L} \qquad n = 0, \pm 1, \pm 2, \ldots \qquad (22.22)$$

It is convenient in this section to let n only take the values $0, 1, 2, \ldots$ and define:

$$p_n = \frac{2\pi n\hbar}{L} \qquad n = 0, 1, 2, \ldots \qquad (22.23)$$

with the understanding that for $n \neq 0$, $\pm p_n$ are the allowed momenta. The allowed values of energy are therefore quantized to

$$E_n = \frac{p_n^2}{2m} = \frac{4\pi^2 n^2 \hbar^2}{2mL^2} \qquad n = 0, 1, 2, \ldots \qquad (22.24)$$

The corresponding wave functions are

$$\psi_{E_n}(x) \equiv \psi_n(x) = A\exp\left[\frac{ip_n x}{\hbar}\right] + B\exp\left[-\frac{ip_n x}{\hbar}\right]$$
$$(n = 1, 2, \ldots) \qquad (22.25)$$
$$= A\exp\left[\frac{2\pi n i x}{L}\right] + B\exp\left[-\frac{2\pi n i x}{L}\right]$$
$$(n = 1, 2 \ldots) \qquad (22.26)$$
$$= A \quad (n = 0). \qquad (22.27)$$

2. Whereas a classical particle of energy E can also have one of two momenta $p = \pm\sqrt{2mE}$, it has to choose one or the other in any situation. The quantum particle on the other hand can be in a state of indefinite momentum displayed in Eqn. 22.19, in which it can yield *either* value with relative probabilities $|A|^2$ and $|B|^2$.

22.1.1 Analysis of energy levels: degeneracy

Consider the fact that a state of momentum is given by a unique function up to an overall normalization

$$\psi_p(x) = Ae^{ipx/\hbar}, \tag{22.28}$$

while a state of definite energy involves *two* independent functions

$$\psi_E(x) \equiv \psi_n(x) = A\exp\left[\frac{ip_n x}{\hbar}\right] + B\exp\left[-\frac{ip_n x}{\hbar}\right]. \tag{22.29}$$

The functions $e^{\pm ip_n x/\hbar}$ are independent in the sense that one cannot be obtained from the other by multiplying by an x-independent constant. Of the two coefficients A and B we may choose one to be equal to 1 using the freedom in the overall scale, but a ratio A/B remains a meaningful parameter that determines the relative odds for p_n versus $-p_n$. This extra parameter in ψ_E makes the computation of probabilities $P(E)$ a little tricky. Here is how we handle it.

Let $\psi(x)$ be the state that is given to us, for which we want the probabilities for various outcomes in an energy measurement. Forget about $\psi_E(x)$ and express the given $\psi(x)$ as a linear combination of $\psi_p(x)$ and group the terms as shown:

$$\psi(x) = A(p=0)e^{2\pi i \cdot 0 \cdot x/L} + A(p_1)e^{2\pi i \cdot 1 \cdot x/L} + A(-p_1)e^{-2\pi i \cdot 1 \cdot rx/L}$$
$$+ A(p_2)e^{2\pi i \cdot 2 \cdot x/L} + A(-p_2)e^{-2\pi i \cdot 2 \cdot x/L} + \ldots \tag{22.30}$$

I am not using normalized $\psi_p(x)$ but that will not matter since scaling them all by $1/\sqrt{L}$ to normalize them will not affect the *relative* probabilities. (We are able to ignore the normalization requirement only because they are all off by the amount $1/\sqrt{L}$. This preserves the "ray.")

In the preceding form, the un-normalized probability for any allowed p is obvious: $P(\pm p_n) = |A(\pm p_n)|^2$. To find the probability for any energy E_n we use the fact that because E_n is quadratic in the momentum, the probabilities for both $\pm p_n$ will contribute to the probability for E_n. For example, both

$$p_3 = \frac{2\pi \cdot 3 \cdot \hbar}{L} \quad \text{and} \tag{22.31}$$

$$-p_3 = \frac{2\pi \cdot (-3) \cdot \hbar}{L} \quad \text{correspond to the same energy} \quad (22.32)$$

$$E_3 = \frac{4\pi^2 3^2 \hbar^2}{2mL^2}. \tag{22.33}$$

The relative probability for this energy to occur is the sum of the probabilities for the two momentum outcomes that correspond to this energy:

$$P'(E_3) = |A(p_3)|^2 + |A(-p_3)|^2. \tag{22.34}$$

The absolute probability is

$$P(E_3) = \frac{|A(p_3)|^2 + |A(-p_3)|^2}{|A(p=0)|^2 + \sum_{n=1,2,\dots} (|A(p_n)|^2 + |A(-p_n)|^2)}. \tag{22.35}$$

Clearly what is true for $n = 3$ is true for any other n except $n = 0$ when there is just one $A(p = 0)$.

When the energy E_n is obtained in a measurement, the wave function in Eqn. 22.30 collapses to the corresponding *two* terms in the expansion of $\psi(x)$ with the corresponding momenta:

$$\psi(x)_{\text{after}} = A(p_n) \exp\left[\frac{ip_n x}{\hbar}\right]$$

$$+ A(-p_n) \exp\left[-\frac{ip_n x}{\hbar}\right]. \tag{22.36}$$

In this collapse the ratio $A(p_n)/A(-p_n)$ is preserved.

On the other hand, if *momentum* was measured and the result p_n (or $-p_n$) was obtained, the state would collapse to $A(p_n)e^{ip_n x/\hbar}$ or $A(-p_n)e^{-ip_n x/\hbar}$.

Go back to a definite energy state

$$\psi_E(x) = A \exp\left[i\frac{\sqrt{2mE}}{\hbar}x\right] + B \exp\left[-i\frac{\sqrt{2mE}}{\hbar}x\right]. \tag{22.37}$$

While any choice of A and B makes it a state of definite energy, two choices have added attractions. Suppose only $A \neq 0$. *This is now a state of definite momentum as well.* We could label it by a pair of numbers

$$(E,p) = \left(E, +\sqrt{2mE}\right). \tag{22.38}$$

The state of opposite momentum could likewise be labeled

$$(E,p) = \left(E, -\sqrt{2mE}\right). \tag{22.39}$$

Unlike the incompatible pair x and p, for which it is impossible to get a state with labels (x, p), we can have the pair (E, p). The reason is that when $V(x) \equiv 0$,

$$E = \frac{p^2}{2m} \tag{22.40}$$

and one can measure p and compute E from it.

For each E (except 0) there are two momentum states with that energy. This is called *degeneracy.* In general an energy level is degenerate if there are two or more independent wave functions with that energy. This situation is depicted in Figure 22.1.

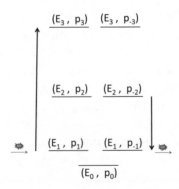

Figure 22.1   The energy states of a particle on a ring labeled by the pair (E, p). There are two states at each allowed energy corresponding to two possible directions of momentum $p = \pm\sqrt{2mE}$, except at $E = p = 0$, which is non-degenerate. The "height" of the levels goes as n^2. The vertical arrows on the sides denote transitions down from $n = 2$ to $n = 1$ by the emission of a photon of energy $\hbar\omega = E_2 - E_1$ and up from $n = 1$ to $n = 3$ by the absorption of a photon of energy $\hbar\omega = E_3 - E_1$.

I want to emphasize that the state

$$\psi_E(x) = A\exp\left[i\frac{\sqrt{2mE}}{\hbar}x\right] + B\exp\left[-i\frac{\sqrt{2mE}}{\hbar}x\right] \qquad (22.41)$$

is not a state of definite momentum: we can get $\pm\sqrt{2mE}$ with probabilities $|A|^2$ and $|B|^2$ respectively. It is, however, a state of definite energy E because E is blind to the sign of p. In other words, the state of definite momentum is also a state of definite energy, but the converse is not true.

If we shine light on this "atom" it can undergo transitions illustrated by the two vertical arrows on the side: down from $n = 2$ to $n = 1$ by the emission of a photon of energy

$$\hbar\omega = E_2 - E_1 = \frac{4\pi^2\hbar^2}{2mL^2}(2^2 - 1^2) \qquad (22.42)$$

and up from $n = 1$ to $n = 3$ by the absorption of a photon of energy

$$\hbar\omega = E_3 - E_1 = \frac{4\pi^2\hbar^2}{2mL^2}(3^2 - 1^2). \qquad (22.43)$$

Only light of the appropriate frequency can induce such transitions and, conversely, by looking at the frequencies that induce such transitions, we can learn about the energy level structure of the "atom."

Experiments at Yale (and elsewhere) have verified the prediction that if you take a small enough metallic ring in a magnetic field it will have a persistent (permanent) current, *which is not due to a battery or superconductivity*. The experiments can measure the tiny current due to one electron. In this context the circumference L, which was an artifact for normalizing wave functions in free space, is a physically significant parameter describing the quantum ring on the scale of a micron.

A real atom is more complicated mathematically than a particle in a ring because the electron is confined by the Coulomb force. But the ideas are the same: only some energies are allowed and they may be degenerate, corresponding usually to different values of angular momentum. (Whereas in a ring the angular momentum can have only two signs, clockwise and counterclockwise, leading to a twofold degeneracy, in three dimensions we also have many possible planes of rotation.) When you studied atoms earlier maybe you encountered shells with 2 electrons or 8 electrons and so on.

These numbers represent degeneracies. From the frequencies of emitted and absorbed light we can deduce the energy level structure of the atom.

It is quantum mechanics that rigidly limits the set of allowed energies and wave functions of bound states, and the corresponding frequencies of emission and absorption. This is why atoms come in a countable number of varieties (H, He, and so on), and atoms of any one kind, say He, are also the same across the universe (given that electrons, protons, and neutrons are the same across the universe, which is amazing in itself). This reproducibility is what allows us to deduce what atoms are contained in distant stars and how fast some galaxies are moving by the Doppler shift of the emitted light. Here is an example. The hydrogen atom has two levels that are close, and the wavelength of the light emitted when it jumps down is 21 centimeters. This is a standard fingerprint of hydrogen anywhere in the universe. If the observed wavelength is not 21 cm but 22 cm, you may say, "I guess it is not hydrogen." Nonetheless, the correct answer is that it is hydrogen, but that galaxy is moving away from you, and its light is Doppler shifted into the red. If the galaxy were coming toward you, the line would be blue shifted. If you believe that hydrogen atoms all over the universe are the same, and the frequency shift is only due to the motion of the galaxy, you find two things: its constitution and its speed of recession. The interpretation of the observations of Edwin Hubble (1889–1953), which related the red shift of galaxies to their distance, was used to demonstrate the expansion of the universe.

Quantization is no less important in biology and the life sciences. It is what ensures that the molecules that play a central role in molecular biology and genetics constitute a discrete set of possibilities, which may be reliably produced over and over again, the way digital music can be copied with no errors, unlike analog music, which cannot.

22.2 Thinking inside the box

Now for a very standard pedagogical exercise, called the *particle in a quantum well*, and a limiting case of the well, *the particle in a box*. It is more representative of quantization than the particle on a ring because the particle is confined by a potential just like the electron is in an atom.

22.2.1 Particle in a well

The particle in the well experiences a potential $V(x)$, two versions of which are shown in the left half of Figure 22.2 by dotted and solid lines. Both

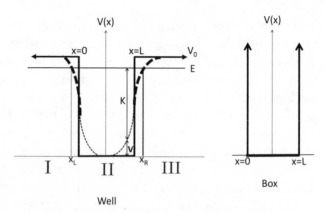

Figure 22.2   Left: A well of finite depth V_0, where the dotted line gives the realistic version and the solid line the artificial one to facilitate the quantum treatment. Right: A well of infinite depth or a *box*.

approach $V(x) = V_0$ for $x \rightarrow \pm\infty$. We want to find the allowed states of definite energy. A potential that changes gradually (dotted lines) is more realistic and also better suited to analyze various kinds of trajectories as a function of energy in the classical description. However, in order to simplify the math in the quantum mechanical treatment, we will use a *square well* whose sides abruptly rise from 0 to V_0 at $x = 0$ and $x = L$, as depicted by the solid lines.

Let us consider the classical dynamics first. The energy of the particle is made up of the kinetic term $K(x)$ and the potential term $V(x)$, the total being an x-independent constant E is shown by a horizontal line. We are interested mainly in the case $E < V_0$ shown in the figure. The particle is then in a *bound state* and can never escape to infinity because the kinetic energy would then be negative: $K(\pm\infty) = E - V(\pm\infty) = E - V_0 < 0$, which is impossible.

If the trapped particle moving to the right hurls itself against the confining potential with some initial velocity, it will climb till it reaches zero velocity at the turning point x_R and then start rolling down into the well. It will do the same at x_L if it is moving the other way. It will rattle back and forth at constant total energy between the two turning points.

For $E > V_0$ the particle can escape to infinity. If launched from the far left, it will speed up on descending into the well, slow down on the way out, and exit with the initial velocity (in this case where $V(\infty) = V(-\infty)$).

We will not spend much time on this case of the unbound particle because its energy is not quantized.

Now for the quantum treatment of the well. To find the allowed energies and wave functions $\psi_E(x)$ we go back to the Schrödinger equation:

$$-\frac{\hbar^2}{2m}\frac{d^2\psi_E(x)}{dx^2} + V(x)\psi_E(x) = E\psi_E(x). \tag{22.44}$$

We were able to solve for the case $V \equiv 0$ in terms of states of definite momentum. For special cases like $V(x) = \frac{1}{2}kx^2$ an analytical solution is possible. For some arbitrary $V(x)$ there is usually no analytical solution.

Consider the case of constant $V = V_0$. It is just as easy to solve as $V = 0$ because we can take the $V_0\psi$ in the left-hand side of the Schrödinger equationto the right-hand side, absorb V_0 into E, and change E to $E - V_0$:

$$-\frac{\hbar^2}{2m}\frac{d^2\psi_E(x)}{dx^2} = (E - V_0)\psi_E(x). \tag{22.45}$$

This is just like the free-particle problem with no V but with E replaced by $E - V_0$. Unfortunately the $V(x)$ describing the well is not a constant. However, we can consider a *square well* in which the potential is *piecewise constant* and jumps from V_0 to 0 and back to V_0 as x increases, as shown by the solid line. This allows us to tame the problem by the following "divide and conquer" procedure.

We divide space into regions *I*, *II*, and *III* as in the figure and solve for $\psi_E(x)$ in each region. Because V is a constant in each region we can make the change $E \to E - V$ and solve for what will look like the free-particle wave equation. *But then we have to glue the solutions together so that ψ and $\frac{d\psi}{dx}$ are continuous at the interfaces between regions I and II and between regions II and III.* I have already argued for the continuity of ψ. The continuity of $\frac{d\psi}{dx}$ follows from the Schrödinger equationitself. If $\frac{d\psi}{dx}$ is discontinuous at some point, $\frac{d^2\psi}{dx^2}$ will blow up there and the equation cannot be satisfied because the other terms, $E\psi$ and $V\psi$, are finite.

Let us begin with ψ_I, the solution in region *I*. It obeys

$$\frac{d^2\psi_I}{dx^2} + \left[\frac{2m(E - V_0)}{\hbar^2}\right]\psi_I = 0. \tag{22.46}$$

I have traded the subscript E (obvious in this context) for I, the region label.

Although we do not know E yet, let us first assume $E < V_0$ and see if there are any solutions in this range. This is the case in which the particle is hopelessly trapped in the well in classical mechanics. Since $E - V_0$ is negative, let us introduce a real positive parameter κ as follows:

$$\kappa = \sqrt{\frac{2m(V_0 - E)}{\hbar^2}}. \tag{22.47}$$

In terms of κ, the Schrödinger equationbecomes

$$\frac{d^2 \psi_I}{dx^2} - \kappa^2 \psi_I = 0, \tag{22.48}$$

with the general solution

$$\psi_I(x) = A e^{\kappa x} + B e^{-\kappa x}, \tag{22.49}$$

where A and B are arbitrary at this point. But look at the B term: it blows up exponentially as $x \to -\infty$. This makes ψ_I non-normalizable. So we choose $B = 0$. The remaining A term vanishes exponentially as we move along the negative x-direction, and it becomes negligible for large and negative κx, or for

$$|x| \gg \frac{\hbar}{\sqrt{2m(V_0 - E)}}. \tag{22.50}$$

So the wave function dies off more and more rapidly as $V_0 - E$ increases. *For future use remember that ψ_I vanishes in all of region I in the limit $V_0 \to \infty$.*

By similar reasoning

$$\psi_{III}(x) = C e^{\kappa x} + D e^{-\kappa x} \qquad x \geq L \tag{22.51}$$

with the same κ as in region I. Now we must choose $C = 0$ to kill the growing exponential. Once again remember that if we let $V_0 \to \infty$, the falling exponential, which is all that is left of $\psi_{III}(x)$, simply vanishes.

A dramatic feature of quantum mechanics is that for finite V_0 the particle has a non-zero probability to be in the classically forbidden regions

I and *III*. While quantum theory does not totally forbid excursions into this region, it does curb the excursions exponentially.

This leaves us with region *II* where $V = 0$. We have just the free-particle solutions of the form $e^{\pm i\sqrt{2mE}x/\hbar}$. With foresight we trade the exponentials for sines and cosines and write

$$\psi_{II}(x) = F\sin\left[\frac{\sqrt{2mE}}{\hbar}x\right] + G\cos\left[\frac{\sqrt{2mE}}{\hbar}x\right]. \qquad (22.52)$$

Of the six possible parameters, two (*B* and *C*) were set to zero to avert blowup at $x = \mp\infty$. The remaining four, $A, D, F,$ and G, seem to be just what we need to satisfy the four conditions of matching ψ and $d\psi/dx$ at the two interfaces. This is an illusion. Consider for example the interface between *I* and *II* at $x = 0$, where these conditions become

$$\psi_I(0) = \psi_{II}(0) \Rightarrow A = G \qquad (22.53)$$

$$\left.\frac{d\psi_I}{dx}\right|_{x=0} = \left.\frac{d\psi_{II}}{dx}\right|_{x=0} \Rightarrow \kappa A = \sqrt{\frac{2mE}{\hbar}}F. \qquad (22.54)$$

The overall scale of ψ, i.e., the overall scale of $A, D, F,$ and G, which I have emphasized has no physical significance, does not help satisfy these conditions. If we could not match the wave function and its slope at the interface for some choice of ψ, rescaling it (and its derivative) everywhere will not help since ψ (and its derivative) appear in both sides of the matching conditions. To make this transparent, we may choose one of the coefficients, say *A*, to be equal to 1, leaving us with just three free parameters. With four conditions and three genuine parameters, we seem doomed. But there is one hidden parameter: the energy *E*, which we took to be an arbitrary real number. *It is found that for special choices of E (the allowed values) the wave function obeys all the continuity equations and also dies off at* $x = \pm\infty$. I refer you to a more advanced text for a graphical but not totally analytic demonstration of this fact. (However, I will show you a case wherein energy quantization for bound states can be demonstrated analytically.)

A typical bound-state solution is shown in Figure 22.3.

This counting of parameters and the existence of normalizable solutions only at some special energies holds even if we go from the single square well to a more complicated one, as long as we are talking about a

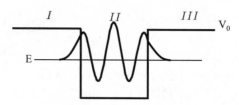

Figure 22.3   A bound-state wave function in a well at one of the allowed energies. As the confining potential $V_0 \to \infty$, and the well becomes the box, the exponential tail outside the well (regions *I* and *III*) shrinks to zero width and $\psi(x)$ is non-zero only in region *II*, the box.

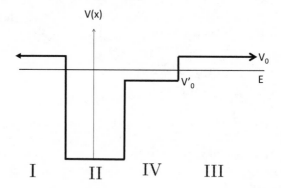

Figure 22.4   A modification of the simple well with an extra segment.

bound state with $E < V(\pm\infty)$. To see this, take the well in Figure 22.3 and add another segment numbered *IV*, between *II* and *III*, with a different constant potential V_0' which then reverts to V_0 in region *III*, as shown in Figure 22.4.

The new region *IV* introduces two extra parameters. (This is true even if $V_0' > E$ because even the rising exponential is allowed in the *finite* region *IV*.) It also introduces one more interface and two more matching conditions. Thus adding more and more segments still leaves us one parameter short. By adding such segments of variable widths and heights we can approximate any given $V(x)$, such as the one sketched in Figure 22.5. We will always have to tune the energy to get a normalizable solution that vanishes as $x \to \pm\infty$.

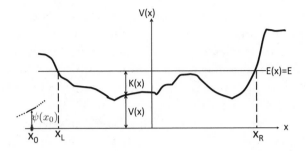

Figure 22.5   Particle of energy E in a potential $V(x)$ for the case $E < V(\pm\infty)$
when it is not allowed to go to $\pm\infty$. Classically its excursions are limited to the
right and left turning points x_R and x_L, where its kinetic energy $K(x) = E - V(x)$
vanishes. (If $V(x)$ exceedes E in points between x_L and x_R, there will be more
turning points and the well will break up into classically disconnected regions.)
Quantum theory allows short forays into the classically forbidden region. At the
far left is an "initial" value of $\psi_E(x_0)$ with some slope (solid line) and its
numerical continuation in either direction (dotted line).

Let us understand in another way how energy quantization for bound
states arises in a generic potential with $E < V(\pm\infty)$. Go back to the
Schrödinger equation

$$\frac{d^2\psi_E}{dx^2} + \left[\frac{2m(E - V(x))}{\hbar^2}\right]\psi_E = 0 \qquad (22.55)$$

and think of x as "time" and $\psi(x)$ as a coordinate that varies with this
"time" as it goes from $-\infty$ to ∞. The Schrödinger equationdetermines
the evolution of a coordinate $\psi_E(x)$ as a function of "time" x in a
"time-dependent" potential $V(x)$, the way Newton's law relates the second
time derivative of the coordinate to the applied force. Let us begin at some
"time" $x_0 \ll 0$, with some initial "coordinate" $\psi(x_0)$ and ("initial velocity")
$(d\psi/dx)_{x_0}$ and solve or integrate the equation numerically on a computer
as a function of "time" using our Newton's law, Eqn. 22.55.

Remember how you "solve it on a computer." You start with some
initial position $x(0)$ and velocity $v(0)$ at time $t = 0$. You use the initial
velocity to get the position a very small time Δt later ($x(\Delta t) = x(0) +
v(0)\Delta t$) and similarly the initial acceleration (from Newton's law) to get

$v(\Delta t)$ and keep inching forward in steps of size Δt. You can repeat the calculation with a smaller value of Δt to get more accurate results.

Unlike in mechanics, we want the solution for "times" past $(x < x_0)$ as well as future times to make sure it is well behaved for all x. If we try to implement our "time" evolution on a computer to earlier "times," we will find that generically the solution blows up as $x \to -\infty$. We saw this explicitly in the square well, which we could solve analytically. Recall the solution in Eqn. 22.51 for region I, which had a piece that blew up as $x \to -\infty$:

$$\psi_I(x) = Ae^{\kappa x} + Be^{-\kappa x}. \tag{22.56}$$

To avert the blowup as $x \to -\infty$ we had to choose $B = 0$. It turns out that we can achieve the same end in another way that is applicable to the numerical solution. *This is done by a judicious choice of initial conditions* $\psi(x_0)$ *and* $(d\psi/dx)_{x_0}$. Although the overall scale of $\psi(x_0)$ and $(d\psi/dx)_{x_0}$ is physically without significance, their ratio is a real degree of freedom. It is in fact the only freedom we have. To see how choosing this ratio judiciously can have the effect of setting $B = 0$, consider the ratio of $(d\psi/dx)$ to $\psi(x)$ of the analytical solution at some point:

$$\left.\frac{\frac{d\psi_I(x)}{dx}}{\psi_I(x)}\right|_{x_0} = \left.\frac{A\kappa e^{\kappa x} - B\kappa e^{-\kappa x}}{Ae^{\kappa x} + Be^{-\kappa x}}\right|_{x_0}. \tag{22.57}$$

Suppose we demand that this ratio equal κ:

$$\left.\frac{A\kappa e^{\kappa x} - B\kappa e^{-\kappa x}}{Ae^{\kappa x} + Be^{-\kappa x}}\right|_{x_0} = \kappa. \tag{22.58}$$

This can happen only if $B = 0$. *Thus there is a magic value of the ratio of* $(d\psi/dx)_{x_0}$ *to* $\psi(x_0)$ *(which happens to be κ in this example, independent of x_0) that corresponds to $B = 0$ in the analytical solution.* This means that if we try to integrate the square well problem numerically the ratio that ensures a well-behaved solution for $x \to -\infty$ will turn out to be κ for any x_0.

This strategy works for the numerical solution in the general case. There too we will find by trial and error that there is always a particular $(x_0$-dependent) ratio of $(d\psi/dx)_{x_0}$ to $\psi(x_0)$ that will kill the growing exponential for $x \to -\infty$. This is reasonable: we are trying to impose one constraint (no growing exponential as $x \to -\infty$) using one free parameter, the ratio of $(d\psi/dx)_{x_0}$ to $\psi(x_0)$.

Having made this choice to avert the blowup for $x \to -\infty$, we now integrate the equation toward increasing x. We have to live with what we get since the only freedom, the ratio of $(d\psi/dx)_{x_0}$ to $\psi(x_0)$, has been used up. We will find oscillations inside the well and exponential growth for $x \gg L$. In other words, the exponentially growing term (present in region *III* for the square well) will now raise its ugly head for large positive x. If, however, we keep varying the energy E in the time-independent Schrödinger equation, we may find that at some special isolated values the growing exponential will be absent as $x \to \infty$ as well, and $\psi(x \to \pm\infty)$ will die exponentially. These will be the allowed or quantized values of the bound-state energies.

Although this problem of generic $V(x)$ cannot be solved in closed form analytically, I hope the preceding analysis has persuaded you that energy has to be quantized for bound states. For those who insist on an analytic example, I will provide one shortly.

But first, a brief comment on the case $E > V_0$. The particle is now free to escape to infinity with positive kinetic energy. The Schrödinger equationwill allow oscillatory (and hence bounded) solutions in all three regions, and there will be no need to kill one coefficient in each of the regions *I* and *III*. We will find a two-parameter family of solutions at every energy, though one parameter may be rescaled to 1 without changing anything. An example of one such well and (the real part of) its wave

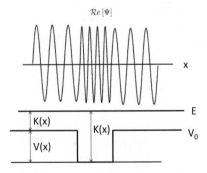

Figure 22.6   Real part of a wave function describing an unbound state $(E > V_0)$. The particle has a larger kinetic energy (and smaller λ) inside the well.

function are given in Figure 22.6. Notice that the wave number k increases inside the well due to the speeding up of the particle.

22.2.2 The box: an exact solution

Let us pass to the box, which is the limit $V_0 \to \infty$ of the well. The quantization of energy is demonstrated easily in this limit. According to the formula

$$\kappa = \sqrt{\frac{2m(V_0 - E)}{\hbar^2}} \tag{22.59}$$

the exponential tail on either side of the well (Figure 22.3) shrinks to zero as $V_0 \to \infty$, and consequently $\psi_I(x)$ is non-zero only inside the box.

Since $V = 0$ inside, we just do what we did with the free particle earlier. I repeat it here:

$$\frac{d^2 \psi_E(x)}{dx^2} + k^2 \psi_E(x) = 0 \quad \text{where} \tag{22.60}$$

$$k^2 = \frac{2mE}{\hbar^2}. \tag{22.61}$$

(Note that E is back as a subscript, there being only one region.)

Before you got into quantum mechanics you might have said (based on the oscillator that satisfies a similar equation with two derivatives in t) that the solutions are of the form

$$\psi_E(x) = A\sin kx + B\cos kx. \tag{22.62}$$

After entering the quantum world and seeing oscillating exponentials everywhere you may be tempted to favor

$$\psi_E(x) = Ce^{ikx} + De^{-ikx} \tag{22.63}$$

where C and D are some constants. Both choices are equally correct because we can write the trigonometric functions in terms of complex exponentials and vice versa. For a given A and B you could get the corresponding C and D using Euler's formula.

It turns out better to use Eqn. 22.62 with sines and cosines here. Once again the first impression that at every E there are two independent solutions with coefficients A and B is an illusion, because only their ratio is of physical importance. With this one free parameter, we have to satisfy the boundary conditions at the edges of the box. Since ψ vanishes identically outside the box, the ψ inside must vanish at the two ends by continuity or single-valuedness.

This boundary condition $\psi(\text{ends}) = 0$ is insensitive to the overall scale of ψ. If a given ψ does not vanish at the ends, neither will a rescaled one. Therefore we have two conditions and one real parameter, which means solutions will exist only at special energies. (What about two more conditions due to matching $\frac{d\psi}{dx}$ at the ends? The answer is that we do not match the slopes and let $\frac{d^2\psi}{dx^2}$ blow up at $x = 0$ and $x = L$. This is permitted on this occasion as V also diverges there.)

Let us see how the boundary conditions restrict A and B and also determine the allowed energies. At $x = 0$ we demand

$$\psi(0) = 0 = A \sin k \cdot 0 + B \cos k \cdot 0 = 0 + B, \tag{22.64}$$

which means $B = 0$. The cosine has to be killed since it refuses to vanish at the left end of the box. At the right end we want

$$\psi(L) = 0 = A \sin kL. \tag{22.65}$$

If we satisfy this by killing A, we would kill the entire solution. So we demand

$$\sin kL = 0. \tag{22.66}$$

This means k is restricted to be

$$k = k_n = \frac{n\pi}{L} \quad n = 1, 2, 3 \ldots \tag{22.67}$$

and the energy is quantized to be

$$E = E_n = \frac{\hbar^2 k_n^2}{2m} = \frac{\hbar^2 \pi^2 n^2}{2mL^2}. \tag{22.68}$$

The corresponding wave functions are

$$\psi_E \equiv \psi_n(x) = A \sin \left[\frac{n\pi x}{L} \right]. \tag{22.69}$$

Since the average value of $\sin^2 \theta$ over half a period is $\frac{1}{2}$, the wave function has an integral

$$\int_0^L |A|^2 \sin^2 \left[\frac{n\pi x}{L} \right] dx = \frac{|A|^2 L}{2}, \tag{22.70}$$

which means the normalized wave function is

$$\psi_n(x) = \sqrt{\frac{2}{L}} \sin \left[\frac{n\pi x}{L} \right]. \tag{22.71}$$

Look at the first few energies and wave functions shown in Figure 22.7. The following features are always true.

- The levels are non-degenerate, unlike on a ring.
- Every ψ has a kink at the ends and so $\frac{d^2\psi}{dx^2}$ diverges there. But that is permissible in this case because V also blows up at the walls. This is why we did not match the slopes at the ends of the box.
- The solution with label n completes n half cycles over the length of the box. These are exactly the functions that arise in the solution to the wave equation of a string clamped at the two ends. These are called its *normal modes*: If we deform the string to take one of these shapes, i.e., $\psi(x,0) = A\sin \left[\frac{n\pi x}{L} \right]$ and let it go, *every part* will go up and down in

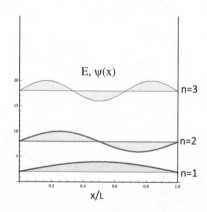

Figure 22.7   First three energy levels and wave function in a box.

step at frequency $\omega_n = k_n v$, where v is the wave velocity:

$$\psi_{string}(x,t) = \psi(x,0)\cos\left[\frac{n\pi\, vt}{L}\right]$$

$$= A\sin\left[\frac{n\pi\, x}{L}\right]\cos\left[\frac{n\pi\, vt}{L}\right]. \qquad (22.72)$$

We will encounter a similar result in quantum dynamics: a state that begins as $\psi_E(x)$ at time zero will preserve its form and simply pick up a time-dependent phase factor in front.

The fact that the quantization of energies has the same origin as the classical quantization of the frequencies of a string is the reason why the Schrödinger equationwas immediately embraced by the community when it was announced. One suddenly understood why energy is quantized: by trying to fit some number of half wavelengths into the box one is fixing the allowed wavelengths or wave numbers k and that translates into allowed energies.

- The allowed values of n do not include 0 or negative integers, whereas these were allowed for free particles on a ring. Here is the reason. On the ring we had

$$\psi_n(x) = A\exp\left[\frac{2\pi\, nix}{L}\right] \qquad n = 0, \pm 1, \pm 2\ldots \qquad (22.73)$$

If we set $n = 0$ above we get $\psi_0 = Ae^{i\cdot 0} = A$, whereas if we set $n = 0$ in $A\sin\frac{n\pi x}{L}$ the corresponding $\psi_0(x) \equiv 0$. Likewise when we change n to $-n$

$$\exp\left[\frac{2\pi\, nix}{L}\right] \rightarrow \exp\left[-\frac{2\pi\, nix}{L}\right], \qquad (22.74)$$

which is an independent function, whereas under the same change

$$A\sin\frac{n\pi\, x}{L} \rightarrow A\sin\left[\frac{-n\pi\, x}{L}\right] = -A\sin\left[\frac{n\pi\, x}{L}\right], \qquad (22.75)$$

which is just -1 times $A\sin\frac{n\pi x}{L}$ and hence not a second, independent solution.

- You can probe the energy levels by shining light and seeing what frequency of light the particle absorbs or emits. When you learn more quantum mechanics, you will see that the wave functions control the *rate* of absorption and emission.

- The probability density in one of these states is

$$P(x) = \frac{2}{L} \sin^2 \left[\frac{n\pi x}{L} \right].$$

(22.76)

- The ground state energy is $E_1 = \frac{\hbar^2 \pi^2}{2mL^2}$. Why is it not zero? If you were in a prison of size L and infinitely high walls what would be your lowest energy state? I know I would just sit on the floor and feel sorry for myself. But such a state of zero momentum and fixed position is not allowed for the particle by the uncertainty principle. Since the particle's position is known to be within the box, its uncertainty is $\Delta x \simeq L$, and its momentum uncertainty bounded below by $\Delta p \gtrsim \frac{\hbar}{L}$, and its energy should be of order $(\Delta p)^2/2m = \frac{\hbar^2}{2mL^2}$. Indeed the ground state energy is of order $\hbar^2/mL^2$ (dropping factors of order unity). Furthermore the wave function $\sin \pi x/L$ *seems to be* an admixture of $p = \pm \frac{\pi \hbar}{L}$, which implies $\Delta p \simeq \frac{\hbar}{L}$. (I say "seems to be" even though $\psi_1(x)$ is clearly a sum of $\exp\left[\pm \frac{i\pi \hbar x}{\hbar L}\right]$, because the wave function $\psi_1(x)$ has this form *only inside the box*. It is identically zero outside. However, such a function *can* be expressed as a superposition of states of definite momentum inside and outside the box. These momentum functions, non-zero on the infinite line, can have a continuous range of allowed p unlike those on a finite ring. Their normalization is very tricky due to the infinite volume. However, the bottom line is that a suitable superposition of such momentum functions, non-zero on all of space, will add up to zero outside the box and to $\psi_1(x)$ inside the box. This superposition will contain not just the two momenta $\pm \frac{\pi \hbar}{L}$, but a continuum of momenta given by a distribution centered at $p = 0$ and of "width" $\Delta p \simeq \hbar/L$.)

   Such uncertainty principle arguments abound in physics. For example, to estimate the lowest kinetic energy of a proton in a nucleus whose size is $\Delta x \simeq 10^{-15} m$ we set $\Delta p \simeq \frac{\hbar}{\Delta x}$ and estimate the ground state kinetic energy to be of order $\frac{1}{2m} \frac{\hbar^2}{\Delta x^2} \simeq 10^7 \, eV$.

The box is the simplest example of quantization of bound-state energies by boundary conditions. It is the caricature for atoms. An electron remains bound to an atom by a deep $1/r$ potential between the electron and the nucleus. We need to solve the Schrödinger equation in three dimensions, which is an obvious generalization of the one-dimensional version. But it is much harder to solve. Even Schrödinger needed help from a mathematician. The solution gives the energy levels, with the right degeneracies and the corresponding wave functions $\psi_E(\mathbf{r})$. The "size" of the wave functions is of the order of the *Bohr radius*

$$a_0 = \frac{4\pi \varepsilon_0 \hbar^2}{m_e e^2} \simeq 10^{-10} m \tag{22.77}$$

where m_e is the electron mass. The estimated kinetic energy based on the uncertainty principle (keeping track of just powers of ten) is around $\frac{\hbar^2}{m_e a_0^2} \simeq 10 eV$. Indeed the electron volt is a natural unit of energy for atomic physics.

Bear in mind that to do the quantum mechanics you need to know the classical potential energy $V(x)$, which you need to stick into the Schrödinger equationto find the allowed energies and ψ_E's. If the system is an oscillator you need the force constant. If it is an atom, you should know the Coulomb potential. If it is a nucleon in the nucleus, you need the nuclear potential, which is usually of the form $\frac{1}{r}e^{-r/r_0}$, where $r_0 \simeq 10^{-15} m = 1 fermi$.

22.3   Energy measurement in the box

It has taken so long to find and analyze the states of definite energy $\psi_E(x) \equiv \psi_n(x)$ that you may have forgotten why we did all this. Let me remind you, just in case. The goal was to find the possible outcomes and their probabilities when an energy measurement was performed on some arbitrary state $\psi(x)$. Now we return to the recipe, but this time fully armed with all its ingredients.

First we expand $\psi(x)$ for a particle in the box, as a superposition

$$\psi(x) = \sum_{n=1}^{\infty} A(n)\psi_n(x) = \sum_{n=1}^{\infty} A(n)\sqrt{\frac{2}{L}}\sin\left[\frac{n\pi x}{L}\right] \tag{22.78}$$

after having determined the $A(n)$ by

$$A(n) = \int_0^L \psi_n^*(x)\psi(x)dx \tag{22.79}$$

$$= \int_0^L \sqrt{\frac{2}{L}} \sin\left[\frac{n\pi x}{L}\right] \psi(x)\, dx. \tag{22.80}$$

The probability for finding the system in a state with label n is

$$P(n) = |A(n)|^2. \tag{22.81}$$

Here is a somewhat artificial example to illustrate this formula. Let us take the normalized ψ to be

$$\psi(x) = \sqrt{\frac{2}{L}} \quad 0 \le x < \frac{L}{2} \tag{22.82}$$

$$= 0 \quad \frac{L}{2} < x \le L. \tag{22.83}$$

(Notice that the ψ we chose to consider also vanishes outside the box just like the ψ_n's did. The reason is that if it did not, it could not be built out of ψ_n's. I asserted earlier that no matter what variable $\mathcal{A}$ we choose, it will be possible to express any given ψ as a linear superposition of the ψ_a. Actually there are some restrictions on the ψ's for which this is true. One of them is that they cannot wander into a region where *all* the ψ_n vanish. Our ψ violates this condition at one point $x = 0$. It should therefore be seen as the limit of a family of functions that drop more and more precipitously as $x \to 0$ from the right.)
Continuing,

$$A(n) = \int_0^{L/2} \sqrt{\frac{2}{L}} \sin\left[\frac{n\pi x}{L}\right] \sqrt{\frac{2}{L}}\, dx \tag{22.84}$$

$$= \frac{2}{L} \frac{L}{n\pi} \left(\cos\left[\frac{n\pi x}{L}\right]\right)\Big|_{L/2}^0 \tag{22.85}$$

$$= \frac{2}{n\pi} \left(1 - \cos\frac{n\pi}{2}\right) \tag{22.86}$$

$$= \frac{4}{n\pi} \sin^2\left[\frac{n\pi}{4}\right] \tag{22.87}$$

$$P(n) = \frac{16}{n^2\pi^2} \sin^4\left[\frac{n\pi}{4}\right]. \tag{22.88}$$

The main features of the result are that $P(n)$ falls with n like $1/n^2$ and that it vanishes whenever n is a multiple of 4. (Try to see why by sketching the first four wave functions on top of $\psi(x)$.)

As a concrete example of Eqn. 22.88 let us compute the probability that the particle will be found in the ground state $n = 1$:

$$P(1) = \frac{16}{\pi^2} \sin^4\left[\frac{\pi}{4}\right] = \frac{4}{\pi^2} \simeq 0.41. \qquad (22.89)$$

CHAPTER 23

Scattering and Dynamics

Consider the following problem in classical mechanics. You fire a particle of mass m and momentum p_0 from the far left on level ground. It then encounters a potential hill that grows smoothly from 0 to a constant V_0 as shown in Figure 23.1.

What happens depends on the energy $E = p_0^2/2m$ of the particle. If $E < V_0$ (upper part of figure), the particle will climb, losing kinetic energy $K(x)$ and gaining potential energy $V(x)$, keeping the total constant and equal to E, shown as a flat line in the figure. It will stop at the turning point x_T where $E = V(x_T)$ and roll back to you. This is called *reflection*. If you crank up the energy and reach $E > V_0$ (lower half of figure), the projectile will go over the top and exit with some positive kinetic energy $K = E - V_0$ and you will never see it again. This is called *transmission*. Even if the hill is invisible from where you shoot the projectile, you can determine V_0: it is the lowest launch energy E at which the projectile fails to come back. Scattering is used in this manner to probe the forces between subatomic particles by shooting them at each other and seeing how they scatter.

23.1 Quantum scattering

Now let us explore the same scattering process in quantum mechanics. The proper way to handle this is to employ the time-dependent Schrödinger equation, which tells us how any initial $\psi(x, 0)$ evolves with time into

524

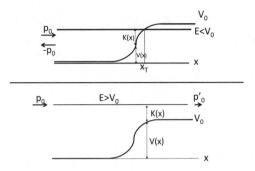

Figure 23.1   Classical scattering off a potential that rises smoothly from 0 at
$x \to -\infty$ to V_0 as $x \to \infty$. If $E > V_0$, transmission is guaranteed and if $E < V_0$,
reflection is guaranteed.

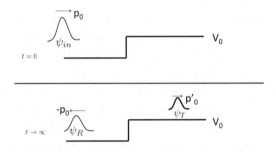

Figure 23.2   Time-dependent view of scattering for the case $E > V_0$. (The
smooth $V(x)$ in the classical discussion has been replaced by the step to facilitate
computation.) A wave packet ψ_{in} of mean momentum p_0 and mean energy
greater than V_0 is incident from the left at $t = 0$. After a long time as $t \to \infty$, it
turns into two packets: a reflected one ψ_R with mean momentum $-p_0$ and a
transmitted one ψ_T with smaller momentum p'_0. Assuming the incident packet is
normalized to unity, the square integrals of the reflected and transmitted packets
give the reflection and transmission coefficients R and T.

$\psi(x, t)$. Look at Figure 23.2. The smooth $V(x)$ in the classical discussion
has been replaced by the step of height V_0 to facilitate computation. For
$\psi(x, 0)$ we choose ψ_{in}, a normalized wave function, called a *wave packet*.
It is well localized in space and made up of states with positive values of p
sharply peaked around some p_0, with Δx and Δp obeying the uncertainty

principle. Then we let the *time-dependent* Schrödinger equationtake over and compute its fate for large future times. (See this calculation as a black box and just consider the results.) What happens depends on the average energy of the incident packet.

23.1.1 Scattering for $E > V_0$

I illustrate in Figure 23.3 the case where the average incident energy $E = \frac{p_0^2}{2m} = \frac{\hbar^2 k^2}{2m} > V_0$, which is what interests us most. (The constant energy line with $E > V_0$ is not shown.) By the time $t \to \infty$, the incident packet would have split into the reflected packet $\psi_R(x, t)$, which moves in region I toward $x = -\infty$ with average momentum $-p_0$, and the transmitted packet $\psi_T(x, t)$, which moves in region II toward $x = \infty$ with average momentum $p_0' = \hbar k'$ where $(\hbar k')^2/2m = E - V_0$. By the basic postulate on probability density, the total areas under $|\psi_R(x, t \to \infty)|^2$ and $|\psi_T(x, t \to \infty)|^2$, respectively, give the probability R that the particle is at the far left asymptotically and thus has been reflected, and the probability T that it is at the far right and has been transmitted.

If we send in a very large number of particles all in the same state ψ_{in}, a fraction R will get reflected and a fraction T will get transmitted. This is what happens when an accelerator shoots out a beam of projectiles at some target and the scattered particles are picked up by many detectors.

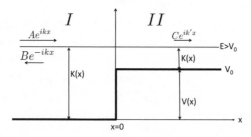

Figure 23.3   Time-independent quantum treatment of a particle that approaches a step potential with energy $E = \frac{\hbar^2 k^2}{2m} > V_0$. The incident and reflected waves *coexist* to the left of the barrier *as waves of opposite momenta* $\pm \hbar k = \pm\sqrt{2mE}$ and the transmitted wave becomes a wave with momentum $\hbar k' = \sqrt{2m(E - V_0)}$. Shown are typical values of kinetic energy $K(x)$, potential energy $V(x)$, and total energy $E > V_0$.

I merely describe this involved computation and skip the details because we have not yet studied the time-dependent Schrödinger equationand because the final answer can be found using just the time-independent Schrödinger equationin a certain limit.

That limit is approached by making the incident packet broader and broader in x and sharper and sharper around the mean momentum $p_0 = \hbar k$. The same broadening happens to the transmitted and reflected packets. At this point they are no longer packets but extended waves. The incident and reflected waves *coexist* to the left of the barrier *as waves of opposite momenta* $\pm p_0$ and the transmitted wave becomes a wave with momentum $p_0' = \sqrt{2m(E - V_0)}$. The reflection and transmission coefficients R and T are fully determined by $p_0 = \hbar k$, the sharply defined initial momentum and V_0. In this limit there is no sense of time, and the values of R and T may be found in terms of p_0 and V_0 by solving the *time-independent* Schrödinger equationwith appropriate boundary conditions.

Here are the details of the time-independent approach to scattering. The treatment of familiar material will be brief. We just have two regions I and II and in each of them we may write as before

$$\psi_I(x) = Ae^{ikx} + Be^{-ikx} \quad \text{where } p_0 = \hbar k = \sqrt{2mE} \qquad (23.1)$$

$$\psi_{II}(x) = Ce^{ik'x} + De^{-ik'x}$$

$$\text{where } \hbar k' = p_0' = \sqrt{2m(E - V_0)}. \qquad (23.2)$$

Note that k' is smaller than k because of the loss of momentum in climbing up the step.

Since there is no exponential blowup as $|x| \to \infty$, we may keep all four coefficients, which allow for incoming (toward the step) and outgoing (away from the step) waves in both regions. However, if we are to describe the scattering process under discussion, *we want an incoming wave only in region I*, producing a reflected wave in region I and a transmitted wave in region II. So we choose $D = 0$ to kill the incoming wave in region II.

You may have noticed that the momentum wave functions have not been normalized by the usual factor $1/\sqrt{L}$ because we are in infinite volume with $L = \infty$. We cannot put the system on a ring either because we do not want the transmitted wave to come back at us from the left! Fortunately we can find R and T using only the ratios of A, B, and C, leaving the overall normalization unspecified. It will not enter R and T. (Although in this example we could finesse the question of normalization in the infinite volume, you should know that there *is* a subtle way to normalize momentum states in infinite volume. It works only because their square integral diverges linearly with the length L of the universe, as compared to wave functions whose square integrals diverge exponentially with L. The latter are simply disallowed.)

Since the overall scale of the wave function is arbitrary, let us choose $A = 1$. The coefficients B and C must ensure that ψ and $\psi' \equiv \frac{d\psi}{dx}$ are continuous at $x = 0$:

$$1 + B = C \quad (\psi_I(0) = \psi_{II}(0)) \tag{23.3}$$

$$ik(1 - B) = ik'C \quad (\psi_I'(0) = \psi_{II}'(0)). \tag{23.4}$$

The solution is

$$B = \frac{k - k'}{k + k'} \tag{23.5}$$

$$C = \frac{2k}{k + k'}. \tag{23.6}$$

How should we define R and T now that there is no sense of time? We cannot define them in terms of the ratios of areas under these wave functions since all such areas are infinite. So let us turn to basics and ask how we would define R and T if we had many projectiles at our disposal. We would fire a large number, say 10000, at the step, and if 6000 come back and 4000 get through we would define

$$R = \frac{6000}{10000} \quad T = \frac{4000}{10000}. \tag{23.7}$$

(We may have to increase all the numbers till the ratios R and T stabilize.)

Look at the incoming, reflected and transmitted wave functions with amplitudes 1, B and C. If there are a large number of projectiles, all given by the same wave functions, their number density will be proportional to the probability density. The particles will have a velocity $v = \frac{p}{m} = \hbar k/m$ in region I and $v' = \frac{\hbar k'}{m}$ in region II. There is a steady stream of particles coming in and getting reflected or transmitted. So we cannot work with the total number of particles (infinite) and must deal with the number arriving and getting scattered or reflected *per second*. That is, we must work with the currents.

Recall from our study of electric currents that

$$j = \rho v \tag{23.8}$$

where j is the current density, ρ is the particle density, and v is the velocity. In one dimension the current (particles crossing some checkpoint per second) is the same as the current density (no area normal to the current to divide by). So

$$j_{\text{incident}} = P(x)v = P(x)\frac{\hbar k}{m} = 1 \cdot \frac{\hbar k}{m}. \tag{23.9}$$

Similarly

$$j_R = |B|^2 \cdot \frac{\hbar k}{m} \tag{23.10}$$

$$j_T = |C|^2 \cdot \frac{\hbar k'}{m}. \tag{23.11}$$

The crucial factor $v' = \frac{\hbar k'}{m}$ in the transmitted current ensures that we take into account not only the density of particles but their velocity in computing their arrival rate at the far right.

So finally we have

$$R = \frac{j_R}{j_{\text{incident}}} = |B|^2 = \left(\frac{k - k'}{k + k'}\right)^2 \tag{23.12}$$

$$T = \frac{j_T}{j_{\text{incident}}} = |C|^2 \frac{k'}{k} = \left(\frac{2k}{k + k'}\right)^2 \frac{k'}{k} = \frac{4kk'}{(k + k')^2}. \tag{23.13}$$

Observe that $R + T = 1$, expressing the conservation of probability. For a steady beam of projectiles, this means the number arriving per second equals the number reflected per second plus the number transmitted per second.

Had we used the time-dependent Schrödinger equationwe would have arrived at exactly these values of R and T in the limit in which the incident packet had a sharply defined momentum p_0.

Now it is time to analyze the results.

- If $V_0 = 0$, i.e., there is no step, we expect no reflection and we do indeed find $R = 0$.
- Even when the incident energy exceeds the step height, the particle has a non-zero probability to bounce back. This happens in quantum theory because the particle is controlled by a wave and waves undergo reflection when there is a change of medium, or, in this case, a change in potential from 0 to V_0.
- If $E \to \infty$, the potential energy is negligible compared to kinetic energy, $k'/k \to 1$, and $T \to 1$, i.e., there is perfect transmission even in quantum theory.

23.1.2 Scattering for $E < V_0$

The wave function that behaves well at spatial infinity is

$$\psi_I(x) = e^{ikx} + B e^{-ikx} \quad \text{where } \hbar k = \sqrt{2mE} \tag{23.14}$$

$$\psi_{II}(x) = C e^{-\kappa x} \quad \text{where } \hbar \kappa = \sqrt{2m(V_0 - E)} \tag{23.15}$$

as indicated in Figure 23.4.

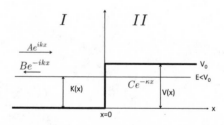

Figure 23.4 Time-independent quantum treatment of a particle that approaches a step potential with energy $E = \frac{\hbar^2 k^2}{2m} < V_0$. The incident and reflected waves *coexist* to the left of the barrier *as waves of opposite momenta* $\pm \hbar k = \pm \sqrt{2mE}$ and the transmitted wave is exponentially damped with $\hbar \kappa = \sqrt{2m(V_0 - E)}$. Shown are typical values of kinetic energy $K(x)$ and potential $V(x)$ energy.

The matching conditions are

$$1 + B = C \tag{23.16}$$

$$ik(1 - B) = -\kappa C \tag{23.17}$$

with the solution

$$B = \frac{k - i\kappa}{k + i\kappa} \tag{23.18}$$

$$C = \frac{2k}{k + i\kappa}. \tag{23.19}$$

The reflected current has a value

$$j_R = |B|^2 \frac{\hbar k}{m} = \left| \frac{k - i\kappa}{k + i\kappa} \right|^2 \frac{\hbar k}{m} = \frac{\hbar k}{m} = j \text{ incident.} \tag{23.20}$$

Thus when the step is taller than the incident energy, there is complete reflection of the incident current: $R = 1$. But if $R = 1$, what does the non-zero $|C|^2$ do to the condition $R + T = 1$? The result $|C|^2 > 0$ only means that there is an exponential tail of probability in the forbidden region, and does not imply a transmitted current. (The formula $j = P(x)v$ does not apply here since the wave function does not describe particles moving with a real momentum. A more advanced definition of current will yield a vanishing current in region II and, in fact, whenever ψ is real.)

23.2 Tunneling

Suppose the barrier, instead of going on forever at the value V_0, dropped to 0 beyond some point, as in region III of Figure 23.5. In region II both $e^{\pm \kappa x}$ are allowed, but the ultimate amplitude for transmission through the barrier ends up being exponentially small if we match ψ and $d\psi/dx$ at the two interfaces. The wave function will always "leak out" into region III as long as the height and width of the barrier are finite. Once the particle leaks to the allowed region, it need not hide; it can go with a real momentum, equal in fact to the incident momentum (assuming V is the same on either side of the barrier). There will also be a non-zero current flowing to the right. This leakage is called *barrier penetration*. It means that if you send a particle with an energy not enough to overcome the barrier in classical

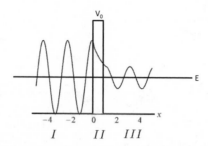

Figure 23.5   A particle coming in from the left with energy less than the barrier height can manage to tunnel to the other side because ψ is non-zero in the barrier. Once it gets to the other side the wave function is once again oscillatory.

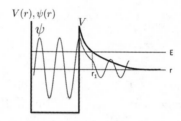

Figure 23.6   Inside the potential well created by the other nucleons, the alpha has an oscillating wave function. It tunnels to r_1 (with exponential suppression) and escapes to infinity (again with an oscillating ψ).

mechanics, quantum mechanics gives it a small chance of being found on the other side.

So no barrier is completely safe in quantum theory. Here is a final survival tip. You are in a prison with walls of finite height and thickness. What's your strategy? I say go ram yourself against the walls as often as you can, because there is a small probability that you will suddenly find yourself on the other side.

This is what happens in *alpha decay*. The alpha particle, which is just the helium nucleus, resides inside a big radioactive nucleus. The attraction between the alpha and the rest of the nucleons creates an attractive well $V(r)$ that keeps the alpha inside for energies below the barrier height, as shown in Figure 23.6. But the barrier tapers off as we move away from the nucleus and at r_1 it falls below the energy E of the alpha. The alpha can then tunnel and come out as a legitimate free particle beyond r_1. The alpha particle does exactly what I told you to do. It goes rattling back and forth inside the nucleus and once in a while it manages to penetrate the barrier and come out. This is the *alpha decay* of the nucleus.

The alpha pounds on the walls with a very high frequency f, which we may estimate as follows. The nucleus has a size Δ, the momentum of the alpha is of the order $\hbar/\Delta$, its velocity is of the order $\hbar/(m\Delta)$, and so it bounces back and forth with a frequency

$$f \simeq \frac{v}{\Delta} = \frac{\hbar}{m\Delta^2}, \tag{23.21}$$

dropping all factors of order unity. For $\hbar \simeq 10^{-34} J \cdot s, m \simeq 10^{-27}$ kg, and $\Delta \simeq 10^{-15} m$ we find $f \simeq 10^{23} Hz$. (A better estimate is $10^{21} Hz$.) That is the good news. The bad news is that the tunneling probability *per attempt* is exponentially small, say $\simeq 10^{-38}$. It could take about a billion years for a successful escape. But you do not have to wait that long for your Geiger counter to click because a very large number of nuclei are simultaneously trying to decay. By comparison, your prison escape is even less likely to succeed, but if you have given up on your lawyer or a presidential pardon, it may be your best bet.

23.3 Quantum dynamics

You want to know the theory of everything? You're almost there, because I'm going to reveal to you the law of (non-relativistic) quantum dynamics. It tells you how things change with time. It is the analog of $F = ma$. It is called the *time-dependent* Schrödinger equation, or simply the Schrödinger equation. It contains Newton's laws as part of it, because if you can do the quantum theory, you can always find hidden in it the classical theory. It describes an astounding number of phenomena around you on this planet and in the cosmos. There are of course some phenomena it cannot describe, but it goes a long long way.

We have seen that the wave function $\psi(x)$ is the analog of (x, p) in Newtonian mechanics. It contains the maximum possible information about the particle. Of course, extracting this information is a lot more difficult in quantum theory than in classical theory. Classically, if you were given (x, p) and wanted to know what you would get if you measured any dynamical variable, some function of coordinates and momenta, like the angular momentum $\mathbf{L} = \mathbf{r} \times \mathbf{p}$, you just entered the values of (x, p) (or its generalization to three dimensions) in the expression for the variable. The answer to the corresponding question in quantum mechanics is so long and tedious you will wish you had not asked: express the given ψ as a sum of functions $\psi_\alpha(x)$ with coefficients $A(\alpha)$ and so forth. Even with all this you just get the probabilities for various outcomes. Let us not go over all that again.

What we want to do is consider the dynamics. How does ψ change with time? If it has a value $\psi(x,0)$ at time $t = 0$, what is $\psi(x,t)$, at a later time t? What is the analog of $F = ma$?

I am just going to write it down: the evolution of $\psi(x,t)$ is determined by the *time-dependent Schrödinger equation*

$$i\hbar\frac{\partial \psi(x,t)}{\partial t} = -\frac{\hbar^2}{2m}\frac{\partial^2 \psi(x,t)}{\partial x^2} + V(x,t)\psi(x,t).$$

$$(23.22)$$

We will restrict ourselves to the case of time-independent potentials $V(x,t) = V(x)$, even though the time-dependent Schrödinger equation is valid for that case as well.

Before looking at solutions to the equation let us explore some obvious and striking features.

The first is that $i = \sqrt{-1}$ enters the very equation of motion. When we used complex numbers as an artifact for solving problems in circuits or oscillations, it was a matter of convenience. The quantity of interest to us was always real, whether it was the coordinate of an oscillator or the current in a circuit. But the i here is deeply embedded, right into the equation of motion. We have already seen that without it we could not write down a state of definite momentum $\psi_p(x) = Ae^{ipx/\hbar}$.

The second is that unlike Newton's laws for $x(t)$, this is first order in time. This means that as our initial condition we need just $\psi(x,0)$. The Schrödinger equationwill tell us what $\partial \psi/\partial t$ is, just like Newton's law determines the acceleration in terms of the force.

The $\psi(x,t)$ in Eqn. 23.22 is not anything particular: *every $\psi(x,t)$ obeys this equation just like every classical trajectory $x(t)$ obeys Newton's second law.*

Do not get confused between this equation and the time-independent Schrödinger equation you saw in the last chapter:

$$-\frac{\hbar^2}{2m}\frac{d^2 \psi_E(x)}{dx^2} + V(x)\psi_E(x) = E\psi_E(x).$$

$$(23.23)$$

In the above there is no time and $\psi_E(x)$ are not generic, but special wave functions corresponding to states of definite energy. However, this equation will reenter the discussion very soon.

How can we calculate the future, given the present using this equation? How do we solve this equation? I'm going to do it at different levels.

1. Just write down a particularly simple solution and verify that it satisfies the equation. Understand a remarkable property of that solution.
2. Understand how that solution could be derived. This is optional.
3. Show how to find the future wave function $\psi(x, t)$ *for any given* initial state $\psi(x, 0)$ building on the simple solution.

23.3.1 A solution of the time-dependent Schrödinger equation

Provided V is time-independent, i.e., $V = V(x)$, the following is a solution to the time-dependent Schrödinger equation:

$$\psi_E(x, t) = \psi_E(x)e^{-\frac{iEt}{\hbar}}. \tag{23.24}$$

On the left-hand side is a particular solution to the time-dependent Schrödinger equation, which carries the label E because it is built out of the solution $\psi_E(x)$ of the time-independent Schrödinger equation. Let us verify the claim, starting with the left-hand side:

$$i\hbar\frac{\partial \psi_E(x, t)}{\partial t} = i\hbar\frac{\partial\left[\psi_E(x)e^{-\frac{iEt}{\hbar}}\right]}{\partial t} \tag{23.25}$$

$$= \psi_E(x)i\hbar\frac{\partial e^{-\frac{iEt}{\hbar}}}{\partial t} \tag{23.26}$$

$$= \psi_E(x)i\hbar\frac{de^{-\frac{iEt}{\hbar}}}{dt} \tag{23.27}$$

$$= E\psi_E(x)e^{-\frac{iEt}{\hbar}} = E\psi_E(x, t). \tag{23.28}$$

I have drawn on the fact that the partial time derivative acts only on the $e^{-\frac{iEt}{\hbar}}$ part of $\psi_E(x, t)$ and does so as the ordinary or total derivative.

Now for the right-hand side:

$$-\frac{\hbar^2}{2m}\frac{\partial^2 \psi_E(x,t)}{\partial x^2} + V(x)\psi_E(x,t)$$

$$= \left(e^{-\frac{iEt}{\hbar}}\right)\left(-\frac{\hbar^2}{2m}\frac{\partial^2 \psi_E(x)}{\partial x^2} + V(x)\psi_E(x)\right) \qquad (23.29)$$

$$= \left(e^{-\frac{iEt}{\hbar}}\right)\left(-\frac{\hbar^2}{2m}\frac{d^2 \psi_E(x)}{dx^2} + V(x)\psi_E(x)\right) \qquad (23.30)$$

$$= e^{-\frac{iEt}{\hbar}}E\psi_E(x) = E\psi_E(x,t). \qquad (23.31)$$

The partial x-derivative acts only on the $\psi_E(x)$ part of $\psi_E(x,t)$ (as a total derivative) and that along with $V(x)\psi_E(x)$ gives $E\psi_E(x)$ because $\psi_E(x)$ is a solution to the time-independent Schrödinger equation 23.23. From Eqns. 23.28 and 23.31 we see the time-dependent Schrödinger equation is satisfied.

23.3.2   Derivation of the particular solution $\psi_E(x,t)$

Suppose we say, "Look, we don't know if we can find every possible solution to the time-dependent Schrödinger equation. So let's begin with the modest goal of looking for solutions of the *product form*

$$\psi(x,t) = X(x)T(t) \qquad (23.32)$$

where $X(x)$ is a function only of x and $T(t)$ is a function only of t." These are by no means the only types of solutions. (We will see examples that are not.) But right now we are desperate for *any* solution, even of the restricted form, because such a tactic has proven fruitful in previous encounters with partial differential equations. To see if even this modest goal can be reached, we take the assumed form and stick it into the equation. Once again we note that the t-derivative acts only on $T(t)$ and the x-derivative only on $X(x)$. Consequently the time-dependent Schrödinger equation tells us that our product form must obey

$$X(x)\left[i\hbar\frac{dT(t)}{dt}\right] = T(t)\left[-\frac{\hbar^2}{2m}\frac{d^2 X(x)}{dx^2} + V(x)X(x)\right]. \quad (23.33)$$

We now divide both sides by $X(x)T(t)$ to arrive at

$$\frac{1}{T(t)}\left[i\hbar\frac{dT(t)}{dt}\right] = \frac{1}{X(x)}\left[-\frac{\hbar^2}{2m}\frac{d^2X(x)}{dx^2} + V(x)X(x)\right].$$
(23.34)

Only the *total* derivatives with respect to t and x appear, because the partial derivatives act only on functions of the corresponding variable.

Look at Eqn. 23.34. On the left-hand side is a function only of t and on the right-hand side is a function only of x. (This is why we required that V have no time-dependence.) Can the left-hand side vary with t? It cannot, because if it did, the right-hand side, which has no t in it, cannot keep up. So the left-hand side must be t-independent. The right-hand side cannot depend on x for the same reason. Both sides must equal a t and x-independent constant, which I will call E for a good reason:

$$\frac{1}{T(t)}\left[i\hbar\frac{dT(t)}{dt}\right] = \frac{1}{X(x)}\left[-\frac{\hbar^2}{2m}\frac{d^2X(x)}{dx^2} + V(x)X(x)\right]$$
$$= E.$$
(23.35)

The original partial differential equation has broken down into two ordinary differential equations *for the product solution*:

$$\left[i\hbar\frac{dT(t)}{dt}\right] = ET(t)$$
(23.36)

$$\left[-\frac{\hbar^2}{2m}\frac{d^2X(x)}{dx^2} + V(x)X(x)\right] = EX(x).$$
(23.37)

The solution to the first equation is obviously

$$T(t) = Ae^{-\frac{iEt}{\hbar}}$$
(23.38)

and $X(x)$, the solution to the second, is the function we have been calling $\psi_E(x)$! Thus the product solution is

$$\psi_E(x,t) = e^{-\frac{iEt}{\hbar}}\psi_E(x).$$
(23.39)

In other words, a product solution exists only if the time-dependent part $T(t)$ is the exponential $e^{-\frac{iEt}{\hbar}}$ and the x-dependent part $X(x)$ is a solution to the time-independent Schrödinger equation corresponding to the energy E. The allowed solutions of the product form will exist only for the allowed energies E.

This completes the derivation of the product solution.

23.4 Special properties of the product solution

Look at the product solution in Eqn. 23.39. If we set $t = 0$ on both sides we find

$$\psi(x,0) = \psi_E(x). \tag{23.40}$$

Thus the product solution begins as $\psi_E(x)$ and as time goes by, all that happens is that it picks up a phase factor $e^{-\frac{iEt}{\hbar}}$ and turns into

$$\psi_E(x,t) = e^{-\frac{iEt}{\hbar}} \psi_E(x,0). \tag{23.41}$$

The x-dependence does not change with time at all! Recall my earlier analogy to a string. If you pluck a string (clamped at $x = 0$ and $x = L$) into some arbitrary shape $\psi(x,0)$ and let it go, it will wiggle and jiggle in a complicated fashion into some $\psi(x,t)$ dictated by the wave equation. If, however, you started it out in the state

$$\psi_n(x,0) = A\sin\left[\frac{n\pi x}{L}\right] \tag{23.42}$$

it will evolve into

$$\psi_n(x,t) = A\sin\left[\frac{n\pi x}{L}\right]\cos\left[\frac{n\pi vt}{L}\right]. \tag{23.43}$$

As time goes by, the profile of the string will change only in its overall scale by the cosine factor. Every part will rise and fall in step.

As in the case of the string, in the product solution of Eqn. 23.41 the ψ at every x oscillates the same way, as $e^{-\frac{iEt}{\hbar}}$. But unlike in the case of the string where the cosine factor changes the appearance of the string with time, *nothing measurable changes with time in $\psi_n(x,t)$.* Consider for example the particle that starts out in state n of the box. It evolves into

$$\psi_n(x,t) = \sqrt{\frac{2}{L}} \sin\left[\frac{n\pi x}{L}\right] \exp\left[-\frac{iE_n t}{\hbar}\right]$$

$$= \sqrt{\frac{2}{L}} \sin\left[\frac{n\pi x}{L}\right] \exp\left[-\frac{i\hbar n^2 \pi^2 t}{2mL^2}\right]. \tag{23.44}$$

The probability density at time t is

$$P_n(x,t) = |\psi_n(x,t)|^2 \tag{23.45}$$

$$= \left|\sqrt{\frac{2}{L}} \sin\left[\frac{n\pi x}{L}\right] \exp\left[-\frac{i\hbar n^2 \pi^2 t}{2mL^2}\right]\right|^2 \tag{23.46}$$

$$= \left|\sqrt{\frac{2}{L}} \sin\left[\frac{n\pi x}{L}\right]\right|^2 \left|\exp\left[-\frac{i\hbar n^2 \pi^2 t}{2mL^2}\right]\right|^2 \tag{23.47}$$

$$= \left|\sqrt{\frac{2}{L}} \sin\left[\frac{n\pi x}{L}\right]\right|^2 \tag{23.48}$$

$$= P_n(x,0). \tag{23.49}$$

Thus the odds of finding the particle at some x does not change with time! The oscillating complex exponential plays a role in the time-dependent Schrödinger equation when $i\hbar\frac{\partial}{\partial t}$ acts on it, but drops out of $|\psi|^2$. It's very interesting. The wave function depends on time and yet in a practical sense the physical properties don't depend on time. This is analogous to what happens in states of definite momentum: they oscillate as $e^{ipx/\hbar}$ (which defines a de Broglie wavelength $\lambda = 2\pi\hbar/p$) but the probability density $P(x)$ is flat.

    If a system is found to be in a state $\psi_E(x)$ after an energy measurement, it stays that way not just for an infinitesimal time but forever. The phase factor $e^{-iEt/\hbar}$ does not affect $P(x,t)$.

Next consider $P(p, t)$, the probability of finding a momentum p in a state that starts out as a state of definite energy $\psi_E(x)$. If at $t = 0$ we expand

$$\psi_E(x, 0) = \sum_p A(p) \psi_p(x), \tag{23.50}$$

then at a later time

$$\psi_E(x, t) = \exp\left[-\frac{iEt}{\hbar}\right] \psi_E(x, 0)$$

$$= \exp\left[-\frac{iEt}{\hbar}\right] \sum_p A(p) \psi_p(x) \tag{23.51}$$

$$= \sum_p A(p) \exp\left[-\frac{iEt}{\hbar}\right] \psi_p(x). \tag{23.52}$$

This means that as time evolves each initial $A(p)$ picks up a phase:

$$A(p, t) = A(p) \exp\left[-\frac{iEt}{\hbar}\right] \tag{23.53}$$

$$|A(p, t)|^2 = |A(p)|^2 \tag{23.54}$$

and the probability for measuring a value p does not change with time:

$$P(p, t) = P(p, 0). \tag{23.55}$$

The same goes for all observables: the probabilities do not change with time. For this reason the product states are called *stationary states*. The little clouds you see in textbooks describing the electronic states of the atom correspond to the time-independent distributions $P_n(\mathbf{r})$ in some definite-energy state of the atom labeled n.

If an electron in such an atomic state does not evolve with time, how does it jump from one state to another and absorb or emit a photon? The answer is that if the atom were truly isolated it would remain in the state ψ_n forever. If, however, we shine light on it, we are applying new forces on the electron. The vector and scalar potentials ϕ and $\mathbf{A}$ describing the $\mathbf{E}$ and $\mathbf{B}$ of the incident electromagnetic wave will enter the time-dependent Schrödinger equation for as long as the radiation is turned on. During this time the initial state with a definite n can evolve into a sum over many such states. At the end, we may be left with the atom in a different state and the electromagnetic field with one more or one less photon.

Actually even an atom with no externally applied electromagnetic field, in a vacuum, can jump to a *lower* level by emitting a photon. This is called *spontaneous emission.* You leave an isolated hydrogen atom in the first excited state, come back a short time later, and find the fellow has come down to the ground state. And you say, "Look, I didn't turn on any electric or magnetic field: $\mathbf{E} = 0, \mathbf{B} = 0$. What made the atom come down?" Where is the field? It turns out that the state $\mathbf{E} = 0, \mathbf{B} = 0$ is like a state $x = p = 0$ of the oscillator, sitting still at the bottom of the potential well. We know that's not allowed in quantum mechanics. You cannot have $x = 0, p = 0$. It turns out in the quantum theory of the electromagnetic field, $\mathbf{E}$ and $\mathbf{B}$ are like x and p. That means the state of definite $\mathbf{E}$ cannot be a state of definite $\mathbf{B}$. In particular $\mathbf{E} = 0, \mathbf{B} = 0$ is impossible. It looks that way in the macroscopic world, because the fluctuations in $\mathbf{E}$ and $\mathbf{B}$ are very small. Therefore, just as the oscillator in its lowest energy state has got some probability to be jiggling back and forth in x and p, the vacuum has its own *vacuum fluctuations* in which we may find $\mathbf{E} \neq 0$ and $\mathbf{B} \neq 0$. These fluctuations can tickle the atom and cause the "spontaneous" emission. There can be no *spontaneous absorption,* because the field is in its lowest energy state and has no energy to give the atom. I promised you the theory of everything, but that interlude was the theory of nothing, the vacuum.

23.5 General solution for time evolution

The product solutions are very special. In general things do change with time because the solutions are generally not of the product form $X(x)T(t)$. It is very easy and instructive to manufacture a non-product solution. If $\psi_1(x,t)$ and $\psi_2(x,t)$ are two solutions to the time-dependent Schrödinger equation, then so is a linear combination

$$\psi_{1+2}(x,t) = A(1)\psi_1(x,t) + A(2)\psi_2(x,t)$$
$$= A(1)\psi_{E_1}(x)e^{-iE_1t/\hbar} + A(2)\psi_{E_2}(x)e^{-iE_2t/\hbar} \quad (23.56)$$

because the time-dependent Schrödinger equation is linear. Since the two exponentials are different, we cannot pull out a common time-dependent factor and the solution above is not of the product form $X(x)T(t)$. One consequence is that measurable quantities like $P(x,t)$ will become time-dependent.

Let us take as an example the superposition of the two lowest energy states in the box:

$$\psi_{1+2}(x,t) = A(1)\sqrt{\frac{2}{L}}\sin\left[\frac{\pi x}{L}\right]e^{-iE_1 t/\hbar}$$

$$+ A(2)\sqrt{\frac{2}{L}}\sin\left[\frac{2\pi x}{L}\right]e^{-iE_2 t/\hbar}. \qquad (23.57)$$

In this state energy measurement can give only two answers, E_1 or E_2, with absolute probabilities:

$$P(n=1) = \frac{|A(1)e^{-iE_1 t/\hbar}|^2}{|A(1)e^{-iE_1 t/\hbar}|^2 + |A(2)e^{-iE_2 t/\hbar}|^2}$$

$$= \frac{|A(1)|^2}{|A(1)|^2 + |A(2)|^2} \qquad (23.58)$$

$$P(n=2) = \frac{|A(2)e^{-iE_2 t/\hbar}|^2}{|A(1)e^{-iE_1 t/\hbar}|^2 + |A(2)e^{-iE_2 t/\hbar}|^2}$$

$$= \frac{|A(2)|^2}{|A(1)|^2 + |A(2)|^2}. \qquad (23.59)$$

Suppose we had chosen $A(1) = 3$ and $A(2) = 4$. Then

$$P(1) = \frac{9}{25} \quad P(2) = \frac{16}{25}. \qquad (23.60)$$

It is wiser in this case with just two A's to get the absolute probabilities from the relative ones by dividing by $|A(1)|^2 + |A(2)|^2 = 25$ instead of normalizing the initial wave function $\psi_{1+2}(x)$ by computing its square integral. You might want to verify that, if you did this, the rescaling factor for $\psi_{1+2}(x)$ would be $\frac{1}{5}$.

Although the odds for different energies do not change with time, this is not so for other observables. The probability density for position is

$$P(x,t) = \left| A(1)\sqrt{\frac{2}{L}} \sin\left[\frac{\pi x}{L}\right] e^{-iE_1 t/\hbar} \right.$$

$$\left. + A(2)\sqrt{\frac{2}{L}} \sin\left[\frac{2\pi x}{L}\right] e^{-iE_2 t/\hbar} \right|^2 \qquad (23.61)$$

$$= |A(1)|^2 \frac{2}{L} \sin^2\left[\frac{\pi x}{L}\right] + |A(2)|^2 \frac{2}{L} \sin^2\left[\frac{2\pi x}{L}\right]$$

$$+ A^*(1)A(2)\frac{2}{L} \sin\left[\frac{\pi x}{L}\right] \sin\left[\frac{2\pi x}{L}\right] e^{-i(E_2 - E_1)t/\hbar}$$

$$+ A^*(2)A(1)\frac{2}{L} \sin\left[\frac{2\pi x}{L}\right] \sin\left[\frac{\pi x}{L}\right] e^{+i(E_2 - E_1)t/\hbar}$$

where $\qquad\qquad\qquad\qquad\qquad\qquad\qquad\qquad\qquad (23.62)$

$$E_2 - E_1 = \frac{\hbar^2 \pi^2}{2mL^2}(2^2 - 1^2). \qquad (23.63)$$

The probability density $P(x,t)$ is evidently time-dependent. For example, if $A(1) = A(2) = 1$,

$$P(x,t) = \frac{2}{L} \sin^2\left[\frac{\pi x}{L}\right] + \frac{2}{L} \sin^2\left[\frac{2\pi x}{L}\right]$$

$$+ \frac{4}{L} \sin\left[\frac{\pi x}{L}\right] \sin\left[\frac{2\pi x}{L}\right] \cos\left(\frac{(E_2 - E_1)t}{\hbar}\right). \quad (23.64)$$

Other densities like $P(p,t)$ also vary with time.

Whereas in a state of definite energy nothing changes with time, in this state, made of two different energies, $P(x,t)$ changes with time. To see appreciable change we must wait at least a time Δt comparable to the time period T of the oscillating cosine:

$$\Delta t \gtrsim T = \frac{2\pi}{\omega} = \frac{2\pi}{(E_2 - E_2)/\hbar} \simeq \frac{\hbar}{\Delta E} \qquad (23.65)$$

where factors like 2 and π have been dropped and $\Delta E = E_2 - E_1$ is the spread in the energy of the state.

This is a special case of the *energy-time uncertainty principle* to be discussed in the next chapter. It states that a system with a spread ΔE in its energy needs a minimum time $\Delta t \gtrsim \hbar/\Delta E$ to show appreciable change.

Let us generalize to a sum over all energy states:

$$\psi_{\text{general}}(x,t) = \sum_{n=1}^{\infty} A(n) \sqrt{\frac{2}{L}} \sin\left[\frac{n\pi x}{L}\right] e^{-iE_n t/\hbar}, \qquad (23.66)$$

which also solves the time-dependent Schrödinger equation by linearity. What sort of initial state did this evolve from? Setting $t = 0$ we find

$$\psi_{\text{general}}(x,0) = \sum_{n=1}^{\infty} A(n) \sqrt{\frac{2}{L}} \sin\left[\frac{n\pi x}{L}\right]. \qquad (23.67)$$

Thus we can predict the future of *any* initial state that may be written in this form. This is, however, no restriction, since the general mathematical theorem alluded to earlier assures us that *any* function $\psi(x,0)$ may be expanded in this form with coefficients

$$A(n) = \int_0^L \sqrt{\frac{2}{L}} \sin\left[\frac{n\pi x}{L}\right] \psi(x,0)\,dx. \qquad (23.68)$$

(There is no need to conjugate $\psi_n(x)$ because it is real.)

We have therefore the following recipe for finding the state $\psi(x,t)$ given the arbitrary initial state $\psi(x,0)$ in any time-independent potential $V(x)$:

1. Express the initial state as

$$\psi(x,0) = \sum_E A(E)\psi_E(x) \qquad (23.69)$$

with coefficients

$$A(E) = \int \psi_E^*(x)\psi(x,0)\,dx. \qquad (23.70)$$

2. The state at later times is obtained by appending a factor $e^{-iEt/\hbar}$ to every $A(E)$:

$$\psi(x,t) = \sum_E A(E) e^{-iEt/\hbar} \psi_E(x). \qquad (23.71)$$

While the mathematical theorems assure us that $\psi(x,0)$ may be expanded in terms of states in which *any* other observable $\mathcal{A}$ has a definite value α,

$$\psi(x,0) = \sum_\alpha A(\alpha) \psi_\alpha(x), \qquad (23.72)$$

with the same rule

$$A(\alpha) = \int \psi_\alpha^*(x) \psi(x,0) dx, \qquad (23.73)$$

the coefficients of the state at a later time, $A(\alpha, t)$, *will not be simply given by* $A(\alpha, 0)$ *times some phase factor.* One can show that instead each $A(\alpha, t)$ will generally be some complicated linear combination of all the $A(\beta, t)$'s. This is the reason that one expends so much time in computing the solutions $\psi_E(x)$ to the time-independent Schrödinger equation: it holds the key to the future.

23.5.1   Time evolution: a more complicated example

In the preceding example we were given the initial state as a combination of box wave functions $\psi_1(x)$ and $\psi_2(x)$ and had to compute its evolution. We just had to append the exponentials $e^{-iE_1 t/\hbar}$ and $e^{-iE_2 t/\hbar}$ due to time evolution to the coefficients $A(1)$ and $A(2)$. Now let us turn to a more complicated situation where we are given $\psi(x,0)$ as a function of x but not written out as a linear combination of $\psi_E(x)$. In this case we first have to find out the coefficients $A(E)$ of the linear combination and *then* attach the exponentials to them.

Consider as an example the following initial state in a box

$$\psi(x,0) = Bx \quad 0 < x < L. \qquad (23.74)$$

(This function does not vanish at $x = L$ and is therefore not a function that can be expanded in terms of box wave functions that do vanish at the ends.

We should therefore see it as the limit of a family of functions that plunge to zero more and more rapidly as $x \to L$. Equivalently, the sum over box functions can approximate it arbitrarily well except at $x = L$, but this will suffice for illustrative purposes.)

Before doing the time development let us probe the initial state a little more, starting with the computation of $P(n)$, the absolute probability of finding the system in energy state n. This is going to be a problem where we will have many (possibly infinite) non-zero $A(n) \equiv A(E_n)$ and normalizing them *after* computing them could be hard. On the other hand the initial state is simple enough to be normalized. So we will do that first, and the $A(n)$ we compute from it will come out normalized. We require

$$\int_0^L B^2 x^2 \, dx = \frac{B^2 L^3}{3} = 1, \tag{23.75}$$

which means the normalized wave function is

$$\psi(x,0) = \sqrt{\frac{3}{L^3}} x \qquad 0 < x < L. \tag{23.76}$$

The coefficients are (for $E = E_n$)

$$A(n) = \int_0^L \sqrt{\frac{2}{L}} \sin\left[\frac{n\pi x}{L}\right] \sqrt{\frac{3}{L^3}} x \, dx \tag{23.77}$$

$$= \frac{\sqrt{6}}{L^2} \left[-\frac{L}{n\pi} x \cos\left[\frac{n\pi x}{L}\right] \right]_0^L$$

$$+ \frac{L}{n\pi} \int_0^L \cos\left[\frac{n\pi x}{L}\right] \right] dx \tag{23.78}$$

$$= \frac{\sqrt{6}}{n\pi}(-\cos n\pi) = (-1)^{n+1} \frac{\sqrt{6}}{n\pi}. \tag{23.79}$$

For example, the probability of finding the system in the ground state of the box is

$$P(1) = |A(1)|^2 = \frac{6}{\pi^2}. \tag{23.80}$$

Here is a brief aside. Since the initial state is normalized we have an interesting mathematical result

$$1 = \sum_{n=1}^{\infty} P(n) = \sum_{n=1}^{\infty} \frac{6}{n^2 \pi^2}, \tag{23.81}$$

which may be written as a celebrated result due to Euler:

$$\sum_{n=1}^{\infty} \frac{1}{n^2} = \frac{\pi^2}{6}. \tag{23.82}$$

Back to the time evolution of the state. Our general formula, applied to this case, gives

$$\psi(x,t) = \sum_{n=1}^{\infty} A(n) e^{\frac{-iE_n t}{\hbar}} \psi_n(x) \tag{23.83}$$

$$= \sqrt{\frac{2}{L}} \sum_{n=1}^{\infty} (-1)^{n+1} \frac{\sqrt{6}}{n\pi} \sin\left[\frac{n\pi x}{L}\right]$$

$$\times \exp\left[-\frac{i\hbar n^2 \pi^2 t}{2mL^2}\right]. \tag{23.84}$$

Figure 23.7 shows the evolution of $P(x,t)$ for parameters $L = \hbar = \frac{\pi^2}{2m} = 1$ and the sum over n truncated at $n = 50$. The times selected are $t = 0, \frac{\pi}{2}, \pi$. I will explain this choice of times a little later.

One can show analytically that

$$\psi(x,t) = -\psi(1-x, t+\pi) \tag{23.85}$$

using properties of the sine and the result

$$e^{-in^2\pi} = (-1)^n. \tag{23.86}$$

(If n is even, so is n^2 and $\exp[-i\pi \times \text{even}] = +1 = (-1)^n$, while if n is odd so is n^2 and $\exp[-i\pi \times \text{odd}] = -1 = (-1)^n$.) Thus after a time π, the wave function $\psi(x,t)$ changes sign (we do not see it in $P(x,t)$) and gets reflected around $x = \frac{1}{2}$, and after another π it comes right back to $\psi(x,0)$.

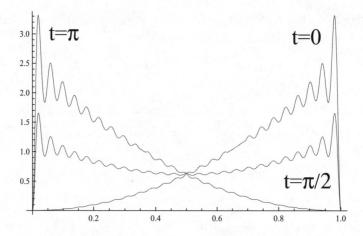

Figure 23.7 Time-dependence of $P(x,t)$ for $\psi(x,0) = \sqrt{\frac{3}{L^3}}\,x$ and $L = \hbar = \frac{\pi^2}{2m} = 1$ at times $0, \frac{\pi}{2}, \pi$. Note the oscillations at $t = 0$ near $x = L = 1$ where the initial wave function plunges to 0 from 1. You can see the symmetry $\psi(x,t) = -\psi(1-x,t+\pi)$. The minus sign is lost in going from $\psi(x,t)$ to $P(x,t)$.

By comparison, the 2π periodicity is obvious because the time-dependent factor $e^{-2\pi i n^2}$ in Eqn. 23.84 equals 1 for any integer n.

Such simple periodic behavior is uncommon and usually occurs only in problems that can be solved analytically. The density $P(x,t)$ generally does not repeat itself. So the only take-away message for you should be that as time goes by, the initial $\psi(x,0)$ and $P(x,0)$ evolve into $\psi(x,t)$ and $P(x,t)$, which we can calculate.

Now to explain why I chose the times $0, \frac{\pi}{2}$, and π and not, say, 0, 1, and 2 seconds in Figure 23.7: the times are chosen in *natural units* arising from the problem itself. Suppose you wanted to show various states of a pendulum of length L and mass m. You should display the pendulum at times comparable to its period so that you can show interesting stages of one or two oscillations. For example, if it had a time period of 10s, it makes sense to show its position every second or two and not every nanosecond or every year. Dimensional analysis gives a way to associate a time constructed out of L, m, and g. We write

$$T = L^a g^b m^c \qquad\qquad (23.87)$$

where the equation aims to balance only the units by suitable choice of the constants a, b, and c. Continuing,

$$T = L^a \left[\frac{L}{T^2} \right]^b m^c \tag{23.88}$$

$$a + b = 0 \tag{23.89}$$

$$-2b = 1 \tag{23.90}$$

$$c = 0, \tag{23.91}$$

which gives us the following natural unit of time

$$T = \sqrt{\frac{L}{g}}. \tag{23.92}$$

Notice that T is not the actual time period of the pendulum (a 2π is missing). In general the motion may not even be periodic. A natural time unit simply avoids the introduction of very large or very small times in discussing the problem in question. For example, in studying planetary motion a year is a natural unit, not a nanosecond.

To find a natural time scale for a quantum problem, we may extract a frequency $\omega = E_1/\hbar$ from its lowest energy, and a corresponding time using $\omega = \frac{2\pi}{T}$. (Using E_2 instead will only change the unit by a numerical factor of order unity.) In our problem, for our choice of parameters, this leads to $\frac{2\pi}{T} = \omega = E_1/\hbar = \frac{\hbar \pi^2}{2mL^2} = 1$, which leads to the natural time unit $T = 2\pi$. I emphasize that in general such a natural time unit does not imply periodic behavior with period T, or even periodic behavior, though both happened to be true in our example.

CHAPTER 24

Summary and Outlook

Now it is time to consolidate everything, to present the subject starting from its postulates. Whereas one could simply say "$F = m\frac{d^2x(t)}{dt^2}$" and launch you into a study of mechanics, there is a lot of groundwork that has to be done in the quantum case. Now that the ground work is behind us, it will be useful to have in one place all the rules of the quantum cookbook. These rules or postulates summarize findings from experiments and cannot be deduced by pure cerebration.

There are many ways to write down the postulates and there can even be arguments about how many there are. What I present below are the postulates appropriate to this course, and they are restricted to a single massive particle in one dimension. Given these, and some mathematical results, you can do any of the problem sets. Following this, I will dig a little deeper into the postulates to unify some of them into a single one. That digression is optional but recommended if you are thinking beyond this course. The chapter will conclude with the study of more than one particle and the *energy-time* uncertainty principle.

24.1 Postulates: first pass

1. **Postulate I.** The complete information on the state of a particle at any fixed time is given by the complex, continuous, normalizable wave function $\psi(x)$. States of definite position and momentum are non-normalizable exceptions that need special treatment.

2. **Postulate II.** The probability density for finding the particle at x is given by

$$P(x) = |\psi(x)|^2.\tag{24.1}$$

If a particle is found at some x_0, the wave function collapses to a spike at x_0. If $P(x)$ is normalized, then

$$\int P(x)\,dx = \int |\psi(x)|^2\,dx = 1.\tag{24.2}$$

Normalization is a convenience and not a requirement because rescaling ψ has no physical effect.

3. **Postulate III (momentum states).** A state guaranteed to yield a momentum p upon measurement is described by

$$\psi_p(x) = A\exp\left[\frac{ipx}{\hbar}\right].\tag{24.3}$$

This function cannot be normalized by any choice of A on the infinite line. If we fold the finite line into a ring of circumference L, we can choose $A = 1/\sqrt{L}$. The requirement of single-valuedness, $\psi(x) = \psi(x+L)$, leads to the quantization of momentum to the values

$$p_n = \frac{2\pi\hbar n}{L} \quad n = 0, \pm 1, \pm 2\ldots\tag{24.4}$$

The quantization of p follows by mathematical reasoning given the requirements on ψ. It is not a postulate.

4. **Postulate IV (energy states).** A state guaranteed to yield the result E upon energy measurement is the solution to the time-independent Schrödinger equation

$$-\frac{\hbar^2}{2m}\frac{d^2\psi_E(x)}{dx^2} + V(x)\psi_E(x) = E\psi_E(x).\tag{24.5}$$

Solving this equation with appropriate boundary conditions will determine the allowed values E and corresponding functions $\psi_E(x)$. Notice that V is assumed to depend only on x and not on t. States of definite energy exist only when V is time-independent. (This is also the condition for a conserved energy E to exist classically.)

Mathematical interlude: Let $\mathcal{A}$ be a dynamical variable, such as momentum or energy, and $\psi_\alpha(x)$ a wave function that describes a state guaranteed to yield an answer α (like p_n or E_n) if $\mathcal{A}$ is measured. Mathematical considerations (not discussed here) assure us that any $\psi(x)$ may be written as a superposition

$$\psi(x) = \sum_\alpha A(\alpha)\psi_\alpha(x) \qquad (24.6)$$

where the coefficients of the expansion are given by

$$A(\alpha) = \int \psi_\alpha^*(x)\psi(x)dx. \qquad (24.7)$$

5. **Postulate V (measurement).** If $\mathcal{A}$ is measured in the state (described by) $\psi(x)$, the only possible outcomes α are the ones that appear in the superposition Eqn. 24.6 and occur with probability

$$P(\alpha) = |A(\alpha)|^2. \qquad (24.8)$$

Right after a measurement yielding the result α_0, the state will collapse (from being a sum over α) to $\psi_{\alpha_0}(x)$. An immediate remeasurement of $\mathcal{A}$ will yield the same value α_0.

Complication due to degeneracy: Sometimes there will be two or more independent wave functions that can correspond to the same value of a variable $\mathcal{A}$. An example is the free particle on a ring: at energy E there are two states of definite momentum $p = \pm\sqrt{2mE}$ described by independent functions $e^{\pm ipx/\hbar}$. Any linear combination of them is a state of definite energy. To find $P(E)$ in this case, it is best to express the given $\psi(x)$ in terms of $\psi_p(x)$, compute the probabilities for $p = \pm\sqrt{2mE}$, and add them to obtain $P(E) = P(p = \sqrt{2mE}) + P(p = -\sqrt{2mE})$.

Complication due to more than one variable: If we are interested in two variables there may not be a state in which both are guaranteed to have definite values. In the case of position and momentum there are no such states. On the other hand, for a free particle on a ring it is possible to have a state with guaranteed E and p.

6. **Postulate VI (time evolution).** The time evolution of the wave function is governed by the time-dependent Schrödinger equation:

$$i\hbar\frac{\partial \psi(x,t)}{\partial t} = -\frac{\hbar^2}{2m}\frac{\partial^2 \psi(x,t)}{\partial x^2} + V(x,t)\psi(x,t). \qquad (24.9)$$

In this equation $V = V(x,t)$ may depend on time.

It may be verified by substitution that if $V = V(x)$, the following is a solution:

$$\psi_E(x,t) = \psi_E(x)e^{-iEt/\hbar} \qquad (24.10)$$

where E and $\psi_E(x)$ are the solutions to the time-independent Schrödinger equation 24.5. It is called a *stationary state* because none of the probabilities $(P(x), P(p), P(\alpha))$ vary with time.

A superposition of such stationary states with arbitrary coefficients $A(E)$,

$$\psi(x,t) = \sum_E A(E)\psi_E(x)e^{-iEt/\hbar}, \qquad (24.11)$$

is also a solution to the time-dependent Schrödinger equation by its linearity. If we ask what kind of initial state corresponds to such a solution, we find, upon setting $t = 0$,

$$\psi(x,0) = \sum_E A(E)\psi_E(x). \qquad (24.12)$$

This is no restriction at all on the initial state, since the mathematics assures us that *any* function $\psi(x,0)$ may be written as a superposition of $\psi_E(x)$. Thus Eqn. 24.11 describes the solution to the most general problem of time evolution one could pose in a time-independent V.

If Eqn. 24.12 is valid at $t = 0$, it is valid at time t, provided of course we choose the coefficients A as a function of time:

$$\psi(x, t) = \sum_E A(E, t)\psi_E(x). \tag{24.13}$$

Comparing this to Eqn. 24.11 we find

$$A(E, t) = A(E)e^{-iEt/\hbar}. \tag{24.14}$$

Thus the coefficients of the expansion have a very simple time evolution *if the general state is written in terms of* $\psi_E(x)$. In other words, although mathematically $\psi(x)$ may be written as a superposition of states $\psi_\alpha(x)$ that have well-defined values for *any* variable $\mathcal{A}$, only the expansion coefficients in terms of states of definite energy have this simple time-dependence.

Do not look for the uncertainty principle among the postulates: it can be deduced given that the particle is described by a wave function and that definite momentum corresponds to definite wavelength.

24.2   Refining the postulates

The postulates as written above would not be found in any book. I gave them to you as a set of rules that would allow you to handle the material in this course. Quantum mechanics is one big recipe but even as recipes go the above list is wanting. There are at least two deficiencies you may have noticed.

1. For every variable, there seems to be a different prescription for finding wave functions with definite values for that variable. For example, I simply gave you $\psi_p(x) = Ae^{ipx/\hbar}$ as the state of definite momentum, while I asked you to solve the time-independent Schrödinger equation to obtain $\psi_E(x)$. Since one can imagine an infinite number of such variables, corresponding to arbitrary functions of x and p, there must be an infinite number of such prescriptions. Are there really an infinite number of such postulates, one for each variable?
2. I treated x differently from any other variable. First, I was evasive about (the wave function for) a state of definite position $x = x_0$, simply

referring to it as a spike at x_0. Next, the rule

$$\psi(x) = \sum_\alpha A(\alpha)\psi_\alpha(x) \qquad (24.15)$$

where

$$A(\alpha) = \int \psi_\alpha^*(x)\psi(x)\,dx \qquad (24.16)$$

was never applied to the case where A was the position: I never wrote $\psi(x)$ as a linear combination of states of definite position with some coefficients $A(x)$ and did not relate the mod-squared of the expansion coefficients to the probability of finding the particle at some x. Instead I gave $P(x) = |\psi(x)|^2$ as a postulate.

24.2.1 Toward a compact set of postulates

I will now remedy these interrelated defects to the extent that is possible within the constraints of this course.

Consider first the momentum states. Following some plausibility arguments based on the double-slit experiment, these were postulated to be

$$\psi_p(x) = A\exp\left[\frac{ipx}{\hbar}\right]. \qquad (24.17)$$

Without changing the substance of this postulate let me rewrite it as follows:

Postulate III. A state of definite momentum p is a solution to the differential equation

$$-i\hbar\frac{d\psi_p(x)}{dx} = p\psi_p(x). \qquad (24.18)$$

You can solve this equation in your head and see that the solutions are indeed the ones in Eqn. 24.17. The arbitrary scale factor A appears because $\psi_p(x)$ appears on both sides. Given one solution, you can get another by rescaling. A common way to choose A is to impose the cosmetic requirement of normalization.

24.2.2 Eigenvalue problem

The familiar differential equation (24.18) and its solution are a simple introduction to the fertile realm of the *eigenvalue problem*. Let us take some time to explore it. Take some arbitrary function $f(x)$ and differentiate it. It will turn into a new function. For example,

$$\frac{d\sin x}{dx} = \cos x \tag{24.19}$$

$$\frac{dx^3}{dx} = 3x^2. \tag{24.20}$$

Let us rewrite these as

$$D[\sin x] = \cos x \tag{24.21}$$

$$D[x^3] = 3x^2, \tag{24.22}$$

which you should take to be a definition of D. One calls D an *operator*. *Just like a function is a recipe that takes in a variable x and spits out a value $f(x)$, an operator takes in a function $f(x)$ and spits out another function.* The function $f(x)$ is always placed to the right of the operator, as in $D[f(x)]$ or simply Df. The thing D does to $f(x)$ is to differentiate it.

The operator D is linear, meaning

$$D[af(x) + \beta g(x)] = aD[f(x)] + \beta D[g(x)], \tag{24.23}$$

which is a familiar property of differentiation. All operators we will consider here will be linear.

It is natural to define the operator D^2 as a D followed by another D and as having the following effect:

$$D^2[f(x)] = \frac{d^2 f}{dx^2} \quad \text{so that, for example,} \tag{24.24}$$

$$D^2[\sin x] = -\sin x. \tag{24.25}$$

Don't let the exponent in D^2 fool you into thinking it is a non-linear operator. After all,

$$D^2\left[\alpha f(x) + \beta g(x)\right] = \alpha \frac{d^2 f}{dx^2} + \beta \frac{d^2 g}{dx^2}$$
$$= \alpha D^2\left[f(x)\right] + \beta D^2\left[g(x)\right]. \qquad (24.26)$$

You can form operators that are sums of various powers of D each multiplied by some constant.

In general operators modify the function they operate on and turn them into other functions. But sometimes an operator may have some privileged functions, called its *eigenfunctions*, on which its effect *is to simply multiply them by a constant, called the* eigenvalue. Let us consider the *eigenvalue equation* for D. Its eigenfunctions must obey

$$D\left[f(x)\right] = \frac{df(x)}{dx} = \kappa f(x) \qquad (24.27)$$

where the constant κ is the eigenvalue. The solution or eigenfunction is clearly

$$f(x) = Ae^{\kappa x}. \qquad (24.28)$$

In other words, although the effect of differentiation by D is usually to transform a function into something else, there are some functions, the exponentials, on which the effect of D is to multiply them by a constant. It is common to label the eigenfunctions by the corresponding eigenvalues as follows:

$$f_\kappa(x) = Ae^{\kappa x}. \qquad (24.29)$$

At this point, there is no restriction on the eigenvalue κ.

In this language we may say that the states of definite momentum $\psi_p(x)$ are eigenfunctions of the operator

$$P = -i\hbar D, \qquad (24.30)$$

called the *momentum operator* in quantum theory, and therefore the solutions to

$$P[\psi_p(x)] \equiv -i\hbar \frac{d\psi_p(x)}{dx} = p\psi_p(x), \tag{24.31}$$

where p is the eigenvalue. In summary,

Postulate III (momentum states). The states of definite momentum p are eigenfunctions of P:

$$P[\psi_p(x)] = p\psi_p(x). \tag{24.32}$$

If the solution $\psi_p(x)$ lives on a ring of circumference L, the single-valued requirement restricts the eigenvalues p to $p_n = \frac{2\pi n\hbar}{L}$.

24.2.3 The Dirac delta function and the operator X

Just one more such operator and we are done. It is called X and this is what it does to any $f(x)$ placed to its right:

$$X[f(x)] = xf(x) \quad \text{so that, for example,} \tag{24.33}$$

$$X[\sin x] = x\sin x. \tag{24.34}$$

Thus the action of X is to take the given function $f(x)$ and change it into the new function $xf(x)$. Evidently X^2 is an X followed by another X and thus

$$X^2[f(x)] = x^2 f(x) \quad \text{so that, for example,} \tag{24.35}$$

$$X^2[\sin x] = x^2 \sin x. \tag{24.36}$$

We can form more complicated operators using X and P, whose action is quite obvious. For example,

$$(3D^2 + 9X^2)[f(x)] = 3\frac{d^2 f(x)}{dx^2} + 9x^2 f(x). \tag{24.37}$$

Consider the eigenfunctions of the operator X, remembering that

$$X[f(x)] = xf(x). \tag{24.38}$$

If it had an eigenfunction $f_{x_0}(x)$ with eigenvalue x_0, it would have to satisfy

$$xf_{x_0}(x) = x_0 f_{x_0}(x). \tag{24.39}$$

But look! Multiplying by x has a different effect at different x, and yet we are looking for a function that when multiplied by x becomes a constant x_0 times that function! How can any function retain its functional form (up to a multiplicative constant) when multiplied by x? And yet there is such a function. It is a little weird as you might expect. It is called a *Dirac delta function* or simply the δ-function. It is a limit of any number of smooth functions, and here is one example. Look at Figure 24.1. It shows three rectangles of decreasing width w and increasing height $1/w$ centered at $x = x_0$. All have unit area. If you take the limit $w \to 0$ you get the Dirac delta function $\delta(x - x_0)$, shown by a vertical arrow going to infinity. *It is infinitely tall at x_0, zero everywhere else, and still has unit area:*

$$\delta(x - x_0) = 0 \quad x \neq x_0 \tag{24.40}$$

$$= \infty \quad x = x_0 \tag{24.41}$$

$$\int_a^b \delta(x - x_0) dx = 1 \quad \text{if } a < x_0 < b; \text{ otherwise 0.} \tag{24.42}$$

The delta function is even, like the one in Figure 24.1 whose limit it is:

$$\delta(x - x_0) = \delta(x_0 - x). \tag{24.43}$$

Let us see how the Dirac delta function $\delta(x - x_0)$ satisfies the eigenvalue equation

$$x\delta(x - x_0) = x_0 \delta(x - x_0). \tag{24.44}$$

Consider first a point $x \neq x_0$. Now both sides vanish due to the $\delta(x - x_0)$. So it does not matter that the δ-function on one side has an x multiplying it and on the other side an x_0. At $x = x_0$, the x on the left-hand side becomes x_0 and the two sides again agree.

Here is another way to say it. The factor x rescales any $f(x)$ by a *variable* amount x, but our eigenfunction lives only at one point $x = x_0$ where it gets rescaled by x_0. So it is correct to say that it gets rescaled everywhere by just one number, x_0.

You can only plot the δ-function before taking the limit $w \to 0$ of the rectangular spike (see the left half of Figure 24.1) or any function with the δ-function as the limit. The limiting function itself has the only two values, zero or infinity. Luckily we never need the function by itself, just some integral within which it appears. Eqn. 24.42 will then tell us exactly how to handle it.

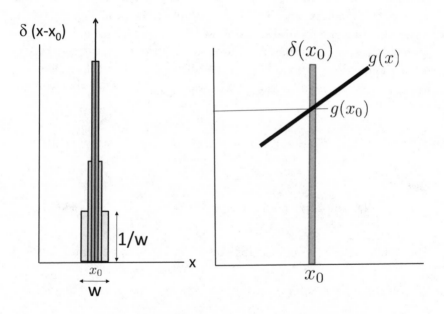

Figure 24.1 Left: Three rectangles centered at $x = x_0$, of width w and height $1/w$ as $w \to 0$. The height and width of the broadest alone are shown. The limit, shown by the arrow, is the delta function $\delta(x - x_0)$. It is even: $\delta(x - x_0) = \delta(x_0 - x)$. Right: The integral of $g(x)$ times $\delta(x - x_0)$ receives a non-zero contribution only infinitesimally close to x_0. Within this interval $g(x) \simeq g(x_0)$ is a constant that can be pulled out of the integral, and the delta function then integrates to 1.

Let $g(x)$ be some smooth function. Consider the integral

$$\int_a^b g(x)\delta(x-x_0)dx \qquad a < x_0 < b \qquad (24.45)$$

just before the limit $w \to 0$ is taken. The integrand vanishes for any x not infinitesimally close to x_0 due to $\delta(x-x_0)$; see Figure 24.1. So the entire integral comes from an infinitesimal region around x_0. We manipulate it as follows:

$$\int_a^b g(x)\delta(x-x_0)dx = \lim_{\varepsilon \to 0} \int_{x_0-\varepsilon}^{x_0+\varepsilon} g(x)\delta(x-x_0)dx \qquad (24.46)$$

$$= g(x_0) \lim_{\varepsilon \to 0} \int_{x_0-\varepsilon}^{x_0+\varepsilon} \delta(x-x_0)dx \qquad (24.47)$$

$$= g(x_0) \cdot 1 \qquad (24.48)$$

where I could pull out $g(x_0)$ from the integral because the smooth function $g(x)$ is essentially constant within the infinitesimal neighborhood of $x = x_0$. Thus $\delta(x-x_0)$ can be used to pull out or *sample* the value of $g(x)$ at x_0:

$$\int_a^b g(x)\delta(x-x_0)dx = g(x_0), \text{ if } a < x_0 < b \qquad (24.49)$$

$$= 0 \text{ if not.} \qquad (24.50)$$

For example,

$$\int_1^{10} x^3\delta(x-5)dx = 5^3 \qquad (24.51)$$

and

$$\int_1^{10} x^3\delta(x-15)dx = 0 \qquad (24.52)$$

and

$$\int_0^\pi \sin x \cdot \delta\left(x - \frac{\pi}{2}\right)dx = \sin\frac{\pi}{2} = 1. \qquad (24.53)$$

Now back to quantum mechanics. Let $g(x)$ be a wave function $\psi(x)$. Its value at x_0 is sampled by $\delta(x - x_0)$:

$$\psi(x_0) = \int_{-\infty}^{\infty} \delta(x - x_0)\psi(x)\,dx. \tag{24.54}$$

Compare this to a mathematical result for extracting the coefficients of expansion of a generic $\psi(x)$ in terms of states of definite value for a general variable $\mathcal{A}$:

$$A(\alpha) = \int \psi_\alpha^*(x)\psi(x)\,dx. \tag{24.55}$$

The correspondence is obvious:

$$\mathcal{A} \to \quad \text{position} \tag{24.56}$$
$$\alpha \to x_0 \tag{24.57}$$
$$\psi_\alpha(x) \to \delta(x - x_0) \tag{24.58}$$
$$A(\alpha) \to \psi(x_0). \tag{24.59}$$

Thus we see that the wave function $\psi(x_0)$ is itself the coefficient in the expansion of $\psi(x)$, the "amount" of $\delta(x - x_0)$ we need in the expansion of $\psi(x)$. By the measurement postulate then

$$P(x_0) = |\psi(x_0)|^2, \tag{24.60}$$

and we see that the rule for x is like that for any other variable such as p or E. The only difference is that since x_0 is a continuous variable, $P(x_0)$ is a probability density whose integral over all x_0 is 1, and not a probability whose sum over all possibilities is 1.

If you are following this closely you will ask, "Where is the analog of

$$\psi(x) = \sum_\alpha A(\alpha)\psi_\alpha(x), \tag{24.61}$$

which says any $\psi(x)$ may be expanded in terms of functions of definite $\mathcal{A}$, which is now position?" Go to

$$\psi(x_0) = \int_{-\infty}^{\infty} \psi(x)\delta(x - x_0)dx \qquad (24.62)$$

and make the exchange $x \leftrightarrow x_0$ to obtain

$$\psi(x) = \int_{-\infty}^{\infty} \psi(x_0)\delta(x_0 - x)dx_0. \qquad (24.63)$$

Comparing this to Eqn. 24.61 we find the correspondence

$$\mathcal{A} \rightarrow \text{ position} \qquad (24.64)$$

$$\alpha \rightarrow x_0 \qquad (24.65)$$

$$\sum_{\alpha} \rightarrow \int dx_0 \qquad (24.66)$$

$$\psi_\alpha(x) \rightarrow \delta(x - x_0) \qquad (24.67)$$

$$A(\alpha) \rightarrow \psi(x_0). \qquad (24.68)$$

Thus I have exhibited a generic $\psi(x)$ as an integral (rather than sum) over states (spikes) of definite position x_0 with coefficients $\psi(x_0)$.

I mentioned earlier that states of definite position and momentum are non-normalizable. In the case of momentum $|\psi_p(x)|^2$ is a constant whose integral over all of space is infinite. For the case of position we find the following square-integral

$$\int_{-\infty}^{\infty} \delta(x - x_0)\delta(x - x_0)dx = \delta(x_0 - x_0) = \delta(0) = \infty. \qquad (24.69)$$

(I have used one of the delta functions to sample the other at $x = x_0$.)

Having seen that the position and momentum can both be analyzed on the same footing, as eigenvalue problems, let us take a second look at states of definite energy, which are the solutions to

$$-\frac{\hbar^2}{2m}\frac{d^2\psi_E(x)}{dx^2} + V(x)\psi_E(x) = E\psi_E(x), \text{ that is,} \qquad (24.70)$$

$$\frac{1}{2m}\left[-i\hbar\frac{d}{dx}\left(\left(-i\hbar\frac{d}{dx}\psi_E(x)\right)\right)\right.$$

$$\left. + V(x)\psi_E(x) = E\psi_E(x), \text{ that is,} \right. \tag{24.71}$$

$$\left(\frac{p^2}{2m} + V(X)\right)[\psi_E(x)] = E\psi_E(x) \tag{24.72}$$

upon using $P\psi_E(x) = -i\hbar\frac{d\psi_E}{dx}$ and the fact that the action of $V(X)$ on $f(x)$ is to replace it by $V(x)f(x)$.

Comparing Eqn. 24.72 to the formula from classical mechanics

$$E = \frac{p^2}{2m} + V(x), \tag{24.73}$$

we see that the states of definite energy are the solutions to the eigenvalue equation

$$E\left(x \to x, p \to -i\hbar\frac{d}{dx}\right)\psi_E(x) = E\psi_E(x), \tag{24.74}$$

or more abstractly,

$$E(x \to X, p \to P)\psi_E(x) = E\psi_E(x). \tag{24.75}$$

In the left-hand side we take the classical expression for energy E as a function of x and p and replace every x by $X = x$, and every p by $P = -i\hbar\frac{d}{dx}$, and we let the result act on the $\psi_E(x)$ sitting to its right.

The combination

$$H = \frac{p^2}{2m} + V(X) \tag{24.76}$$

is called the *Hamiltonian operator* or simply Hamiltonian. It depends on the potential $V(x)$. For example, in the case of the harmonic oscillator it is

$$H = \frac{p^2}{2m} + \frac{1}{2}kX^2. \tag{24.77}$$

This means that the states of definite energy for the quantum oscillator are the normalizable solutions to

$$-\frac{\hbar^2}{2m}\frac{d^2\psi_E(x)}{dx^2} + \frac{1}{2}kx^2\psi_E(x) = E\psi_E(x). \tag{24.78}$$

The time-independent Schrödinger equation may be written in terms of H as

$$H\psi_E(x) = E\psi_E(x). \tag{24.79}$$

24.3   Postulates: Final

We are now ready to combine the postulates into the following set, which is more compact and free of the defects in the initial set.

1. **Postulate I.** The complete information on the state of a particle is given by a complex, continuous wave function $\psi(x)$ which is normalizable except for states of definite x or p.
2. **Postulate II.** Let $A(x,p)$ be a dynamical variable, such as momentum or position or a function thereof, like energy. Then its allowed values α, and the corresponding $\psi_\alpha(x)$, are the normalizable (except in the case of position and momentum), single-valued solution to

$$A\left(x \to x, p \to -i\hbar\frac{\partial}{\partial x}\right)\psi_a(x) = \alpha\,\psi_\alpha(x). \tag{24.80}$$

(I use partial derivatives of x in anticipation of additional coordinates y and z.)

Mathematical interlude: A mathematical result assures us that any $\psi(x)$ may be written as

$$\psi(x) = \sum_\alpha A(\alpha)\psi_\alpha(x) \tag{24.81}$$

where

$$A(\alpha) = \int \psi_\alpha^*(x)\psi(x)dx. \tag{24.82}$$

3. **Postulate III.** If $\mathcal{A}$ is measured in the state $\psi(x)$, the only possible outcomes are the α's that appear in the superposition above and the probability for each α is

$$P(\alpha) = |A(\alpha)|^2. \tag{24.83}$$

The state right after measurement will collapse from the sum over α to the single term corresponding to the value of α obtained. An immediate remeasurement of $\mathcal{A}$ will yield the same value.

4. **Postulate IV.** The time evolution of the wave function is governed by the time-dependent Schrödinger equation:

$$i\hbar \frac{\partial \psi(x,t)}{\partial t} = -\frac{\hbar^2}{2m} \frac{\partial^2 \psi(x,t)}{\partial x^2} + V(x,t)\psi(x,t)$$

$$\equiv H\psi(x,t). \tag{24.84}$$

In this equation the classical potential $V = V(x,t)$ may depend on time.

24.4 Many particles, bosons, and fermions

What does quantum mechanics of more than one particle look like? There are some obvious consequences like more coordinates and some real surprises of quantum origin.

First let the two particles be different, say a proton and an electron. Now each of them has its position, say x_1 and x_2, and these appear in the two-particle wave function $\psi(x_1, x_2)$. The probability density for finding particle 1 at x_1 and particle 2 at x_2 is

$$P(x_1, x_2) = |\psi(x_1, x_2)|^2. \tag{24.85}$$

With three particles you will have a $\psi(x_1, x_2, x_3)$ and so on, but I will stop with two because you can learn some profound things in about fifteen minutes just by exploring this case.

Imagine both particles are in a box and the electron is in state $n = 3$ and the proton is in state $n = 5$. The corresponding $\psi(x_1, x_2)$ is of the *product form*

$$\psi_{3,5}(x_1, x_2) = \psi_3(x_1)\psi_5(x_2)$$

$$= \sqrt{\frac{2}{L}}\sin\left[\frac{3\pi x_1}{L}\right]\sqrt{\frac{2}{L}}\sin\left[\frac{5\pi x_2}{L}\right]. \qquad (24.86)$$

The probability density for finding the electron at $x = 4$ and the proton at $x = 8$ is given by

$$P_{3,5}(x_1 = 4, x_2 = 8) = |\psi_3(4)|^2|\psi_5(8)|^2 \qquad (24.87)$$

$$= \frac{4}{L^2}\sin^2\left[\frac{3\pi \cdot 4}{L}\right]\sin^2\left[\frac{5\pi \cdot 8}{L}\right]. \qquad (24.88)$$

If instead we ask for the probability density for finding the electron at $x = 8$ and the proton at $x = 4$, the probability density would be

$$P_{3,5}(x_1 = 8, x_2 = 4) = \frac{4}{L^2}\sin^2\left[\frac{3\pi \cdot 8}{L}\right]\sin^2\left[\frac{5\pi \cdot 4}{L}\right], \qquad (24.89)$$

which is quite different. For example, if the box had a size $L = 40$, then $P_{3,5}(x_1 = 4, x_2 = 8)$ would vanish while $P_{3,5}(x_1 = 8, x_2 = 4)$ would not.

That's perfectly okay, because they are two different probabilities for two different outcomes: finding the electron here and the proton there is not the same as finding the electron there and the proton here. To verify the probabilities, I take many many boxes with electrons in the state $n = 3$ and protons in the state $n = 5$ and measure their positions and tally my findings in the form of a histogram in two dimensions labeled by x_1 and x_2. Following each measurement I can unambiguously assign the measured positions to the electron or the proton. In the end the histogram should agree with $P_{3,5}(x_1, x_2)$. If I find the electron at $x = 8$ and the proton at $x = 4$, that would be consistent with (but not fully confirm) the predictions, but if I found the electron at $x = 4$ and the proton at $x = 8$ even once that would deal a fatal blow to the theory because the proton should not be found at $x = 8$, which is a zero of its wave function.

24.4.1 *Identical versus indistinguishable*

Something very dramatic happens if the two particles are identical. The words "identical particles" have a connotation in quantum mechanics

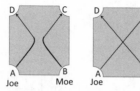

Figure 24.2   Two identical twins Joe and Moe enter doors *A* and *B* and exit via doors *C* and *D*, along two possible classical paths. In quantum theory we cannot say which of the two things depicted happens. Had these been identical macroscopic twins Moe and Joe we could, based on continuous observation.

that is very different from that in classical mechanics. Consider identical twins. I mean absolutely identical. They are separated at birth and they are moving around. Even though they look identical in every way, we can still follow them. We know this is Joe and that is Moe. We can keep track of them continuously. Consider the following experiment involving these twins, depicted in Figure 24.2. There are four doors in a room and Joe enters from door *A* and Moe enters from door *B*, and both are headed for the center of the room. There are now two options. Either they exit via the doors in front of their entry door (left half of figure) or cross over and Joe exits via door *C* and Moe via door *D*. Now suppose you saw them entering the room in the beginning, and you left the room briefly and came back in time to see them leaving the room. You cannot tell whether they crossed or not, because you just see two identical twins at these doors. But *somebody* knows what has happened, somebody who was watching them at all times. So even though they are *identical*, they are *distinguishable*. They cannot swap roles without *someone* knowing.

But imagine now that these are not classical twins but quantum particles, like electrons, which do not have a definite trajectory between observations. You know an electron was emitted at door *A* and another at door *B* and they were eventually detected at doors *C* and *D*. You cannot tell who really went where. Was it this guy or was it that guy? There's no way to tell. So when you have identical particles whose trajectories you cannot follow, when you catch a particle here and a particle there, you cannot say Joe was here and Moe was there. It's not allowed, because you're not following them continuously. You can only say, "I found a particle here, and I found a particle there" and not " I found Joe here and Moe there." Therefore the theory cannot assign different probabilities for finding particle 1 here and particle 2 there, and particle 2 there and particle 1 here, because the two outcomes are indistinguishable. *It must give the same odds for two indistinguishable outcomes:*

$$P(x_1, x_2) = P(x_2, x_1). \tag{24.90}$$

We saw this was not true for the product function written above for the electron-proton system. It vanished when the electron was at $x = 4$ and the proton was at $x = 8$ but not the other way around. Product functions cannot describe two electrons in a box.

However, we can cook up a function that respects the indistinguishability of the particles by superposing the two alternatives;

$$\psi_{3,5,S}(x_1, x_2) = \psi_3(x_1)\psi_5(x_2) + \psi_5(x_1)\psi_3(x_2) \tag{24.91}$$

$$= \sqrt{\frac{2}{L}} \sin\left[\frac{3\pi x_1}{L}\right] \sqrt{\frac{2}{L}} \sin\left[\frac{5\pi x_2}{L}\right]$$

$$+ \sqrt{\frac{2}{L}} \sin\left[\frac{5\pi x_1}{L}\right] \sqrt{\frac{2}{L}} \sin\left[\frac{3\pi x_2}{L}\right]. \tag{24.92}$$

In this superposition of two product states, one has particle 1 in state $n = 3$ and particle 2 in state $n = 5$ and the other has particles with exchanged states. The subscript S stands for *symmetric*, meaning that because we have added the two possible product states related by particle exchange, the two particles now play symmetric roles. You can only infer from this wave function that there is *one* particle in $n = 3$ and *one* in $n = 5$ and not that particle 1 is in $n = 3$ and 2 is in $n = 5$.

Formally, this means the symmetric wave function is insensitive to the exchange of particle coordinates:

$$\psi_{3,5,S}(x_1, x_2) = \psi_3(x_1)\psi_5(x_2) + \psi_5(x_1)\psi_3(x_2) \tag{24.93}$$

$$\psi_{3,5,S}(x_2, x_1) = \psi_3(x_2)\psi_5(x_1) + \psi_5(x_2)\psi_3(x_1)$$

$$= \psi_{3,5,S}(x_1, x_2). \tag{24.94}$$

If we exchange the coordinates x_1 and x_2, the two terms in the symmetric wave function exchange roles and their sum is unaffected. The labels 1 and 2 in the quantum wave function no longer refer to the individual particles, which do not have a specific identity anymore.

The symmetric function is also unaffected if we leave x_1 and x_2 alone and *swap the state labels*:

$$\psi_{3,5,S}(x_1, x_2) = \psi_{5,3,S}(x_1, x_2). \tag{24.95}$$

This is an equivalent way of saying that all we know is that there is *a* particle in $n = 3$ and *a* particle in $n = 5$.

In any event, the probability density, which is simply the mod-squared of ψ, has the requisite symmetry

$$P_{3,5,S}(x_1, x_2) = P_{3,5,S}(x_2, x_1). \tag{24.96}$$

Figure 24.3 will help you visualize the situation. On the left is the state with particle 1 in $n = 3$ and particle 2 in $n = 5$, and in the middle is the state with particle 1 in $n = 5$ and particle 2 in $n = 3$. These are the two product states. Both these states are allowed in quantum theory if we are talking about a proton and an electron in the box, and they are counted as distinct states. But if they are two identical particles, the labeling makes a distinction that is meaningless in quantum theory. There is only one allowed state, the symmetric one depicted on the right. We just see two particles, one in $n = 3$ and one in $n = 5$, with no labels.

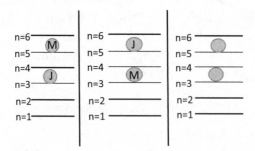

Figure 24.3   At the left is a state with particle 1 (Joe) in state $n = 3$ and particle 2 (Moe) in state $n = 5$ and in the middle a state with the particles exchanged. These states are permitted in quantum theory if the particles are different and count as two different possible outcomes. If they are identical, only the depiction at the right, which carries no labels, is allowed.

More generally for any two quantum states a and b (not just box states $n = 3$ and $n = 5$ as in our example) the allowed wave function is

$$\psi_{a,b,S}(x_1, x_2) = \psi_a(x_1)\psi_b(x_2) + \psi_a(x_2)\psi_b(x_1). \qquad (24.97)$$

I can get the symmetric state by adding to the product state, a state in which x_1 and x_2 are exchanged keeping a and b fixed or vice versa. Both reflect the fact that the particles have no identity. The probability density is

$$P_{a,b,S}(x_1, x_2) = |\psi_{a,b,S}(x_1, x_2)|^2$$
$$= |\psi_a(x_1)\psi_b(x_2) + \psi_a(x_2)\psi_b(x_1)|^2. \qquad (24.98)$$

In the symmetric wave function we finally seem to have found a way for describing two electrons, one in $n = 3$ and one in $n = 5$, respecting the requirement of indistinguishabilty. However, this is not true. It works for two identical pi-mesons or pions but not two electrons. But what else could we do besides symmetrize the product wave function?

There is another allowed combination called the *antisymmetric wave function* where we *subtract* the product state with the exchanged ordering:

$$\psi_{a,b,A}(x_1, x_2) = \psi_a(x_1)\psi_b(x_2) - \psi_a(x_2)\psi_b(x_1). \qquad (24.99)$$

If we now exchange the coordinates we find $\psi_{a,b,A}$ *changes sign*:

$$\psi_{a,b,A}(x_2, x_1) = \psi_a(x_2)\psi_b(x_1) - \psi_a(x_1)\psi_b(x_2)$$
$$= -\psi_{a,b,A}(x_1, x_2). \qquad (24.100)$$

This seems to violate the premise that exchanging identical particles should not make any difference. However, in quantum theory ψ itself is not directly observable (remember ψ and $-\psi$ are the same state) and only quantities quadratic in ψ such as $P = |\psi|^2$ are. Indeed we find that

$$P_{a,b,A}(x_1, x_2) = |\psi_a(x_1)\psi_b(x_2) - \psi_a(x_2)\psi_b(x_1)|^2$$

$$\text{while} \qquad\qquad\qquad\qquad\qquad\qquad (24.101)$$

$$P_{a,b,A}(x_2, x_1) = |\psi_a(x_2)\psi_b(x_1) - \psi_a(x_1)\psi_b(x_2)|^2 \qquad (24.102)$$

$$= |(-1)(\psi_a(x_1)\psi_b(x_2)$$

$$- \psi_a(x_2)\psi_b(x_1))|^2 \qquad (24.103)$$

$$= P_{a,b,A}(x_1, x_2). \qquad (24.104)$$

So in quantum mechanics, there are two options for identical particles. Either you can take the product function and add to it the product with the exchanged coordinates to obtain the symmetric wave function ψ_S, or subtract the product with the exchanged coordinates to obtain the antisymmetric wave function ψ_A. Remarkably every particle in the universe goes with one camp or the other. Particles called *bosons* always choose the symmetric wave functions, and particles called *fermions* always choose the antisymmetric wave function. Every particle is either boson or fermion. Pions are bosons. Electrons are fermions. Quarks are fermions. Photons and gravitons are bosons. For example, two pions of definite momentum p_1 and p_2 will be symmetric under exchange. We cannot say which one has momentum p_1 and which has momentum p_2. We can only say there is a pion with p_1 and a pion with p_2. If you put two identical bosons in a box, their symmetric wave function will remain the same when you exchange them. If you put two identical fermions in a box, their wave function will change sign if you exchange them. (If you put non-identical particles, say an electron and a proton, you may use a product wave function. If you exchange the coordinates, you generally get a different product state, and not ± 1 times the original.)

The previous discussion assumes the states a and b are different. Let us see what happens if $a = b$. For bosons we find

$$\psi_{a,a,S}(x_1, x_2) = \psi_a(x_1)\psi_a(x_2) + \psi_a(x_2)\psi_a(x_1)$$

$$= 2\psi_a(x_1)\psi_a(x_2) \qquad (24.105)$$

where the overall factor of 2 is physically unimportant. So bosons have no problem being in the same state, and if we consider more bosons we will find that they love being in the same state as others, a feature exploited in the laser, and one which we must reluctantly skip.

We are more interested in the dramatic case of two identical fermions. Can they both be in the same quantum state? If we set $a = b$

in the antisymmetric function, we find

$$\psi_{a,a,A}(x_2, x_1) = \psi_a(x_1)\psi_a(x_2) - \psi_a(x_2)\psi_a(x_1) \equiv 0. \quad (24.106)$$

This is the famous *Pauli exclusion principle*, which says *two identical fermions cannot be in the same quantum state.*

Notice also that even if $a \neq b$ the two fermions cannot occupy the same position: if we set $x_1 = x_2 = x$ we find ψ_A vanishes:

$$\psi_{a,b,A}(x, x) = \psi_a(x)\psi_b(x) - \psi_a(x)\psi_b(x) \equiv 0. \quad (24.107)$$

(This is not for the trivial reason that two particles cannot sit at the same point. Quantum theory allows two pions to be at the same point.) Since $\psi_A(x_1, x_2)$ vanishes when $x_1 = x_2$, by continuity it is also small when x_1 and x_2 approach each other. Thus two identical fermions avoid each other, not due to any repulsive forces between them, but due to the Pauli principle.

What does the Pauli principle say when we have three fermions? If we demand that the wave function change sign whenever we exchange any two fermions, we come up with

$$\begin{aligned}
\psi_{a,b,c,A}(x_1, x_2, x_3) = {} & \psi_a(x_1)\psi_b(x_2)\psi_c(x_3) - \psi_a(x_2)\psi_b(x_1)\psi_c(x_3) \\
& - \psi_a(x_1)\psi_b(x_3)\psi_c(x_2) - \psi_a(x_3)\psi_b(x_2)\psi_c(x_1) \\
& + \psi_a(x_2)\psi_b(x_3)\psi_c(x_1) + \psi_a(x_3)\psi_b(x_1)\psi_c(x_2).
\end{aligned}$$

I invite you to verify that in addition if you set any two coordinates equal or any two state labels equal, $\psi_{a,b,c,A}(x_1, x_2, x_3)$ vanishes. There is a way to write down such *totally antisymmetric wave functions* for any number of identical fermions. If you know *determinants*, here is the answer for three particles:

$$\psi_{a,b,c,A}(x_1, x_2, x_3) = \begin{vmatrix} \psi_a(x_1) & \psi_b(x_1) & \psi_c(x_1) \\ \psi_a(x_2) & \psi_b(x_2) & \psi_c(x_2) \\ \psi_a(x_3) & \psi_b(x_3) & \psi_c(x_3) \end{vmatrix}.$$

You can rely on the theory of determinants or verify by explicit computation that this wave function vanishes whenever two of the rows or columns are equal, that is, when two of the coordinates or state labels become equal. It also changes sign when two rows or columns are exchanged. For more particles, you just need a bigger determinant.

We need the symmetric and antisymmetric states only if the particles in question are identical. Even if there is the slightest difference between two particles, they will be treated as distinguishable and described by product wave functions. What makes the formalism worthwhile is that *there are in nature many many particles that are absolutely identical.* Every electron is identical to every other electron. One could have been produced in an accelerator on the earth and the other in another galaxy. You put those two guys in a box or an atom and they will obey the Pauli principle. It is remarkable how nature manages to churn out exactly identical particles in such widely separated regions of the universe.

24.4.2 Implications for atomic structure

Let us try to work out the structure of atoms based on what we know. First we have to compute the stationary states of an electron in the field of a nucleus of charge Ze, where Z is the number of protons.

This means solving the Schrödinger equationin three dimensions with a potential

$$V(r) = -\frac{Ze^2}{4\pi\varepsilon_0 r}. \tag{24.108}$$

From the classical expression for energy

$$E = \frac{p_x^2 + p_y^2 + p_z^2}{2m} - \frac{Ze^2}{4\pi\varepsilon_0\sqrt{x^2 + y^2 + z^2}} \tag{24.109}$$

and the final Postulate II, we know $\psi_E(x,y,z)$ obeys the Schrödinger equation

$$-\frac{\hbar^2}{2m}\left[\frac{\partial^2\psi_E}{\partial x^2} + \frac{\partial^2\psi_E}{\partial y^2} + \frac{\partial^2\psi_E}{\partial z^2}\right]$$

$$-\frac{Ze^2}{4\pi\varepsilon_0\sqrt{x^2 + y^2 + z^2}}\psi_E = E\psi_E. \tag{24.110}$$

The solution, which I will skip, gives the following spectrum. The allowed energies are

$$E_n = -\frac{Z^2 m e^4}{32\pi^2 \varepsilon_0^2 \hbar^2 n^2} = -\frac{13.6Z^2}{n^2} eV \quad n = 1, 2, 3\ldots \quad (24.111)$$

The levels are degenerate: there are n^2 levels at a given n. In addition, the electron has a twofold degree of freedom called *spin* corresponding to an internal angular momentum $\pm\frac{1}{2}\hbar$ not connected with motion and not discussed so far. So the real degeneracy is $2n^2$, which takes on values $2, 8, 18, \ldots$. The probability densities $P(r)$ for a particle in level n and with maximum allowed angular momentum, being found in a spherical shell between r and $r + dr$, are functions peaked at a radius $n^2 a_0$ where

$$a_0 = \frac{4\pi \varepsilon_0 \hbar^2}{m e^2} \quad (24.112)$$

is the Bohr radius. For this reason the states at any n are often called shells. In some books these are depicted as orbits or clouds of that radius.

Given the spectrum of an atom we can predict the frequencies of light it will emit or absorb when the value of n changes:

$$\omega_{n_1 \to n_2} = \left| \frac{E(n_1) - E(n_2)}{\hbar} \right|. \quad (24.113)$$

We can even compute the rate at which it will absorb or emit light, but this will require invoking the wave functions ψ_E.

Combined with the Pauli principle, we can understand a lot of chemistry by asking what the electrons will be doing in a given atom.

Hydrogen has just one electron, which we may place in the $n = 1$ state with spin $\frac{1}{2}\hbar$ or spin $-\frac{1}{2}\hbar$. Helium has $Z = 2$, and its two electrons occupy the $n = 1$ level with opposite spins. Lithium has $Z = 3$ and its third electron has to go to one of the eight $n = 2$ states. (At this point we may have to include the fact that the two electrons in the inner states $n = 1$ may screen some of the nuclear charge seen by the $n = 2$ electron.) We keep going till we hit Neon whose 10 electrons fill the $n = 1$ and $n = 2$ shells. If we add one more electron we need to go to the next level $n = 3$. This is the case for Na (sodium), which has 11 electrons. When the 11th electron looks in toward the nucleus it sees a charge $1e$, since the 10 inner electrons screen the rest of the nuclear charge. Its binding energy is a low $5.1 eV$. The atom then looks longingly at F (fluorine), which has 9 electrons. Its seven $n = 2$ electrons have a huge binding energy of $17.46 eV$ each. There is room for one more electron in its $n = 2$ shell. If the Na could unload its lone $n = 3$ electron to the vacancy in the $n = 2$ shell of F, the two atoms could lower their combined energy. This is what they do given a chance. But after this transfer, the Na atom will be positively charged and the F atom negatively charged. The two will be electrostatically bound by the *ionic bond* to form the NaF molecule.

The pattern is clear. Atoms with filled shells (like He or Ne) will have no incentive to talk to anyone else. Atoms with a lone electron in the outermost shell (*valance electron*) will try to unload it on atoms with a vacancy in their outermost shell. (The same goes for more than one transferred electron.) As the shells get filled this behavior will repeat. This explains the *periodic table*. Given that maxim "Happiness is a filled shell," we can anticipate who will be interested in whom. There are, however, some surprises and anomalies in the many-electron atoms that we cannot get into here.

Our belief in our description of the quantum world is based on very different considerations compared to the classical world. For example, if Newton says, "I can show using my laws that the planetary orbits are ellipses," this can be confirmed by direct observation. (In this case, the observation had already been done by Kepler before Newton.) For atoms, on the other hand, all we have are the energy levels and corresponding wave functions. Using these we can predict the structure of atoms and their interaction with each other and with the electromagnetic field. It is the spectacular agreement between theory and experiment that corroborates our faith in quantum mechanics as the way to describe the atomic world to which we do not have direct sensory access.

24.5   Energy-time uncertainty principle

We now consider the *energy-time uncertainty principle*

$$\Delta E \, \Delta t \geq \frac{\hbar}{2}. \tag{24.114}$$

This inequality presumes a particular definition of Δt to be described later and may have to be replaced by $\Delta E \, \Delta t \gtrsim \hbar$ or $\Delta E \, \Delta t \simeq \hbar$. This is because even if ΔE is the precisely defined uncertainty (see Eqn. 24.128), there is no unique definition of Δt. This is because time is not a dynamical variable with a probability distribution, instead it is a parameter on which dynamical variables like $x(t)$ and $\psi(t)$ depend. We all know exactly what the time is by looking at a clock and Δt is *not* the uncertainty in time.

What does Δt mean? For what definition of Δt is Eqn. 24.114 valid? What do Eqn. 24.114 and its variants signify?

They often reflect the fact that in order for a phenomenon to be ascribed a well-defined period, it must complete many cycles.

Suppose, upon observing you for some time, I assert that you go from New Haven to New York City and back once a day. I plot your distance $x(t)$ from New Haven as a function of time and find you complete a full cycle in one day. For me to say with absolute confidence that the frequency of your visits is once a day, I need to have seen you do this for many days. If you have been doing this just two days in a row, it is not enough, though after ten days I become more certain. I am never really sure because you may stop any time. To be absolutely positively sure, I have to wait an infinite time. But what can I say after a finite period of observation? I would like to say I know f, the frequency of your visits, with some uncertainty Δf, which should decrease with the observation time Δt. But what is Δf?

Suppose I have collected data over a time Δt (not necessarily small). The slice of time Δt will typically enclose a non-integer number of trips because you will typically be somewhere in the course of your round trip at the beginning and at the end of the interval Δt. Thus N, the number of trips you made in this time Δt, will be uncertain by an amount of order 1. So the estimated frequency will be

$$f \simeq \frac{N \pm 1}{\Delta t} = \frac{N}{\Delta t} \pm \frac{1}{\Delta t} \equiv f_0 + \Delta f \qquad (24.115)$$

and the uncertainty in f will be

$$\Delta f \simeq \frac{1}{\Delta t} \qquad (24.116)$$

$$\Delta f \Delta t \simeq 1. \qquad (24.117)$$

There is a more technical definition of Δf. If I use a Fourier transform to express your $x(t)$ *during the observation period* Δt as a sum of truly periodic waves that last for all time, it will be a sum over a continuum

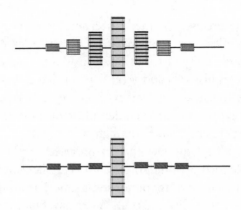

Figure 24.4   A series of reeds ordered by resonant frequency and pointing out of the page are shown end on. The top shows their early response to a frequency f_0 equal to that of the central reed. (The rectangles show the range of motion of each reed as seen end on.) The bottom shows the response after many periods.

of frequencies, with coefficients peaked at $f_0 = (24\text{ hrs})^{-1}$, and a width of order $1/\Delta t$, once again leading to Eqn. 24.117.

Here is a mechanical example of this phenomenon. Look at Figure 24.4, which shows a line of reeds arranged according to their resonant frequencies, with one end fixed, and the other end pointing out of the page and free to vibrate up and down. If we now stimulate them with a mechanical vibrator at some f_0, we may expect only the reed of that f_0 to respond strongly. But we will find that when we turn on the vibrator with its dial set at f_0, many of the reeds near the one at f_0 also respond substantially, as indicated in the upper half of the figure. The rectangles show the range of motion of each reed as seen end on.

This occurs because the reeds do not care what the dial on the vibrator says: they go by what they have experienced up to a time Δt, which is a finite wave train of a periodic stimulus of length Δt. However, as time goes by they will get the message that we are applying a periodic force, and eventually only the reed at f_0 will show any appreciable response, as shown in the lower half of the figure.

All this has nothing to do with quantum mechanics and merely reflects the fact that to measure the period (or frequency) of something you need to wait a few cycles, and that the longer you wait, the better will be your determination of the frequency.

Now for quantum mechanics. Suppose we have many identical atoms in their ground states (with energy E_0) and we want to find out their higher levels. To this end we turn on laser light of some frequency f_0 and see if it gets absorbed. If it does, we know there is a state at energy $E_0 + hf_0 \equiv E_0 + \hbar\omega_0$. However, what we will find is that initially the atoms will make transitions not only to states separated by $hf_0 = \hbar\omega_0$, but also several states on either side. Once again the dial on the laser may read f_0 or ω_0, but the atoms are going with the data they have over the time Δt. They will respond to the frequencies of waves that make up this *finite* (in time) train whose expansion Fourier coefficients are given by a distribution peaked at f_0 with a width $\Delta f \simeq 1/\Delta t$. Thus the spread in energy of the incoming photons and of the final atomic states will be

$$\Delta E = \hbar \Delta f \simeq \hbar \frac{1}{\Delta t}, \quad \text{which means } \Delta E \cdot \Delta t \simeq \hbar. \quad (24.118)$$

The meaning of ΔE is not the amount by which energy conservation is violated. It is the *range of possible energies* that could be absorbed by the atom if the energy transfer from the field has been going on for time Δt. However, once the atom absorbs a photon with one of these energies, the radiation field would have lost an equal amount of energy.

This is analogous to the Δp of the photon entering the Heisenberg microscope. It is not the amount by which momentum conservation is violated: instead Δp gives the range of momenta the photon *could have* upon measurement. Once one of these values is measured, you can be sure that the scattered electron will have just the right momentum to satisfy momentum conservation. After scattering, the combined photon-electron system will be in a superposition of product states with photon momentum p, electron momentum $P - p$, and total momentum P, shown below in obvious notation:

$$\psi^{e\gamma} = \sum_p A(p)\psi_p^\gamma \psi_{P-p}^e. \quad (24.119)$$

We are given that the coefficients $A(p)$ are significant only within a width Δp of the average p_0. Even though the photon can be detected with a range of momenta p, in every case the electron will have the missing momentum $P - p$. Measuring the photon momentum collapses the sum over product states to the one that was observed.

Consider next a system in some stationary state of initial energy E_i. If we turn on a constant potential V_0 at time $t = 0$, the system will jump to final states of energy

$$E_f = E_i \pm \frac{\hbar}{\Delta t} \tag{24.120}$$

for a time Δt. This is so because the applied potential is not really a constant, but a step function that jumped from 0 to V_0 at $t = 0$. This process violates the conservation of the energy *of the system* by $\Delta E \simeq \frac{\hbar}{\Delta t}$. However, the energy of the system and the external agency that suddenly imposed the potential V_0 will be conserved.

Next consider an atom that has been sitting in an excited state of energy E_n for some time Δt before it decays to its ground state. *During the time it is excited* its wave function varies with time as $e^{-iE_n t/\hbar}$. The Fourier transform of this function (in existence only for the time Δt) will have a width $\hbar/\Delta t$. Consequently the light emitted by this atom when it relaxes to the ground state will have a spectrum centered around $\omega_{n,0} = (E_n - E_0)/\hbar$ and of width (called the *line width*) $\Delta E \simeq \hbar/\Delta t$. Once again this spread ΔE only means that the atom and radiation field begin in a superposition of energy states with the atom having some energy and the field the rest of the conserved energy. The energy of the atom and the field will always equal a fixed conserved value, just like the total momentum P in the Eqn. 24.119. The statement that "a system that has been in existence for a finite time cannot be assigned a definite energy" has to be understood as above and not as a violation of energy conservation.

There is another way to derive and interpret $\Delta E \Delta t \gtrsim \hbar$. Let us begin with an analogy. Suppose the grade distribution of a class is some bell-shaped curve with some average $\langle G \rangle$ and width ΔG. Now some educator comes along with a scheme to improve the average. If the benefits of the strategy are to be convincing, the center of the distribution $\langle G \rangle$ has to move by at least the width ΔG when the changes are implemented.

Now carry this idea to the quantum problem. Consider a wave packet with mean momentum p_0 and an uncertainty Δp. Its width in x must be

$$\Delta x \geq \frac{\hbar}{2\Delta p}. \tag{24.121}$$

The center of the packet moves by a detectable amount when it moves by at least the uncertainty Δx. With a mean velocity $v = \frac{p_0}{m}$ the time required is

$$\Delta t = \frac{\Delta x}{v} \geq \frac{\hbar}{2\Delta p \cdot v} = \frac{\hbar}{2\Delta p \cdot (p_0/m)}. \tag{24.122}$$

Now consider ΔE, the spread in the energy of the particle due to Δp, the spread in the momentum of the wave packet:

$$E = \frac{p^2}{2m} = \frac{(p_0 \pm \Delta p)^2}{2m}$$

$$= \frac{p_0^2}{2m} \pm \frac{p_0 \Delta p}{m} + (\text{Order } \Delta p)^2 \tag{24.123}$$

$$= E_0 + \Delta E. \tag{24.124}$$

This tells us the spread in the energy of the wave packet is

$$\Delta E = \frac{p_0 \Delta p}{m}. \tag{24.125}$$

Substituting in Eqn. 24.122 we find $\Delta E \cdot \Delta t \geq \frac{\hbar}{2}$.

To summarize, if a particle is in a state with uncertainty ΔE in energy, the time it takes to move by Δx, the uncertainty in its position, obeys $\Delta t \geq \frac{\hbar}{2\Delta E}$. We have a precise inequality here because both ΔE and Δt are precisely defined.

We have already seen one example of this in the last chapter. In a state made of two box states of energies E_1 and E_2, the minimum time Δt over which one could see appreciable change in $P(x,t)$ (see Eqns. 23.64 and 23.66) was given by

$$\Delta t \gtrsim \frac{\hbar}{\Delta E} \tag{24.126}$$

where $\Delta E \simeq E_2 - E_1$.

This argument is very general. Consider a quantum state and a variable $\mathcal{A}$ that is measured. There will be a range of possible outcomes with probability $P(\alpha)$, expectation value

$$\langle \mathcal{A} \rangle = \sum_\alpha P(\alpha)\alpha, \tag{24.127}$$

and uncertainty

$$\Delta \mathcal{A} = \sqrt{\sum_\alpha P(\alpha)(\langle \mathcal{A} \rangle - \alpha)^2}. \tag{24.128}$$

In a stationary state of some energy E, neither $\langle \mathcal{A} \rangle$ nor $\Delta \mathcal{A}$ will change with time because $P(\alpha)$ will not.

Suppose now that the system starts out in a *superposition* of energy states. Then $P(\alpha)$ and $\langle \mathcal{A} \rangle$ can change with time, the way $P(x, t)$ did when we started off the particle in a box state $\psi(x, 0) = Ax$ made of many energy levels. Let ΔE be the range of energies in the superposition. *Let Δt be the time over which $\langle \mathcal{A} \rangle$ changes by an amount equal to the uncertainty in $\mathcal{A}$.* In other words Δt is given by

$$\frac{d\langle \mathcal{A} \rangle}{dt} \cdot \Delta t = \Delta \mathcal{A} \tag{24.129}$$

$$\Delta t = \frac{\Delta \mathcal{A}}{\frac{d\langle \mathcal{A} \rangle}{dt}}. \tag{24.130}$$

From the time-dependent Schrödinger equation one can compute the rate of change of $\mathcal{A}$ and establish a precise inequality

$$\Delta t \geq \frac{\hbar}{2\Delta E}. \tag{24.131}$$

In other words, if a system has an energy uncertainty ΔE in its wave function, the minimum time Δt it takes a variable $\mathcal{A}$ to change by the width in its probability distribution (precisely defined by its uncertainty $\Delta \mathcal{A}$) is $\frac{\hbar}{2\Delta E}$. Since a quantity with an intrinsic uncertainty $\Delta \mathcal{A}$ in its value has to change by at least that amount for us to know it has changed, Δt is the minimum time for this change to be detectable. The minimum time depends on the choice of variable $\mathcal{A}$. The smallest of such times defines the natural timescale for the system, which is the time it takes to notice *any* kind of change in the system with respect to *any* observable. This minimum time grows as the energy content of the state becomes sharper and sharper. Finally, for a state of definite energy, we have to wait forever to witness any change.

Let us apply this argument to a problem we have already analyzed: an atom that lives in an excited state for a short time τ and decays to the ground state. The initial atomic state is clearly not a stationary state because things change; the atom relaxes to the ground state. So the initial state must have been a superposition of many energy states. We know from the preceding argument that the spread ΔE is related to the time Δt over which something noticeable happens by

$$\Delta E \gtrsim \frac{\hbar}{\Delta t}. \tag{24.132}$$

Since the atom decays in a time τ, called the *lifetime*, and since its decay is certainly a noticeable event, it is fair to assert that $\Delta t \simeq \tau$. This leads to

$$\Delta E \cdot \tau \simeq \hbar. \tag{24.133}$$

This is what leads to the statement that a system with a lifetime τ has an uncertainty in its energy $\Delta E \simeq \frac{\hbar}{\tau}$. What this means is that such an unstable state is a state in a superposition of ψ_E's with the values of E spread over a width ΔE.

24.6 What next?

The end of this book is just the beginning of your journey into physics. There is so much to learn on every front. For example, the quantum mechanics I described here is based on the Schrödinger equation and begins with the Newtonian expression for kinetic energy

$$E = \frac{p^2}{2m}. \tag{24.134}$$

This works if the particles involved have kinetic energies small compared to the rest energy mc^2. When this condition is not met, we need the relativistic wave equation due to P. A. M. Dirac, who begins with the relativistic expression

$$E^2 = c^2 p^2 + m^2 c^4. \tag{24.135}$$

At some point, the Dirac theory too becomes inadequate. For example, two energy levels of hydrogen that are supposed to be degenerate in the Dirac theory were found to differ slightly by what is called the Lamb shift.

To explain this we need quantum field theory, which treats matter and radiation in accordance with the laws of quantum mechanics and relativity. Quantum field theory has proven to be a powerful way to describe and possibly unify electromagnetic, weak and strong interactions into one big gauge theory. But this too has some problems: there are unwanted infinities at intermediate stages in the calculation of finite answers. There is a recipe for fixing these infinities and extracting the finite answers (which agree exceptionally well with experiment, say in describing the Lamb shift), but the fix does not work for gravity, which we would want to include.

String theory is a potential solution to all the woes of field theory: no infinities appear, gravity is seamlessly incorporated, and even the number of space-time dimensions, which could be anything in field theory, is fixed to be 10. All this is very seductive. However, there are some technical complications at the present time, and, more significantly, the real differences due to strings appear only at the unimaginably small distance called the Planck length $\simeq 10^{-35} m$, which is 10^{-20} times the proton size. (For comparison, the radius of the atom is roughly 10^{-20} times the radius of the earth's orbit around the sun.) The energy required to test string physics would be 10^{15} times that of the Large Hadron Collider, which operates at $10^{12} eV$. So even if string theory is right, it may be hard to verify that it is so, based on what we can experimentally probe today. But strings should play a significant role in the very early universe, and that is why people are looking for remnants of "stringy physics."

So you have a lot to learn and you better get started!

Constants

$$G = 6.7 \cdot 10^{-11} \, m^3 \cdot kg^{-1} \cdot s^{-2} \quad \text{gravitational constant}$$

$$e = 1.6 \cdot 10^{-19} \, C \quad \text{proton charge}$$

$$m_e = 9.1 \cdot 10^{-31} \, kg \quad \text{electron mass}$$

$$m_p = 1.7 \cdot 10^{-27} \, kg \quad \text{proton mass}$$

$$\frac{1}{4\pi\varepsilon_0} = 9 \cdot 10^9 \, N \cdot m^2 \cdot C^{-2}$$

$$\frac{\mu_0}{4\pi} = 10^{-7} \, N \cdot s^2 \cdot C^{-2}$$

$$\hbar = 1.05 \cdot 10^{-34} \, J \cdot s \quad \text{Planck's constant}$$

$$a_0 = \frac{4\pi\varepsilon_0 \hbar^2}{m_e e^2} = 0.53 \cdot 10^{-10} \, m$$

$$1 \, Amp = 1 \cdot C/s$$

Index